DYNAMICAL SYSTEM MODELS

DYNAMICAL SYSTEM MODELS

by

A. G. J. MacFarlane
B.Sc., Ph.D., D.Sc., F.I.E.E.

Professor of Control Engineering
The University of Manchester Institute of Science and Technology

GEORGE G. HARRAP & CO. LTD
London   Toronto   Wellington   Sydney

First published in Great Britain 1970
by GEORGE G. HARRAP & CO. LTD
182 High Holborn, London, W.C.1

© A. G. J. MacFarlane 1970

All rights reserved. No part of this publication may be reproduced in any form or by any means without the prior permission of George G. Harrap & Co. Ltd

SBN 245 50404 4

*Composed in Times Monophoto and printed by J. W. Arrowsmith Ltd.
Bristol*

Made in Great Britain

Preface

The aim of this book is to give a complete, self-contained treatment of all the types of deterministic dynamical system model which are currently used by engineers and applied mathematicians in the fields of automatic control, systems engineering, network analysis and feedback system theory. Dynamical system models have been grouped into four types: transform models, Hamiltonian models, network models, and state models. Each specific type of model is described in a single chapter which is self-contained, when taken together with the basic physical and mathematical background provided in the first two chapters. No particular significance is therefore attached to the order in which Chapters Three to Six are presented. For each type of model discussed, all the definitions and proofs required are given in the text. Those mathematical results from the theory of linear vector spaces and matrix algebra which are repeatedly used in the later chapters have all been collected into Chapter Two for reference purposes.

The reason for writing the book is the increasing use of dynamical system models in engineering studies. This increase has led to a proliferation of specialized textbooks which threatens to obscure the fact that all the topics treated have a common basis in general dynamical theory. A few examples may be cited of the resulting difficulties faced by students and teachers. Many textbooks on electrical network theory deal only with linear phenomena, and give no indication of how to tackle non-linear or non-electrical systems. Conventional treatments of analytical mechanics, although applicable to non-linear mechanical systems, do not use the concept of co-energy. This leads to acute difficulty in attempts to apply conventional analytical mechanical theory to non-linear electrical systems. Treatments of the recently-developed theory of optimal control usually plunge students, with no previous exposure to classical Hamiltonian mechanics, straight into generalized Hamiltonian techniques. Discussions of state-variable techniques are usually presented at a very abstract level with little indication of how they are related to other basic methods of dynamical system analysis. All these difficulties arise from the fragmentation of dynamical theory among a wide variety of specialized courses.

The advantages of a comprehensive treatment, such as is attempted here, are two-fold. In the first place, the essential unity of the subject becomes apparent. An awareness of this enables the engineering dynamicist to handle a variety of non-linear and linear physical system models with insight and understanding, and to turn confidently from one type of model representation to another. Secondly, the central role played in modern dynamical studies by algebraic theory becomes apparent. This is particularly true of the interplay between algebraic and dynamical duality. The systematic use of algebraic methods in dynamical systems analysis is the subsidiary theme of this book. The wide range of topics presented here has never before been collected together and given a unified presentation in one volume.

During the past ten years in which I have been lecturing on and developing the material presented here, I have consulted and used a large number of textbooks and papers. Certain books and papers have had a great effect on the way in which I have developed several parts of the work, and it is appropriate to make particular acknowledgement of them here. In the treatment of Components in Chapter One, the presentation by Shearer, Murphy and Richardson [S3] was very helpful. I was much influenced by teaching this material, during a brief spell in the Mechanical Engineering Department of the Massachusetts Institute of Technology. The treatment of Hamiltonian Models in Chapter Four owes much to Lanczos [L1], and in preparing this chapter I also found helpful the treatment by Crandall and his colleagues of mechanical and electromechanical systems [C4]. The expanding wavefront treatment of Pontryagin's theory of optimal control is based on Halkin's work [H1].

In the presentation of network models in Chapter Five I have leaned heavily on Bryant's approach to linear network theory [B9], [B10], [B11] and Brayton and Moser's study of non-linear networks [B7]. The extensive use of co-energy and related duality concepts in dynamical theory is based on the original work of Cherry [C3] and Millar [M9]. In the algebraic treatment of linear system stability I found notes and papers supplied by Dr P. C. Parks of Warwick University most helpful, particularly with regard to Hermite's work.

Much of my enthusiasm for dynamical theory results from my brief spell at M.I.T.; during this time I had many stimulating conversations with Professor H. M. Paynter on general dynamical model theory. I am also grateful to many past and present students of the University of Manchester Institute of Science and Technology who helped me to clarify and correct early lecture treatments, and in particular to Mr C. C. Arcasoy who carefully read the proofs and helped to correct them.

A. G. J. MacFarlane

Contents

Chapter One   Components

1.1	Introduction	1
1.2	Translational Mechanical Variables	3
1.3	Translational and Rotational Mechanical Components	18
	1.3.1 Translational Spring	24
	1.3.2 Translational Mass	27
	1.3.3 Translational Converters	29
	1.3.4 Representation of Translational Mechanical Systems by Linear Graphs	33
	1.3.5 Relationships between Translational and Rotational Mechanical Interactions	39
	1.3.6 Rotational Mechanical Components	41
	1.3.7 Mechanical Transformers	51
1.4	Fluid Components	52
	1.4.1 Fluid Capacitance	53
	1.4.2 Fluid Inertance	59
	1.4.3 Fluid Dissipatance	61
	1.4.4 Representation of Lumped Fluid Systems by Linear Graphs	63
	1.4.5 Wave Propagation in Fluid Systems	64
1.5	Thermal System Components	66
1.6	Electrical System Components	74
	1.6.1 Capacitor	77
	1.6.2 Inductor	80
	1.6.3 Converters	83
	1.6.4 Representation of Lumped Electrical Systems by Linear Graphs	85
	1.6.5 Wave Propagation in Electrical Systems	86

Chapter Two   Spaces

2.1	Sets	90
2.2	Metric Spaces	92
2.3	Linear Vector Spaces	92
2.4	Linear Independence and Bases	94
2.5	Inner Product	96
2.6	Matrix Representation of Linear Operators in an n-dimensional Linear Vector Space	98

2.7	Dual Basis and the Projection Theorem	100
2.8	Matrix Representation of Change of Basis in an n-dimensional Euclidean Vector Space	101
2.9	Coordinate Transformations	102
2.10	Representation of an Operator Matrix in a Different Basis	102
2.11	Eigenvalues and Eigenvectors of a Linear Operator Matrix Representation	103
2.12	Scalar-valued Functions of Vectors	109
2.13	Range Space, Null Space, Rank and Nullity of a Linear Operator	110
2.14	Vector Differentiation	112
2.15	Differentiation and Integration of Matrices and Determinants	114
2.16	Direct Sum of Matrices	115
2.17	Extremum Values of Vector Functions Subject to Constraints	116
2.18	Extremum Characteristics of Eigenvalues	120
2.19	Matrix Functions	124
2.20	Miscellaneous Notes	125

Chapter Three Transform Models

3.1	Laplace Transform	126
3.2	Transfer Function Relationships	143
	3.2.1 Convolution	145
	3.2.2 Asymptotic Relations between Weighting Function and Transfer Function	149
	3.2.3 Response to Sinusoidal Input	150
	3.2.4 Determination of System Response from Transfer Function Pole and Zero Distribution	151
	3.2.5 Relationships between Real and Imaginary Part of Transfer Function	154
	3.2.6 Minimum-phase Transfer Functions	158
3.3	Block Diagrams	159
	3.3.1 Block Diagram Conventions	159
	3.3.2 Block Diagram Manipulations	163
3.4	Signal-flow Graphs	170
	3.4.1 Signal-flow Graph Conventions	170
	3.4.2 Signal-flow Graph Manipulations	172
	3.4.3 Mason's Circuit Rule	177
3.5	Nyquist's Stability Criterion	184
	3.5.1 Complex Plane Mappings	184
	3.5.2 Open- and Closed-loop Transfer Function Relationships	187

		3.5.3	Basic Closed-loop Stability Theorem	188

        3.5.3 Basic Closed-loop Stability Theorem 188
        3.5.4 Simple Form of Nyquist's Criterion 189
        3.5.5 General Form of Nyquist's Criterion 191
        3.5.6 Relative Stability Criteria 192
3.6 Root Locus Method 194
        3.6.1 General Rules for Construction of Root Loci 197
3.7 Optimal Linearization and the Describing Function 201
3.8 z-transforms for Discrete Systems 206
        3.8.1 Basis Sequences 206
        3.8.2 Discrete Operators 206
        3.8.3 z-transform 207
        3.8.4 Convolution of Sequences 214
        3.8.5 z-transfer Functions 215

Chapter Four   Hamiltonian Models

4.1 Fundamental Processes of the Calculus of Variations 217
        4.1.1 Conditions for Stationary Values of Definite Integral 220
        4.1.2 Stationary Value of Integral with Fixed End-points and Several Dependent Variables 223
4.2 Generalized Coordinates 226
        4.2.1 Generalized Velocities 226
        4.2.2 Generalized Forces 227
4.3 Primal Form of Hamilton's Postulate and the Set of Lagrangian Equations 227
        4.3.1 Lagrangian Equation Set 232
4.4 Generalized Momenta 238
4.5 Dual Form of Hamilton's Postulate and the Set of Co-Lagrangian Equations 239
4.6 Conservation of Energy and Momentum 241
4.7 Hamilton's Equations 243
        4.7.1 Hamilton–Jacobi Equation 244
        4.7.2 Liouville's Theorem 245
4.8 Hamiltonian Principles for Electrical Networks 246
4.9 Pontryagin's Equations 248
        4.9.1 Optimal Control Problem 251
        4.9.2 Event Vector and Event Space 251
        4.9.3 The Set of Possible Events 252
        4.9.4 Expanding Wave-fronts in the State Space 253
        4.9.5 Analogy with Huygens' Principle in Geometrical Optics 254
        4.9.6 Pontryagin's Maximum Principle 255
        4.9.7 Generation of Optimal Trajectory 255

4.9.8	Derivation of Pontryagin's Equations	257
4.9.9	Derivation for the Case When the Control Inputs are Unrestricted	262
4.9.10	Minimization of an Integral Functional of System Motion	265
4.10	Maximal-effort or "Bang-bang" Systems	270

Chapter Five   Network Models

5.1	Basic Definitions for Linear Graphs	285
5.2	Interconnective Constraints on Power Variables	287
5.3	Topological Relationships between Network Variables	290
5.4	Tellegen's Theorem	296
5.5	The Dynamical Transformation Matrix	298
5.6	Analogues, Duals and Dualogues	301
5.7	Circuit, Vertex and Mixed Transform Analysis Methods for Linear Electrical Networks	304
	5.7.1 Circuit Method	308
	5.7.2 Vertex Method	313
	5.7.3 Mixed Method	318
5.8	Systems Matrix Analysis of Networks	319
5.9	Lagrangian Equations for Networks	323
5.10	Co-Lagrangian Equations for Networks	325
5.11	Special Variational Principles for Networks	329
5.12	Formulation of Canonical Equation Sets for Linear Networks	338
5.13	Formulation of State Space Equations for Nonlinear Networks	363
	5.13.1 Integral Invariants	369
	5.13.2 Derivation of Canonical Equation Set	371
	5.13.3 Construction of Scalar Functions	377

Chapter Six   State Models

6.1	Analytical Aspects of State Space Equation Sets	385
	6.1.1 Existence and Uniqueness of Solutions	385
	6.1.2 Singular Points and the Liapunov First-Approximation Matrix	389
	6.1.3 Simple Trajectory Properties in the State Plane	393
	6.1.4 Analytical Solutions of Linear Equation Sets	396
6.2	Stability	420
	6.2.1 Liapunov Stability Theory	424
6.3	Modality	434

6.4	Discrete Model Approximation of Linear Constant Coefficient Systems		437
6.5	Functional Matrices		439
	6.5.1	Equivalent Free-motion Systems for Network Impulse, Step and Ramp Response	444
6.6	Generalized Mohr Circles and their Use in Feedback Design		446
	6.6.1	Generalized Mohr Circles	447
	6.6.2	Properties of Generalized Mohr Circles	449
6.7	Controllability and Observability		455
	6.7.1	Controllability	456
	6.7.2	Observability	459
	6.7.3	Decomposition of State Space Systems	461
	6.7.4	Duality and Adjoint Systems	462
	6.7.5	Determination of the Controllable Part of a Given Representation	466
	6.7.6	Determination of the Observable Part of a Given Representation	469
	6.7.7	Determination of Transfer Function Matrix Representations	471
6.8	Reduction		476
	6.8.1	Optimal Orthogonal Projection on to a Subspace	480
	6.8.2	Optimal Projection Along an Invariant Subspace	483

Appendix A   486
References   490
Index   497

CHAPTER ONE

Components

1.1 Introduction

A system may be broadly defined as any ordered set of interrelated physical or abstract objects. A dynamical system model is an ordered set of interrelated abstract objects used to predict the behaviour of some natural or engineering system. It is related to the set of systems shown in Figure 1–1.

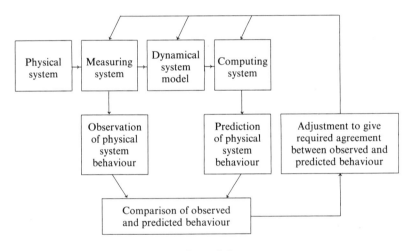

Figure 1–1

In Figure 1–1, the arrow directions define the sequence in which the various systems involved are built up and modified. The evolution of a satisfactory dynamical system model is obviously an iterative procedure. When a sufficiently close agreement exists between the predicted and observed behaviour of the physical system of interest, the dynamical system model thus obtained may be used to represent the physical system in theoretical investigations. In any such investigations however it must never be forgotten that any dynamical system model is usually evolved for some quite specific purpose, and represents a vastly simplified and highly abstract system whose validity can only be justified in terms of a comparison of observed and predicted physical system behaviour.

In discussing any dynamical system model we are constantly dealing with:

(a) Sets of objects such as ideal masses, springs, capacitors, inductors, levers, transformers, etc.
(b) Sets of variables such as forces, velocities, voltages, currents, etc.
(c) Sets of relationships between variables.
(d) Interactions between objects.

Since at least two things are needed to specify a relationship, relationships between pairs of variables play a dominant part in dynamical systems analysis. An object is defined in terms of relationships between variables and thus the simplest and most basic objects in dynamical systems analysis are those defined in terms of the relationship between a pair of variables. Since at least two objects are required for an interaction, and since the specification of an interaction must involve at least a pair of variables, the simplest and most basic interactions in dynamical analysis are those specified in terms of pairs of object variables.

In constructing abstract models to predict physical system behaviour we work in terms of the three fundamental intuitive concepts of space, matter and time. The essence of a dynamical system is that its present behaviour is influenced by its past history; dynamical system objects cannot therefore be specified simply in terms of an instantaneous relationship between 'input' and 'output' variables. An extra set of variables is required which are used to take into account the past behaviour of the system. Such a set of variables will be termed *state* variables, and the simplest dynamical system models may be described from a causal point of view in terms of input variables, state variables and output variables. In mathematical terms the state variables of a dynamical system are functionals of the input variables, i.e. they associate numbers, the values of the state variables, with functions of time, the past behaviour of the input variables with time. The simplest functional relationship is the integral relationship for a single object:

$$\text{value of state variable} = \int_{-\infty}^{t} (\text{input function of time}) \, d(\text{time})$$

In this case

$$\frac{d(\text{state})}{d(\text{time})} = \text{input function of time}$$

and thus dynamical systems whose state is characterized in this way are described by sets of ordinary differential equations. It is important to remember that this is merely the result of using the simplest functional relationship possible. More complicated functional relationships have been introduced and studied by Volterra [V1].

The detailed treatment of dynamical system models is best introduced by considering the first type of system studied in the historical development of the subject: the translational mechanical system.

1.2 Translational Mechanical Variables

In mechanical systems analysis we are first of all concerned with the three intuitive aspects of the physical world: space, time and matter. It is convenient to describe those objects and variables which are concerned with purely spatial aspects of the analysis as *geometrical*; to describe those which are concerned with purely spatial and temporal aspects as *kinematical*; and to describe those which are also concerned with the material aspects as *physical*. In order to describe interactions between objects, we add to the kinematical variables used in mechanical systems analysis, such as displacement and strain, the physical variables such as force and stress.

Interaction is the fundamental physical concept upon which all dynamical analysis is based. The exact science of dynamics originated with Newton's enunciation of his celebrated Laws of Motion. These contain the essence of all quantitative treatments of interaction. In particular they show how the formalism of dynamics has a triadic structure by virtue of its relationships to the three fundamental concepts of space, time and matter, and to the input, state, output description of object behaviour. It is interesting to look briefly at Newton's three laws from these points of view.

(NLM 1) "Every body perseveres in its state of rest or of moving uniformly in a straight line, except in so far as it is made to change that state by external forces."

This law introduces the concept of the *state* of uniform translational motion. It is also a statement about a *material* property of an object, namely its inertia. Finally it introduces the intuitive concept of *force* as a quantitative measure of the *interaction* of an object with its environment.

(NLM 2) "The rate of change of momentum of a body is equal to the impressed force and takes place in the direction in which the force is impressed." (In modern terminology we have substituted "rate of change of momentum" for the original "change of motion".)

This is one of the functional concepts from which the differential equations of mechanics arise. The momentum is the quantitative measure of the state of motion and it is related by a *temporal* rate of change to an input force variable.

(NLM 3) "Action and reaction are always equal and opposite, that is to say the actions of two bodies on each other are always equal and opposite."

This is essentially concerned with *spatial relationships* between forces, and highlights the concept of force as the key physical variable characterizing an interaction.

Primal Interaction Experiment

The type of analytical structure which may be set up on these three laws, and their equivalents for other types of discrete physical systems, underlies the whole of what we may call the causal approach to dynamical analysis, and is thus worthy of fairly detailed consideration at this point. The first step in a systematic study of mechanical systems would be a series of qualitative experiments which would show that material objects interact mechanically by virtue of deformation or relative motion. The next step would be a quantitative experiment on mechanical interaction; the particular one we will consider here is illustrated in Figure 1–2. Consider a pair of material particles at each end of a light spring, and suppose the resulting motions are confined to a straight line along the axis of the spring and constrained in such a way that gravitational effects on the motion are excluded. Suppose we first experiment with the *material* property of inertia and, for this purpose, take a given light spring S and a set of different massive particles $^\#i$, $^\#j$ and $^\#s$. (When letters and numbers are used as identifying labels this will be explicitly denoted by means of the 'label symbol' $^\#$.) Let the particles interact in pairs via the spring S and consider the trio of interactions:

$$^\#i \quad \text{with} \quad ^\#s$$
$$^\#s \quad \text{with} \quad ^\#j$$
$$^\#i \quad \text{with} \quad ^\#j$$

Denote the consequent particle accelerations by a_i, a_s, a_j; a_i', a_s', a_j'; a_i'', a_s'', a_j'', where the primes denote the different accelerations associated with different interactions. From careful observation we are led to *postulate* that the accelerations are independent of the particle motion, and we may therefore introduce a set of real, positive interaction coefficients T_{is}, T_{sj} and T_{ij} such that the experiments of Figure 1–2 give, with the positive direction of acceleration taken as that of a_i:

$$a_i = -T_{is}a_s \quad \text{for} \quad ^\#i \text{ interacting with } ^\#s$$
$$a_s' = -T_{sj}a_j' \quad \text{for} \quad ^\#s \text{ interacting with } ^\#j \qquad (1\text{--}1)$$
$$a_i'' = -T_{ij}a_j'' \quad \text{for} \quad ^\#i \text{ interacting with } ^\#j$$

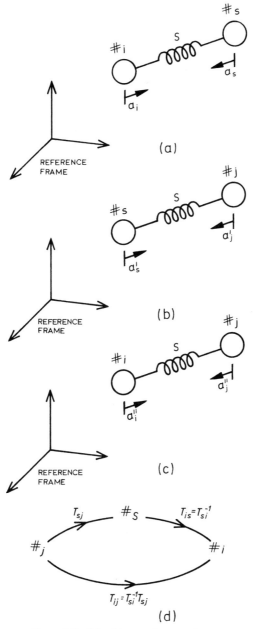

Figure 1–2 Primal interaction experiment.

For the first of these interactions we may introduce the further coefficient T_{si} such that

$$a_s = -T_{si}a_i$$
$$a_i = -T_{si}^{-1}a_s \qquad (1-2)$$

Group postulate. We now make the fundamental postulate that these interaction coefficients *form a group*. Thus we have that [B3]

$$T_{ij} = T_{is}T_{sj} = T_{si}^{-1}T_{sj} \qquad (1-3)$$

This postulate is of fundamental importance in the structure of dynamical systems analysis since it enables one to make a fundamental factorization of the interaction coefficient T_{ij} for two interacting material particles $^\#i$ and $^\#j$ into an interaction coefficient for $^\#i$ with a *standard dynamical object* $^\#s$ and one for $^\#s$ with $^\#j$, as depicted in the scheme of Figure 1–2(d). Its importance arises from the fact that we may now allocate to *each individual* material particle a quantity, called its inertial mass, which is characteristic of the way in which a *single* particle reacts with what we may call its environment. This is done by defining

$$M_i = T_{si} \triangleq \text{inertial mass of } ^\#i \text{ (with reference to } ^\#s)$$
$$M_j = T_{sj} \triangleq \text{inertial mass of } ^\#j \text{ (with reference to } ^\#s)$$

In turn, since we have

$$a_i'' = -T_{si}^{-1}T_{sj}a_j'' \qquad (1-4)$$

this enables us to put

$$M_i a_i'' = -M_j a_j'' \qquad (1-5)$$

and introduce the variables

$$f_i = M_i a_i'' \triangleq \text{force on } ^\#i$$
$$f_j = M_j a_j'' \triangleq \text{force on } ^\#j$$

for which

$$f_i = -f_j \qquad (1-6)$$

in accordance with Newton's Third Law of Motion above. This force equation essentially describes a spatial aspect of the interaction. The force variable has been introduced by reference of the interactions to a standard object, namely the unit inertial mass; *it is common to the pair of interacting objects by virtue of this reference.*

If we let v_i, v_j = velocities of material points $^\#i$ and j respectively, we have

$$f_i = M_i \frac{dv_i}{dt} = \frac{d}{dt}(M_i v_i)$$
$$f_j = M_j \frac{dv_j}{dt} = \frac{d}{dt}(M_j v_j) \qquad (1-7)$$

The momenta of the material particles are defined as

$$p_i \triangleq M_i v_i = \text{momentum of particle } {}^\#i$$
$$p_j \triangleq M_j v_j = \text{momentum of particle } {}^\#j$$

This gives

$$f_i = \frac{dp_i}{dt}$$
$$f_j = \frac{dp_j}{dt}$$
(1-8)

In the restricted context of this idealized experiment, Newton's First Law would follow from observations of particle motions following the severance of the spring; Newton's Second Law from equations (1–8) and Newton's Third Law from equation (1–6).

Ideal spring postulate. We now add the further postulate that the forces exerted at each end of an ideal spring are always equal and opposite, proportional to the extension, and independent of the velocity.

Summary of fundamental interaction experiment. From the discussion above, the fundamental interaction experiment leads to the following conclusions:

(FE 1) Each material particle is associated with a triad of variables: force, momentum and velocity.

(FE 2) Each light spring is associated with a triad of variables: force, displacement and velocity.

(FE 3) Between the momentum and velocity variables of a material particle there is a material or *constitutive* relationship

$$p = Mv \qquad (1-9)$$

(FE 4) Between the force and displacement variables of a light spring there is a material or constitutive relationship

$$f = \phi(q) \qquad (1-10)$$

(FE 5) Between the force and momentum variables of a material particle there is a temporal or *dynamic* relationship

$$f = \frac{dp}{dt} \qquad (1-11)$$

(FE 6) Between the velocity and the displacement variables of a light spring there is a temporal or dynamic relationship

$$v = \frac{dq}{dt} \qquad (1-12)$$

8   Components

(FE 7) A set of spatial or *interconnective* relationships exist between the forces and velocities of the interconnected material particles and spring. For the positive directions shown in Figure 1–3, we have, where f_s is the force on the spring:

$$f_i = f_s, \text{ the force balance for particle } ^\#i$$
$$f_s = f_j, \text{ the force balance for particle } ^\#j \quad (1\text{–}13)$$

$$v_i + v_s + v_j = 0 \quad (1\text{–}14)$$

Equation (1–14) is a velocity balance round a mechanical circuit which includes a point on an inertial reference frame.

(FE 8) The system may be regarded as an interconnected set of discrete, isolatable components. Equations (1–13) and (1–14) define the interconnective relationships between the isolatable components. The behaviour of each isolatable component is defined by a pair of constitutive and dynamic relationships.

Dual interaction experiment. The interaction experiment shown in Figure 1–4 is said to be *dual* to that of Figure 1–2; the role played by material particles and light springs has been interchanged in the two experiments. In this dual experiment we have a set of light linear springs $^\#i$, $^\#j$ and $^\#s$ and we vary the material property associated with the light spring, allowing the three springs to interact in turn via the same mass M at their common free end. If we introduce a set of real, constant interaction coefficients U_{is}, U_{sj} and U_{ij}, the results of the experiment will give:

$$\frac{df_i}{dt} = U_{is}\frac{df_s}{dt} \text{ for } ^\#i \text{ and } ^\#s$$

$$\frac{df'_s}{dt} = U_{sj}\frac{df'_j}{dt} \text{ for } ^\#j \text{ and } ^\#s \quad (1\text{–}15)$$

$$\frac{df''_i}{dt} = U_{ij}\frac{df''_j}{dt} \text{ for } ^\#i \text{ and } ^\#j$$

where the three interaction coefficients form a group, so that:

$$U_{ij} = U_{is}U_{sj} = U_{si}^{-1}U_{sj} \quad (1\text{–}16)$$

We could now proceed to repeat in exact detail all the arguments of the primal interaction experiment discussion, defining a spring compliance as

U_{si} = compliance of spring $^\#i$ with respect to a standard spring $^\#s$

and so on. There is little point in doing this however; all that is important about the dual interaction experiment is the concept of *mechanical*

Translational Mechanical Variables 9

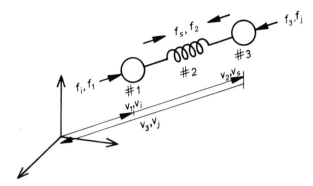

Figure 1-3 Identifiers and positive reference directions for analysis of primal interaction experiment.

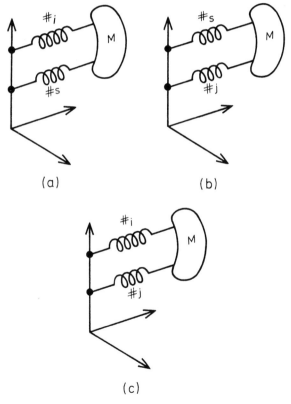

Figure 1-4

duality. It is important to realize that, from the point of view of causal mechanical theory, there is no *a priori* reason why one set of mechanical variables should be regarded as in any way 'more fundamental' than another. Both velocity, which is a visual-intuitive concept, and force, which is a tactile-intuitive concept, are of equal standing in intuition and in the formal structure of practical dynamics. It is natural, however, to prefer the visual-intuitive concepts of distance, velocity and acceleration to the tactile-intuitive concept for the purposes of standardization. Added to this the *linearity*, for most practical purposes, of the mass constitutive relationship completes the reasons why the triad of mass, length and time are taken as the basic quantities for the dimensional classification of mechanical variables.

Mechanical duality. Interactions take place between sets of objects, and the simplest interactions on a study of which the methodology of dynamics is based are those in which pairs of objects of one sort interact via an object of a different sort. For any such study we will always have a primal and a dual interaction experiment, one being obtained from the other by interchanging the roles played by the different types of object. The two properties of matter involved are said to be dual properties. Thus elasticity or the property of resisting deformation is the mechanical property dual to inertia or the property of resisting change of motion. This in turn leads to the primal and dual classification of the relations between variables:

(a) *Constitutive Relations*

The dual constitutive relations are those between momentum and velocity

$$p = Mv$$

and between force and displacement

$$f = \phi(q)$$

(b) *Dynamic Relations*

The dual dynamic relations are between force and momentum

$$f = \frac{dp}{dt}$$

and between velocity and displacement

$$v = \frac{dq}{dt}$$

(c) *Interconnective Relations*
The dual interconnective relations are those between forces

$$\sum_i f_i = 0 \quad \text{(D'Alembert's Principle)}$$

and between velocities

$$\sum_j v_j = 0 \quad \text{(Continuity of Space Law)}$$

In discussing such dual relationships we may refer, as convenient, to one relationship as primal and the other as dual. It is most important to realize that, from a dynamical point of view, there is no *a priori* significance in which relation we call primal and which dual. The duality in dynamics arises because two kinds of objects are required for an interaction, and for this reason alone.

While we are almost entirely concerned here with engineering systems in which mechanical interactions arise from the effects of elasticity and inertia, it is most important to remember that there is another type of mechanical interaction and therefore a further mechanical duality, namely that associated with gravitationally-interacting masses.

It has been pointed out by Hill [H4] that gravitationally-interacting masses may be treated on a purely kinematical basis by introducing the following postulate.

Gravitational postulate for mass points. Any celestial object P has associated with it a positive scalar magnitude μ_p, called its *gravitational mass*. Any other celestial object Q has an acceleration μ_p/r_{PQ}^2 towards P, where r_{PQ} is the distance between P and Q, additional to and independent of whatever acceleration Q would have in the absence of P.

Since the relative accelerations of a set of interacting celestial objects are all thus defined, such problems can be dealt with on a purely kinematical basis, without invoking the physical concept of force. If however we introduce the force concept into the analysis then the force exerted on objects by virtue of their interaction becomes proportional to their gravitational mass and we get a symmetrical description of the interaction relationships into which gravitational mass and inertial mass enter as dual concepts.

Analysis of Primal Interaction Experiment
The dynamical behaviour of the objects involved in the primal interaction experiment may now be analysed. We first choose a set of positive reference directions for the velocities and forces as shown in Figure 1–3. The component-labelling scheme adopted here is the identification of each component by a different index number. The parameters and variables associated with each component are then

identified by using this number as a subscript. With this convention, the system relationships may now be summarized as follows:

(a) *Constitutive Relationships*

$$p_1 = M_1 v_1$$
$$p_3 = M_3 v_3 \qquad (1\text{--}17)$$
$$f_2 = K_2 q_2$$

(b) *Dynamical Relationships*

It is convenient to have these in both differential and integral form

$$f_1 = \frac{dp_1}{dt}$$
$$f_3 = \frac{dp_3}{dt} \qquad (1\text{--}18)$$
$$v_2 = \frac{dq_2}{dt}$$

$$p_1 = [p_1]_{t_0} + \int_{t_0}^{t} f_1 \, dt$$
$$p_3 = [p_3]_{t_0} + \int_{t_0}^{t} f_3 \, dt \qquad (1\text{--}19)$$
$$q_2 = [q_2]_{t_0} + \int_{t_0}^{t} v_2 \, dt$$

(c) *Interconnective Relationships*

$$v_1 + v_2 + v_3 = 0$$
$$f_1 = f_2 \qquad (1\text{--}20)$$
$$f_2 = f_3$$

These three sets of relationships completely define the dynamical system behaviour. To obtain from them an appropriate set of differential equations defining system behaviour, we first combine the constitutive and dynamical relationships into:

(d) *Component Relationships*

$$f_1 = M_1 \frac{dv_1}{dt}$$

$$f_3 = M_3 \frac{dv_3}{dt} \qquad (1\text{-}21)$$

$$v_2 = K_2^{-1} \frac{df_2}{dt}$$

where M_1 and M_3 are the inertial masses of the material particles and K_2 is the stiffness of the spring, i.e. the force per unit deflection.

Combining these component relationships with the interconnective relationships gives:

$$\frac{dv_1}{dt} = M_1^{-1} f_1 = M_1^{-1} f_2$$

$$\frac{dv_3}{dt} = M_3^{-1} f_3 = M_3^{-1} f_2 \qquad (1\text{-}22)$$

$$\frac{df_2}{dt} = K_2 v_2 = -K_2(v_1 + v_3)$$

We thus obtain the system differential equation set in matrix and vector form as

$$\frac{d}{dt}\begin{bmatrix} v_1 \\ v_3 \\ f_2 \end{bmatrix} = \begin{bmatrix} 0 & 0 & M_1^{-1} \\ 0 & 0 & M_3^{-1} \\ -K_2 & -K_2 & 0 \end{bmatrix} \begin{bmatrix} v_1 \\ v_3 \\ f_2 \end{bmatrix} \qquad (1\text{-}23)$$

This set of equations completely defines the system behaviour when taken together with the auxiliary equations relating displacement to velocity, etc. A consideration of the implications of equation (1–23) will show that for known, fixed values of inertial masses and spring stiffness, the future system behaviour is completely determined for any given initial set of values of v_1, v_3 and f_2, i.e. the two mass velocities and the spring force. Such a set of system variables is called a set of *state* variables.

The structural relationships between the dynamical system variables are clearly revealed by combining the constitutive, dynamical and interconnective relationships into the block diagram of Figure 1–5. From such a diagram it is obvious that the set of integrator outputs also constitute a set of system state variables. (The use of block diagrams is treated in detail in Chapter 3.)

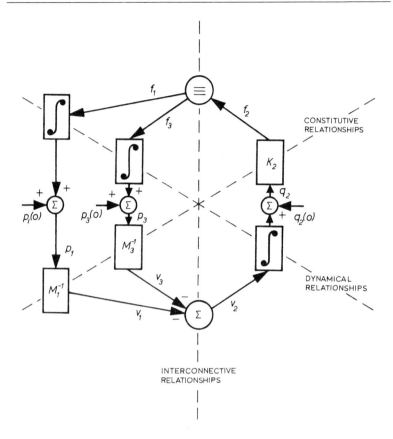

Figure 1–5 Computational block diagram for primal interaction experiment.

Analysis of Dual Interaction Experiment

For the dual interaction experiment, let the positive reference directions and identifying index numbers be as shown in Figure 1–6.

The system relationships may now be summarized as:

(a) *Constitutive Relationships*

$$f_1 = K_1 q_1$$
$$f_3 = K_3 q_3 \qquad (1\text{–}24)$$
$$p_2 = M_2 v_2$$

where K_1 and K_3 are the stiffnesses of the springs, i.e. the forces per unit deflection, and M_2 is the inertial mass of the material particle.

Translational Mechanical Variables 15

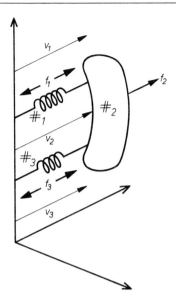

Figure 1–6 Identifiers and positive reference directions for analysis of dual interaction experiment.

(b) *Dynamical Relationships*

$$v_1 = \frac{dq_1}{dt}$$

$$v_3 = \frac{dq_3}{dt} \qquad (1\text{--}25)$$

$$f_2 = \frac{dp_2}{dt}$$

$$q_1 = [q_1]_{t_0} + \int_{t_0}^{t} v_1 \, dt$$

$$q_3 = [q_3]_{t_0} + \int_{t_0}^{t} v_3 \, dt \qquad (1\text{--}26)$$

$$p_2 = [p_2]_{t_0} + \int_{t_0}^{t} f_2 \, dt$$

(c) *Interconnective Relationships*

$$v_1 = v_2 = v_3$$
$$f_1 + f_2 + f_3 = 0 \qquad (1\text{--}27)$$

Combining the constitutive and dynamical relationships

(d) *Component Relationships*

$$\frac{df_1}{dt} = K_1 v_1$$

$$\frac{df_3}{dt} = K_3 v_3 \qquad (1\text{--}28)$$

$$\frac{dv_2}{dt} = M_2^{-1} f_2$$

To obtain a set of differential equations describing the system behaviour, the component and interconnective relationships are combined to give:

$$\frac{df_1}{dt} = K_1 v_1 = K_1 v_2$$

$$\frac{df_3}{dt} = K_3 v_3 = K_3 v_2 \qquad (1\text{--}29)$$

$$\frac{dv_2}{dt} = M_2^{-1} f_2 = -M_2^{-1}(f_1 + f_3)$$

from which we obtain

$$\frac{d}{dt}\begin{bmatrix} f_1 \\ f_3 \\ v_2 \end{bmatrix} = \begin{bmatrix} 0 & 0 & K_1 \\ 0 & 0 & K_3 \\ -M_2^{-1} & -M_2^{-1} & 0 \end{bmatrix} \begin{bmatrix} f_1 \\ f_3 \\ v_2 \end{bmatrix} \qquad (1\text{--}30)$$

Alternatively, using the integral form of the dynamical relationships, we obtain the block diagram of Figure 1–7.

Power and Energy

Consider the interconnective relationships for the pair of interaction experiments:

$$\left.\begin{array}{r} v_1 + v_2 + v_3 = 0 \\ f_1 = f_2 = f_3 \end{array}\right\} \text{primal}$$

$$\left.\begin{array}{r} v_1 = v_2 = v_3 \\ f_1 + f_2 + f_3 = 0 \end{array}\right\} \text{dual}$$

In each case a combination of the force and velocity variables gives

$$f_1 v_1 + f_2 v_2 + f_3 v_3 = 0 \qquad (1\text{--}31)$$

Translational Mechanical Variables

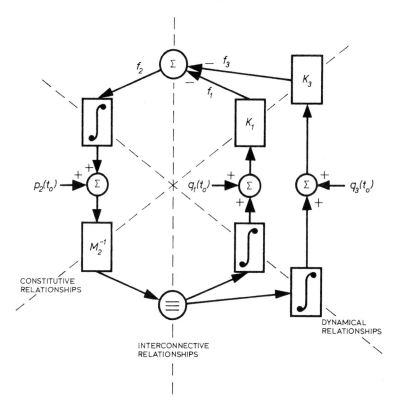

Figure 1-7 Computation block diagram for dual interaction experiment.

The quantity $f_i v_i$ is a measure of the instantaneous part which component #i is taking in the interaction. This quantitative measure of interaction is defined to be the component *power*. The system interconnective relationships must always be such that the power summed over all components involved in an interaction is identically zero. If we define the component energies E_i by

$$E_i \triangleq \int_{t_0}^{t} f_i v_i \, dt + [E_i]_{t_0} \tag{1-32}$$

$$\frac{dE_i}{dt} = f_i v_i \tag{1-33}$$

then

$$\sum_{i=1}^{3} E_i(t) = \text{constant} \tag{1-34}$$

by virtue of equation (1-31).

For an inertial mass component

$$\frac{dE_m}{dt} = vf = v\frac{dp}{dt}$$

giving, by definition, the *work increment* for an inertial mass component as

$$dW_m = dE_m = v\,dp \tag{1-35}$$

Work is defined as the quantitative measure of changing the state of a physical object.

For a spring component

$$\frac{dE_s}{dt} = fv = f\frac{dq}{dt}$$

giving, by definition, the work increment for a spring component as

$$dW_s = dE_s = f\,dq \tag{1-36}$$

Thus for an increase in energy from an initial reference value of zero we have for an inertial mass component

$$E_m(t) = \int_{t_0}^{t} v\frac{dp}{dt}\,dt = \int_{0}^{p} v\,dp = \int_{0}^{v} vM\,dv = \tfrac{1}{2}Mv^2 \tag{1-37}$$

and for a *linear* spring component

$$E_s(t) = \int_{t_0}^{t} f\frac{dq}{dt}\,dt = \int_{0}^{q} f\,dq = \int_{0}^{f} fK^{-1}\,df = \tfrac{1}{2}K^{-1}f^2 = \tfrac{1}{2}Kq^2 \tag{1-38}$$

1.3 Translational and Rotational Mechanical Components

Suppose we take a lump of metal and experiment with it under three different spatial constraints.

(a) Let it move freely in a near vacuum. Under these conditions we will find that it will behave like an ideal inertial mass, maintaining constant velocity in the absence of applied forces and accelerating in a manner linearly proportional to applied force.

(b) Let it be fixed at one end solidly to a massive object. Under these conditions we will find that it will behave like an ideal spring, maintaining a constant shape for a steady applied force, and deforming by an amount proportional (though not necessarily linearly) to applied force.

(c) Let it move freely through a heavy viscous liquid. Under these conditions it will require a constant force to maintain a constant velocity, and the resultant velocity will be proportional (though not necessarily linearly) to applied force. It is said to behave like an ideal dissipator.

Thus the *same* piece of matter would require three entirely different abstract dynamical model components to describe its behaviour under different forms of spatial constraint. If the spatial constraints were a mixture of the above then several ideal components would have to be combined into a composite model for the physical object.

Now consider the *transmission of power through* given mechanical objects. If we assume that the effect of dissipatance can be ignored we can identify two 'limiting cases'.

(a) The object 'absorbs force' in accelerating its inertial mass. The object may be represented by a compact mass with input and output forces transmitted by light stiff rods as shown in Figure 1–8(a).

The input and output rod velocities are the same, and the power absorbed in accelerating the inertial mass is $(f_1 - f_2)v$. The relationships between the variables involved are shown in the block diagram of Figure 1–8(b).

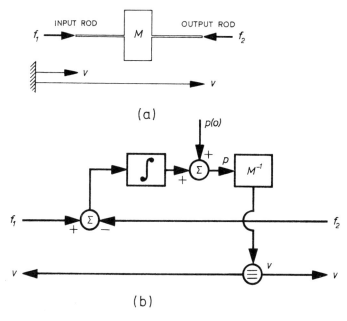

Figure 1-8

(b) The object 'absorbs velocity' in changing its shape. The object may be represented by a conventionally coiled long flexible spring as shown in Figure 1–9(a). The input and output rod forces are the same, and the power absorbed in changing the object shape is $(v_1 - v_2)f$.

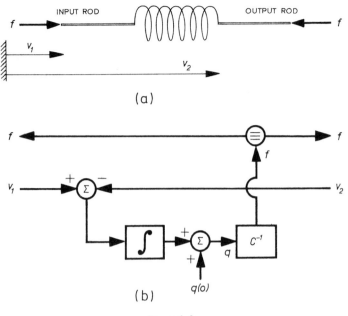

Figure 1-9

The relationships between the variables involved are shown in Figure 1-9(b), where C is the spring compliance, i.e. the deflection per unit force.

In general both these effects will be present together. We now consider the one-dimensional spatial aspects of the transmission of power through mechanical objects in order to get some quantitative idea of the validity of procedures which ignore the spatial variation of mechanical variables [M12]. Consider the *longitudinal* transmission of power through a relatively long bar of density ρ, uniform cross-sectional area A, and Bulk Modulus of Elasticity Y_B (bulk Young's Modulus). Let a small element of length Δx at x be translated to a small element of length $\Delta x'$ at x' in a time Δt, as illustrated by Figure 1-10. Let the corresponding force and velocity respectively change from f to $(f+\Delta f)$ and v to $(v+\Delta v)$, over a space Δx and in a time Δt.

For stretching we have

$$\text{compression per unit area} = -Y_B \frac{\partial x'}{\partial x}$$

$$\left[\frac{\text{compression force}}{A}\right]\Delta x = -Y_B \Delta x'$$

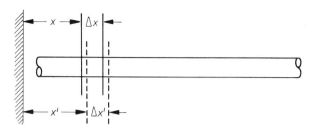

Figure 1–10

If the force transmitted at x is f then

$$f\Delta x = -AY_B \Delta x'$$

so that

$$\Delta x' = \frac{-f}{AY_B}\Delta x \qquad (1\text{–}39)$$

Let C be defined as the spring compliance per unit length of the bar, i.e. the deflection per unit force per unit length, then:

$$\Delta x' = -(C\Delta x)f$$

where

$$C = \frac{1}{AY_B}$$

Let I be defined as the bar mass per unit length. Thus

$$\text{mass of length } \Delta x = A\rho\Delta x = I\Delta x$$

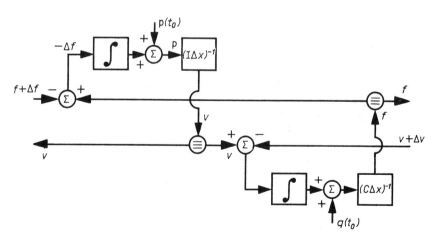

Figure 1–11

For a given 'micro-element' of bar of length Δx, the relationships between the forces and velocities involved may be illustrated by Figure 1–11.

From an inspection of Figure 1–11 we have

$$\int_{t_0}^{t} \Delta f \, dt + [p]_{t_0} = -I\Delta x v \tag{1-40}$$

$$\int_{t_0}^{t} \Delta v \, dt + [q]_{t_0} = -C\Delta x f \tag{1-41}$$

Partially differentiating (1–40) and (1–41) with respect to t gives

$$\Delta f = -I\Delta x \frac{\partial v}{\partial t}$$

$$\Delta v = -C\Delta x \frac{\partial f}{\partial t}$$

Proceeding to the limit in the normal way as Δx, Δf and Δv become arbitrarily small, we get the pair of first-order partial differential equations:

$$\frac{\partial f}{\partial x} = -I\frac{\partial v}{\partial t} \tag{1-42}$$

$$\frac{\partial v}{\partial x} = -C\frac{\partial f}{\partial t} \tag{1-43}$$

from which we may obtain the second-order equations:

$$\frac{\partial^2 f}{\partial x^2} = IC\frac{\partial^2 f}{\partial t^2} \tag{1-44}$$

$$\frac{\partial^2 v}{\partial x^2} = IC\frac{\partial^2 v}{\partial t^2} \tag{1-45}$$

for which it may easily be verified that

$$\psi(x+ct)$$

is a solution if

$$c = \frac{1}{\sqrt{(IC)}} = \frac{1}{\sqrt{[\rho A \cdot (1/AY_B)]}} = \sqrt{\frac{Y_B}{\rho}} \tag{1-46}$$

These solutions represent *travelling waves* of longitudinal force and velocity whose speed of propagation is $\sqrt{(Y_B/\rho)}$.

For various materials of interest we have:

Table 1–1

	Y_B N m^{-2}	ρ kg m^{-3}	c m s^{-1}
Rubber	1×10^9	0.93×10^3	1 040
Copper	18.4×10^{10}	8.9×10^3	4 560
Steel	27.5×10^{10}	7.8×10^3	5 940
Aluminium	10.8×10^{10}	2.7×10^3	6 320

This table shows that the velocity of propagation of longitudinal waves in such materials is of the order of several kilometres a second. Such results provide a guide to physical dimensions and speeds of variation for which it is reasonable to ignore propagation effects.

For the most detailed discussion of dynamical phenomena, both space and time variables must be used in the analysis as independent variables. In such an analysis, all physical variables must be determined both as functions of time and also as functions of space variables. In many situations of engineering interest, the system behaviour may be described sufficiently accurately if the space variable aspect of the analysis is neglected; this is equivalent to assuming that changes in certain variables are instantaneously propagated to all parts of the system. Such an assumption will be valid if the disturbances in the variables are propagated a distance much greater than the largest dimension of the physical system, in the smallest time interval of interest in the investigation. The limit for a reasonable approximation of system behaviour would correspond to a ratio of these distances of about ten to one.

For such a physical system representation, the corresponding dynamical model will consist of a finite number of abstract dynamical objects interacting in a finite number of ways. Such a model is said to be discrete or 'lumped', since its behaviour is defined in terms of discrete spatial concepts, and it thus represents the lumped-space behaviour of the physical system. The relationships between distributed and lumped systems will be further considered when we discuss electrical systems below.

Mechanical vertex. A mechanical vertex is a designated point in space with reference to which measurements defining mechanical system behaviour are made.

Mechanical terminal. A mechanical vertex at which a power measurement is made will be termed a mechanical terminal.

In what follows, for the purposes of defining the elements of lumped translational mechanical models, the speeds of time variation of the variables involved and the dimensions of the physical objects involved are taken to be such that an instantaneous propagation of the appropriate variables may be assumed.

1.3.1 Translational Spring

A translational spring is associated with a physical object where there is a spatial variation of relative velocity and a *uniform distribution of force at all times*. This uniform distribution of force at all times implies a negligible inertia as there is no 'absorption of force' through the object due to the acceleration of inertia. An ideal spring is defined as having negligible mass and dissipation, and as having a known functional relationship between deformation and applied force; a linear ideal spring will have a deformation linearly proportional to force. Since the ideal spring has zero mass, a force at one end must always be exactly balanced by an equal and opposite force at the other end. Thus force is *propagated directly through* a spring as shown in the schematic representation of Figure 1–12.

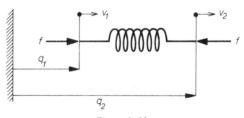

Figure 1–12

The ideal spring is associated with the triad of variables

velocity v
force f
displacement q

Interconnective relationships exist between the spring power variables, v and f, and other model element power variables which govern the interchange of power with the spring.

A *dynamical* relationship exists between the displacement and velocity variables

$$q_{21} = \int_{t_0}^{t} v_{21}\, dt + [q_{21}]_{t_0}$$
$$\frac{dq_{21}}{dt} = v_{21}$$
(1–47)

where $[q_{21}]_{t_0}$ is the equilibrium length of the spring at time t_0.

A *constitutive* relationship exists between the force and displacement variables

$$q_{21} - [q_{21}]_{t_0} = \phi(f) \tag{1-48}$$

For an ideal linear spring

$$f = Kq$$
$$q = K^{-1}f = Cf \tag{1-49}$$

where K is defined as the spring stiffness and C is defined as the spring compliance. For an ideal linear spring we thus have the component relationship between force and velocity variables:

$$\frac{df}{dt} = Kv \tag{1-50}$$

Classification of Nonlinear Springs
Displacement-controlled spring. A spring is said to be displacement-controlled if, for all values of time and all finite values of displacement, the spring force is a single-valued function of displacement.
Force-controlled spring. A spring is said to be force-controlled if, for all values of time and all finite values of force, the spring displacement is a single-valued function of spring force.
One-to-one spring. A spring is said to be one-to-one if it is both displacement-controlled and force-controlled.

Examples of constitutive characteristics for displacement-controlled, force-controlled, and one-to-one springs are given in Figures 1–13(a), (b) and (c) respectively. Figure 1–14 shows a linear spring constitutive characteristic.

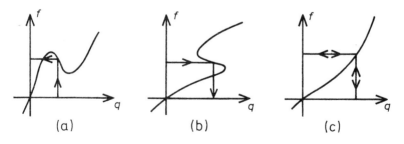

Figure 1–13

The work increment for a spring component has been shown to be

$$dW_s = f\, dq$$

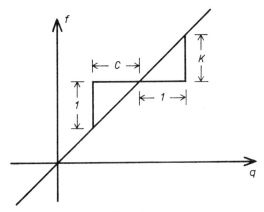

Figure 1–14

so that the quantity

$$\int_0^q f\, dq$$

has the dimensions of energy. For a one-to-one spring, let the spring displacement be changed by an amount q from the equilibrium condition. Then the corresponding *stored potential energy* is defined as

$$U_{tm} \triangleq \int_0^q f\, dq \qquad (1\text{--}51)$$

and is shown as a shaded area on the constitutive relationship of Figure 1–15. The complementary integral [C3], [S3], [C4]

$$U_{tm}^* \triangleq \int_0^f q\, df \qquad (1\text{--}52)$$

is defined as the *potential co-energy* of the spring.

Thus, if for a given spring the potential energy U_{tm} is a known function of the displacement q, the force corresponding to the given displacement may be evaluated as

$$f = \frac{\partial U_{tm}}{\partial q} \qquad (1\text{--}53)$$

and if the potential energy U_{tm}^* is a known function of the force f then the displacement for a given force may be evaluated as

$$q = \frac{\partial U_{tm}^*}{\partial f} \qquad (1\text{--}54)$$

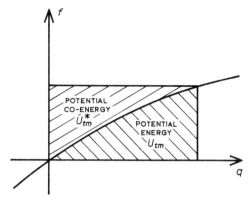

Figure 1-15

This gives the component relationship between force and velocity for an ideal spring as

$$v = \frac{d}{dt}\left(\frac{\partial U_{tm}^*}{\partial f}\right) \qquad (1\text{-}55)$$

Legendre transformation. We note that, for an ideal spring, if we are working in terms of U_{tm} and q we can find U_{tm}^* and f by

$$U_{tm}^* = qf - U_{tm}$$
$$f = \frac{\partial U_{tm}}{\partial q} \qquad (1\text{-}56)$$

and, if we are working in terms of U_{tm}^* and f we can find U_{tm} and q by

$$U_{tm} = fq - U_{tm}^*$$
$$q = \frac{\partial U_{tm}^*}{\partial f} \qquad (1\text{-}57)$$

The completely symmetrical relationships of equations (1-56) and (1-57) are a simple example of Legendre's Dual Transformation, which plays an important role in classical mechanics and which occurs in Chapter 4 in connection with the transformation between Lagrangian and Hamiltonian equations of motion.

1.3.2 Translational Mass

A translational mass is associated with a physical object for which there is no appreciable variation of relative velocity (i.e. a rigid object whose parts all 'move together'). An ideal mass is associated with the triad of variables:

velocity v
force f
momentum p

Interconnective relationships exist between the mass power variables, v and f, and other model element power variables which govern the interchange of power with the mass.

A dynamical relationship exists between the momentum and force variables

$$p = [p]_{t_0} + \int_{t_0}^{t} f \, dt$$

$$f = \frac{dp}{dt}$$
(1–58)

A constitutive relationship exists between the momentum and velocity variables

$$p = Mv \qquad (1\text{–}59)$$

where M is the inertial mass. In the Special Theory of Relativity [C4] this is modified to

$$p = \frac{Mv}{\sqrt{[1-(v^2/c^2)]}} \qquad (1\text{–}60)$$

but we shall treat the mass as an essentially linear element since only velocities small compared with c, the velocity of light waves, will be considered for mechanical systems. The linear constitutive characteristic for a translational mass is shown in Figure 1–16. If a mass is

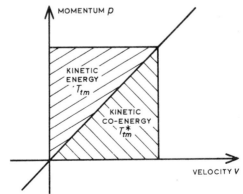

Figure 1–16

accelerated from rest at time t_0 to a time t, then the translational mechanical stored kinetic energy is

$$T_{tm} = \int_{t_0}^{t} fv\, dt = \int_{t_0}^{t} v\frac{dp}{dt}\, dt = \int_{0}^{p(t)} v\, dp \qquad (1\text{--}61)$$

where $p(t)$ is the final value of momentum. This kinetic energy is shown as a shaded area on Figure 1–16. The complementary integral

$$\int_{0}^{v(t)} p\, dv = T_{tm}^* \qquad (1\text{--}62)$$

is defined as the translational mechanical kinctic co-energy. If the mass starts from rest it is the shaded area shown in Figure 1–16.

The linearity of the constitutive relationship for an inertial mass means that the kinetic energy and kinetic co-energy of a translational mass are always equal. This is of some importance in considering the historical development of mechanics, since Lagrange had no need to distinguish between them in formulating his original equations of motion. For the mass we have

$$v = \frac{\partial T_{tm}}{\partial p} \quad \text{and} \quad p = \frac{\partial T_{tm}^*}{\partial v} \qquad (1\text{--}63)$$

when the kinetic energy T_{tm} and the kinetic co-energy T_{tm}^* are known functions of the momentum and velocity respectively. This gives the component relationship for a mass as

$$f = \frac{d}{dt}\left(\frac{\partial T_{tm}^*}{\partial v}\right) \qquad (1\text{--}64)$$

which, since the mass constitutive characteristic will always be linear in a practical engineering system, simply reduces to

$$f = M\frac{dv}{dt}$$

1.3.3 Translational Converters

An ideal translational mechanical converter is defined as an object which exhibits no inertial or spring effects and for which the force exerted by the object is related in a known way to the object terminal velocities. A translational mechanical converter is thus defined in terms of a constitutive characteristic of the type shown in Figure 1–17(a). Power is defined as positive when it flows into the converter. The first and third quadrants of the converter constitutive characteristic thus correspond to power flowing into the converter, and the second and fourth quadrants correspond to a flow of power out of the converter.

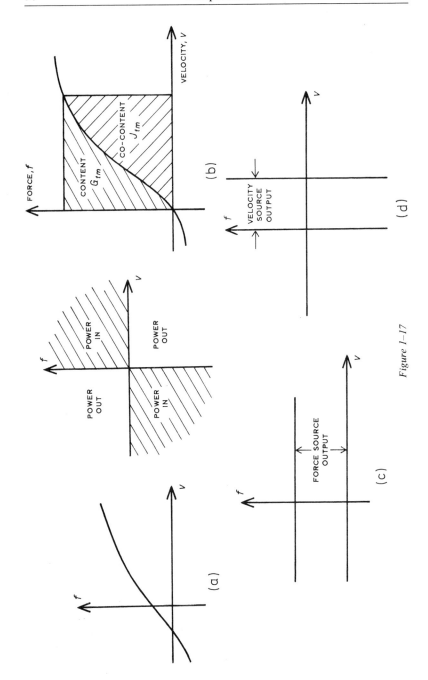

Figure 1–17

The converters are conveniently divided into two types:

(a) *Dissipators.* The constitutive characteristic of a dissipator passes through the origin and is confined to the first and third quadrants. It therefore always absorbs power and is used in network models to represent the irreversible conversion of energy to thermal form. A translational dissipator characteristic is shown in Figure 1–17(b).

(b) *Sources.* A source is defined as a converter in which one power variable, which may be any specified function of time, is independent of the other. The constitutive characteristics for constant force and constant velocity sources are shown in Figures 1–17(c) and (d) respectively.

Any arbitrary converter may be represented by a combination of source and dissipator; consequently we will only deal with these two generic types of converter. For a similar reason, we will only deal with passive storage element constitutive characteristics which pass through the origin since any active storage element may be represented in terms of a source combined with a passive element.

An ideal translational mechanical dissipator is defined as an object which exhibits no inertial mass or spring effects and which exerts equal and opposite forces at its ends which are a function of the relative velocity between its two terminals. The dissipator forces always oppose the relative motion of its two terminals. It is defined by a constitutive relationship of the form

$$f = \phi(v) \tag{1–65}$$

as illustrated by Figure 1–17(b).

If the constitutive relationship is nonlinear then we can define force-controlled, velocity-controlled and one-to-one dissipators in a manner analogous to the definition of force-controlled, displacement-controlled and one-to-one springs. The power absorbed in a one-to-one dissipator is

$$fv = \int_0^v f\, dv + \int_0^f v\, df = J_{tm} + G_{tm} \tag{1–66}$$

where

$$G_{tm} \triangleq \int_0^f v\, df \tag{1–67}$$

is defined as the dissipator content. If the content is a known function of the dissipator force, then the corresponding velocity is given by

$$v = \frac{\partial G_{tm}}{\partial f} \tag{1–68}$$

The complementary quantity

$$J_{tm} \triangleq \int_0^v f \, dv \qquad (1\text{-}69)$$

is defined as the dissipator co-content. If the co-content is a known function of the dissipator velocity, then the corresponding force is given by

$$f = \frac{\partial J_{tm}}{\partial v} \qquad (1\text{-}70)$$

For an ideal linear dissipator we have

$$f = Bv \qquad (1\text{-}71)$$

where B is defined as the viscous damping coefficient of the dissipator.

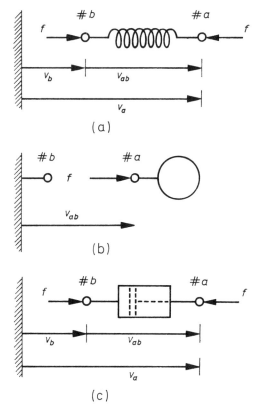

Figure 1–18

1.3.4 Representation of Translational Mechanical Systems by Linear Graphs [T1], [S3]

Schematic diagrams of the three idealized physical objects corresponding to the mass, spring and dissipator translational mechanical components are shown in Figure 1–18. In each case the component behaviour is defined in terms of a pair of points in space, labelled $^\#a$ and $^\#b$ on the diagrams. On schematic diagrams the orientation arrows for forces show the positive direction of force exerted on objects; for velocity orientations an arrow pointing from $^\#b$ to $^\#a$ shows the positive direction of the relative velocity of $^\#a$ with respect to $^\#b$.

Network diagram elements. A network diagram is a form of linear graph used for the representation of lumped dynamical system models. Each vertex of the graph corresponds to a point in space with respect to which actual or conceptual measurements of physical variables are carried out. Each line segment is a coded symbol which represents the sets of relationships between measured variables which define an ideal component. The network diagram elements corresponding to the idealized objects of Figure 1–18 are shown in Figure 1–19.

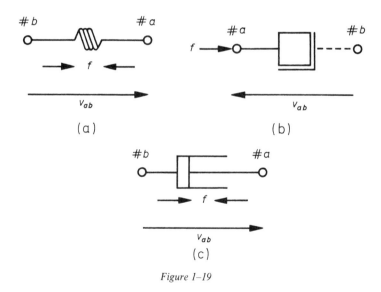

Figure 1–19

The same force and velocity orientation conventions are adopted for network diagrams as for the schematic diagrams. The network diagram symbols for the spring and the dissipator are self-explanatory. The object terminal on a mass symbol represents a specific point on the physical object modelled; the reference terminal represents a fixed point on an inertial reference system. The *L*-shaped guide associated with the reference terminal emphasises the fact that the symbol represents the storage of kinetic energy in an object constrained to have only one degree of freedom in translational movement or the energy corresponding to one degree of freedom of a less constrained object. The dotted line from the guide to the reference terminal emphasizes that no force is propagated to the reference terminal, all the force being absorbed in accelerating the inertia.

Oriented line segment representation. For each of the components in Figure 1–18, the power absorbed by an element may be conceptually determined by a simultaneous measurement of force and velocity. For example, the power for a spring may be conceptually measured by an ideal force-measuring instrument of zero mass through which force is applied to the spring, and an ideal velocity-measuring instrument connected across the spring, as shown in Figure 1–20(a). In order to determine the algebraic sign to be allocated to the power, both measuring instruments must be allocated a polarity.† In Figure 1–20(a) positive reference directions for velocity and force applied by the spring are indicated and we could say that the velocity meter will read positive when $^\#a$ is moving away from $^\#b$ and the force meter will read positive when the spring is being extended. If a consistent set of conventions is adopted for every pair of meters measuring element power in a system, then the pair of measurements may be represented by a single oriented line segment as shown in Figure 1–20(b).

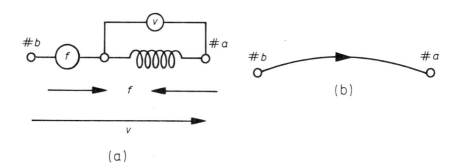

Figure 1–20

† Force orientations may be taken as showing forces exerted on objects or by objects. Either interpretation may be used, providing that it is consistently applied to the whole diagram. Normally, positive force and positive velocity mean a power flow into an object.

Translational and Rotational Mechanical Components

For an oriented linear graph representation of a translational mechanical system, the orientations of the various line segments involved must be interpreted with respect to a single arbitrary positive reference direction. The fact that we can use a single reference direction follows immediately from the fact that we are, by definition, only considering motions along a single linear direction. The positive direction of power is taken to be into an object, i.e. positive work is done on an object. The conventions adopted for interpreting orientated line segments representing a pair of power variable measurements are shown in Figure 1–21.

The oriented line segment conventions adopted for the spring and the dissipator are the same. The mass oriented line segment convention is shown in Figure 1–22. The only difference is that the line segment is

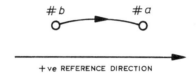

(1) v_{ab} is positive, i.e. $\#a$ is instantaneously moving away from $\#b$.

(2) The positive directions of force exerted *on* the object are:

(3) The positive directions of force exerted *by* the object are:

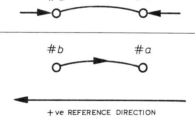

(1) v_{ab} is negative, i.e. $\#a$ is instantaneously moving towards $\#b$.

(2) The positive directions of force exerted *on* the object are:

(3) The positive directions of force exerted *by* the object are:

Figure 1–21 Interpretation of linear segment orientations.

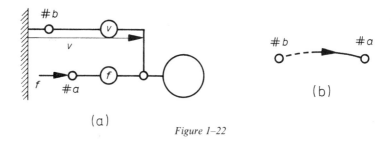

Figure 1-22

dashed at the reference terminal end to emphasize again that no force is propagated to the reference terminal. [S3]

Direct Representation of Simple Systems by Networks

Certain simple physical systems can be regarded as a set of distinct, interacting physical objects whose behaviour may be defined in terms of pairs of measurements carried out with respect to pairs of points in space. Consider the simple translational mechanical system consisting of two springs and two masses shown in Figure 1–23(a), and assume that the motion is not affected by gravitational effects or by any significant dissipation. The system is attached to an inertial reference at point $^\#a$ and only horizontal motions of the system are to be considered.

First consider each part of the system in turn *in isolation from the remainder of the system*. When the rest of the system is removed, the dynamical behaviour of spring $^\#1$ may be defined in terms of a measured relationship between the displacement between points $^\#a$ and $^\#b$ and the equal and opposite forces applied at points $^\#a$ and $^\#b$. Similarly the isolated behaviour of spring $^\#3$ may be defined in terms of a measured relationship between the displacement between points $^\#b$ and $^\#c$ and the equal and opposite forces applied at points $^\#b$ and $^\#c$. Taking point $^\#a$ as the inertial reference point, the dynamical behaviour of the isolated mass $^\#2$ may be defined in terms of a measured relationship between a force applied at point $^\#b$ and the resultant acceleration between points $^\#a$ and $^\#b$; similar measurements with respect to points $^\#a$ and $^\#c$ define the isolated dynamical behaviour of mass $^\#4$. A complete description of the dynamical behaviour of the physical system requires the specification of:

(a) the relationships between the measured variable pairs and their associated variables, and
(b) the way in which the objects comprising the system are connected together. The way in which the connected system behaves would be investigated by connecting the appropriate set of instruments to the assembled system. Such an assembled set of pairs of measuring instruments may be represented by the *network diagram* of Figure 1–23(b).

Translational and Rotational Mechanical Components

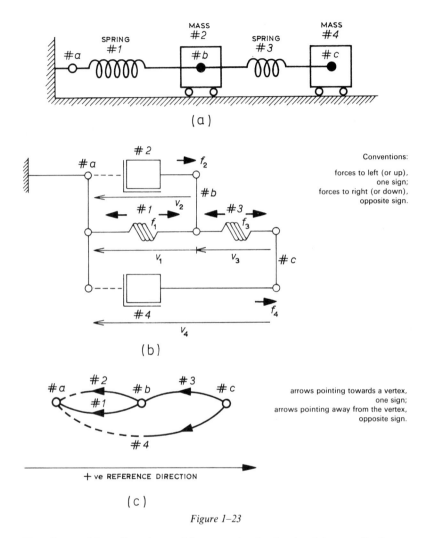

Figure 1–23

For the positive directions of force and velocity (and hence displacement and acceleration) orientations shown on Figure 1–23(b) we have the system oriented linear graph shown in Figure 1–23(c).

Mechanical systems are analysed with respect to a specified inertial reference framework. A mechanical vertex is defined as a specified point in space, usually the junction point of several components.
Mechanical circuit. A mechanical circuit is defined as a closed path in space which includes one point on the inertial reference framework.

Translational mechanical system postulates. The analysis of translational mechanical systems is based on the following pair of postulates:
Mechanical vertex postulate. The sum of all component forces exerted on a mechanical vertex is identically zero.
Mechanical circuit postulate. The sum of all component velocities taken round a mechanical circuit is identically zero.

These postulates define the spatial or interconnective relationships among the system power variables. For any translational mechanical system we therefore have the following sets of relationships:

Constitutive relationships $\Big\}$ Component relationships between
Dynamical relationships   power variables.

Interconnective relationships $\Big\{$ Vertex relationships
between power variables.   Circuit relationships

Example. The procedure for the analysis of mechanical systems may now be introduced by the simple example of the free motion of the system of Figure 1–23. We proceed as follows:

(a) Write down the constitutive relationships for each component. For the purposes of this example we take all the components to be linear.

$$f_1 = K_1 q_1$$
$$f_3 = K_3 q_3$$
$$p_2 = M_2 v_2$$
$$p_4 = M_4 v_4$$

The variable and parameter identifying scheme adopted is to give each component an identifying number which is used as a subscript on all its variables and parameters.

(b) Write down the dynamical relationships for each component.

$$v_1 = \frac{dq_1}{dt} \qquad f_2 = \frac{dp_2}{dt}$$
$$v_3 = \frac{dq_3}{dt} \qquad f_4 = \frac{dp_4}{dt}$$

(c) Combine the constitutive and dynamical relationships into component relationships between the object power variables.

$$\frac{df_1}{dt} = K_1 v_1 \qquad \frac{df_3}{dt} = K_3 v_3$$
$$f_2 = M_2 \frac{dv_2}{dt} \qquad f_4 = M_4 \frac{dv_4}{dt}$$

(d) Write down the interconnective vertex and circuit relationships between the power variables.

$$f_1 + f_2 - f_3 = 0$$
$$f_3 + f_4 = 0$$
$$v_3 + v_1 - v_4 = 0$$
$$v_1 - v_2 = 0$$

(e) Select an *independent* set of power variables whose derivatives occur in the component relationship set. In this case it is obviously f_1, f_3, v_2 and v_4.

(f) Use the interconnective relationships together with the component relationships to form a set of first-order differential equations in the set of independent power variables.

$$\frac{df_1}{dt} = K_1 v_1 = K_1 v_2$$

$$\frac{df_3}{dt} = K_3 v_3 = K_3(-v_2 + v_4)$$

$$\frac{dv_2}{dt} = M_2^{-1} f_2 = M_2^{-1}(-f_1 + f_3)$$

$$\frac{dv_4}{dt} = M_4^{-1} f_4 = M_4^{-1}(-f_3)$$

Or, in matrix and vector form:

$$\frac{d}{dt}\begin{bmatrix} f_1 \\ f_3 \\ v_2 \\ v_4 \end{bmatrix} = \begin{bmatrix} 0 & 0 & K_1 & 0 \\ 0 & 0 & -K_3 & K_3 \\ -M_2^{-1} & M_2^{-1} & 0 & 0 \\ 0 & -M_4^{-1} & 0 & 0 \end{bmatrix} \begin{bmatrix} f_1 \\ f_3 \\ v_2 \\ v_4 \end{bmatrix}$$

1.3.5 Relationships between Translational and Rotational Mechanical Interactions

The work done in changing the mechanical state of a rotational object is simply calculated in terms of translational system variables. Suppose, as illustrated in Figure 1–24, we have an object $^\#i$ of mass δM_i attached to a rigid rod of negligible inertia in such a way as to rotate freely about a fixed axis and in a fixed plane. Let a force f be applied to the object at the point A and let f_r be the component of this applied force perpendicular to the line OA. Let δs be the distance moved by the point of application of the force f_r for a small rotation $\delta\theta$.

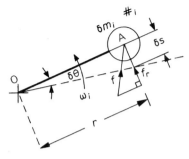

Figure 1–24

Then

$$\text{work done on system} = \delta W = f_r \delta s = f_r r \delta\theta = \tau \delta\theta$$

where the quantity τ is defined as the *torque* exerted on the rotational object.

The result of applying a force to the object $^\#i$ is to accelerate it, and we thus have:

$$f_r = \delta M_i \frac{d^2 s}{dt^2} = \delta M_i \frac{d^2(r\theta)}{dt^2} = \delta M_i r \frac{d^2\theta}{dt^2}$$

$$= \delta M_i r \frac{d\omega_i}{dt}$$

where

$$\omega_i = \frac{d\theta_i}{dt}$$

is defined as the angular velocity of the object $^\#i$.

We thus have:

$$\delta W = f_r r \delta\theta_i$$

$$= \left[\delta M_i r \frac{d\omega_i}{dt}\right] r \delta\theta_i$$

$$= \delta M_i r^2 \frac{d\omega_i}{dt} \delta\theta_i$$

$$= \lim_{\delta t \to 0} \delta M_i r^2 \frac{\delta\omega_i}{\delta t} \delta\theta_i$$

$$= \lim_{\delta t \to 0} \delta M_i r^2 \frac{\delta\theta_i}{\delta t} \delta\omega_i$$

$$= \delta M_i r^2 \omega_i \delta\omega_i$$

$$= \omega_i [\delta M_i r^2 \delta\omega_i]$$

If we now define the *angular momentum* of the object #i to be

$$\sigma_i \triangleq \delta M_i r^2 \omega_i$$

we have $\quad\quad\quad\quad \delta W = \omega_i \delta \sigma_i \quad\quad\quad\quad$ (1-72)

The *rotational inertia* of the object #i is defined to be

$$I_i = r^2 \delta M_i \quad\quad\quad\quad (1\text{-}73)$$

The unit standard of rotational inertia is simply related to the unit standards of mass and length by defining unit rotational inertia to be the equivalent of unit mass rotating at unit distance in a plane about a fixed axis. If a more complex object is composed of n elements of mass at known radial distances from a common axis of rotation, then its moment of inertia is defined to be

$$I = \sum_{i=1}^{n} r_i^2 \delta M_i \quad\quad\quad\quad (1\text{-}74)$$

For more general objects of simple continuous shape the moment of inertia is evaluated by simple integration. [C4]

Since the incremental work done in changing the state of a rotational mechanical system is $\tau \delta \theta$ we have:

$$\text{rotational component power} = \tau \frac{d\theta}{dt} = \tau \omega \quad\quad (1\text{-}75)$$

From this we deduce the dynamical relationship for a rotational inertia. Since the increment of work may be expressed as

$$\delta W_{ri} = \omega \delta \sigma$$

we have $\quad\quad\quad\quad \text{power} = \omega \dfrac{d\sigma}{dt} = \omega \tau$

so that $\quad\quad\quad\quad \tau = \dfrac{d\sigma}{dt} \quad\quad\quad\quad$ (1-76)

We thus have for the stored kinetic energy of a rotational inertia accelerated from rest

$$E_{ri}(t) = \int_0^t \omega \frac{d\sigma}{dt} dt = \int_0^\sigma \omega \, d\sigma = \int_0^\omega I\omega \, d\omega = \tfrac{1}{2} I \omega^2 \quad (1\text{-}77)$$

1.3.6 Rotational Mechanical Components

If wave equations are derived for the propagation of torsional waves it is found that the propagation of angular velocity and torque changes are governed by [M12]:

$$\frac{\partial^2 \omega}{\partial x^2} = \frac{\rho}{C} \frac{\partial^2 \omega}{\partial t^2} \qquad (1\text{--}78)$$

$$\frac{\partial^2 \tau}{\partial x^2} = \frac{\rho}{C} \frac{\partial^2 \tau}{\partial t^2} \qquad (1\text{--}79)$$

where ρ is the material density and C is the torsional rigidity (or shear modulus) which, for thin rods, is related to Young's modulus Y_0 by

$$C = \frac{Y_0}{2(\bar{\sigma} + 1)}$$

where

$$Y_0 \approx \tfrac{3}{4} Y_B$$

for most stiff materials and $\bar{\sigma}$, the Poisson ratio is in a range of approximately $\tfrac{1}{3}$ to $\tfrac{1}{4}$ for most stiff materials.

The velocity of propagation of torsional waves is thus

$$v = \sqrt{\frac{C}{\rho}} \approx \sqrt{\frac{3 Y_B}{10\rho}} \qquad (1\text{--}80)$$

and we may thus estimate the speed of propagation of torsional waves using the data in Table 1–1.

Rotational spring. A rotational spring is associated with a physical object where there is a spatial variation of relative angular velocity and a *uniform distribution of torque at all times*. This uniform distribution of torque at all times implies that there is a negligible rotational inertia since there is no 'absorption of torque' through the element due to the angular acceleration of rotational inertia. An ideal rotational spring is defined as having negligible rotational inertia and dissipation, and as having a known functional relationship between angular deformation and applied torque; a linear rotational spring will have an angular deformation linearly proportional to torque. For a linear rotational spring, rotated from its equilibrium configuration, the stored energy

$$E_{rs}(t) = \int_0^t \tau \frac{d\theta}{dt} dt = \int_0^\theta \tau\, d\theta = \int_0^\tau K_r^{-1} \tau\, d\tau = \tfrac{1}{2} K_r^{-1} \tau^2 = \tfrac{1}{2} K_r \theta^2 \qquad (1\text{--}81)$$

where K_r is defined as the rotational stiffness of the linear rotational spring, i.e. the torque exerted per unit angular deflection.

Since the ideal rotational spring has zero rotational inertia, a torque at one end must always be exactly balanced by an equal and opposite torque at the other end. Thus torque is propagated directly through a rotational spring. The ideal rotational spring is associated with the triad of variables

> angular velocity ω
> torque τ
> angular displacement θ

Interconnective relationships exist between the rotational spring power variables, ω and τ, and other model element power variables which govern the interchange of power with the rotational spring.

A dynamical relationship exists between the angular displacement and angular velocity variables:

$$\theta_{21} = \int_0^t \omega_{21}\, dt + [\theta_{21}]_{t_0}$$

$$\frac{d\theta_{21}}{dt} = \omega_{21}$$

(1-82)

where $[\theta_{21}]_{t_0}$ is the equilibrium angle of the spring at time t_0.

A constitutive relationship exists between the torque and angular displacement variables

$$\theta_{21} - [\theta_{21}]_{t_0} = \phi(\tau) \tag{1-83}$$

For an ideal linear spring

$$\left.\begin{array}{l} \tau = K_r \theta \\ \theta = K_r^{-1}\tau = C_r \tau \end{array}\right\} \tag{1-84}$$

where K_r is the rotational stiffness and C_r is the rotational compliance. For an ideal linear spring we thus have the component relationship between torque and angular velocity variables:

$$\frac{d\tau}{dt} = K_r \omega \tag{1-85}$$

The constitutive characteristics for rotational springs may be classified as torque-controlled, displacement-controlled, and one-to-one in a manner exactly analogous to the classification of translational spring characteristics.

The work increment for a rotational spring component has been shown to be

$$dW_{rs} = \tau\, d\theta$$

The stored rotational mechanical potential energy is defined as

$$U_{rm} = \int_0^\theta \tau\, d\theta \tag{1-86}$$

for a rotation θ from the equilibrium condition.

The corresponding rotational mechanical potential co-energy is defined as

$$U_{rm}^* = \int_0^\tau \theta \, d\tau \tag{1-87}$$

For a one-to-one spring displacement from its equilibrium condition, these quantities are shown as areas in Figure 1–25. If the rotational potential energy is a known function of the angular displacement, the corresponding torque is given by

$$\tau = \frac{\partial U_{rm}}{\partial \theta} \tag{1-88}$$

If the rotational potential co-energy is a known function of the torque, the corresponding angular displacement is given by

$$\theta = \frac{\partial U_{rm}^*}{\partial \tau} \tag{1-89}$$

This gives the component relationship between torque and angular velocity as

$$\omega = \frac{d}{dt}\left(\frac{\partial U_{rm}^*}{\partial \tau}\right) \tag{1-90}$$

Rotational Inertia

A rotational inertia is associated with a rigid body rotating about an axis; it is associated with the triad of variables:

- angular velocity ω
- torque τ
- angular momentum σ

Interconnective relationships exist between the rotational inertia power variables, ω and τ, and other model element power variables which govern the interchange of power with the inertia.

A dynamical relationship exists between the angular momentum and torque variables:

$$\sigma = [\sigma]_{t_0} + \int_{t_0}^t \tau \, dt$$

$$\tau = \frac{d\sigma}{dt} \tag{1-91}$$

A constitutive relationship exists between the angular momentum and angular velocity variables:

$$\sigma = I\omega \tag{1-92}$$

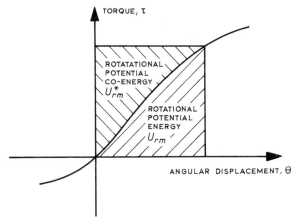

Figure 1-25

where I is the rotational inertia. If the rotational inertia is accelerated from rest then:

$$\text{stored rotational kinetic energy} = T_{rm} \triangleq \int_0^\sigma \omega \, d\sigma \qquad (1\text{-}93)$$

and the corresponding rotational mechanical kinetic co-energy is defined as

$$T_{rm}^* \triangleq \int_0^\omega \sigma \, d\omega \qquad (1\text{-}94)$$

If the rotational kinetic energy is a known function of angular momentum, the corresponding angular velocity is given by

$$\omega = \frac{\partial T_{rm}}{\partial \sigma} \qquad (1\text{-}95)$$

Similarly, if the rotational kinetic co-energy is a known function of the angular velocity, the corresponding angular momentum is given by

$$\sigma = \frac{\partial T_{rm}^*}{\partial \omega} \qquad (1\text{-}96)$$

This gives the component relationship for a rotational inertia as

$$\tau = \frac{d}{dt}\left(\frac{\partial T_{rm}^*}{\partial \omega}\right) \qquad (1\text{-}97)$$

or

$$\tau = I \frac{d\omega}{dt} \qquad (1\text{-}98)$$

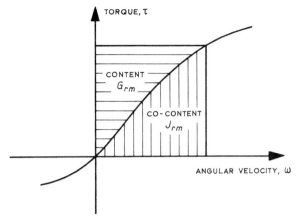

Figure 1-26

Rotational Dissipator

An ideal rotational dissipator is an object which exhibits no rotational inertia or rotational spring effects and which exerts equal and opposite torques at its ends which are a function of the relative angular velocity between its input and output shafts. It is defined by a constitutive relationship of the form

$$\tau = \phi(\omega) \tag{1-99}$$

as illustrated in Figure 1-26. If the constitutive relationship is nonlinear then we can define torque-controlled, angular-velocity controlled, and one-to-one rotational dissipators. The power absorbed in a one-to-one rotational dissipator is

$$\omega\tau = \int_0^\omega \tau \, d\omega + \int_0^\tau \omega \, d\tau = J_{rm} + G_{rm}$$

where
$$G_{rm} \triangleq \int_0^\tau \omega \, d\tau \tag{1-100}$$

is defined as the rotational dissipator content. If the content is a known function of the dissipator torque, then the corresponding angular velocity is given by

$$\omega = \frac{\partial G_{rm}}{\partial \tau} \tag{1-101}$$

The corresponding quantity

$$J_{rm} \triangleq \int_0^\omega \tau \, d\omega \tag{1-102}$$

is defined as the rotational dissipator co-content.

If the co-content is a known function of the dissipator angular velocity, then the corresponding torque is given by

$$\tau = \frac{\partial J_{rm}}{\partial \omega} \qquad (1\text{--}103)$$

For an ideal linear dissipator we have

$$\tau = B_r \omega \qquad (1\text{--}104)$$

where B_r is defined as the rotational viscous damping coefficient of the dissipator.

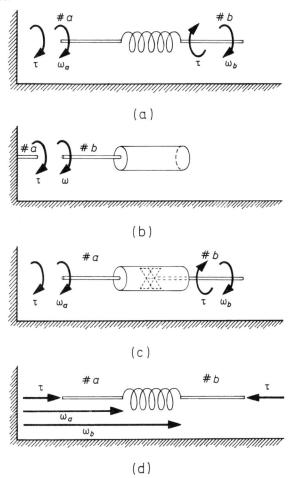

Figure 1-27

Representation of Rotational Mechanical Systems by Linear Graphs

Schematic diagrams of the three idealized physical objects corresponding to the rotational spring, rotational inertia and rotational damper are shown in Figure 1–27. On such diagrams the positive directions of angular velocity and of the torque applied to an object are indicated by circulatory arrows showing the positive rotational directions involved. Such circulatory arrow conventions may be replaced by an *axial* vector convention as shown in Figure 1–27(d) in which the positive direction of rotation is defined as that in which a right-handed screw would have to be rotated to advance in the direction of the arrow. This axial vector convention is adopted in linear graph representations of rotational mechanical systems in distinction to the *polar* vector convention used in the representation of translational systems.

The network diagram elements corresponding to these objects are shown in Figure 1–28 where the positive directions of torque applied to the objects are indicated by the axial vector orientation. If an axial velocity vector ω_{ab} points from $^{\#}a$ to $^{\#}b$ this means that $^{\#}a$ is rotating faster in the positive direction than $^{\#}b$. The vertices on a rotational mechanical network diagram represent lines (or axes) in space about which object rotations are specified.

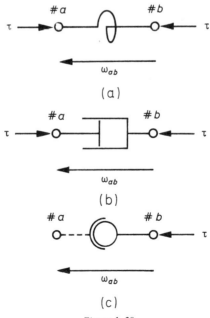

Figure 1–28

Translational and Rotational Mechanical Components 49

The oriented line segment conventions are shown in Figure 1–29. A similar convention to that for the mass is adopted for the line segment representation of a rotational inertia, the segment being dashed at the reference terminal end.

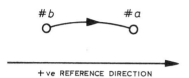

(1) ω_{ab} is positive, i.e. angular velocity at #a in +ve reference direction exceeds angular velocity at #b.

(2) The positive directions of torque exerted *on* the object are:

(3) The positive directions of torque exerted *by* the object are:

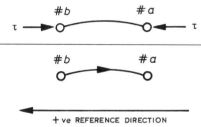

(1) ω_{ab} is negative i.e. angular velocity about #b in +ve reference direction exceeds angular velocity about #a.

(2) The positive directions of torque exerted *on* the object are:

(3) The positive directions of torque exerted *by* the object are:

Figure 1–29

Figure 1–30 shows a simple example of the representation of a rotational mechanical system by network diagram and oriented linear graph.

Sources. The constitutive characteristics for constant torque and constant angular velocity sources are shown in Figures 1–31 and 1–32. In general the torque and angular velocities may be any stipulated

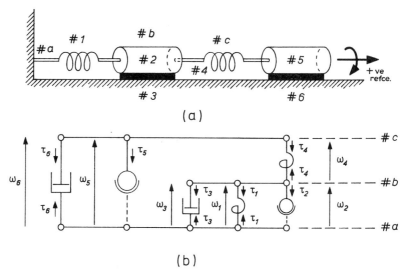

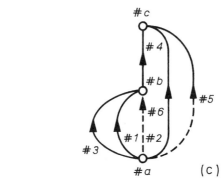

Figure 1–30

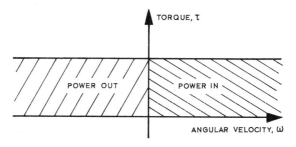

Figure 1–31

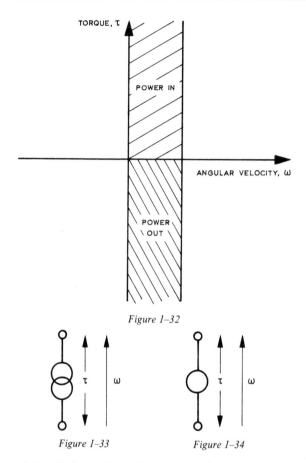

Figure 1-32

Figure 1-33   Figure 1-34

functions of time independent of angular velocity and torque respectively. The corresponding network diagram symbols are shown in Figures 1-33 and 1-34.

1.3.7 Mechanical Transformers

An ideal mechanical transformer is an element which exhibits no spring, inertia or dissipator effects but which changes torques, forces and velocities in such a way that the net power flow into the element is zero. The simplest examples are:

(a) Rotational transformer or gear train for which

$$\frac{\omega_i}{\omega_0} = \frac{-\tau_0}{\tau_i} = n \triangleq \text{gear ratio}$$

where

ω_i = input shaft angular velocity
ω_0 = output shaft angular velocity
τ_i = input shaft torque
τ_0 = output shaft torque

(b) Translational transformer or lever for which

$$\frac{v_i}{v_0} = \frac{-f_0}{f_i} = n \triangleq \text{lever ratio}$$

where

v_i = input point velocity
v_0 = output point velocity
f_i = input point force
f_0 = output point force

(c) Rotational to translational transformer or rack and pinion for which

$$\frac{\omega_i}{v_0} = \frac{-f_0}{\tau_i} = n \triangleq \text{transduction ratio}$$

where

ω_i = input pinion shaft angular velocity
v_0 = output rack translational velocity
τ_i = input shaft torque
f_0 = output point force

1.4 Fluid Components

In certain cases some simple fluid systems may be treated as lumped systems with enough accuracy for practical purposes. In such systems, the fluid is transported in long pipes between large storage vessels. The pipes are associated with the kinetic state of the fluid in motion between the storage vessels; the fluid is stored in the vessels which are thus associated with the potential energy of the fluid. In treating the energy stored in the static state, liquids and gases must be considered separately; fluid systems are therefore subdivided into hydraulic and pneumatic systems. In order to have a quantitative assessment of when it is reasonable to make lumped approximations, it is necessary to consider wave propagation in fluids; this is done below after the basic components have been considered.

Fluid pressure. Fluid pressure is that physical quantity which causes or results from fluid flow or sustains the weight of a column of fluid in a gravitational field. In an analysis, one is always concerned with the difference in pressure between two points; in many cases one of these points will be an implied reference point and not explicitly indicated. Pressure is quantitatively defined as the *normal* force per unit area, F_n, acting on a surface; the tangential force per unit area is called the shear stress. The total force exerted on the cross-sectional area A of some pipe containing fluid at a pressure P is

$$F = \int_A dF_n = \int_A P\, dA$$

If the pressure distribution is uniform over the area, or if we take P as the average pressure over the area, then

$$F = PA \qquad (1\text{–}105)$$

The work done in moving a small plug of fluid through a distance dx is

$$dW = F_n\, dx = PA\, dx \qquad (1\text{–}106)$$

and the quantity of fluid moved past A is

$$dV = A\, dx \qquad (1\text{–}107)$$

so that
$$P = \frac{dW}{dV} \qquad (1\text{–}108)$$

which shows that pressure is the work done in passing a unit volume of fluid across the area on which it acts. In this sense it is directly analogous to electrical potential difference which is the work done in moving unit quantity of charge through the distance across which the potential difference acts. Pressure will be taken positive for compression.

Volume flow rate. If in a time dt, volume dV of fluid crosses A, the volume flow rate is defined as

$$Q = \frac{dV}{dt} \qquad (1\text{–}109)$$

If the fluid is compressible and the pipe is rigid, the flow into a pipe is obviously equal to the flow leaving the pipe and we may thus speak of the flow *through* the pipe.

1.4.1 Fluid Capacitance

A fluid capacitor is a fluid system element for which kinetic and dissipation effects are negligible, and in which the energy stored is a function of pressure.

Fluid reservoir. The simplest form of fluid capacitor represents a physical object consisting of a uniform cross-sectional open tank or

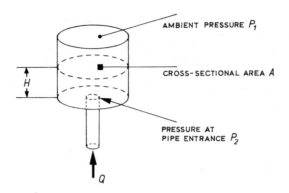

Figure 1-35

reservoir, located in a gravitational field, which is supplied with liquid through a pipe in the bottom as shown in Figure 1-35.

Let:

ρ = mass density of liquid kg/m³
Q = volume flow rate through connecting pipe m³/s
A = uniform cross-sectional area of tank m²
H = height of liquid level in tank m

Then conservation of mass gives that:

$$\rho Q = \frac{d}{dt}(\rho A H)$$

Taking the liquid to be incompressible gives

$$\rho = \text{constant}$$

and thus

$$Q = A \frac{dH}{dt} \qquad (1\text{-}110)$$

For relatively slow motions of the liquid in the tank for which no pressure differences arise from the acceleration of the fluid mass, the pressure difference between top and bottom of the tank

$$P_2 - P_1 = P_{21}$$

supports the weight of the fluid so that

$$P_{21} = \frac{\rho g H A}{A} = \rho g H \qquad (1\text{-}111)$$

where g is the acceleration of a free particle in the gravitational field.

Combining equations (1–110) and (1–111) we have

$$Q = \frac{A}{\rho g} \frac{dP_{21}}{dt} \qquad (1\text{--}112)$$

The quantity

$$C_f = \frac{A}{\rho g} \left(\frac{m^4 s^2}{kg} \right) \qquad (1\text{--}113)$$

is defined as the fluid capacitance of the reservoir.

Pressurized tank. Another simple form of fluid capacitance is associated with a physical object consisting of a closed container of volume V with rigid walls completely filled with fluid of density ρ, for which pressure variations due to the gravitational field are ignored and the only pressure variations considered arise from the fluid forced into or out of the container. Let P_2 be the containing tank pressure, P_1 the fixed reference pressure and $P_{21} = P_2 - P_1$.

Conservation of mass gives that

$$\rho Q = \frac{d}{dt}(\rho V) = V \frac{d\rho}{dt} \qquad (1\text{--}114)$$

Hydraulic tank capacitance. For hydraulic systems

$$d\rho = \frac{\rho}{B} dP_{21} \qquad (1\text{--}115)$$

where B is the bulk modulus of elasticity. For some oils used in hydraulic servo-work B could have a value of about 2×10^9 N m^{-2}. From equations (1–114) and (1–115) we have

$$\rho Q = \frac{V\rho}{B} \frac{dP_{21}}{dt}$$

thus

$$Q = \frac{V}{B} \frac{dP_{21}}{dt} \qquad (1\text{--}116)$$

Thus for suitably small flows into and out of the tank its behaviour may be represented by a fluid capacitor of capacitance

$$C_f = \frac{V}{B} \left(\frac{m^4 s^2}{kg} \right) \qquad (1\text{--}117)$$

Pneumatic tank capacitance. In the pneumatic case two particular values of pneumatic capacitance may be derived, one on the assumption that the gas is compressed adiabatically, and the other on the assumption that it is compressed isothermally.

Let:

V = volume of containing vessel m³
P_{21} = pressure of fluid in vessel N m⁻²
Q = volume flow rate into vessel m³ s⁻¹

Suppose that the storage vessel is supplied from a single pipe, and consider the intrusion of a small amount of fluid from the pipe into the storage vessel.

Let:

v = volume of fluid in the containing vessel plus fluid up to some specified movable reference cross-section in the supply pipe as shown in Figure 1–36.

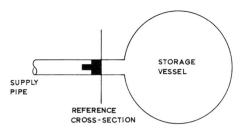

Figure 1–36

Suppose the fluid between the movable reference cross-section and the storage vessel is now intruded into the vessel. Then for adiabatic compression

$$P_{21} v^\gamma = \text{constant} \qquad (1-118)$$

where

$$\gamma = \frac{C_p}{C_v}$$

is the ratio of the gas specific heats at constant pressure and constant volume. Differentiating (1–118) with respect to v gives

$$\frac{dP_{21}}{dv} v^\gamma + P_{21} \gamma v^{\gamma-1} = 0$$

so that

$$\frac{dP_{21}}{dv} v^\gamma + \frac{P_{21}}{v} \gamma v^\gamma = 0$$

and thus

$$\frac{dP_{21}}{dv} = -\frac{P_{21} \gamma}{v}$$

giving

$$dv = -\frac{v}{P_{21} \gamma} dP_{21} \qquad (1-119)$$

If the volume flow rate Q is defined to be positive out of the storage vessel then

$$Q = -\frac{dv}{dt} = \frac{v}{P_{21}\gamma}\frac{dP_{21}}{dt}$$

Thus for sufficiently small volume flow rates so that $v \approx V$ we have

$$Q = \frac{V}{\gamma P_{21}}\frac{dP_{21}}{dt} \qquad (1\text{--}120)$$

giving the pneumatic capacitance for a storage vessel of volume V containing gas at a mean pressure P_{21} and subject to adiabatic (i.e. rapid) compression as

$$C_f = \frac{V}{\gamma P_{21}} \left(\frac{m^4 s^2}{kg}\right) \qquad (1\text{--}121)$$

If the compression is isothermal (i.e. slow) then

$$Pv = \text{constant}$$

and the element power variables are thus related by

$$Q = \frac{V}{P_{21}}\frac{dP_{21}}{dt}$$

where

$$C_f = \frac{V}{P_{21}} \qquad (1\text{--}122)$$

is the pneumatic capacitance for isothermal compression of the gas at a mean pressure P_{21} in a storage vessel of volume V. In general for pneumatic systems we will have

$$d\rho = \frac{\rho}{kP_{21}} dP_{21}$$

where k is a constant in the range $1\cdot 0 \rightarrow 1\cdot 4$ depending on the rapidity with which the fluid is compressed. If pressure variations in a gas are small it may be considered to have a bulk modulus of elasticity

$$B \approx kP_{21}$$

General fluid capacitance. An ideal general fluid capacitance is associated with the triad of variables:

volume flow-rate Q
pressure P
volume V

Interconnective relationships exist between the fluid capacitor power variables, Q and P, and other model element power variables. These relationships govern the interchange of power with the fluid capacitor.

A dynamical relationship exists between the volume and volume flow rate variables

$$V = \int_{t_0}^{t} Q\, dt + [V]_{t_0}$$
$$\frac{dV}{dt} = Q \tag{1-123}$$

where $[V]_{t_0}$ is the initial volume of the fluid storage region.

A constitutive relationship exists between pressure and volume variables

$$V = \phi(P)$$

as shown in Figure 1–37.

The input power to a fluid capacitor is

$$\text{power} = PQ \tag{1-124}$$

so that the input fluid potential energy stored in an ideal fluid capacitor is

$$U_f = \int_{t_0}^{t} \text{power}\, dt = \int_{t_0}^{t} PQ\, dt = \int_{t_0}^{t} P\frac{dV}{dt}\, dt = \int_{0}^{V} P\, dV \tag{1-125}$$

shown as a shaded area on Figure 1–37. The complementary integral

$$U_f^* = \int_{0}^{P} V\, dP \tag{1-126}$$

is defined as the fluid potential co-energy.

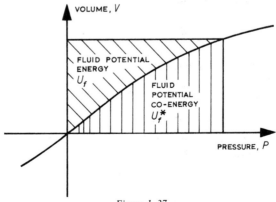

Figure 1–37

For a linear ideal fluid capacitor, volume is directly proportional to pressure

$$V = C_f P \tag{1-127}$$

$$U_f = \int_0^P C_f P\, dP = \tfrac{1}{2} C_f P^2 = \frac{V^2}{2C_f} \tag{1-128}$$

where C_f is the fluid capacitance.

1.4.2 Fluid Inertance

A fluid inertance is a fluid system element for which potential and dissipation effects are negligible, and in which the energy stored is a function of fluid flow rate. It is associated with the kinetic state of the fluid in a relatively long length of pipe in which the fluid moves as a unit (a so-called 'plug' of fluid) with uniform velocity, the fluid being considered as incompressible. If the pipe has a constant area and the velocity of the fluid is uniform across all cross-sections of the pipe, every fluid particle will have the same acceleration.

Let:

L = length of pipe m
ρ = mean density of fluid in pipe kg m^{-3}
A = cross-sectional area of pipe m^2
Q = volume flow rate m^3 s^{-1}
P_2, P_1 = pressures at each end of pipe N m^{-2}
$P_{21} = P_2 - P_1$
v = observed velocity of fluid past a reference cross-section m s^{-1}

Then, accelerating force on fluid in pipe = AP_{21}
= [mass of fluid in pipe] × [acceleration]
= $\rho A L \dfrac{dv}{dt}$

where

$$v = \frac{Q}{A}$$

thus

$$P_{21} = \frac{\rho L}{A} \frac{dQ}{dt} = I \frac{dQ}{dt} \tag{1-129}$$

where

$$I = \frac{\rho L}{A} \left(\frac{\text{kg}}{\text{m}^4}\right)$$

is defined as the fluid inertance of the flow in the pipe.

A fluid *pressure-momentum* variable, Γ, is defined by the dynamical relationship

$$\Gamma_{21} \triangleq \int_{t_0}^{t} P_{21} \, dt + [\Gamma_{21}]_{t_0}$$
$$P_{21} = \frac{d\Gamma_{21}}{dt}$$
(1–130)

An ideal fluid inertance is associated with the triad of variables

volume flow rate Q
pressure P
pressure-momentum Γ

Interconnective relationships exist between the fluid inertance power variables, Q and P, and other model element power variables which govern the interchange of power with the inertance.

The dynamical relationship of equation (1–130) exists between pressure and pressure-momentum variables.

A constitutive relationship will exist between pressure-momentum and flow rate

$$\Gamma = \phi(Q) \tag{1–131}$$

as shown in Figure 1–38.

The fluid kinetic energy stored in a fluid inertance will be

$$T_f = \int_{t_0}^{t} QP \, dt = \int_{t_0}^{t} Q \frac{d\Gamma}{dt} \, dt = \int_{0}^{\Gamma} Q \, d\Gamma \tag{1–132}$$

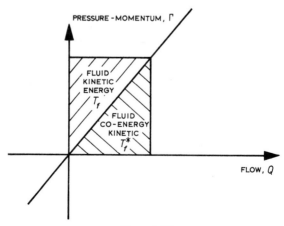

Figure 1–38

Fluid Components

when the fluid is accelerated from rest. The complementary integral

$$T_f^* = \int_0^Q \Gamma \, dQ \tag{1-133}$$

is defined as the fluid kinetic co-energy. For an ideal linear fluid inertance

$$\Gamma = IQ \tag{1-134}$$

and

$$T_f = \frac{1}{2}IQ^2 = \frac{1}{2}\frac{\Gamma^2}{I} \tag{1-135}$$

where I is the fluid inertance.

1.4.3 Fluid Dissipatance

An ideal fluid dissipator is associated with fluid system objects in which inertance and capacitance effects are negligible and for which the pressure drop is determined solely by the fluid flow rate. There are several situations in which dissipatance effects are important. If only relatively small pressure changes are involved, the analysis will be the same for liquids and gases since we can consider the fluid as incompressible. The Reynolds Number, Re, of the flow is

$$Re = \frac{4\rho Q}{\pi d \mu} \tag{1-136}$$

where d is the pipe diameter and μ the absolute viscosity of the fluid. The flow will almost certainly be turbulent for $Re > 5000$ in which case

$$P_{21} \approx kQ|Q|^{\frac{3}{4}} \tag{1-137}$$

where k is a constant whose value depends on the flow rate at which the flow becomes turbulent, the pipe dimensions d and L, the surface finish of the inside of the pipe, and on the fluid properties ρ and μ. It is best found by experiment if required accurately for a particular situation. Incompressible turbulent flow through a long pipe may therefore be represented by a nonlinear dissipator whose constitutive characteristic is of the form of equation (1–137).

If the Reynolds Number is small, say $Re < 100$, then the corresponding laminar flow through a long capillary pipe is governed by the Hagen–Poiseuille Law

$$P_{21} = \frac{128 \mu L}{\pi d^4} Q$$

or

$$P_{21} = R_f Q$$

where

$$R_f \triangleq \frac{128 \mu L}{\pi d^4} \tag{1-139}$$

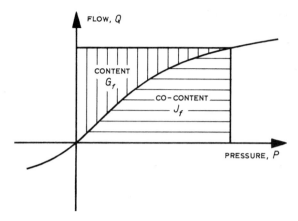

Figure 1–39

is defined as the fluid dissipatance for the pipe. In a similar fashion the incompressible flow through a porous plug or similar obstruction will be governed by a relationship of form (1–138). In this case R_f will depend on the geometry of the plug and the size of the pressure drop, and is best found by experiment.

Incompressible flow through an orifice may be represented by a fluid dissipator whose constitutive characteristic is

$$P_{21} = \frac{\rho}{2C_d^2 A_0^2} Q|Q| \qquad (1\text{--}140)$$

where:

C_d = discharge coefficient of the orifice (0·62 for a sharp-edged orifice)

A_0 = orifice area

In general an ideal fluid dissipator is defined in terms of a constitutive relationship between volume flow rate and pressure drop

$$Q = \phi(P_{21}) \qquad (1\text{--}141)$$

as shown in Figure 1–39. In this general case we define:

$$\text{fluid dissipator content} = G_f = \int_0^Q P_{21}\, dQ \qquad (1\text{--}142)$$

$$\text{fluid dissipator co-content} = J_f = \int_0^P Q\, dP_{21} \qquad (1\text{--}143)$$

Fluid Components

For a linear fluid dissipatance

$$Q = \frac{1}{R_f} P_{21} \qquad (1\text{-}144)$$

and   power absorbed $= P_{21} Q = \dfrac{P_{21}^2}{R_f} = R_f Q^2 \qquad (1\text{-}145)$

where R_f is the fluid resistance.

1.4.4 Representation of Lumped Fluid Systems by Linear Graphs

The network and oriented linear graph conventions for the representation of fluid system elements are shown in Figure 1–40. On the network diagram elements the pressure orientation arrow points from lower to higher pressure and the volume flow rate arrow points in the positive direction of flow. On an oriented line segment for a fluid system the arrow points in the positive direction of pressure rise and the positive direction of flow is such that positive pressure and positive flow give a power flow into the element. The positive direction of flow is thus opposite to the orientation arrow on the line segment. The essential point to remember is that while on a network diagram element we can indicate the flow and pressure positive orientations separately we must use a *single* consistent representation for both on an oriented line segment. The link from the fluid capacitance symbol to the pressure reference point is dashed to emphasize that no liquid flow is propagated through to the reference point; the corresponding oriented line segment is dashed at the reference terminal end for the same reason.

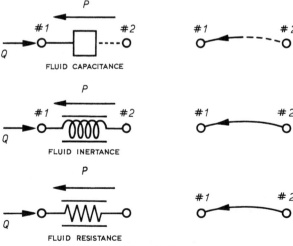

Figure 1–40

1.4.5 Wave Propagation in Fluid Systems

We consider the gaseous case first. In the transmission of an acoustic wave through a gas there is insufficient time between compression and expansion regimes for heat to diffuse so that the expansion is essentially adiabatic. Consider the propagation of a plane-wave through the gas associated with a one-dimensional motion of gas particles in the x-direction. The spatial relationship involved are shown in Figure 1–41.

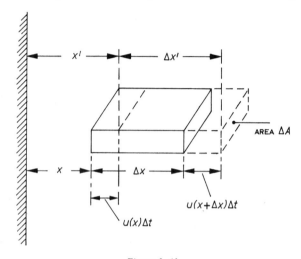

Figure 1–41

This depicts the volume, V_0, say, of a fixed small mass of gas at pressure P_0 which at time t is contained in a box of cross-sectional area ΔA lying between x and $(x+\Delta x)$. In a time Δt this mass of gas shifts to then lie between x' and $(x'+\Delta x')$. Let the velocity of a gas particle in the x-direction at x be $u(x)$ and the velocity of a gas particle at $(x+\Delta x)$ be $u(x+\Delta x)$.

Then

$$V_0 = \Delta x \Delta A \tag{1-146}$$

and the change in volume in time Δt is

$$\Delta V_0 = [u(x+\Delta x)\Delta t - u(x)\Delta t]\Delta A \tag{1-147}$$

Since the expansion is adiabatic, the argument above between equations (1–118) and (1–119) gives

$$\Delta V_0 = -\frac{V_0}{P_0 \gamma}\Delta P_0 \tag{1-148}$$

Equations (1–146), (1–147) and (1–148) thus give

$$[u(x+\Delta x)\Delta t - u(x)\Delta t]\Delta A = -\frac{\Delta x \Delta A}{P_0 \gamma}\Delta P_0$$

whence
$$\frac{u(x+\Delta x)-u(x)}{\Delta x} = -\frac{1}{\gamma P_0}\frac{\Delta P_0}{\Delta t}$$

and, in the limit as Δx and Δt become arbitrarily small,

$$\frac{\partial u}{\partial x} = -\frac{1}{\gamma P_0}\frac{\partial P_0}{\partial t}$$

or, letting P denote a small pressure difference from an ambient pressure P_a, i.e. $P_0 = P + P_a$

$$\frac{\partial u}{\partial x} = -\frac{1}{\gamma P_a}\frac{\partial P}{\partial t} \qquad (1\text{–}149)$$

If we now consider the acceleration of the inertia of the gaseous mass in the x-direction, we have

$$[P(x) - P(x+\Delta x)]\Delta A = \rho \Delta A \Delta x \frac{\partial^2 x'}{\partial t^2} \qquad (1\text{–}150)$$

where x' is the instantaneous measure of position. This gives

$$\frac{P(x+\Delta x)-P(x)}{\Delta x} = -\rho\frac{\partial^2 x'}{\partial t^2} = -\rho\frac{\partial u}{\partial t}$$

so that in the limit as Δx becomes arbitrarily small

$$\frac{\partial P}{\partial x} = -\rho\frac{\partial u}{\partial t}$$

Combining equations (1–149) and (1–151) in the usual way gives

$$\frac{\partial^2 P}{\partial x^2} = \frac{\rho}{\gamma P_a}\frac{\partial^2 P}{\partial t^2}$$

and
$$\frac{\partial^2 u}{\partial x^2} = \frac{\rho}{\gamma P_a}\frac{\partial^2 u}{\partial t^2}$$

so that the velocity of propagation of plane acoustic waves is seen to be

$$c = \sqrt{\frac{\rho}{\gamma P_a}}$$

In the liquid case, we have
$$V_0 = -B^{-1}V_0 \Delta P_0$$
where B is the bulk modulus of elasticity ($4{\cdot}78 \times 10^{10}$ for water at normal temperatures and pressures). Combining equations (1–146), (1–147) and (1–150) then gives

$$[u(x+\Delta x)\Delta t - u(x)\Delta t]\Delta A = -B^{-1}\Delta x \Delta A \Delta P_0$$

so that
$$\frac{u(x+\Delta x)-u(x)}{\Delta x} = -B^{-1}\frac{\Delta P_0}{\Delta t}$$

and, in the limit as Δx and Δt become arbitrarily small,

$$\frac{\partial u}{\partial x} = -B^{-1}\frac{\partial P_0}{\partial t}$$

Letting P again denote a small pressure difference from ambient pressure we have

$$\frac{\partial u}{\partial x} = -B^{-1}\frac{\partial P}{\partial t} \qquad (1\text{–}151)$$

An exactly similar consideration of acceleration effects as for the gaseous case gives the second transmission equation as

$$\frac{\partial P}{\partial x} = -\rho\frac{\partial u}{\partial t} \qquad (1\text{–}152)$$

Combining equations (1–151) and (1–152) we get

$$\frac{\partial^2 P}{\partial x^2} = \rho B^{-1}\frac{\partial^2 P}{\partial t^2} \qquad (1\text{–}153)$$

$$\frac{\partial^2 u}{\partial x^2} = \rho B^{-1}\frac{\partial^2 u}{\partial t^2} \qquad (1\text{–}154)$$

so that the velocity of propagation of plane acoustic waves in a liquid is given by

$$c = \sqrt{(\rho B^{-1})} \qquad (1\text{–}155)$$

Typical values of the velocity of propagation are shown in Table 1–2.

1.5 Thermal System Components

In order to distinguish between the material relationships which are studied in mechanical, electrical and thermal systems we must consider the underlying structure of matter. In very simple terms any portion of solid matter may be considered as a vast assemblage of molecules, and

Table 1–2

	Temp. °C	Density kg m^{-3}	Vel. of prop. m s^{-1}
Air at 760 cm of mercury pressure	0	1·29	331
Steam at 760 cm of mercury pressure	100	0·60	405
Alcohol	0	801	1 240
Water	15	1 000	1 450

each molecule may be crudely modelled in terms of an assembly of massive particles representing the atoms, and springs which represent the cohesive interaction forces between the atoms of a molecule. Each minute dynamical system will have a set of characteristic patterns of vibration, or modes, and the magnitudes of the forces and masses involved are such that one cycle of any characteristic vibration pattern will last for the order of a billionth of a second. In observing the macroscopic behaviour of matter we can only observe what survives of the characteristic vibrations after averaging over the whole assembly of microscopic systems in the sample of matter involved. The mechanical properties of the solid, such as elasticity, arise from certain microscopic modes of vibration which survive the statistical averaging process involved in considering the macroscopic piece of matter because they are associated with an additive property of volume change. The electrical properties of the solid are associated with certain other characteristic microscopic modes of vibration which give a bulk electrical effect because they are associated with an additive property of dipole moment change.

The thermal properties of matter are associated with the macroscopic consequences of those characteristic vibrations which, by virtue of the averaging process, do not give rise to explicitly mechanical or electrical effects in the macroscopic description of a piece of matter. In gases and liquids the individual molecules are in a greater or lesser degree of relative motion. The mechanical effects such as pressure are then associated with the mean kinetic energy of the constituent molecules in translational motion and the thermal effects such as temperature with the microscopic vibrations of the individual molecules.

Such concepts together with simple qualitative experiments with material objects lead to an intuitive concept which we may call 'hotness.' Associated with changes in the hotness of a body are related physical changes, particularly in the linear dimensions of objects and their colour. Such related, reproducible physical changes may be used to establish a quantitative measure of the thermal state of an object. This is the *temperature* of the object, the quantity measured by a thermometer. A mercury bulb thermometer measures temperature in terms of the length of a mercury column in a capillary tube; the temperature scale is obtained by dividing into equal increments the change in length of the column between the two conditions obtained by immersing the thermometer in melting ice and boiling water. The detailed considerations of thermodynamics show that an absolute scale of temperature may be defined in a manner independent of the physical characteristics of material substances. In what follows T will be used to denote temperature measured in degrees centigrade.

Thermal interaction experiment. When two solid material objects, having initially different temperatures, are brought into physical contact their temperatures are observed to vary monotonically with lapse of time, one temperature decreasing and the other increasing, until both objects attain an equal equilibrium temperature. The objects are said to interact thermally. Further experiments will show that we can only find one such form of pure thermal interaction and thus we *do not have a dual thermal interaction experiment*; no oscillatory changes of temperature can arise from *purely* thermal interactions; a little reflection on the simple microscopic considerations presented above will show why this is so.

Consider a set of material objects $^\#i, ^\#j, \ldots, ^\#s$, etc., having known initially different temperatures, and let them interact thermally in pairs by bringing them into physical contact.

For the trio of interactions:

$$
\begin{array}{ccc}
^\#i & \text{with} & ^\#s \\
^\#s & \text{with} & ^\#j \\
^\#i & \text{with} & ^\#j
\end{array}
$$

experiment will show that, under suitable conditions and at least for small excursions in temperature, there exists a set of real positive interaction coefficients H_{is}, H_{sj} and H_{ij} such that:

$$\frac{dT_i}{dt} = -H_{is}\frac{dT_s}{dt} \quad \text{for } ^\#i \text{ and } ^\#s$$

$$\frac{dT'_s}{dt} = -H_{sj}\frac{dT'_j}{dt} \quad \text{for } ^\#s \text{ and } ^\#j \qquad (1\text{--}156)$$

$$\frac{dT''_i}{dt} = -H_{ij}\frac{dT''_j}{dt} \quad \text{for } ^\#i \text{ and } ^\#j$$

Thermal System Components

We *postulate* that these interaction coefficients *form a group* so that

$$H_{ij} = H_{is}H_{sj} = H_{si}^{-1}H_{sj}$$

giving
$$H_{si}\frac{dT_i''}{dt} = -H_{sj}\frac{dT_j''}{dt} \qquad (1\text{--}157)$$

for an interaction between $^\#i$ and $^\#j$ referred to a standard object $^\#s$.

The common quantity for the two bodies which is characteristic of their thermal interaction as exemplified by equation (1–157) is called the *heat-flow*; its integral with respect to time which is characteristic of the capacity of the object to interact thermally with other objects is called *heat*.

If we select some specific material object $^\#s$ as a standard thermal object we may define

$H_{is} \triangleq C_{t_i} \triangleq$ thermal capacitance of object $^\#i$ with reference to standard thermal object $^\#s$

$H_{js} \triangleq C_{t_j} \triangleq$ thermal capacitance of object $^\#j$ with reference to standard thermal object $^\#s$

We then have
$$C_{t_i}\frac{dT_i}{dt} = -C_{t_j}\frac{dT_j}{dt} \qquad (1\text{--}158)$$

for $^\#i$ interacting thermally with $^\#j$, and define

$$q_i \triangleq C_{t_i}\frac{dT_i}{dt} \triangleq \text{heat-flow of object } ^\#i$$

$$q_j \triangleq C_{t_j}\frac{dT_j}{dt} \triangleq \text{heat-flow of object } ^\#j$$

so that
$$q_i + q_j \equiv 0 \qquad (1\text{--}159)$$

We then define

$$Q_i = \int_{t_0}^{t} q_i \, dt + [Q_i]_{t_0} = \text{heat of body } ^\#i$$

$$\frac{dQ_i}{dt} = q_i$$
$$\qquad (1\text{--}160)$$
$$Q_j = \int_{t_0}^{t} q_j \, dt + [Q_j]_{t_0} = \text{heat of body } ^\#j$$

$$\frac{dQ_j}{dt} = q_j$$

and note that we have no grounds for attributing an absolute value of heat to a body but must deal in relative amounts of heat and that, by definition, heat will be conserved in a purely thermal interaction.

The capacity of a material object to do work or to heat by virtue of its thermal state will be termed *internal energy*. Only a fraction of this internal energy is available to do work; if a heat-flow q is put into an ideal reversible thermal-to-mechanical transducer operating between absolute temperatures T_2 and T_1 where $T_2 > T_1$, the available mechanical power is $q(T_2 - T_1)/T_2$ [S9]

Thermal capacitance. In the basic interaction experiment the standard thermal object is taken to be a unit mass of water at standard temperature and pressure. The thermal capacitance of other objects is then found to depend on their mass and the material of which they are composed, and is defined in terms of a material property called specific heat, C_p, by putting

$$C_t = C_p M \qquad (1\text{-}161)$$

where M is the mass of the object and C_p is defined as the specific heat of the object. If heat-flow is taking place into an object at temperature T_2 relative to a constant reference temperature T_1, and $T_{21} = T_2 - T_1$ then

$$q = C_p M \frac{dT_2}{dt} = C_p M \frac{dT_{21}}{dt} \qquad (1\text{-}162)$$

In the general case specific heat will be a function of temperature, and we define a general thermal capacitance to be characterized by a material relationship between heat and temperature

$$Q = \phi(T) \qquad (1\text{-}163)$$

Thermal inductance. At ordinary temperatures there is no known dual material property to thermal capacitance. For temperatures near absolute zero there is some evidence that some purely thermal form of 'heat wave' may exist; such considerations are completely outside the scope of the simple approach adopted here.

Thermal resistance. The spatial flow of heat through a solid material is due to the microscopic transference of the vibrational kinetic energy of the molecules within the material. It is governed by Fourier's Law

$$q = \frac{\sigma_c A (T_2 - T_1)}{L} \qquad (1\text{-}164)$$

where

q = heat-flow rate
A = cross-sectional area normal to heat-flow
L = length of object
T_2, T_1 = temperatures at ends of object
σ_c = thermal conductivity of material

where thermal conductivity is defined by equation (1–164).
If
$$T_{21} = T_2 - T_1$$
then we have
$$q = \left[\frac{\sigma_c A}{L}\right] T_{21}$$
$$q = G_t T_{21} \qquad (1\text{–}165)$$
where
$$G_t = \frac{\sigma_c A}{L}$$
is defined as the thermal conductance of the material. Alternatively
$$T_{21} = R_t q \qquad (1\text{–}166)$$
$$R_t = \frac{L}{\sigma_c A} \qquad (1\text{–}167)$$

where R_t is defined as the *thermal resistance* of the material. In the general case we could define thermal resistance in terms of the material relationship
$$T = \phi(q) \qquad (1\text{–}168)$$

The flow of heat through a fluid is only governed by Fourier's Law when the fluid is not in motion; when fluid motion occurs the heat transfer is mostly by convection and is difficult to infer analytically. Turbulent fluid motion reduces the resistance to heat-flow through a fluid and in practice the major portion of thermal resistance is associated with thin films of near-stationary fluids at boundary walls; such a resistance is usually described in terms of a film coefficient of heat transfer. An *overall coefficient of heat transfer* for an arrangement of physical objects may be defined as
$$C_h = \frac{q}{(T_2 - T_1)A} \qquad (1\text{–}169)$$
giving the thermal resistance in terms of the coefficient of heat transfer as
$$R_t = \frac{1}{C_h A} \qquad (1\text{–}170)$$

Heat may be also transferred to or from an object by *radiation*; this is governed by the Stefan–Boltzmann Law
$$q = C_r(T_2^4 - T_1^4) \qquad (1\text{–}171)$$
where T_2 and T_1 are the *absolute* temperatures of the surfaces emitting

and receiving radiation respectively, and the radiation coefficient C_r depends on the shape, area and surface characteristics of the objects involved. For radiation from a spherical black body to a black environment

$$C_r = \sigma_{SB} A \qquad (1\text{–}172)$$

where A is the surface area of the radiating sphere in m^2 and the Stefan–Boltzmann constant is

$$\sigma_{SB} = 6\cdot 595 \times 10^{-2}\,\text{J}\,\text{m}^{-2}\,\text{s}^{-1}\,\text{K}^{-4}$$

When considering practical cases of the determination of thermal resistance arising from thermal conduction, it is important to remember that, when two solid objects are placed in contact, the total thermal resistance will normally greatly exceed the sum of the appropriate thermal resistances, due to the presence of oxide films and thin layers of entrapped air.

Although the thicknesses of air and oxide involved are small both are very poor conductors of heat.

Representation of Thermal Systems by Linear Graphs

The conventions adopted for the representation of thermal systems by oriented linear graphs are shown in Figure 1–42. On a network diagram the orientation arrow for heat-flow denotes the positive direction of heat-flow and the orientation arrow for temperature points from

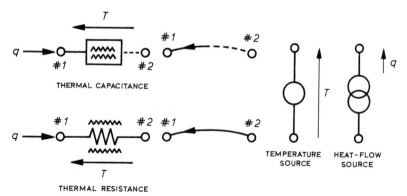

Figure 1–42

lower to higher temperature. The connection from the thermal capacitance symbol to the reference terminal is dashed to emphasize that no heat-flow is propagated through to the reference terminal point. The reference terminal end of the oriented line segment corresponding to a thermal capacitance is dashed for the same reason. Figure 1–43

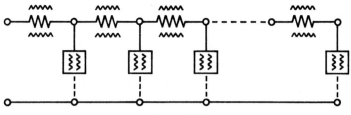

Figure 1-43

shows a simple lumped thermal representation of heat transmission down a bar.

Thermal sources. Figure 1-42 shows the network diagram symbols adopted for ideal temperature and heat sources.

Thermal transformer. There is no such thing as a purely thermal transformer.

Wave propagation in thermal systems. Consider the one-dimensional propagation of heat through a small rectangular piece of material as shown in Figure 1-44. Let $q(x)$ be the rate of heat-flow into the face of

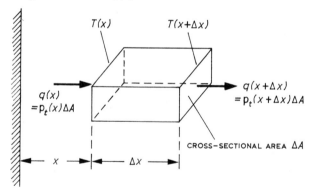

Figure 1-44

area ΔA located at x and $q(x+\Delta x)$ the rate of heat-flow out of the face of area ΔA located at $(x+\Delta x)$. Define

$$\frac{q(x)}{\Delta A} = p_t(x) \quad \text{and} \quad \frac{q(x+\Delta x)}{\Delta A} = p_t(x+\Delta x)$$

as the corresponding *thermal flux densities* in W m^{-2} for the two faces. Then applying Fourier's Law of heat conduction to the small element of material we have

$$q = -\frac{\sigma_c \Delta A \Delta T}{\Delta x}$$

so that
$$p_t \Delta A = -\frac{\sigma_c \Delta A \Delta T}{\Delta x}$$

giving
$$\frac{\Delta T}{\Delta x} = -\frac{1}{\sigma_c} p_t$$

and in the limit as Δx and ΔT become arbitrarily small

$$\frac{\partial T}{\partial x} = -\frac{1}{\sigma_c} p_t \qquad (1\text{-}173)$$

Again, we have that the heat in volume ΔV corresponding to the temperature difference ΔT is $C_p \rho \Delta V \Delta T$, so that

$$[p_t(x+\Delta x) - p_t(x)]\Delta A \Delta t = -C_p \rho \Delta V \Delta T = -C_p \rho \Delta A \Delta x \Delta T$$

thus
$$\frac{p_t(x+\Delta x) - p_t(x)}{\Delta x} = -C_p \rho \frac{\Delta T}{\Delta t}$$

and in the limit, as Δx, ΔT and Δt become arbitrarily small,

$$\frac{\partial p_t}{\partial x} = -C_p \rho \frac{\partial T}{\partial t} \qquad (1\text{-}174)$$

Combining equations (1-173) and (1-174) we have

$$\frac{\partial^2 T}{\partial x^2} = \frac{C_p \rho}{\sigma_c} \frac{\partial T}{\partial t} = \frac{1}{D_t} \frac{\partial T}{\partial t} \qquad (1\text{-}175)$$

$$\frac{\partial^2 p_t}{\partial x^2} = \frac{C_p \rho}{\sigma_c} \frac{\partial p_t}{\partial t} = \frac{1}{D_t} \frac{\partial p_t}{\partial t} \qquad (1\text{-}176)$$

where $D_t = \sigma_c / C_p \rho$ is defined as the thermal diffusivity of the material.

A quantitative idea of the velocities involved in thermal propagation may be obtained from the fact that the phase velocity for a sinusoidal wave of frequency ω is $\sqrt{(2\omega D_t)}$ and using the data contained in Table 1-3 (for 0°C).

1.6 Electrical System Components

The fundamental electrical quantity is *charge*, q, measured in coulombs. Charge is transported on charged material particles such as electrons or ions. Certain materials, called conductors, may sustain a continuous flow of charge. The continuous flow of charge past a reference point is defined as a current, i, and measured in coulombs per second or amperes, i.e.

$$i = \frac{dq}{dt} \qquad (1\text{-}177)$$

Table 1–3

Material	Thermal conductivity σ_c	Specific heat C_p	Density ρ	Thermal diffusivity D_t
Silver	418	234	10.5×10^3	1.7×10^{-4}
Copper	387	380	8.94×10^3	1.14×10^{-4}
Aluminium	203	870	2.71×10^3	0.86×10^{-4}
Steel	44.6	460	7.85×10^3	1.24×10^{-5}
Asbestos	0.15	1 046	0.58×10^3	2.48×10^{-7}
Glass	0.88	670	2.6×10^3	5.1×10^{-7}
Concrete	2.42	920	2.47×10^3	1.07×10^{-6}
	$\text{J s}^{-1} \text{m}^{-1} {}^\circ\text{C}^{-1}$	$\text{J kg}^{-1} {}^\circ\text{C}^{-1}$	kg m^{-3}	$\text{m}^2 \text{s}^{-1}$

Positive current may be defined as the flow of a positive charge in a stipulated positive direction (or negative charge in the opposite direction). Charges whose relative motion is very small compared with the velocity of light exert forces on each other which are defined by Coulomb's Law, illustrated in Figure 1–45.

$$f_{21} = -f_{12} = \frac{kq_1q_2}{d^2} \frac{\mathbf{r}}{\|\mathbf{r}\|} \qquad (1\text{–}178)$$

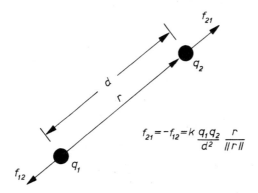

Figure 1–45

where
- $\mathbf{r}$ = vector displacement between charges
- q_1, q_2 = charges, coulomb
- d = distance between charges
- k = constant depending on medium in which the charges are immersed

If a charged particle experiences forces exerted by a constellation of other fixed charges then the resulting net force is determined by the vectorial summation of the forces exerted by each charge as determined by equation (1–178). In all cases the force is a function of the charged particle position and work must therefore be done on a particle as it is moved about in the field of force exerted by a set of other particles. For a displacement of the charged particle from $\mathbf{r}_1$ to $\mathbf{r}_2$ the work done is equal to a change in potential energy of

$$\Delta U = \int_{r_1}^{r_2} \mathbf{f}(\mathbf{r}) \cdot \mathbf{dr} \qquad (1\text{–}179)$$

where $\mathbf{r}$ is the charged particle position relative to a reference frame in which all the other particles remain fixed. Since $\mathbf{f}$ is under all circumstances proportional to the charge, q, of the moving particle we can define a *potential difference* or *voltage* function, v, in terms of *potential energy difference per unit charge*

$$v(r) = v(r_2) - v(r_1) = \int_{r_1}^{r_2} \frac{\mathbf{f}(\mathbf{r}) \cdot \mathbf{dr}}{q} \qquad (1\text{–}180)$$

where the voltage v is measured in newton-metres per coulomb or volts. Note that only differences in potential are of significance so that any arbitrary constant could be added in the definition of equation (1–180). In terms of the potential difference the change in potential energy is

$$\Delta U = qv \qquad (1\text{–}181)$$

Now suppose we have a continuous flow of charge constituting a current i flowing along a path from $\mathbf{r}_1$ to $\mathbf{r}_2$. In a time t the amount of charge transferred between $\mathbf{r}_1$ and $\mathbf{r}_2$ is

$$\Delta q = i \Delta t \qquad (1\text{–}182)$$

associated with a change in potential energy of

$$\Delta U = v \Delta q = vi(\Delta t)$$

The rate of increase in potential energy per unit time is the instantaneous power which must be supplied by the environment in order to effect the charge flow

$$\text{power} = iv \qquad (1\text{–}183)$$

The power is measured in joules per second, volt-amps or watts. In equation (1–183) v denotes the *increase* in potential through which the charges move; if v is positive power is supplied from the environment; if v is negative power is supplied to the environment.

1.6.1 Capacitor

The concept of capacitance is associated with physical objects consisting of a pair of conductors of large area in close proximity, as shown schematically in Figure 1–46. For such an object it is found experimentally that the voltage difference between the separate conducting areas is a function of the difference in quantity of charge on them. In general the conducting areas may be taken to have equal and opposite charges on them, and the charge of the capacitor is then defined as the quantity added to one area and taken away from the other, starting from a condition in which there was no potential difference between the plates.

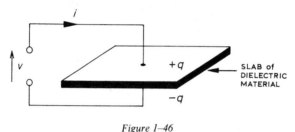

Figure 1–46

An ideal capacitor is associated with the triad of variables:

$$\text{voltage } v$$
$$\text{current } i$$
$$\text{charge } q$$

Interconnective relationships exist between the capacitor power variables, v and i, and other model element power variables which govern the interchange of power with the capacitor.

A dynamical relationship exists between the charge and current variables

$$q = \int_{t_0}^{t} i \, dt + [q]_{t_0}$$

$$\frac{dq}{dt} = i$$

where $[q]_{t_0}$ is the charge on the capacitor at time t_0.

A constitutive relationship exists between the charge and voltage variables
$$v = \phi(q)$$
as illustrated by Figure 1–47. The capacitor constitutive characteristics may be classified as follows:

Charge-controlled capacitor. A capacitor is said to be charge-controlled if, for all values of time and all finite values of charge, the capacitor voltage is a single-valued function of the charge.

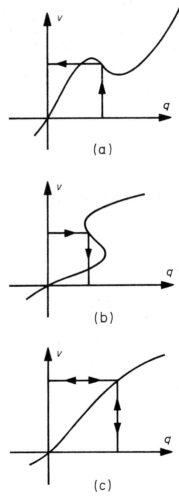

Figure 1–47

Voltage-controlled capacitor. A capacitor is said to be voltage-controlled if, for all values of time and all finite values of voltage, the capacitor charge is a single-valued function of the voltage.

One-to-one capacitor. A capacitor is said to be one-to-one if it is both charge-controlled and voltage-controlled.

Examples of constitutive characteristics for charge-controlled, voltage-controlled and one-to-one capacitors are shown in Figures 1–47 (a), (b) and (c) respectively.

The stored energy of a capacitor is equal to the work done in charging it to a charge q in a time t

$$U_e = \int_0^t vi\, dt = \int_0^t v\frac{dq}{dt}dt = \int_0^q v\, dq$$

which is shown as an area on the one-to-one constitutive characteristic of Figure 1–48. If the constitutive characteristic is linear then

$$q = Cv$$

where C is defined as the capacitance of the capacitor, measured in coulombs per volt or farads. For a linear capacitor the stored energy is

$$U_e = \int_0^q \frac{q}{C}dq = \frac{1}{2}\frac{q^2}{C} = \frac{1}{2}Sq^2 = \frac{1}{2}Cv^2$$

where $S = C^{-1}$ is defined as the susceptance of the capacitor, measured in reciprocal farads or darafs.

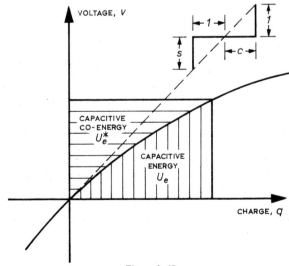

Figure 1–48

The complementary integral

$$U_e^* = \int_0^v q\, dv$$

is defined as the capacitor co-energy. If the capacitive co-energy is a known function of the capacitor voltage, then the corresponding value of charge may be determined from

$$q = \frac{\partial U_e^*}{\partial v} \qquad (1\text{--}184)$$

If the capacitive energy is a known function of the capacitor charge, then the corresponding value of voltage may be determined from

$$v = \frac{\partial U_e}{\partial q}$$

Equation (1–184) gives the component relationship for a capacitor as

$$i = \frac{d}{dt}\left(\frac{\partial U_e^*}{\partial v}\right)$$

If the capacitor is linear $U_e^* = \tfrac{1}{2}Cv^2$ and the component relationship becomes

$$i = C\frac{dv}{dt}$$

where C is the capacitance.

1.6.2 Inductor

The concept of inductance is associated with physical objects consisting of one or more loops of conducting material. When current flows through a coil it is found experimentally that forces are exerted on current-carrying elements placed in its vicinity. The coil is said to have an associated magnetic field which is quantitatively characterized at a point in space by a flux of magnetic induction. Since the neighbouring current-carrying elements of a physical object consisting of a large number of turns of wire in close proximity exercise a repulsive force on each other, energy must be expended to deform an initially long loop of wire into such a shape, and we would therefore expect the flow of current in such an object to be associated with a storage of energy. We therefore have to supply power to any such object in order to change the value of current flowing in it. The associated voltage is determined by Faraday's Law

$$v = \frac{d\lambda}{dt}$$

where λ is defined as the inductor flux-linkages and characterizes the effect of the magnetic field in lumped terms. Flux-linkage is measured in volt-seconds or webers.

An ideal inductor is associaed with the triad of variables:

> voltage v
> current i
> flux-linkage λ

Interconnective relationships exist between the inductor power variables, v and i, and other model element power variables which govern the interchange of power with the inductor.

A dynamical relationship exists between the voltage and flux-linkage variables

$$\lambda = \int_{t_0}^{t} v \, dt + [\lambda]_{t_0}$$

$$\frac{d\lambda}{dt} = v$$

where $[\lambda]_{t_0}$ is the flux-linkage at time t_0.

A constitutive relationship exists between the flux-linkage and current variables

$$\lambda = \phi(i)$$

as illustrated in Figure 1–49. These relationships may be classified

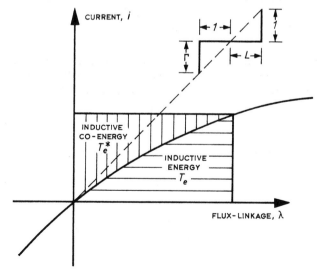

Figure 1–49

into flux-controlled, current-controlled and one-to-one in a manner analogous to the previous classification of constitutive characteristics.

The work done in establishing a flux-linkage λ in a coil in a time t is the stored magnetic energy

$$T_e = \int_0^t iv \, dt = \int_0^t i\frac{d\lambda}{dt} dt = \int_0^\lambda i \, d\lambda \qquad (1\text{-}185)$$

which is shown as an area on the one-to-one constitutive characteristic of Figure 1-49. If the constitutive characteristic is linear then

$$\lambda = Li \qquad (1\text{-}186)$$

where L is defined as the inductance of the inductor, measured in volt-seconds per ampere or henries.

For a linear inductor the stored energy is

$$T_e = \int_0^\lambda i \, d\lambda = \int_0^\lambda \frac{\lambda}{L} d\lambda = \frac{1}{2}\frac{\lambda^2}{L} = \frac{1}{2}\Gamma\lambda^2 = \frac{1}{2}Li^2 \qquad (1\text{-}187)$$

where Γ is the reciprocal inductance.

If the inductive energy is a known function of the inductor flux-linkage, then the corresponding current may be determined from

$$i = \frac{\partial T_e}{\partial \lambda}$$

The complementary integral

$$T_e^* = \int_0^i \lambda \, di \qquad (1\text{-}188)$$

is defined as the inductive co-energy. If the inductive co-energy of an inductor is a known function of the inductor current, the corresponding value of inductor flux-linkage may be determined from

$$\lambda = \frac{\partial T_e^*}{\partial i} \qquad (1\text{-}189)$$

This gives the component relationship for an inductor as

$$v = \frac{d}{dt}\left(\frac{\partial T_e^*}{\partial i}\right) \qquad (1\text{-}190)$$

If the inductor is linear, then

$$T_e^* = \tfrac{1}{2}Li^2$$

so that

$$\frac{\partial T_e^*}{\partial i} = Li$$

and
$$v = L\frac{di}{dt} \tag{1-191}$$

where L is the inductance of the inductor.

1.6.3 Converters

An ideal electrical converter is defined as an object which exhibits no capacitative or inductive effects and for which the terminal voltage is a known function of the current flowing between its terminals. An electrical converter is thus defined in terms of a constitutive characteristic as shown in Figure 1-50. Power is defined as positive when it flows into the converter; the first and third quadrants of the converter constitutive characteristic thus correspond to power flowing into the converter, and the second and fourth quadrants to a flow of power out of the converter. Electrical converters are conveniently divided into three types.

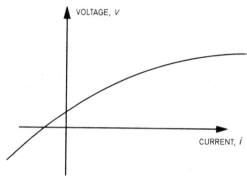

Figure 1-50

(a) *Resistors.* The constitutive characteristic of a resistor passes through the origin and is confined to the first and third quadrants. A resistor therefore always absorbs power and is used in lumped electrical circuits to represent the irreversible conversion of electrical into thermal energy. The resistor characteristics may be divided, in the usual way, into current-controlled, voltage-controlled and one-to-one types. The resistor power is

$$vi = \int_0^v i\,dv + \int_0^i v\,di = J_e + G_e$$

where

$$G_e \triangleq \int_0^i v\,di \tag{1-192}$$

84 Components

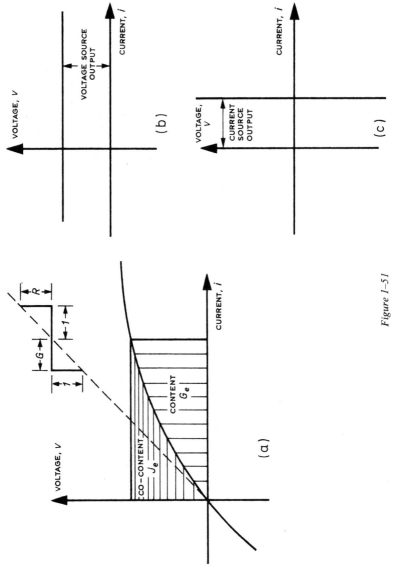

Figure 1-51

Electrical System Components

is defined as the resistor content, shown as an area for the one-to-one case in Figure 1–51(a). If the content for a given resistor is a known function of the resistor current, the corresponding voltage may be determined by

$$v = \frac{\partial G_e}{\partial i} \tag{1-193}$$

The quantity

$$J_e \triangleq \int_0^v i \, dv \tag{1-194}$$

is defined as the resistor co-content, shown as an area in Figure 1–51(a). If the co-content for a given resistor is a known function of the resistor voltage, the corresponding current may be determined by

$$i = \frac{\partial J_e}{\partial v} \tag{1-195}$$

If the resistor is linear then

$$\begin{aligned} v &= Ri \\ i &= Gv \end{aligned} \tag{1-196}$$

where R and G are respectively defined as the resistance and conductance of the resistor. For a linear resistor the dissipated power is

$$vi = i^2 R = \frac{v^2}{G} \tag{1-197}$$

(b) *Voltage source.* A voltage source is a converter for which the voltage is a specified function of time independent of the current. The characteristic for a constant voltage source is shown in Figure 1–51(b).
(c) *Current source.* A current source is a converter for which the current is a specified function of time independent of the voltage. The characteristic for a constant current source is shown in Figure 1–51(c).

1.6.4 Representation of Lumped Electrical Systems by Linear Graphs

The network and oriented linear graph conventions for the representation of electrical system elements are shown in Figure 1–52. On a network diagram the current orientation arrows point in the positive direction of current flow and the voltage orientation arrows point from lower to higher potential. On an oriented line segment representation of an electrical element the orientation arrow points in the direction of positive rise in potential and the corresponding positive current orientation is taken to be in the opposite direction.

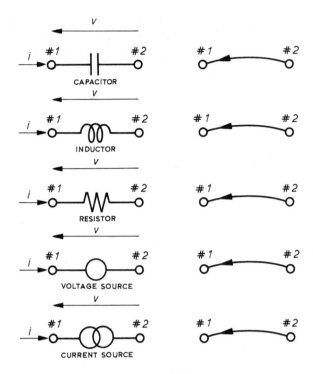

Figure 1-52

1.6.5 Wave Propagation in Electrical Systems

Consider a one-dimensional transmission line having the following parameters:

L = inductance per unit length H m^{-1}
C = capacitance per unit length F m^{-1}
R = series resistance per unit length Ω m^{-1}
G = shunt conductance per unit length Ω^{-1} m^{-1}

A length Δx of line may be modelled by the lumped approximating circuit of Figure 1-53.

Use of Kirchhoff's Voltage Law (see Chapter Five) and the various component characteristics gives

$$v(x+\Delta x) = v(x) - (R\Delta x)i(x) - (L\Delta x)\frac{\partial i(x)}{\partial t} \qquad (1\text{-}198)$$

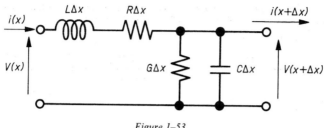

Figure 1-53

Use of Kirchhoff's Current Law and the various component characteristics gives

$$i(x+\Delta x) = i(x) - (G\Delta x)v(x+\Delta x) - (C\Delta x)\frac{\partial v(x+\Delta x)}{\partial t} \quad (1\text{-}199)$$

Expanding $v(x+\Delta x)$ in a Taylor series and neglecting terms of higher than second order in small quantities, we have

$$v(x+\Delta x) = v(x) + \frac{\partial v(x)}{\partial x}\Delta x + \cdots \quad (1\text{-}200)$$

so that

$$\frac{\partial v(x+\Delta x)}{\partial t} = \frac{\partial v(x)}{\partial t} + \frac{\partial v(x)}{\partial t \partial x}\Delta x + \cdots \quad (1\text{-}201)$$

Substituting from equations (1-200) and (1-201) into equation (1-199) we get

$$i(x+\Delta x) = i(x) - (G\Delta x)\left[v(x) + \frac{\partial v(x)}{\partial x}\Delta x\right] - C\Delta x\left[\frac{\partial v(x)}{\partial t} + \frac{\partial v(x)}{\partial t \partial x}\Delta x\right]$$

$$= i(x) - (G\Delta x)v(x) - C\Delta x\frac{\partial v(x)}{\partial t} + \text{terms in } (\Delta x)^2$$

Thus

$$\frac{i(x+\Delta x) - i(x)}{\Delta x} = -Gv(x) - C\frac{\partial v(x)}{\partial t}$$

so that in the limit, as Δx becomes arbitrarily small, we have

$$\frac{\partial i}{\partial x} = -Gv - C\frac{\partial v}{\partial t} \quad (1\text{-}202)$$

From equation (1-198) we have

$$\frac{v(x+\Delta x) - v(x)}{\Delta x} = -Ri(x) - L\frac{\partial i(x)}{\partial t}$$

In all cases the orientation arrow on a linear graph segment signifies the positive orientation of the across variable (see classification in Chapter 5).

Translational Mechanism

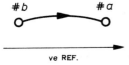

v_{ab} positive when $v_a > v_b$, i.e. #a moving away from #b in the positive direction.

Rotational Mechanism

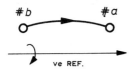

ω_{ab} positive when $\omega_a > \omega_b$

Fluid

Pressure at #a > pressure at #b

Thermal

Temperature of #a > temperature of #b

Electrical

Voltage at #a > voltage at #b

In all cases the positive direction of the through-variable is such that simultaneously positive across- and through-variables imply a power flow *into* the object.

Figure 1–54 Linear graph orientation conventions.

so that in the limit, as Δx becomes arbitrarily small we have

$$\frac{\partial v}{\partial x} = -Ri - L\frac{\partial i}{\partial t} \tag{1-203}$$

Combining equations (1–202) and (1–203) we have

$$\frac{\partial^2 v}{\partial x^2} = RGv + (RC+LG)\frac{\partial v}{\partial t} + LC\frac{\partial^2 v}{\partial t^2} \tag{1-204}$$

$$\frac{\partial^2 i}{\partial x^2} = RGi + (RC+LG)\frac{\partial i}{\partial t} + LC\frac{\partial^2 i}{\partial t^2} \tag{1-205}$$

If the line losses are negligible

$$\frac{\partial^2 v}{\partial x^2} = LC\frac{\partial^2 v}{\partial t^2} \qquad (1-206)$$

$$\frac{\partial^2 i}{\partial x^2} = LC\frac{\partial^2 i}{\partial t^2} \qquad (1-207)$$

and waves of current and voltage propagate with a speed

$$c = \frac{1}{\sqrt{(LC)}} \qquad (1-208)$$

For an open transmission line this velocity will be of the order of 3×10^8 m s^{-1}.

Summary of linear graph orientation conventions. The various linear graph orientation conventions are summarized in Figure 1–54.

CHAPTER TWO

Spaces

The treatment of dynamical system models given in later chapters makes considerable use of the theory of linear vector spaces. The main results required are collected in this chapter for reference purposes.

2.1 Sets

A set (or class) of objects may be intuitively thought of as an arbitrary collection of objects selected according to some rule. It is normal to use the word 'class' to denote a set of other sets, but this usage is not mandatory. Examples of sets are the set of all even integers greater than 16; the set of all points equidistant from a fixed point; the set of all points on the complex plane. The concepts of a set and the associated selection rule are equivalent to the intuitive concept of membership. If an element x is a member of (or belongs to, or is contained in) a set X, this is denoted symbolically by

$$x \in X$$

Two sets X and Y are equal if and only if they both consist of exactly the same elements.

Inclusion. Consider the following two sets: the set A of all points on the complex plane and the set B of all points on the imaginary axis in the complex plane. All members of the set B are included in the set A. The set B is said to be a proper subset of the set A; this is denoted by writing

$$B \subset A$$

This is analogous to the relation $<$ between real variables. By analogy with the real variable relation $\leqslant$ it is convenient to provide for the case when sets are equal. We then say that a set X is a subset of a set Y, symbolically

$$X \subseteq Y \qquad (2\text{--}1)$$

if and only if

$$x \in X \quad \text{implies} \quad x \in Y$$

The expression (2–1) is read alternatively as: "X is contained in (or included in) Y" or "X is a subset of Y". The symbol may be reversed as

$$Y \supseteq X$$

Sets

which is read as "Y contains (or includes) X". Obviously

(a) $X \subseteq Y$ and $Y \subseteq Z$ implies $X \subseteq Z$
(b) $X \subseteq Y$ and $Y \subseteq X$ implies $X = Y$

A given set is usually defined by specifying the membership rule or property which is common to all the elements. The notation used for this is

$$X = \{x : \phi(x)\}$$

where $\phi(x)$ is an explicit statement of the common property defining the elements. For example

$$X = \{x : x \text{ is an even integer} > 16\}$$

Operations on sets

Certain simple operations on sets are extremely important. These are union, intersection, difference and complement. They are defined as follows:

Union. The union of any two sets X and Y is the set denoted by $X \cup Y$ and defined by

$$Z = X \cup Y = \{z : z \in X \text{ or } z \in Y\} \qquad (2\text{--}2)$$

Intersection. The intersection of any two sets X and Y is the set denoted by $X \cap Y$ and defined by

$$Z = X \cap Y = \{z : z \in X \text{ and } z \in Y\} \qquad (2\text{--}3)$$

Difference. The difference of any two sets X and Y is the set denoted by $X \sim Y$ and defined by

$$Z = X \sim Y = \{z : z \in X \text{ and } z \notin Y\} \qquad (2\text{--}4)$$

where the symbol $\notin$ denotes 'is not a member of'.

Complement. The complement X' of any set X is the set defined by

$$Z = X' = \{z : z \notin X\} \qquad (2\text{--}5)$$

In order to give logical completeness to the discussion of sets it is convenient to include a *universal set* which is chosen for any particular class of sets to include all sets under discussion, and an *empty set* or *null set* which has no members.

Cartesian product. If X and Y are any two sets whose elements are denoted by x and y then the Cartesian product of sets X and Y is denoted by $X \times Y$ and defined by

$$X \times Y = \{(x, y) : x \in X \text{ and } y \in Y\}$$

and consists of all possible ordered pairs of elements from X and Y. The name arises because we can specify the Cartesian coordinates of a point in this way by reference to a set of rectangular axes.

2.2 Metric Spaces

A set endowed with a certain structure is said to be *a space*; by structure we mean a defined set of operations on, or relationships between, the elements of the set.

Metric. A metric or distance measure defined on a nonempty set X is a real function d of pairs of elements of X such that the following conditions hold when

$$x \in X \qquad y \in X \qquad z \in X$$

(a) $d(x, y) \geq 0$ for all $x \in X$ and all $y \in X$ and $d(x, y) = 0$ implies $x = y$
(b) $d(x, y) = d(y, x)$
(c) $d(x, y) \leq d(x, z) + d(z, y)$ (This is usually called the triangle inequality.)

Metric space. A metric space is any nonempty set together with a metric defined on it.

Many different metrics may be defined on any given nonempty set; the different metrics turn the same set into different metric spaces.

Sets of n-tuple numbers. In dynamical systems analysis we are constantly discussing classes of sets whose elements are ordered sets of n real or complex numbers.

n-tuple number or n-tuplet. An n-tuple number or n-tuplet is defined as an ordered set of n real or complex numbers. If $x_1, x_2, \ldots, x_n$ is a set of n numbers, the corresponding n-tuple number is written as

$$(x_1, x_2, \ldots, x_n)$$

The sum of a pair of n-tuplets

$$(x_1, x_2, \ldots, x_n) + (y_1, y_2, \ldots, y_n)$$

is defined to be the n-tuplet

$$(x_1 + y_1, x_2 + y_2, \ldots, x_n + y_n)$$

R^n and C^n. The set of all real n-tuplets is conventionally denoted by R^n and the set of all complex n-tuplets by C^n. These sets become metric spaces when endowed with the metric

$$d(x, y) = \left\{ \sum_{i=1}^{n} |x_i - y_i|^2 \right\}^{\frac{1}{2}} \tag{2-6}$$

This metric is called the Euclidean metric.

2.3 Linear Vector Spaces [H2], [B3]

The archetypal vector is a rectilinear displacement in the familiar Euclidean spaces of two and three dimensions. For such vectors one can define the binary operations of addition and multiplication by a scalar. If two displacements are carried out sequentially, the resultant displacement is regarded as the sum of the two displacements to which it is equivalent. Multiplication by the real scalar factor α is interpreted as changing the magnitude of the displacement by the factor α; if α is negative the direction of the displacement is reversed. The consideration of linear combinations of objects of mathematical investigation in terms of a pair of binary operations of the above sort is so frequent and fundamental that an abstract algebraic system has been evolved for this purpose; this system is called a linear vector space. The mathematical objects comprising the system are scalars and abstract objects termed vectors, together with the binary relationships of addition and scalar multiplication.

The scalars are real or complex numbers and the vectors may represent a wide range of mathematical objects or physical variables. A set of objects **X** is said to be a linear vector space if a pair of binary relationships of addition and scalar multiplication are defined on **X** which are such that:

(a) The addition is commutative

$$\mathbf{a}+\mathbf{b} = \mathbf{b}+\mathbf{a} \text{ for any } \mathbf{a} \in \mathbf{X} \text{ and any } \mathbf{b} \in \mathbf{X}$$

and associative

$$\mathbf{a}+(\mathbf{b}+\mathbf{c}) = (\mathbf{a}+\mathbf{b})+\mathbf{c} \text{ for any } \mathbf{a}, \mathbf{b}, \mathbf{c} \in \mathbf{X}$$

(b) The scalar multiplication is associative

$$\alpha(\beta \mathbf{a}) = (\alpha\beta)\mathbf{a}$$

and distributive

$$\alpha(\mathbf{a}+\mathbf{b}) = \alpha\mathbf{a}+\alpha\mathbf{b}$$

$$(\alpha+\beta)\mathbf{a} = \alpha\mathbf{a}+\beta\mathbf{a}$$

for all scalars α, β and any $\mathbf{a}, \mathbf{b} \in \mathbf{X}$

(c) $1\mathbf{a} = \mathbf{a}$ for all $\mathbf{a} \in \mathbf{X}$

(d) For all vectors $\mathbf{a} \in \mathbf{X}$ there exists a unique vector **0**, the zero element, such that

$$\mathbf{a}+\mathbf{0} = \mathbf{a} \quad \text{for any} \quad \mathbf{a} \in \mathbf{X}$$

(e) For any element $\mathbf{a} \in \mathbf{X}$ there exists an element $-\mathbf{a} \in \mathbf{X}$, the inverse of **a**, such that

$$-\mathbf{a}+\mathbf{a} = \mathbf{0}$$

The linear vector space which is most used in discussions of dynamical system behaviour is the space of real n-tuple numbers. The product of a scalar α and an n-tuple number $(x_1, x_2, \ldots, x_n)$ is defined by

$$\alpha(x_1, x_2, \ldots, x_n) \triangleq (\alpha x_1, \alpha x_2, \ldots, \alpha x_n)$$

We thus have

$$-1(x_1, x_2, \ldots, x_n) = (-x_1, -x_2, \ldots, -x_n)$$

so that $(x_1, x_2, \ldots, x_n) - (x_1, x_2, \ldots, x_n) = (0, 0, \ldots, 0)$

It thus follows that the real n-tuple numbers together with the scalar numbers and the two binary operations of addition and scalar multiplication constitute a linear vector space, since all the requirements laid down above are satisfied. The numbers $x_1, x_2, \ldots, x_n$, will be termed the components of the n-tuple number $(x_1, \ldots, x_n)$ and this n-tuplet will be conveniently referred to as the vector **x**.

A linear vector space [H2] is an algebraic system for the consideration of relationships between members of a set of objects. It is an abstraction and generalization of vector algebra which was developed for the study of algebraic relationships between geometrical objects representing physical variables in Euclidean spaces of two and three dimensions. There are two distinct aspects of this generalization. In the first place the vectors of a general linear vector space are *any* abstract objects which fulfil the requirements laid down above. Such objects may have little or no direct geometrical interpretation in themselves as, for example, the set of $m \times n$ matrices, which together with the scalar set of complex numbers and a pair of defined binary operations of addition and scalar multiplication, constitute a linear vector space. However, the way in which such abstract objects *combine linearly* is essentially the same as the way in which objects combine linearly which have a direct geometrical interpretation, such as rectilinear displacements in two- and three-dimensional Euclidean space. In considering the way in which linear combinations of such objects behave, we may consequently appeal to geometrical intuition as a guide. This is of very real value as our geometrical intuition can thus help in a visualization of complicated relationships which may subsequently be investigated by algebraic means. Secondly, the dimensions of the vectors, that is the number of independent objects in terms of which the vector may be expressed, is not restricted. It may be any finite number, or even infinite if suitable care is taken in definition. With high-dimensional spaces there is no direct geometrical interpretation in terms of intuitive understanding of some physical space but we may still be guided by geometrical intuition provided that each n-dimensional object is defined in algebraic terms and all manipulations and proofs are carried out by algebraic means.

2.4 Linear Independence and Bases

Subspace. If **V** is a linear vector space and **X** is some subset of the set of objects comprising **V**, then **X** is said to be a subspace of **V** if

(a) $\quad\quad\quad \mathbf{x}_1, \mathbf{x}_2 \in \mathbf{X} \quad$ implies that $\quad \mathbf{x}_1 + \mathbf{x}_2 \in \mathbf{X}$

(b) $\quad\quad\quad \alpha \in R^1 \quad$ and $\quad \mathbf{x} \in \mathbf{X} \quad$ implies that $\quad \alpha \mathbf{x} \in \mathbf{X}$

That is, a subset of **V** is a subspace if it is itself a linear vector space with respect to the binary operations of addition and scalar multiplications.

Linear independence. A given set of vectors $\{\mathbf{a}, \mathbf{b}, \mathbf{c}, \ldots, \mathbf{n}\}$ is said to be linearly independent if no combination of scalars $\alpha, \beta, \gamma, \ldots, \nu$, not all zero, exists such that

$$\alpha \mathbf{a} + \beta \mathbf{b} + \gamma \mathbf{c} + \cdots + \nu \mathbf{n} = \mathbf{0}$$

Putting this in another way, we say that a given set of vectors is linearly independent if no one member of the set can be expressed as a linear combination of the remainder.

Dimension of a vector space. A linear vector space is said to be of dimension n or n-dimensional if every vector in the space can be expressed as a linear combination of n basic independent vectors in the space.

Basis. A set of basic independent vectors in terms of which all the vectors in a linear vector space may be expressed is termed a basis for the space. The basis set may also be referred to as a reference frame or a coordinate vector set.

Coordinates. If a given set of vectors, $\{\mathbf{a}, \mathbf{b}, \mathbf{c}, \ldots, \mathbf{n}\}$ constitute a basis for an n-dimensional linear vector space, and a vector **x** is expressed in terms of this basis as

$$\mathbf{x} = x_1 \mathbf{a} + x_2 \mathbf{b} + x_3 \mathbf{c} + \cdots + x_n \mathbf{n}$$

then the set of scalars $x_1, x_2, \ldots, x_n$ are termed the coordinates of the vector **x** in this reference frame. The individual vectors $x_1 \mathbf{a}, x_2 \mathbf{b}, \ldots, x_n \mathbf{n}$ are termed the component vectors of the vector **x** in the reference frame.

Standard basis in R^n. The set of n-tuples

$$\mathbf{e}_1 = (1, 0, 0, 0, \ldots 0, 0)$$
$$\mathbf{e}_2 = (0, 1, 0, 0, \ldots 0, 0)$$
$$\mathbf{e}_3 = (0, 0, 1, 0, \ldots 0, 0)$$
$$\vdots$$
$$\mathbf{e}_n = (0, 0, 0, 0, \ldots 0, 1)$$

forms a basis for the linear vector space of real n-tuplets since its members are linearly independent, and any n-tuplet can be expressed in terms of them as a sum of scalar multiples of the basis set. They will be termed the standard (or natural or canonical) basis for the space of real n-tuplets, R^n.

2.5 Inner Product

Let **V** be a linear vector space and let P be a function whose domain is the Cartesian product space $\mathbf{V} \times \mathbf{V}$ and whose range is R^1. P is said to be an inner product or scalar product on **V** if the following conditions are satisfied:

(a) $P(\mathbf{x}_1 + \mathbf{x}_2, \mathbf{x}_3) = P(\mathbf{x}_1, \mathbf{x}_3) + P(\mathbf{x}_2, \mathbf{x}_3)$ for all $\mathbf{x}_1, \mathbf{x}_2, \mathbf{x}_3 \in \mathbf{V}$

(b) $P(\alpha \mathbf{x}_1, \mathbf{x}_2) = \alpha P(\mathbf{x}_1, \mathbf{x}_2)$ for all $\alpha \in R^1$ and all $\mathbf{x}_1, \mathbf{x}_2 \in \mathbf{V}$

(c) $P(\mathbf{x}_1, \mathbf{x}_2) = P(\mathbf{x}_2, \mathbf{x}_1)$ for all $\mathbf{x}_1, \mathbf{x}_2 \in \mathbf{V}$

Suppose P is an inner product on **V** and that $\{\mathbf{u}_1, \mathbf{u}_2, \ldots, \mathbf{u}_n\}$ is a basis for **V**. Let

$$P(\mathbf{u}_i, \mathbf{u}_j) = q_{ij} \qquad i, j = 1, 2, \ldots, n$$

and let two vectors **x** and **y** be expressed in terms of this basis set as

$$\mathbf{x} = \sum_{i=1}^{n} x_i \mathbf{u}_i \qquad \mathbf{y} = \sum_{j=1}^{n} y_j \mathbf{u}_j$$

We then have

$$P(\mathbf{x}, \mathbf{y}) = P\left[\sum_{i=1}^{n} x_i \mathbf{u}_i, \mathbf{y}\right]$$

$$= \sum_{i=1}^{n} x_i P(\mathbf{u}_i, \mathbf{y}) \qquad \text{on using conditions (a) and (b)}$$

$$= \sum_{i=1}^{n} x_i P\left(\mathbf{u}_i, \sum_{j=1}^{n} y_j \mathbf{u}_j\right)$$

$$= \sum_{i=1}^{n} x_i \sum_{j=1}^{n} y_j P(\mathbf{u}_i, \mathbf{u}_j)$$

$$= \sum_{i=1}^{n} x_i \sum_{j=1}^{n} y_j q_{ij}$$

$$= \sum_{i=1}^{n} \sum_{j=1}^{n} x_i q_{ij} y_j$$

$$= \mathbf{x}^t \mathbf{Q} \mathbf{y}$$

where **Q** is the matrix with element q_{ij} at the intersection of the ith row and jth column. **Q** is called the matrix of the inner product with respect to the basis set $\{\mathbf{u}_1, \mathbf{u}_2, \ldots, \mathbf{u}_n\}$; condition (c) ensures that **Q** will be a symmetric matrix. One particular inner product is used more than any others in dynamical systems work; this is the inner product on R^n whose matrix with respect to the standard basis set $\{\mathbf{e}_1, \mathbf{e}_2, \ldots, \mathbf{e}_n\}$ is the unit matrix **I**. This inner product on R^n will be called the standard scalar product on R^n, and distinguished by the special notation $\langle \mathbf{x}, \mathbf{y} \rangle$ for $P(\mathbf{x}, \mathbf{y})$. If $\mathbf{x}$ and $\mathbf{y}$ are the n-tuplets

$$\mathbf{x} = (x_1, x_2, \ldots, x_n)$$
$$\mathbf{y} = (y_1, y_2, \ldots, y_n)$$

then
$$\langle \mathbf{x}, \mathbf{y} \rangle \triangleq \sum_{i=1}^{n} x_i y_i \qquad (2\text{–}8)$$

Euclidean space. An n-dimensional Euclidean vector space, denoted by E^n, is defined as the linear vector space of n-tuple real numbers together with the standard basis and the standard scalar product as defined above.

Norm. If V is an inner product space with P as inner product then $P(\mathbf{x}, \mathbf{x})$ is termed a norm (or modulus, or length) of the element $\mathbf{x}$ of V and denoted by

$$\|\mathbf{x}\|_P = \sqrt{[P(\mathbf{x}, \mathbf{x})]} \qquad (2\text{–}9)$$

For the standard product in R^n the corresponding norm is called the Euclidean norm of $\mathbf{x}$ and we write

$$\|\mathbf{x}\|_E = \sqrt{\langle \mathbf{x}, \mathbf{x} \rangle} \qquad (2\text{–}10)$$

The norm has the following properties:

(a) $\|\mathbf{x}\|_P \geq 0$ for all $\mathbf{x} \in V$ and $\|\mathbf{x}\|_P = 0$ if and only if $\mathbf{x} = \mathbf{0}$
(b) $\|\mathbf{x}_1 + \mathbf{x}_2\|_P \leq \|\mathbf{x}_1\|_P + \|\mathbf{x}_2\|_P$ for all $\mathbf{x}_1, \mathbf{x}_2 \in V$
(c) $\|\alpha \mathbf{x}\|_P = |\alpha| \|\mathbf{x}\|_P$ for all $\alpha \in R^1$ and all $\mathbf{x} \in V$

In terms of the Euclidean distance measure previously introduced the Euclidean distance between $\mathbf{x}$ and $\mathbf{y}$ is the Euclidean norm of $(\mathbf{x} - \mathbf{y})$.

Angle between vectors. The angle θ between two vectors is defined by expressing the standard scalar product as

$$\langle \mathbf{x}, \mathbf{y} \rangle = \|\mathbf{x}\|_E \|\mathbf{y}\|_E \cos \theta \qquad (2\text{–}11)$$

giving
$$\cos \theta = \frac{\langle \mathbf{x}, \mathbf{y} \rangle}{\|\mathbf{x}\|_E \|\mathbf{y}\|_E} \qquad (2\text{–}12)$$

If $\langle \mathbf{x}, \mathbf{y} \rangle = 0$ then $\cos \theta = 0$ and the vectors are said to be orthogonal.

Orthonormal basis. A set of linearly independent vectors $\{\mathbf{u}_1, \mathbf{u}_2, \ldots, \mathbf{u}_n\}$ is said to form an orthonormal basis for the n-dimensional Euclidean vector space E^n if

$$\langle \mathbf{u}_i, \mathbf{u}_j \rangle = \delta_{ij} \qquad i, j = 1, 2, \ldots, n \qquad (2\text{–}13)$$

where δ_{ij}, termed the Kronecker delta, is defined by

$$\delta_{ij} = 1 \quad \text{for} \quad i = j \qquad \delta_{ij} = 0 \quad \text{for} \quad i \neq j$$

Gram–Schmidt Orthonormalization Procedure

Given a set of linearly independent vectors $\{\mathbf{a}_1, \mathbf{a}_2, \ldots, \mathbf{a}_n\}$ an orthonormal set $\{\mathbf{c}_1, \mathbf{c}_2, \ldots, \mathbf{c}_n\}$ may be constructed in the following way, via an intermediate orthogonal set $\{\mathbf{b}_1, \mathbf{b}_2, \ldots, \mathbf{b}_n\}$. Put

$$\mathbf{b}_1 = \mathbf{a}_1$$

$$\mathbf{c}_1 = \frac{\mathbf{b}_1}{\|\mathbf{b}_1\|_E}$$

where $\mathbf{c}_1$ is obtained by normalizing $\mathbf{b}_1$ by dividing it by its own norm. Then put

$$\mathbf{b}_2 = \mathbf{a}_2 - \langle \mathbf{c}_1, \mathbf{a}_2 \rangle \mathbf{c}_1 \tag{2-14}$$

$$\mathbf{c}_2 = \frac{\mathbf{b}_2}{\|\mathbf{b}_2\|_E} \tag{2-15}$$

Now taking the scalar product of both sides of equation (2–14) with $\mathbf{c}_1$ gives, using equation (2–15)

$$\langle \mathbf{c}_1, \mathbf{c}_2 \rangle = \left\langle \mathbf{c}_1, \frac{\mathbf{a}_2}{\|\mathbf{b}_2\|_E} \right\rangle - \langle \mathbf{c}_1, \mathbf{a}_2 \rangle \left\langle \frac{\mathbf{c}_1}{\|\mathbf{b}_2\|_E}, \mathbf{c}_1 \right\rangle$$

$$= 0 \qquad \qquad \text{since } \langle \mathbf{c}_1, \mathbf{c}_1 \rangle = 1$$

If we proceed in this way it is easily verified that an orthonormal set of vectors is constructed

$$\mathbf{b}_3 = \mathbf{a}_3 - \langle \mathbf{c}_1, \mathbf{a}_3 \rangle \mathbf{c}_1 - \langle \mathbf{c}_2, \mathbf{a}_3 \rangle \mathbf{c}_2$$

$$\mathbf{c}_3 = \frac{\mathbf{b}_3}{\|\mathbf{b}_3\|_E}$$

with the general terms given by

$$\mathbf{b}_r = \mathbf{a}_r - \sum_{k=1}^{r-1} \langle \mathbf{c}_k, \mathbf{a}_r \rangle \mathbf{c}_k$$

$$\mathbf{c}_r = \frac{\mathbf{b}_r}{\|\mathbf{b}_r\|_E} \tag{2-16}$$

2.6 Matrix Representation of Linear Operators in an n-dimensional Linear Vector Space

The set of $n \times 1$ matrices, or column vectors, together with the set of scalar numbers and the usual matrix definitions of the binary operations

Matrix Representation of Linear Operators

of addition and multiplication by a scalar, constitute a linear vector space. A reciprocal one-to-one correspondence or *isomorphism* thus exists between the linear vector space of $n \times 1$ matrices and the linear vector space of *n*-tuple real numbers. We may thus identify the *n*-tuple real number vectors with the $n \times 1$ matrices or column vectors and thus deploy the powerful tools of matrix algebra for the discussion of relationships in the vector space. This convention is accordingly adopted from here on and the vector **x** having components $x_1, x_2, \ldots, x_n$ will henceforth be represented by the $n \times 1$ matrix.

$$\mathbf{x} = \begin{bmatrix} x_1 \\ x_2 \\ x_3 \\ \vdots \\ x_n \end{bmatrix}$$

The transpose of vector **x** is denoted by $\mathbf{x}^t$ and is the $1 \times n$ matrix or row vector

$$\mathbf{x}^t = [x_1, x_2, x_3, \ldots, x_n]$$

The standard scalar product of two vectors may now be written alternatively as

$$\mathbf{x}^t \mathbf{y} \equiv \langle \mathbf{x}, \mathbf{y} \rangle$$

An alternative notation often used for the standard scalar product is the bold medial dot

$$\langle \mathbf{x}, \mathbf{y} \rangle \equiv \mathbf{x} \cdot \mathbf{y}$$

Note that if two vectors are both of unit length then the cosine of the angle between them is equal to their scalar product.

An *operator* on a vector space converts any given vector **x** into another vector **y**; it thus maps the linear vector space into itself. Suppose an operator is represented by L so that

$$\mathbf{y} = L\mathbf{x} \tag{2-17}$$

An operator on a space V is said to be linear if it is such that

(a) $L(\mathbf{x}_1 + \mathbf{x}_2) = L\mathbf{x}_1 + L\mathbf{x}_2$ for all $\mathbf{x}_1, \mathbf{x}_2 \in V$
(b) $L\alpha\mathbf{x} = \alpha L\mathbf{x}$ for all $\alpha \in R^1$ and all $\mathbf{x} \in V$

Any linear operator which is bounded, so that no **y** of infinite norm may be obtained from any **x** of finite norm, may be represented by a matrix.

To show this, let $\{\mathbf{u}_i : i = 1, 2, \ldots, n\}$ be an orthonormal basis in the space and, in terms of this basis set, let

$$\mathbf{x} = \sum_{i=1}^{n} x_i \mathbf{u}_i \qquad \mathbf{y} = \sum_{i=1}^{n} y_i \mathbf{u}_i$$

Now consider the action of the operator L on the basis vector $\mathbf{u}_j$. It will convert it into some other vector which may be expressed in terms of the chosen basis as

$$L\mathbf{u}_j = \sum_{i=1}^{n} l_{ij} \mathbf{u}_i \qquad j = 1, 2, \ldots, n$$

We thus have

$$L\mathbf{x} = \sum_{j=1}^{n} x_j L\mathbf{u}_j = \sum_{j=1}^{n} x_j \sum_{i=1}^{n} l_{ij} \mathbf{u}_i = \sum_{i=1}^{n} \sum_{j=1}^{n} l_{ij} x_j \mathbf{u}_i = \sum_{i=1}^{n} y_i \mathbf{u}_i$$

and so we may put

$$\mathbf{y} = L\mathbf{x} \qquad (2\text{--}18)$$

where $\mathbf{y}$ and $\mathbf{x}$ are column vectors having components $y_1, y_2, \ldots, y_n$ and $x_1, x_2, \ldots, x_n$ respectively and L is a matrix having an element l_{ij} at the intersection of the ith row and the jth column.

2.7 Dual Basis and the Projection Theorem

In a dynamical system investigation, the components of the vectors considered will be some set of n independent dynamical system variables. These vectors are regarded as expressed in terms of the standard basis in the n-dimensional Euclidean vector space. It is often convenient to choose some other set of n independent vectors as a new basis set, and then refer all vectors of the space to this new choice of basis. In the general case the new basis vectors will not be an orthogonal set when referred to the standard basis, and it is then convenient to introduce a related vector set known as the *dual* (or reciprocal) basis.

Dual basis. If a set of vectors $\{\mathbf{u}_1, \mathbf{u}_2, \ldots, \mathbf{u}_n\}$ form a basis for an n-dimensional Euclidean vector space, the dual basis is the set of vectors $\{\mathbf{v}_1, \mathbf{v}_2, \ldots, \mathbf{v}_n\}$ defined by

$$\langle \mathbf{u}_i, \mathbf{v}_j \rangle = \delta_{ij} \qquad i, j = 1, 2, \ldots, n \qquad (2\text{--}19)$$

where δ_{ij} is the Kronecker delta defined above. To show that these vectors are linearly independent, suppose some set of scalars $\alpha_1, \alpha_2, \ldots, \alpha_n$ exist such that

$$\sum_{k=1}^{n} \alpha_k \mathbf{v}_k = \mathbf{0}$$

Multiplying both sides of this equation by $\mathbf{u}_i$ for $i = 1, 2, \ldots, n$ and using equation (2-19) above gives

$$\sum_{k=1}^{n} \alpha_k \langle \mathbf{v}_k, \mathbf{u}_i \rangle = \alpha_i = 0 \quad \text{for} \quad i = 1, 2, \ldots, n$$

Thus the set of vectors $\{\mathbf{v}_1, \mathbf{v}_2, \ldots, \mathbf{v}_n\}$ is linearly independent and may therefore be taken as an alternative basis for the vector space.

The set of dual basis vectors may be calculated in a straightforward way. Inspection of the set of equations which define the dual basis shows that if $\mathbf{U}$ is a matrix whose *columns* are the vectors $\mathbf{u}_i$ ($i = 1, 2, \ldots, n$), then its inverse $\mathbf{U}^{-1}$ is a matrix whose *rows* are the transposed vectors $\mathbf{v}_i^t$ ($i = 1, 2, \ldots, n$). The dual basis is thus simply calculated by formation and inversion of the matrix $\mathbf{U}$.

Projection theorem. Let $\mathbf{u}_i$ ($i = 1, 2, \ldots, n$) be a set of independent vectors, referred to the standard basis in an n-dimensional Euclidean vector space, and let $\mathbf{v}_i$ ($i = 1, 2, \ldots, n$) be the corresponding dual basis. If a vector $\mathbf{x}$ is expressed as a linear combination of the set $\mathbf{u}_i$ it is said to have been projected into this reference frame. Let

$$\mathbf{x} = \bar{x}_1 \mathbf{u}_1 + \bar{x}_2 \mathbf{u}_2 + \cdots + \bar{x}_n \mathbf{u}_n \tag{2-20}$$

where $\bar{x}_1, \bar{x}_2, \ldots, \bar{x}_n$ are the components of vector $\mathbf{x}$ in the new basis set $\mathbf{u}_i$ ($i = 1, 2, \ldots, n$). To determine the components $\bar{x}_i$, the dual basis is used in the following way. Take the scalar product of both sides of equation (2-20) with the reciprocal vector set $\mathbf{v}_i$ ($i = 1, 2, \ldots, n$).

Then, using equation (2-19)

$$\langle \mathbf{x}, \mathbf{v}_i \rangle = \sum_{j=1}^{n} \langle \bar{x}_j \mathbf{u}_j, \mathbf{v}_i \rangle = \bar{x}_i \quad i = 1, 2, \ldots, n$$

The components $\bar{x}_i$ ($i = 1, 2, \ldots, n$) are thus simply determined by taking the scalar product of $\mathbf{x}$ with the appropriate member of the dual basis giving

$$\mathbf{x} = \sum_{j=1}^{n} \langle \mathbf{x}, \mathbf{v}_j \rangle \mathbf{u}_j \tag{2-21}$$

This useful result will be referred to as the Projection theorem.

2.8 Matrix Representation of Change of Basis in an n-dimensional Euclidean Vector Space

If a vector whose coordinates are known in the standard basis is expressed in terms of a set of new basis vectors $\mathbf{u}_i$ ($i = 1, 2, \ldots, n$), the coordinates of the vector $\mathbf{x}$ in the new basis are given by

$$\bar{x}_i = \langle \mathbf{v}_i, \mathbf{x} \rangle = \sum_{j=1}^{n} v_{ij} x_j \quad i = 1, 2, \ldots, n$$

where v_{ij} for $j = 1, 2, \ldots, n$ are the coordinates of the dual basis vector $\mathbf{v}_i$ referred to the standard basis. This set of equations may be written in matrix form as

$$\bar{\mathbf{x}} = V\mathbf{x} \qquad (2\text{--}22)$$

where $\bar{\mathbf{x}}$ is a column vector whose elements are the coordinates of vector $\mathbf{x}$ in the new basis, $\mathbf{x}$ is a column vector whose elements are the coordinates of the vector $\mathbf{x}$ in the standard basis, and V is a matrix whose *rows* are the components of the transposed dual basis vectors $\mathbf{v}_i^t$ ($i = 1, 2, \ldots, n$) referred to the standard basis.

2.9 Coordinate Transformations

A transformation of the coordinates $x_1, x_2, \ldots, x_n$ of a vector $\mathbf{x}$ is a set of rules or relations which associate each n-tuplet with a new n-tuplet. Such coordinate transformations may be regarded from two distinct points of view.

(a) *The active or 'alibi' interpretation.* The equation (2–18) describes an *operation* (or mapping) associating a *new* mathematical object $\mathbf{y}$ with each given object $\mathbf{x}$.

(b) *The passive or 'alias' interpretation.* The equation (2–22) describes a *new description* (or relabelling) of the *same* object in terms of a new reference frame. Coordinate transformations of this type arise in physical systems when we change the scheme of measurement used. Such transformations of coordinates are usually restricted to reversible transformations for which the corresponding transformation matrix is non-singular. For such cases the inverse transformation corresponding to equation (2–22) is

$$\mathbf{x} = V^{-1}\bar{\mathbf{x}}$$

2.10 Representation of an Operator Matrix in a Different Basis

Consider a change of reference basis with an associated coordinate transformation matrix $\mathbf{P}$. Let $\mathbf{x}, \mathbf{y}$ be column vectors whose elements are the coordinates of a pair of vectors in the standard basis, and let $\bar{\mathbf{x}}, \bar{\mathbf{y}}$ be column vectors whose elements are the coordinates of these same vectors in the new reference basis. Thus we have

$$\bar{\mathbf{x}} = P\mathbf{x} \qquad \bar{\mathbf{y}} = P\mathbf{y}$$

Let $\mathbf{L}$ be the matrix representation of a linear operator in the standard basis which transforms the vector $\mathbf{x}$ into the vector $\mathbf{y}$, so that

$$\mathbf{y} = L\mathbf{x}$$

This operator may be represented by a matrix in the new basis. To find this new representation we have

giving
$$\mathbf{x} = \mathbf{P}^{-1}\bar{\mathbf{x}} \qquad \mathbf{y} = \mathbf{P}^{-1}\bar{\mathbf{y}}$$
$$\mathbf{P}^{-1}\bar{\mathbf{y}} = \mathbf{L}\mathbf{P}^{-1}\bar{\mathbf{x}}$$

whence
$$\bar{\mathbf{y}} = \mathbf{PLP}^{-1}\bar{\mathbf{x}}$$

This shows that the linear operator, whose matrix representation was $\mathbf{L}$ in the old basis, has the representation $\mathbf{PLP}^{-1}$ in the new basis.

It is frequently convenient, after a transformation of vectors and operators has been carried out in changing from the standard basis in a given vector space to a difference reference basis in the same space, to regard the transformed vector coordinates as the coordinates of vectors referred to a standard basis in some *new* vector space.

2.11 Eigenvalues and Eigenvectors of a Linear Operator Matrix Representation

If $\mathbf{A}$ is the matrix representation of a linear operator on a Euclidean vector space referred to the standard basis, it will transform a given vector $\mathbf{x}$ into a related vector $\mathbf{y}$ such that

$$\mathbf{y} = \mathbf{A}\mathbf{x}$$

Suppose some vector $\mathbf{u}$ exists such that the operator represented by $\mathbf{A}$ alters its Euclidean norm $\|\mathbf{u}\|_E$ in the ratio of $\lambda:1$ and leaves the *relative* values of the individual components unchanged. Then

$$\mathbf{A}\mathbf{u} = \lambda\mathbf{u} \equiv \lambda\mathbf{I}\mathbf{u}$$

where $\mathbf{I}$ is the unit matrix of order n. This gives

$$[\mathbf{A} - \lambda\mathbf{I}]\mathbf{u} = \mathbf{0}$$

which represents a set of homogeneous equations in the components of the vector $\mathbf{u}$. A solution of this set of equations will exist if, and only if

$$\det[\mathbf{A} - \lambda\mathbf{I}] = 0 \qquad (2\text{--}23)$$

This is termed the *characteristic equation* of the matrix $\mathbf{A}$, and the corresponding n roots are defined as the *eigenvalues* of the matrix $\mathbf{A}$. The eigenvalues may be real or complex; if they are complex, they must occur in complex conjugate pairs since the coefficients in the equation (2–23) are all real.

Corresponding to each eigenvalue λ_i there is a nonzero eigenvector $\mathbf{u}_i$ such that

$$\mathbf{A}\mathbf{u}_i = \lambda_i\mathbf{u}_i \qquad i = 1, 2, \ldots, n$$

If $\mathbf{u}_i$ satisfies this equation, then so will $\alpha\mathbf{u}_i$ where α is any real constant.

The norm of an eigenvector is thus indeterminate, and it is usually convenient to normalize the eigenvector set by adjusting the relative size of each eigenvector component by dividing through each by the norm of the corresponding eigenvector. This gives a normalized eigenvector set of unit norms. For simple low-order matrices the most convenient way to solve for an eigenvector is to arbitrarily set one of the components of each $\mathbf{u}_i$ equal to unity and then solve the resulting determinate equation set for the remaining components. For matrices of order greater than about three, it is advisable to use a digital computer to perform the eigensystem analysis. The determination of complex eigenvectors is considered in Chapter 6.

Let $\mathbf{U}$ be a matrix whose columns are the eigenvectors $\mathbf{u}_1, \mathbf{u}_2, \ldots, \mathbf{u}_n$ of a matrix $\mathbf{A}$ and let $\mathbf{V}$ be a matrix whose rows are the transposed dual vector set to $\{\mathbf{u}_1, \mathbf{u}_2, \ldots, \mathbf{u}_n\}$, defined as the dual (or reciprocal) eigenvector set $\{\mathbf{v}_1, \mathbf{v}_2, \ldots, \mathbf{v}_n\}$.

Let $\mathbf{x}$ and $\mathbf{y}$ be vectors referred to the standard basis and let $\bar{\mathbf{x}}, \bar{\mathbf{y}}$ be the same vectors referred to the eigenvectors set $\{\mathbf{u}_1, \mathbf{u}_2, \ldots, \mathbf{u}_n\}$ as a basis. Let $\mathbf{A}$ have distinct eigenvalues.

Then

$$\bar{x}_i = \langle \mathbf{v}_i, \mathbf{x} \rangle \qquad \bar{y}_i = \langle \mathbf{v}_i, \mathbf{y} \rangle$$

and thus

$$\bar{\mathbf{x}} = \mathbf{U}^{-1}\mathbf{x} = \mathbf{V}\mathbf{x} \qquad \bar{\mathbf{y}} = \mathbf{V}\mathbf{y}$$

where $\mathbf{V}$ is defined as above, and also

$$\mathbf{x} = \mathbf{V}^{-1}\bar{\mathbf{x}} = \mathbf{U}\bar{\mathbf{x}} \tag{2-24}$$

$$\mathbf{y} = \mathbf{U}\bar{\mathbf{y}} \tag{2-25}$$

Now consider the matrix $\mathbf{A}$ operating on a vector $\mathbf{x}$ to give a related vector $\mathbf{y}$

$$\mathbf{y} = \mathbf{A}\mathbf{x}$$

Using equations (2-24) and (2-25) we have

$$\mathbf{U}\bar{\mathbf{y}} = \mathbf{A}\mathbf{U}\bar{\mathbf{x}}$$

$$\bar{\mathbf{y}} = \mathbf{U}^{-1}\mathbf{A}\mathbf{U}\bar{\mathbf{x}}$$

$$= \mathbf{V}\mathbf{A}\mathbf{U}\bar{\mathbf{x}}$$

where $\mathbf{V}\mathbf{A}\mathbf{U}$ is the representation of the operator $\mathbf{A}$ in the basis consisting of its eigenvectors. Now we may partition $\mathbf{U}$ into its constituent columns to get

$$\mathbf{V}\mathbf{A}\mathbf{U} = \mathbf{V}\mathbf{A}[\mathbf{u}_1 \mid \mathbf{u}_2 \mid \cdots \mid \mathbf{u}_n] = \mathbf{V}[\lambda_1 \mathbf{u}_1 \mid \lambda_2 \mathbf{u}_2 \mid \cdots \mid \lambda_n \mathbf{u}_n]$$

and, further, partition **V** into its constituent rows to get

$$\mathbf{VAU} = \begin{bmatrix} \mathbf{v}_1^t \\ \mathbf{v}_2^t \\ \vdots \\ \mathbf{v}_n^t \end{bmatrix} [\lambda_1 \mathbf{u}_1 \vdots \lambda_2 \mathbf{u}_2 \vdots \cdots \vdots \lambda_n \mathbf{u}_n] \equiv \begin{bmatrix} \lambda_1 & 0 & 0 & \cdots & 0 \\ 0 & \lambda_2 & 0 & \cdots & 0 \\ 0 & 0 & \lambda_3 & \cdots & 0 \\ \vdots & \vdots & \vdots & & \vdots \\ 0 & 0 & 0 & & \lambda_n \end{bmatrix}$$

Thus we have

$$\mathbf{VAU} = \Lambda$$

where Λ is a diagonal matrix of distinct eigenvalues $\lambda_1, \lambda_2, \ldots, \lambda_n$. Again we have

$$\mathbf{A} = \mathbf{V}^{-1} \Lambda \mathbf{U}^{-1} = \mathbf{U} \Lambda \mathbf{V}$$

Now

$$\mathbf{VA} = \Lambda \mathbf{V}$$

and thus

$$\mathbf{A}^t \mathbf{V}^t = \mathbf{V}^t \Lambda^t$$

which, after suitable partitioning and multiplication gives

$$[\mathbf{A}^t \mathbf{v}_1 \vdots \mathbf{A}^t \mathbf{v}_2 \vdots \cdots \vdots \mathbf{A}^t \mathbf{v}_n] = [\lambda_1 \mathbf{v}_1 \vdots \lambda_2 \mathbf{v}_2 \vdots \cdots \vdots \lambda_n \mathbf{v}_n]$$

so that

$$\mathbf{A}^t \mathbf{v}_i = \lambda_i \mathbf{v}_i \qquad i = 1, 2, \ldots, n$$

Thus the normalized eigenvector set and the normalized dual eigenvector set will coincide for a symmetrical matrix for which $\mathbf{A} = \mathbf{A}^t$. It follows from this that for a *symmetrical* matrix, the eigenvectors will form a mutually orthogonal set and we will have

$$\mathbf{U}^{-1} = \mathbf{V} = \mathbf{U}^t$$
$$\mathbf{V}^{-1} = \mathbf{U} = \mathbf{V}^t$$

so that for a *symmetrical* matrix **A**

$$\mathbf{U}^t \mathbf{A} \mathbf{U} = \Lambda \qquad \mathbf{A} = \mathbf{U} \Lambda \mathbf{U}^t$$

Transposing both sides of equation (2–26) gives

$$\mathbf{v}_i^t \mathbf{A} = \lambda_i \mathbf{v}_i^t \qquad i = 1, 2, \ldots, n$$

For this reason the row vectors $\mathbf{v}_i^t$ ($i = 1, 2, \ldots, n$) are sometimes called the *left-hand eigenvectors* of **A** in distinction to the vectors $\mathbf{u}_i$ for which

$$\mathbf{A} \mathbf{u}_i = \lambda_i \mathbf{u}_i \qquad i = 1, 2, \ldots, n$$

and which are thus sometimes called the *right-hand eigenvectors* of **A**.

Eigenvalue shift theorem. Let $\{\mathbf{u}_i : i = 1, 2, \ldots, n\}$ be a set of eigenvectors for a matrix $\mathbf{A}$ with a corresponding distinct set of eigenvalues $\{\lambda_i : i = 1, 2, \ldots, n\}$. Then

$$[\mathbf{A} + k\mathbf{I}]\mathbf{u}_i = \mathbf{A}\mathbf{u}_i + k\mathbf{u}_i$$
$$= \lambda_i \mathbf{u}_i + k\mathbf{u}_i$$
$$= (\lambda_i + k)\mathbf{u}_i$$

where k is a real constant and $\mathbf{I}$ is a unit matrix of order n. This shows that the eigenvectors of $\mathbf{A}$ are eigenvectors of $[\mathbf{A} + k\mathbf{I}]$ and that the eigenvalues of $[\mathbf{A} + k\mathbf{I}]$ are

$$(\lambda_i + k) \qquad i = 1, 2, \ldots, n$$

Reciprocal matrix spectrum. If

$$\mathbf{A}\mathbf{u}_i = \lambda_i \mathbf{u}_i \qquad i = 1, 2, \ldots, n$$

Then
$$\mathbf{A}\lambda_i^{-1}\mathbf{u}_i = \mathbf{u}_i \qquad i = 1, 2, \ldots, n$$

so that, multiplying both sides by the inverse of $\mathbf{A}$

$$\lambda_i^{-1}\mathbf{u}_i = \mathbf{A}^{-1}\mathbf{u}_i \qquad i = 1, 2, \ldots, n$$

This shows that the eigenvectors of a matrix are also eigenvectors of the inverse matrix (if an inverse exists) and that the eigenvalues of the inverse matrix are the inverses of the matrix eigenvalues.

Invariance of eigenvalues under a change of basis. Let $\mathbf{A}$ be the matrix representation of a linear operator referred to the standard basis in a Euclidean space of n dimensions, and let $\mathbf{P}$ be the coordinate transformation matrix associated with a change of basis. Then, as shown in Section 2.10, the operator is represented in the new basis by a matrix $\mathbf{PAP}^{-1}$. Consider the characteristic equation of this matrix. We have

$$\det[\lambda\mathbf{I} - \mathbf{PAP}^{-1}] = \det[\mathbf{P}\lambda\mathbf{IP}^{-1} - \mathbf{PAP}^{-1}]$$
$$= \det\{\mathbf{P}[\lambda\mathbf{I} - \mathbf{A}]\mathbf{P}^{-1}\}$$
$$= \det(\mathbf{P})\det(\lambda\mathbf{I} - \mathbf{A})\det(\mathbf{P}^{-1})$$
$$= \det(\lambda\mathbf{I} - \mathbf{A})$$

which shows that the eigenvalues of the matrix representation of a linear operator $\mathbf{A}$ are invariant under a simple change of basis.

Location bounds on eigenvalues. The determination of the eigenvalues of a matrix requires a considerable investment in computation and it is important to deduce as much as possible about the location of matrix eigenvalues before embarking on an eigenvalue calculation. It is convenient to denote the eigenvalues of a matrix $\mathbf{A}$ by

$$\lambda_i(\mathbf{A}) \qquad i = 1, 2, \ldots, n$$

The sum of the diagonal elements of a matrix $\mathbf{A}$ is termed the trace of the matrix and is denoted by

$$\text{trace}(\mathbf{A}) = \sum_{i=1}^{n} a_{ii}$$

It follows from the invariance of the characteristic equation under a change of basis that

$$\text{trace}(\mathbf{A}) = \sum_{i=1}^{n} \lambda_i(\mathbf{A})$$

so that the arithmetic mean of the eigenvalue distribution, which gives the 'centre of gravity' of the eigenvalue set, is given by

$$\text{arithmetic mean of eigenvalue set} = \text{trace}(\mathbf{A})/n$$

It also follows that the determinant of a matrix is equal to the product of the eigenvalues, so that

$$\text{geometric mean of eigenvalue set} = [\det(\mathbf{A})]^{1/n}$$

A most useful theorem due to Gershgorin states that the eigenvalues of a matrix $\mathbf{A}$ must be inside, or on the boundary of, the union of the circles of centre a_{ii} and radius

$$\sum_{\substack{i \neq j \\ j=1}}^{n} |a_{ij}| \qquad i = 1, 2, \ldots, n$$

If m of these circles intersect to form a connected domain disjoint from the remaining circles, this domain will contain m eigenvalues. Since the eigenvalues of the transposed matrix $\mathbf{A}^t$ equal those of $\mathbf{A}$, a further union of circles may be constructed for the transposed matrix. We may thus locate the eigenvalues of a given matrix in an easily constructed union of circles. The union of circles so constructed for a stable system may still include part of the right-half of the complex variable plane, so that no simple test for stability is implied. However, if the union of circles does not include any part of the right half-plane, the corresponding dynamical system is certainly stable.

It is straight forward to split a matrix $\mathbf{A}$ into its symmetric and skew-symmetric parts, which will be denoted by $\mathbf{A}_+$ and $\mathbf{A}_-$ respectively, and are given by

$$\mathbf{A}_+ = \tfrac{1}{2}(\mathbf{A} + \mathbf{A}^t) \qquad \mathbf{A}_- = \tfrac{1}{2}(\mathbf{A} - \mathbf{A}^t)$$

where the superscript t denotes transposition. A useful set of bounds exists [M10] which relate the real parts of the eigenvalues of $\mathbf{A}$ to the eigenvalues of $\mathbf{A}_+$. Let λ_i and μ_i denote the eigenvalues of $\mathbf{A}$ and $\mathbf{A}_+$ respectively, and suppose that all the eigenvalues have been ordered such that the subscripts n and 1 denote the largest and smallest eigenvalues respectively.

Let $\mathbf{Q}$ be a symmetric matrix with distinct eigenvalues. Then there exists a matrix $\mathbf{U}$ whose columns are an orthogonal eigenvector set such that

$$\mathbf{\Lambda}^* = \mathbf{U}^t\mathbf{Q}\mathbf{U} \qquad \mathbf{Q} = \mathbf{U}\mathbf{\Lambda}^*\mathbf{U}^t \qquad \mathbf{U}\mathbf{U}^t = \mathbf{I}$$

where $\mathbf{\Lambda}^*$ is a diagonal matrix whose elements are the eigenvalues of $\mathbf{Q}$, and $\mathbf{I}$ is a unit matrix.

Let
$$\mathbf{y} = \mathbf{U}^t\mathbf{x} \qquad \mathbf{x} = \mathbf{U}\mathbf{y}$$

If the eigenvalues of $\mathbf{Q}$ are so ordered that

$$\lambda_n^* \geqslant \lambda_{n-1}^* \geqslant \cdots \geqslant \lambda_1^*$$

we must have that

$$\lambda_n^*\bar{\mathbf{y}}^t\mathbf{y} \geqslant \bar{\mathbf{y}}^t\mathbf{\Lambda}^*\mathbf{y} \geqslant \lambda_1^*\bar{\mathbf{y}}^t\mathbf{y}$$

where $\bar{\mathbf{y}}$ denotes the complex conjugate of $\mathbf{y}$. Thus

$$\lambda_n^*(\mathbf{U}^t\bar{\mathbf{x}})^t(\mathbf{U}^t\mathbf{x}) \geqslant (\mathbf{U}^t\bar{\mathbf{x}})^t\mathbf{\Lambda}^*(\mathbf{U}^t\mathbf{x}) \geqslant \lambda_1^*(\mathbf{U}^t\bar{\mathbf{x}})^t(\mathbf{U}^t\mathbf{x})$$

where $\bar{\mathbf{x}}$ denotes the complex conjugate of $\mathbf{x}$, giving

$$\lambda_n^*\bar{\mathbf{x}}^t\mathbf{U}\mathbf{U}^t\mathbf{x} \geqslant \bar{\mathbf{x}}^t\mathbf{U}\mathbf{\Lambda}^*\mathbf{U}^t\mathbf{x} \geqslant \lambda_1^*\bar{\mathbf{x}}^t\mathbf{U}\mathbf{U}^t\mathbf{x}$$

so that
$$\lambda_n^*\bar{\mathbf{x}}^t\mathbf{x} \geqslant \bar{\mathbf{x}}^t\mathbf{Q}\mathbf{x} \geqslant \lambda_1^*\bar{\mathbf{x}}^t\mathbf{x} \qquad (2\text{-}26)$$

Now let $\mathbf{u}_i$ be an eigenvector of unit length corresponding to the eigenvalue λ_i for a general real matrix $\mathbf{A}$ and let $\bar{\mathbf{u}}_i$, $\bar{\lambda}_i$ denote the complex conjugates of $\mathbf{u}_i$, λ_i respectively. We then have

$$\lambda_i\mathbf{u}_i = \mathbf{A}\mathbf{u}_i \qquad \bar{\lambda}_i\bar{\mathbf{u}}_i = \mathbf{A}\bar{\mathbf{u}}_i$$

The numbers λ_i and $\bar{\lambda}_i$ may therefore be expressed in the following way

$$\lambda_i = \lambda_i\bar{\mathbf{u}}_i^t\mathbf{u}_i = \bar{\mathbf{u}}_i^t\lambda_i\mathbf{u}_i = \bar{\mathbf{u}}_i^t\mathbf{A}\mathbf{u}_i$$
$$\bar{\lambda}_i = \bar{\lambda}_i\mathbf{u}_i^t\bar{\mathbf{u}}_i = \mathbf{u}_i^t\bar{\lambda}_i\bar{\mathbf{u}}_i = \mathbf{u}_i^t\mathbf{A}\bar{\mathbf{u}}_i$$

If Re (λ_i) denotes the real part of λ_i we have:

$$\begin{aligned}
\text{Re}\,(\lambda_i) &= \tfrac{1}{2}(\lambda_i + \bar{\lambda}_i) \\
&= \tfrac{1}{2}\bar{\mathbf{u}}_i^t\mathbf{A}\mathbf{u}_i + \tfrac{1}{2}\mathbf{u}_i^t\mathbf{A}\bar{\mathbf{u}}_i \\
&= \tfrac{1}{2}\bar{\mathbf{u}}_i^t\mathbf{A}\mathbf{u}_i + \tfrac{1}{2}\bar{\mathbf{u}}_i^t\mathbf{A}^t\mathbf{u}_i \\
&= \tfrac{1}{2}\bar{\mathbf{u}}_i^t(\mathbf{A} + \mathbf{A}^t)\mathbf{u}_i \\
&= \bar{\mathbf{u}}_i^t\mathbf{A}_+\mathbf{u}_i
\end{aligned}$$

If μ_n and μ_1 are the largest and smallest eigenvalues of $\mathbf{A}_+$ we have from equation (2–26), since $\mathbf{u}_i$ is of unit length

$$\mu_1 \leq \mathbf{\bar{u}}_i^t \mathbf{A}_+ \mathbf{u}_i \leq \mu_n$$

Inserting the expression found for $\mathrm{Re}\,(\lambda_i)$ then gives the inequalities

$$\mu_1 \leq \mathrm{Re}\,(\lambda_i) \leq \mu_n \qquad i = 1, 2, \ldots, n$$

showing that the real parts of the eigenvalues of a general real matrix are bounded by the eigenvalues of its symmetric part.

Some further useful bounds make use of the matrix norm

$$\|\mathbf{A}\| = \left[\sum_{i,j=1}^{n} |a_{ij}|^2\right]^{\frac{1}{2}}$$

These are

$$0 \leq \sum_{i=1}^{n} |\lambda_i|^2 \leq \|\mathbf{A}\|^2$$

$$0 \leq \sum_{i=1}^{n} |\mathrm{Re}\,(\lambda_i)|^2 \leq \|\mathbf{A}_+\|^2$$

$$0 \leq \sum_{i=1}^{n} |\mathrm{Im}\,(\lambda_i)|^2 \leq \|\mathbf{A}_-\|^2$$

2.12 Scalar-valued Functions of Vectors

In the application of vector space methods to the analysis of dynamical systems, the components of the vectors involved are physical variables associated with some system of interest. Certain entities such as total stored energy or total power are then scalar-valued functions of the vectors involved. Thus the consideration of such scalar-valued functions of vectors is of great importance in the dynamical application of vector space methods. The scalar-valued function most used is the quadratic form in the components of a vector.

A quadratic form in the n variables $x_1, x_2, \ldots, x_n$ is defined as a homogeneous polynomial of the second degree of the form

$$\sum_{i=1}^{n} \sum_{j=1}^{n} x_i q_{ij} x_j \quad \text{with} \quad q_{ij} = q_{ji}$$

If the variables $x_1, x_2, \ldots, x_n$ are the coordinates of a column vector $\mathbf{x}$, then the quadratic form may be written as

$$\mathbf{x}^t \mathbf{Q} \mathbf{x} \quad \text{or} \quad \langle \mathbf{x}, \mathbf{Q} \mathbf{x} \rangle$$

where $\mathbf{Q}$ is a symmetric matrix of real coefficients. Since a scalar is equal to its own transpose we have

$$y^t Q x = (y^t Q x)^t = x^t Q^t y$$

giving the useful relationship

$$\langle y, Q x \rangle = \langle Q^t y, x \rangle \qquad (2\text{–}27)$$

which is used repeatedly in the manipulations carried out in later chapters.

Consider a change of reference basis with an associated transformation matrix P, such that

$$P \bar{x} = x$$

A quadratic form $x^t Q x$ in the old basis thus becomes the quadratic form

$$(P\bar{x})^t Q (P\bar{x}) = \bar{x}^t P^t Q P \bar{x} \quad \text{in the new basis.}$$

Thus under the coordinate transformation P associated with a change of reference basis, the coefficient matrix Q of a given quadratic form is transformed into the coefficient matrix $P^t Q P$ for the new basis.

2.13 Range Space, Null Space, Rank and Nullity of a Linear Operator

In Section 2.6 it has been shown that a linear operator may be represented by a matrix L in a given basis $u_1, u_2, \ldots, u_n$ and that the element of L at the intersection of ith row and jth column is such that

$$L u_j = \sum_{i=1}^{n} l_{ij} u_i \qquad j = 1, 2, \ldots, n$$

Writing this out in full we have

$$L u_1 = l_{11} u_1 + l_{21} u_2 + \cdots + l_{n1} u_n$$
$$L u_2 = l_{12} u_1 + l_{22} u_2 + \cdots + l_{n2} u_n$$
$$\vdots$$
$$L u_n = l_{1n} u_1 + l_{2n} u_2 + \cdots + l_{nn} u_n$$

Thus we see that

$$\alpha L u_1 + \beta L u_2 + \cdots + v L u_n$$
$$= [\alpha l_{11} + \beta l_{12} + \cdots + v l_{1n}] u_1$$
$$+ [\alpha l_{21} + \beta l_{22} + \cdots + v l_{2n}] u_2$$
$$+ \cdots$$
$$+ \cdots + [\alpha l_{n1} + \beta l_{n2} + \cdots + v l_{nn}] u_n \qquad (2\text{–}28)$$

An inspection of equation (2–28) shows that the vector set $\{\mathbf{Lu}_1, \mathbf{Lu}_2, \ldots, \mathbf{Lu}_n\}$ will be linearly dependent if and only if some of the columns of the matrix $\mathbf{L}$ are linearly dependent.

A linear operator transforms the vector space on which it is defined into its *range space*. The above considerations show that if any of the columns of the matrix representing the operator are linearly dependent, then the range space will be of lower dimension than the space on which the operator is defined. Further reflection will show that the dimension of the range space is equal to the number of linearly independent columns of the matrix $\mathbf{L}$. The number of linearly independent columns of a matrix is called its column rank and the number of linearly independent rows is called its row rank. Both column and row rank are equal to the order of the highest order non-zero minor which may be formed from the matrix and which is therefore termed the *rank* of the matrix. To establish this, suppose a matrix has m columns and the column rank is $k \leqslant m$. Then k and not more than k columns are linearly independent so that there exists a $k \times k$ non-zero minor which may be formed from these columns. This will be the highest-order non-zero minor which may be formed from the columns since any of higher order will have at least two dependent columns. If the matrix has l rows and the row rank is $r \leqslant l$, then r and not more than r rows are linearly independent. Thus there exists an $r \times r$ non-zero minor which may be formed from these rows and this will be the highest-order non-zero minor which may be formed from the rows since any higher-order minor will have at least two dependent rows. But the order of the highest-order non-vanishing minor is unique, and so the column rank and row rank are equal, and equal to this order.

The set of all vectors $\mathbf{x}$ such that

$$\mathbf{Lx} = \mathbf{0}$$

is called the *null space* of the operator $\mathbf{L}$, and the dimension of the null space is called the *nullity* of $\mathbf{L}$.

Suppose the nullity of $\mathbf{L}$ is s. The null space will then have a basis, say $\mathbf{v}_1, \mathbf{v}_2, \ldots, \mathbf{v}_s$, of s elements which can be extended to a basis $\mathbf{v}_1, \mathbf{v}_2, \ldots, \mathbf{v}_s, \mathbf{v}_{s+1}, \ldots, \mathbf{v}_{s+r}$ for the whole space. Since

$$\mathbf{Lv}_i = \mathbf{0} \qquad i = 1, 2, \ldots, s$$

the vectors $\mathbf{v}_{s+1}, \ldots, \mathbf{v}_{s+r}$ must span the range space of $\mathbf{L}$. Now the relation

$$\alpha_1 \mathbf{Lv}_{s+1} + \alpha_2 \mathbf{Lv}_{s+2} + \cdots + \alpha_r \mathbf{Lv}_{s+r} = \mathbf{0}$$

where $\alpha_1, \alpha_2, \ldots, \alpha_r$ is a set of constants, implies that $[\alpha_1 \mathbf{Lv}_{s+1} + \cdots + \alpha_r \mathbf{Lv}_{s+r}]$ lies in the null space of $\mathbf{L}$. Since this is not so we must have

$$\alpha_1 = \alpha_2 = \cdots = \alpha_r = 0$$

Thus the vectors $\mathbf{v}_{s+1}, \ldots, \mathbf{v}_{s+r}$ must be linearly independent and therefore form a basis for the range space of $\mathbf{L}$ so that

$$n = s + r$$

In words, that is

Operator Rank + Operator Nullity = Order of vector space

2.14 Vector Differentiation

Suppose a vector $\mathbf{f}$ having components $f_1, f_2, \ldots, f_m$ is a function of another vector $\mathbf{x}$ having components $x_1, x_2, \ldots, x_n$, and that the derivatives $\partial f_i/\partial x_j$, $\partial^2 f_i/\partial x_j \partial x_k$ exist for $i, j, k = 1, 2, \ldots, n$. The multivariable form of Taylor's theorem then gives that, in a sufficiently small neighbourhood of a given set of values of the components of $\mathbf{x}$

$$f_i(x_1 + \Delta x_1, x_2 + \Delta x_2, \ldots, x_n + \Delta x_n)$$
$$= f_i(x_1, x_2, \ldots, x_n) + \sum_{r=1}^{n} \left[\frac{\partial f_i}{\partial x_r}\right] \Delta x_r$$
$$+ \frac{1}{2!} \sum_{r=1}^{n} \sum_{s=1}^{n} \left[\frac{\partial^2 f_i}{\partial x_r \partial x_s}\right] \Delta x_r \Delta x_s + \text{terms of order } (\Delta x_i)^3$$

$$i = 1, 2, \ldots, m$$

This may be written in vector notation as

$$\mathbf{f}(\mathbf{x} + \Delta \mathbf{x}) = \mathbf{f}(\mathbf{x}) + \frac{\partial \mathbf{f}}{\partial \mathbf{x}} \Delta \mathbf{x} + \cdots$$

where the matrix $\partial \mathbf{f}/\partial \mathbf{x}$ may be called the derivative of vector $\mathbf{f}$ with respect to vector $\mathbf{x}$ and is defined as having coefficients given by

$$\frac{\partial \mathbf{f}}{\partial \mathbf{x}} \triangleq \begin{bmatrix} \frac{\partial f_1}{\partial x_1} & \frac{\partial f_1}{\partial x_2} & \cdots & \frac{\partial f_1}{\partial x_n} \\ \frac{\partial f_2}{\partial x_1} & \frac{\partial f_2}{\partial x_2} & \cdots & \frac{\partial f_2}{\partial x_n} \\ \vdots & & & \vdots \\ \frac{\partial f_m}{\partial x_1} & \frac{\partial f_m}{\partial x_2} & \cdots & \frac{\partial f_m}{\partial x_n} \end{bmatrix} \quad (2\text{–}29)$$

If ϕ is a scalar function of the vector $\mathbf{x}$ we therefore have that

$$\frac{\partial \phi}{\partial \mathbf{x}} = \begin{bmatrix} \frac{\partial \phi}{\partial x_1} & \frac{\partial \phi}{\partial x_2} & \cdots & \frac{\partial \phi}{\partial x_n} \end{bmatrix}$$

Vector Differentiation

and so the *gradient vector* of ϕ with respect to **x** is denoted by

$$\left(\frac{\partial \phi}{\partial \mathbf{x}}\right)^t \triangleq \begin{bmatrix} \dfrac{\partial \phi}{\partial x_1} \\ \dfrac{\partial \phi}{\partial x_2} \\ \vdots \\ \dfrac{\partial \phi}{\partial x_n} \end{bmatrix} \tag{2-30}$$

Certain derivatives of scalar functions occur frequently and may usefully be evaluated in matrix terms. We have

$$\left[\frac{\partial(\mathbf{x}^t \mathbf{P} \mathbf{y})}{\partial \mathbf{x}}\right]^t = \begin{bmatrix} \dfrac{\partial}{\partial x_1} \sum_{i=1}^{n}\sum_{j=1}^{m} x_i p_{ij} y_j \\ \dfrac{\partial}{\partial x_2} \sum_{i=1}^{n}\sum_{j=1}^{m} x_i p_{ij} y_j \\ \vdots \\ \dfrac{\partial}{\partial x_n} \sum_{i=1}^{n}\sum_{j=1}^{m} x_i p_{ij} y_j \end{bmatrix} = \begin{bmatrix} \sum_{j=1}^{m} p_{1j} y_j \\ \sum_{j=1}^{m} p_{2j} y_j \\ \vdots \\ \sum_{j=1}^{m} p_{nj} y_j \end{bmatrix} = \mathbf{Py} \tag{2-31}$$

$$\left[\frac{\partial(\mathbf{x}^t \mathbf{P} \mathbf{y})}{\partial \mathbf{y}}\right]^t = \begin{bmatrix} \dfrac{\partial}{\partial y_1} \sum_{i=1}^{n}\sum_{j=1}^{m} x_i p_{ij} y_j \\ \dfrac{\partial}{\partial y_2} \sum_{i=1}^{n}\sum_{j=1}^{m} x_i p_{ij} y_j \\ \vdots \\ \dfrac{\partial}{\partial y_m} \sum_{i=1}^{n}\sum_{j=1}^{m} x_i p_{ij} y_j \end{bmatrix} = \begin{bmatrix} \sum_{i=1}^{n} p_{i1} x_i \\ \sum_{i=1}^{n} p_{i2} x_i \\ \vdots \\ \sum_{i=1}^{n} p_{in} x_i \end{bmatrix} = \mathbf{P}^t \mathbf{x} \tag{2-32}$$

$$\left[\frac{\partial(\mathbf{x}^t \mathbf{Q} \mathbf{x})}{\partial \mathbf{x}}\right]^t = \begin{bmatrix} \dfrac{\partial}{\partial x_1} \sum_{i=1}^{n}\sum_{j=1}^{n} x_i q_{ij} x_j \\ \dfrac{\partial}{\partial x_2} \sum_{i=1}^{n}\sum_{j=1}^{n} x_i q_{ij} x_j \\ \vdots \\ \dfrac{\partial}{\partial x_n} \sum_{i=1}^{n}\sum_{j=1}^{n} x_i q_{ij} x_j \end{bmatrix} = \begin{bmatrix} 2\sum_{j=1}^{n} q_{1j} x_j \\ 2\sum_{j=1}^{n} q_{2j} x_j \\ \vdots \\ 2\sum_{j=1}^{n} q_{nj} x_j \end{bmatrix} = 2\mathbf{Qx} \tag{2-33}$$

Transformation of Gradient Vectors

A certain relationship between gradient vectors is used repeatedly in later chapters, particularly in Chapter 5. To derive this relationship, suppose $\phi(\mathbf{y})$ is a scalar-valued function of a vector $\mathbf{y}$. Let $\mathbf{x}$ be some other vector derived from vector $\mathbf{y}$ under the action of a known transformation matrix $\mathbf{T}$. That is let

$$\mathbf{x} = \mathbf{Ty}$$

The scalar function ϕ may now be expressed as a function of the components of the vector $\mathbf{x}$. We may thus compute the corresponding column vectors $(\partial \phi / \partial \mathbf{x})^t$ and $(\partial \phi / \partial \mathbf{y})^t$ and then form

$$\left\langle \mathbf{x}, \left(\frac{\partial \phi}{\partial \mathbf{x}}\right)^t \right\rangle \quad \text{and} \quad \left\langle \mathbf{y}, \left(\frac{\partial \phi}{\partial \mathbf{y}}\right)^t \right\rangle$$

Since this merely corresponds to computing the same scalar magnitude in two different reference systems we must have

$$\left\langle \mathbf{x}, \left(\frac{\partial \phi}{\partial \mathbf{x}}\right)^t \right\rangle = \left\langle \mathbf{y}, \left(\frac{\partial \phi}{\partial \mathbf{y}}\right)^t \right\rangle \tag{2-34}$$

Now we have

$$\left\langle \mathbf{x}, \left(\frac{\partial \phi}{\partial \mathbf{x}}\right)^t \right\rangle = \left\langle \mathbf{Ty}, \left(\frac{\partial \phi}{\partial \mathbf{Ty}}\right)^t \right\rangle$$
$$= \left\langle \mathbf{y}, \mathbf{T}^t \left(\frac{\partial \phi}{\partial \mathbf{Ty}}\right)^t \right\rangle \tag{2-35}$$

Comparing equations (2–34) and (2–35) we thus get

$$\mathbf{T}^t \left(\frac{\partial \phi}{\partial \mathbf{Ty}}\right)^t = \left(\frac{\partial \phi}{\partial \mathbf{y}}\right)^t \tag{2-36}$$

2.15 Differentiation and Integration of Matrices and Determinants

In the manipulations used in later chapters, it is frequently necessary to differentiate and integrate, with respect to an independent variable t, matrices whose elements are all functions of t. The derivative of a matrix $\mathbf{A}$ with respect to t is defined as the matrix obtained by differentiating each of the elements with respect to t; that is if

$$\mathbf{A} = [a_{ij}] \quad \text{then} \quad \frac{d\mathbf{A}}{dt} = \left[\frac{da_{ij}}{dt}\right] \tag{2-37}$$

Similarly for integration

$$\int_\alpha^\beta \mathbf{A}\, dt = \left[\int_\alpha^\beta a_{ij}\, dt\right] \tag{2-38}$$

The determinant of such a matrix may also be differentiated with respect to time. The determinant may be defined by

$$\det(\mathbf{A}) = \sum \pm a_{1\alpha} a_{2\beta} \ldots a_{n\nu}$$

where the summation (of $n!$ terms) is taken over all permutations $(\alpha\beta \ldots \nu)$ of the column suffixes of the elements a_{ij} with a plus or minus sign according as the permutation is even or odd. The product rule for differentiation thus gives

$$\frac{d[\det(\mathbf{A})]}{dt} = \sum \pm \frac{da_{1\alpha}}{dt} a_{2\beta} \cdots a_{n\nu} + \sum \pm a_{1\alpha} \frac{da_{2\beta}}{dt} \cdots a_{n\nu}$$

$$+ \cdots + \sum \pm a_{1\alpha} a_{2\beta} \cdots \frac{da_{n\nu}}{dt}$$

Thus we have

$$\frac{d[\det(\mathbf{A})]}{dt} = \det(\mathbf{A}_1) + \det(\mathbf{A}_2) + \cdots + \det(\tilde{\mathbf{A}}_n) \qquad (2\text{--}39)$$

where

$$\det(\mathbf{A}_j) = \begin{vmatrix} a_{11} & a_{12} & \cdots & a_{1n} \\ a_{21} & a_{22} & \cdots & a_{2n} \\ \vdots & & & \\ \dfrac{da_{j1}}{dt} & \dfrac{da_{j2}}{dt} & \cdots & \dfrac{da_{jn}}{dt} \\ \vdots & & & \\ a_{n1} & a_{n2} & \cdots & a_{nn} \end{vmatrix}$$

so that the derivative of a determinant is a sum of determinants in each of which an appropriate single row has been differentiated.

2.16 Direct Sum of Matrices

The direct sum of a set of matrices $\mathbf{A}_1, \mathbf{A}_2, \ldots, \mathbf{A}_m$ is defined as the partitioned matrix

$$\mathbf{A} = \begin{bmatrix} \mathbf{A}_1 & 0 & 0 & \cdots & 0 \\ 0 & \mathbf{A}_2 & 0 & \cdots & 0 \\ 0 & 0 & \mathbf{A}_3 & \cdots & 0 \\ \vdots & & & & \\ 0 & 0 & 0 & \cdots & \mathbf{A}_m \end{bmatrix}$$

having the matrices $\mathbf{A}_1, \mathbf{A}_2, \ldots, \mathbf{A}_m$ along its diagonal and zero matrices

of appropriate order everywhere else. The direct sum is usually denoted by
$$\mathbf{A} = \mathbf{A}_1 \oplus \mathbf{A}_2 \oplus \cdots \oplus \mathbf{A}_m$$
It follows immediately from the definitions of trace and determinant that
$$\text{trace}(\mathbf{A}) = \sum_{i=1}^{m} \text{trace}(\mathbf{A}_i)$$
$$\det(\mathbf{A}) = \prod_{i=1}^{m} \det(\mathbf{A}_i)$$

2.17 Extremum Values of Vector Functions Subject to Constraints

Let $V(x_1, x_2, \ldots, x_n)$ be a known scalar function, and suppose we have to find those values of $\mathbf{x}$ for which V is stationary subject to a given set of constraints

$$\begin{aligned} g_1(x_1, x_2, \ldots, x_n) &= 0 \\ g_2(x_1, x_2, \ldots, x_n) &= 0 \\ &\vdots \\ g_m(x_1, x_2, \ldots, x_n) &= 0 \end{aligned} \quad (2\text{-}40)$$

where $m < n$. There are two ways of tackling this problem:
(a) a direct solution using differentiation and the solution of simultaneous equations,
(b) by the method of Lagrange multipliers.

Both methods will be considered in turn; the technique of Lagrange multipliers is the more important of the two and is used in later chapters.

Direct Solution

At a stationary value of V
$$dV = \frac{\partial V}{\partial x_1} dx_1 + \frac{\partial V}{\partial x_2} dx_2 + \cdots + \frac{\partial V}{\partial x_n} dx_n \quad (2\text{-}41)$$
for differentials $dx_1, \ldots, dx_n$ which are subject to the m constraints found by differentiating the constraining equations, namely

$$\begin{aligned} dg_1 &= \frac{\partial g_1}{\partial x_1} dx_1 + \frac{\partial g_1}{\partial x_2} dx_2 + \cdots + \frac{\partial g_1}{\partial x_n} dx_n = 0 \\ dg_2 &= \frac{\partial g_2}{\partial x_1} dx_1 + \frac{\partial g_2}{\partial x_2} dx_2 + \cdots + \frac{\partial g_2}{\partial x_n} dx_n = 0 \\ &\vdots \\ dg_m &= \frac{\partial g_m}{\partial x_1} dx_1 + \frac{\partial g_m}{\partial x_2} dx_2 + \cdots + \frac{\partial g_m}{\partial x_n} dx_n = 0 \end{aligned} \quad (2\text{-}42)$$

Extremum Values of Vector Functions Subject to Constraints 117

Only $(n-m)$ of the n variables $x_1, x_2, \ldots, x_n$ may be varied independently. Let these be $x_{(m+1)}, x_{(m+2)}, \ldots, x_n$. For this set of variables for which we may take independent differentials put

$$dx_{(m+2)} = dx_{(m+3)} = \cdots = dx_n = 0 \qquad (2\text{-}43)$$

and vary $x_{(m+1)}$. Then, solving for the dependent differentials in terms of this single independent differential $dx_{(m+1)}$ gives, using equations (2-41), (2-42) and (2-43),

$$\begin{bmatrix} \dfrac{\partial V}{\partial x_1} & \dfrac{\partial V}{\partial x_2} & \cdots & \dfrac{\partial V}{\partial x_m} & \dfrac{\partial V}{\partial x_{(m+1)}} \\ \dfrac{\partial g_1}{\partial x_1} & \dfrac{\partial g_1}{\partial x_2} & \cdots & \dfrac{\partial g_1}{\partial x_m} & \dfrac{\partial g_1}{\partial x_{(m+1)}} \\ \vdots & \vdots & & \vdots & \vdots \\ \dfrac{\partial g_m}{\partial x_1} & \dfrac{\partial g_m}{\partial x_2} & \cdots & \dfrac{\partial g_m}{\partial x_m} & \dfrac{\partial g_m}{\partial x_{(m+1)}} \end{bmatrix} \begin{bmatrix} dx_1 \\ dx_2 \\ \vdots \\ dx_{(m+1)} \end{bmatrix} = 0 \qquad (2\text{-}44)$$

Now this set of equations has a non-trivial solution if, and only if, the determinant of the matrix vanishes. This determinant is called the Jacobian determinant of the set of equations involved and is normally denoted by

$$\frac{\partial(V, g_1, g_2, \ldots, g_{(m-1)}, g_m)}{\partial(x_1, x_2, x_3, \ldots, x_m, x_{(m+1)})}$$

Thus we have the condition

$$\frac{\partial(V, g_1, \ldots, g_m)}{\partial(x_1, x_2, \ldots, x_{(m+1)})} = 0$$

In a similar manner by putting

$$dx_{(m+1)} = 0 \qquad dx_{(m+3)} = dx_{(m+4)} = \cdots = dx_n = 0$$

and varying $x_{(m+2)}$ alone we get

$$\frac{\partial(V, g_1, \ldots, g_{(m-1)}, g_m)}{\partial(x_1, x_2, \ldots, x_m, x_{(m+2)})} = 0$$

Proceeding with this line of argument, we have that the constrained stationary point is determined by the solution of the following equations:

(i) The m constraint equations

$$g_1 = 0 \qquad g_2 = 0 \ldots g_m = 0 \qquad (2\text{-}45)$$

(ii) The $(n-m)$ Jacobian determinant equations

$$\frac{\partial(V, g_1, \ldots, g_{(m-1)}, g_m)}{\partial(x_1, x_2, \ldots, x_m, x_{(m+1)})} = 0$$

$$\vdots$$

$$\frac{\partial(V, g_1, \ldots, g_{(m-1)}, g_m)}{\partial(x_1, x_2, \ldots, x_m, x_n)} = 0$$

(2-46)

giving a total of $m + (n-m) = n$ equations in the n unknowns $x_1, x_2, \ldots, x_n$.

Lagrange Multiplier Method

Consider the $(n-m)$ Jacobian determinants obtained in the direct method of solution. The value of these determinants is unaltered by an interchange of rows and columns so that we may write the Jacobian determinant set as

$$\begin{vmatrix} \frac{\partial V}{\partial x_1} & \frac{\partial g_1}{\partial x_1} & \frac{\partial g_2}{\partial x_1} & \ldots & \frac{\partial g_m}{\partial x_1} \\ \frac{\partial V}{\partial x_2} & \frac{\partial g_1}{\partial x_2} & \frac{\partial g_2}{\partial x_2} & \ldots & \frac{\partial g_m}{\partial x_2} \\ \vdots & & & & \vdots \\ \frac{\partial V}{\partial x_m} & \frac{\partial g_1}{\partial x_m} & \frac{\partial g_2}{\partial x_m} & \ldots & \frac{\partial g_m}{\partial x_m} \\ \frac{\partial V}{\partial x_{(m+1)}} & \frac{\partial g_1}{\partial x_{(m+1)}} & \frac{\partial g_2}{\partial x_{(m+1)}} & \ldots & \frac{\partial g_m}{\partial x_{(m+1)}} \end{vmatrix}$$

$$\begin{vmatrix} \text{same first set of } m \text{ rows as above} \\ \frac{\partial V}{\partial x_{(m+2)}} \quad \frac{\partial g_1}{\partial x_{(m+2)}} \quad \frac{\partial g_2}{\partial x_{(m+2)}} \quad \ldots \quad \frac{\partial g_m}{\partial x_{(m+2)}} \end{vmatrix}$$

and so on until

$$\begin{vmatrix} \text{same first set of } m \text{ rows} \\ \text{as above} \\ \frac{\partial V}{\partial x_n} \quad \frac{\partial g_1}{\partial x_n} \quad \frac{\partial g_2}{\partial x_n} \quad \ldots \quad \frac{\partial g_m}{\partial x_n} \end{vmatrix}$$

The values of these determinants will be unaltered if we add any multiple of a column to any other column. We may thus multiply the 2nd, 3rd, 4th, ... $(m+1)$th column of *each* of these Jacobian determinants by a set of constants $\lambda_1, \lambda_2, \ldots, \lambda_m$ and add them to the first column of each determinant, leaving the values of the determinants unaltered.

Now each of the determinants obtained in this way will vanish if we choose these constants $\lambda_1, \lambda_2, \ldots, \lambda_m$ in such a way that the first column of each determinant vanishes, that is if we make them such that

$$\frac{\partial V}{\partial x_1} + \lambda_1 \frac{\partial g_1}{\partial x_1} + \lambda_2 \frac{\partial g_2}{\partial x_1} + \cdots + \lambda_m \frac{\partial g_m}{\partial x_1} = 0$$

$$\frac{\partial V}{\partial x_2} + \lambda_1 \frac{\partial g_1}{\partial x_2} + \lambda_2 \frac{\partial g_2}{\partial x_2} + \cdots + \lambda_m \frac{\partial g_m}{\partial x_2} = 0$$

$$\vdots$$

$$\frac{\partial V}{\partial x_m} + \lambda_1 \frac{\partial g_1}{\partial x_m} + \lambda_2 \frac{\partial g_2}{\partial x_m} + \cdots + \lambda_m \frac{\partial g_m}{\partial x_m} = 0 \qquad (2\text{-}47)$$

$$\frac{\partial V}{\partial x_{(m+1)}} + \lambda_1 \frac{\partial g_1}{\partial x_{(m+1)}} + \lambda_2 \frac{\partial g_2}{\partial x_{(m+1)}} + \cdots + \lambda_m \frac{\partial g_m}{\partial x_{(m+1)}} = 0$$

$$\vdots$$

$$\frac{\partial V}{\partial x_n} + \lambda_1 \frac{\partial g_1}{\partial x_n} + \lambda_2 \frac{\partial g_2}{\partial x_n} + \cdots + \lambda_m \frac{\partial g_m}{\partial x_n} = 0$$

Now the set of equations (2–47) will be recognized as the necessary conditions for a stationary value of the function

$$\phi = V + \lambda_1 g_1 + \lambda_2 g_2 + \cdots + \lambda_m g_m = V + \sum_{i=1}^{m} \lambda_i g_i \qquad (2\text{-}48)$$

The constrained extremum point is fixed by the Jacobian determinant equations plus the constraint equations. Furthermore the necessary conditions for a stationary value of the augmented function ϕ defined in equation (2–48) will guarantee the vanishing of the $(n-m)$ Jacobian determinants obtained in the direct method of solution. Thus if a set of constants $\lambda_1, \lambda_2, \ldots, \lambda_m$ exists, and a point $x_1, x_2, \ldots, x_n$ exists, which satisfy the set of $(n+m)$ equations

$$\frac{\partial}{\partial x_i}\left[V + \sum_{i=1}^{m} \lambda_i g_i\right] = 0 \qquad i = 1, 2, \ldots, n$$

$$g_k = 0 \qquad k = 1, 2, \ldots, m \qquad (2\text{-}49)$$

then the point $x_1, x_2, \ldots, x_n$ must give a stationary value of V subject to the constraint equations since

(a) the Jacobian determinants vanish, and
(b) the constraining equations are satisfied.

Example

    Minimize $\qquad V = x_1^2 + x_2^2 + x_3^2$

subject to the constraint
$$ax_1 + bx_2 + cx_3 + d = 0$$
In geometrical terms this may be interpreted as finding the minimum distance from the origin to a given plane. Take the augmented function as
$$\phi = x_1^2 + x_2^2 + x_3^2 + \lambda(ax_1 + bx_2 + cx_3 + d)$$
The necessary conditions for a minimum of this augmented function are
$$\frac{\partial \phi}{\partial x_1} = 2x_1 + a\lambda = 0$$
$$\frac{\partial \phi}{\partial x_2} = 2x_2 + b\lambda = 0$$
$$\frac{\partial \phi}{\partial x_3} = 2x_3 + c\lambda = 0$$
which we solve together with
$$ax_1 + bx_2 + cx_3 + d = 0$$
to obtain x_1, x_2, x_3 and λ.

This gives
$$\lambda = 2\alpha \quad x_1 = -a\alpha \quad x_2 = -b\alpha \quad x_3 = -c\alpha$$
where
$$\alpha = \frac{d}{a^2 + b^2 + c^2}$$
and the corresponding minimum value of V is
$$\frac{|d|}{\sqrt{(a^2 + b^2 + c^2)}}$$

The constants $\lambda_1, \lambda_2, \ldots, \lambda_m$ are called *Lagrange multipliers*. The apparent disadvantage of the Lagrange multiplier method of introducing more unknowns is completely offset by the simpler and more symmetrical form of the equations obtained.

2.18 Extremum Characterization of Eigenvalues [D4]

Consider the real quadratic form
$$Q(x_1, x_2, \ldots, x_n) = \mathbf{x}^t \mathbf{A} \mathbf{x}$$
where $\mathbf{A}$ is a real symmetric matrix. On dividing Q by $\mathbf{x}^t\mathbf{x}$ (which is positive unless $\mathbf{x}$ vanishes) we obtain a function
$$f(x_1, x_2, \ldots, x_n) = \frac{\mathbf{x}^t \mathbf{A} \mathbf{x}}{\mathbf{x}^t \mathbf{x}} = \frac{\langle \mathbf{x}, \mathbf{A}\mathbf{x} \rangle}{\langle \mathbf{x}, \mathbf{x} \rangle} \qquad (2\text{-}50)$$

If **u** is an eigenvector of **A** having components $u_1, u_2, \ldots, u_n$ and corresponding to an eigenvalue λ then

$$f(u_1, u_2, \ldots, u_n) = \frac{\mathbf{u}^t \mathbf{A} \mathbf{u}}{\mathbf{u}^t \mathbf{u}} = \frac{\mathbf{u}^t \lambda \mathbf{u}}{\mathbf{u}^t \mathbf{u}} = \lambda \quad (2\text{-}51)$$

Now consider small changes about **u**. We have

$$df = \frac{\partial f}{\partial \mathbf{u}} d\mathbf{u} = \frac{\partial}{\partial \mathbf{u}} \left[\frac{\mathbf{u}^t \mathbf{A} \mathbf{u}}{\mathbf{u}^t \mathbf{u}} \right] d\mathbf{u}$$

$$= \frac{1}{(\mathbf{u}^t \mathbf{u})^2} \left[(\mathbf{u}^t \mathbf{u}) \frac{\partial}{\partial \mathbf{u}} (\mathbf{u}^t \mathbf{A} \mathbf{u}) - (\mathbf{u}^t \mathbf{A} \mathbf{u}) \frac{\partial}{\partial \mathbf{u}} (\mathbf{u}^t \mathbf{u}) \right] d\mathbf{u}$$

$$= \left(\frac{1}{\mathbf{u}^t \mathbf{u}} \right) \left[2 \mathbf{A} \mathbf{u} - \left(\frac{\mathbf{u}^t \mathbf{A} \mathbf{u}}{\mathbf{u}^t \mathbf{u}} \right) 2 \mathbf{u} \right] d\mathbf{u}$$

$$= \left(\frac{1}{\mathbf{u}^t \mathbf{u}} \right) \left[2 \lambda \mathbf{u} - \left(\frac{\mathbf{u}^t \lambda \mathbf{u}}{\mathbf{u}^t \mathbf{u}} \right) 2 \mathbf{u} \right] d\mathbf{u}$$

$$= \left(\frac{1}{\mathbf{u}^t \mathbf{u}} \right) [2 \lambda \mathbf{u} - 2 \lambda \mathbf{u}] d\mathbf{u}$$

$$= 0$$

This shows that f is equal to an eigenvalue when evaluated for an eigenvector and is stationary for small changes about the eigenvector. We are therefore led to seek means of characterizing the eigenvalues of **A** in terms of the stationary values of f. For a real symmetric matrix **A** there will be n eigenvalues $\lambda_1, \lambda_2, \ldots, \lambda_n$. Suppose they are ordered in the following way

$$\lambda_1 \leqslant \lambda_2 \leqslant \lambda_3 \leqslant \cdots \leqslant \lambda_n \quad (2\text{-}52)$$

and suppose the corresponding eigenvectors $\mathbf{u}_1, \mathbf{u}_2, \ldots, \mathbf{u}_n$ are independent and normalized to unit length. Any vector **x** may be expressed as a linear combination of the eigenvectors as

$$\mathbf{x} = \sum_{i=1}^{n} c_i \mathbf{u}_i \quad (2\text{-}53)$$

Substituting from equation (2–53) into equation (2–50) gives

$$f(\mathbf{x}) = \frac{\left(\sum_{i=1}^{n} c_i \mathbf{u}_i^t \right) \mathbf{A} \left(\sum_{j=1}^{n} c_j \mathbf{u}_j \right)}{\left(\sum_{k=1}^{n} c_k \mathbf{u}_k^t \right) \left(\sum_{m=1}^{n} c_m \mathbf{u}_m \right)} = \frac{\sum_{i=1}^{n} \sum_{j=1}^{n} c_i \mathbf{u}_i^t c_j \lambda_j \mathbf{u}_j}{\sum_{k=1}^{n} \sum_{m=1}^{n} c_k c_m \mathbf{u}_k^t \mathbf{u}_m} \quad (2\text{-}54)$$

Now it has been shown in Section 2.11 above that the eigenvectors of a symmetric matrix form a mutually orthogonal set. Since we have stipulated that the eigenvectors are normalized to unit length we must therefore have

$$\mathbf{u}_i^t \mathbf{u}_j = 1 \quad \text{if} \quad i = j$$
$$\mathbf{u}_i^t \mathbf{u}_j = 0 \quad \text{if} \quad i \neq j \tag{2-55}$$

Using the relationships (2–55) in equation (2–54) we thus have

$$f(\mathbf{x}) = \frac{\sum_{i=1}^{n} \sum_{j=1}^{n} \lambda_j c_i c_j \delta_{ij}}{\sum_{k=1}^{n} \sum_{m=1}^{n} c_k c_m \delta_{km}} = \frac{\sum_{i=1}^{n} \lambda_i c_i^2}{\sum_{k=1}^{n} c_k^2} \tag{2-56}$$

where δ_{ij}, δ_{km} denote the Kronecker delta defined in Section 2.5 above.

Now we may write the eigenvalue λ_1 in the form

$$\lambda_1 = \frac{\sum_{i=1}^{n} \lambda_1 c_i^2}{\sum_{k=1}^{n} c_k^2}$$

and the eigenvalue λ_n in the form

$$\lambda_n = \frac{\sum_{i=1}^{n} \lambda_n c_i^2}{\sum_{k=1}^{n} c_k^2}$$

We therefore have

$$\lambda_1 - f(\mathbf{x}) = \frac{\sum_{i=1}^{n} (\lambda_1 - \lambda_i) c_i^2}{\sum_{k=1}^{n} c_k^2} \leq 0$$

and

$$\lambda_n - f(\mathbf{x}) = \frac{\sum_{i=1}^{n} (\lambda_n - \lambda_i) c_i^2}{\sum_{k=1}^{n} c_k^2} \geq 0$$

in view of the ordering imposed by inequalities (2–52). We therefore have

$$\lambda_1 \leq f(\mathbf{x}) \leq \lambda_n \tag{2-57}$$

for any arbitrary vector **x**. It immediately follows from this that

$$\lambda_1 = \underset{\mathbf{x}}{\text{minimum}}\,[f(\mathbf{x})] \text{ with minimum at } \mathbf{u}_1$$
$$\lambda_n = \underset{\mathbf{x}}{\text{maximum}}\,[f(\mathbf{x})] \text{ with maximum at } \mathbf{u}_n$$
(2–58)

The remaining eigenvalues can be expressed in this way by successively restricting the regions over which the vector **x** is allowed to vary. Let **x** be confined to the subspace of vectors orthogonal to $\mathbf{u}_1$, the eigenvector corresponding to λ_1. In this subspace, **x** can be expressed as a linear combination of the remaining eigenvectors $\mathbf{u}_2, \mathbf{u}_3, \ldots, \mathbf{u}_n$ and we then have

$$f(\mathbf{x}) = \frac{\sum_{i=2}^{n} \lambda_i c_i^2}{\sum_{k=2}^{n} c_k^2} \quad \text{for} \quad \{\mathbf{x} : \langle \mathbf{x}, \mathbf{u}_1 \rangle = 0\}$$

This gives

$$\lambda_2 \leqslant f(\mathbf{x}) \leqslant \lambda_n \quad \text{for} \quad \langle \mathbf{x}, \mathbf{u}_1 \rangle = 0$$

so that
$$\lambda_2 = \underset{\{\mathbf{x}:\langle \mathbf{x},\mathbf{u}_1\rangle = 0\}}{\text{minimum}} [f(\mathbf{x})] \text{ with minimum at } \mathbf{u}_2$$

Proceeding in this way, λ_3 can be found by minimizing $f(\mathbf{x})$ over the subspace orthogonal to $\mathbf{u}_1$ and $\mathbf{u}_2$. Continuing this iterative process gives all the eigenvectors and eigenvalues as solutions to simple variational problems.

It is instructive to look at this problem from the point of view of a constrained minimization using Lagrange multipliers. From this angle we consider finding stationary values of

$$Q(\mathbf{x}) = \mathbf{x}^t \mathbf{A} \mathbf{x}$$

subject to the constraint

$$\mathbf{x}^t \mathbf{x} - 1 = 0$$

Using a Lagrange multiplier λ we therefore consider the stationary values of

$$\phi(\mathbf{x}) = \mathbf{x}^t \mathbf{A} \mathbf{x} - \lambda \mathbf{x}^t \mathbf{x}$$

Differentiating with respect to **x** gives

$$\left(\frac{\partial \phi}{\partial \mathbf{x}}\right)^t = 2\mathbf{A}\mathbf{x} - 2\lambda \mathbf{x} = 0$$

so that we get the usual characteristic equation for the eigenvalues, namely

$$\det(\mathbf{A} - \lambda \mathbf{I}) = 0$$

The eigenvectors are then determined, as usual, from

$$\mathbf{A}\mathbf{u}_i = \lambda_i \mathbf{u}_i$$
$$\langle \mathbf{u}_i, \mathbf{u}_i \rangle = 1 \qquad i = 1, 2, \ldots, n$$

2.19 Matrix Functions

Many functions of a scalar variable are defined by Taylor series expansions, such as

$$\exp(x) = \sum_{n=0}^{\infty} \frac{x^n}{n!}$$

$$\sin(x) = \sum_{n=0}^{\infty} (-1)^n \frac{x^{2n+1}}{(2n+1)!}$$

$$\cos(x) = \sum_{n=0}^{\infty} (-1)^n \frac{x^{2n}}{(2n!)}$$

If we are given a scalar power series defining a scalar function f as

$$f(x) = \sum_{k=0}^{\infty} a_k x^k$$

then the matrix-valued function $\mathbf{f}(\mathbf{A})$ of the *square* matrix $\mathbf{A}$ is *defined* as

$$\mathbf{f}(\mathbf{A}) = \sum_{k=0}^{\infty} a_k \mathbf{A}^k \quad \text{with} \quad \mathbf{A}^0 = \text{unit matrix}$$

for all matrices $\mathbf{A}$ for which the series converges. An infinite series of matrices

$$\sum_{k=0}^{\infty} \mathbf{B}^{(k)} = \mathbf{B}^{(0)} + \mathbf{B}^{(1)} + \cdots + \mathbf{B}^{(m)} + \cdots = \mathbf{B}$$

of order m by n is said to converge to a matrix $\mathbf{B}$ if and only if the sums of all the elements of the constituent matrices converge to the appropriate element of $\mathbf{B}$. Thus the convergence of a matrix power series requires the convergence of mn scalar power series, one for each of the mn elements in $\mathbf{B}$. The Cayley–Hamilton theorem, proved in Chapter 6, makes it possible to avoid calculating matrix powers of higher than n for a power series of order n. Some examples of matrix functions defined by power series are

$$\exp(\mathbf{X}) = \sum_{k=0}^{\infty} \frac{\mathbf{X}^k}{k!}$$

$$\cos(\mathbf{A}) = \sum_{k=0}^{\infty} (-1)^k \frac{\mathbf{A}^{2k}}{(2k)!}$$

2.20 Miscellaneous Notes

Matrices over a field. When the elements of a matrix are chosen from a group of elements comprising a field, the matrix is said to be defined over this field. In the matrices used here which have scalar elements, the fields over which the elements are defined are the field of real numbers and the field of complex numbers.

Idempotency. Any nonzero matrix **A** is said to be idempotent if and only if

$$\mathbf{A}^2 = \mathbf{A} \quad \text{where} \quad \mathbf{A}^2 = \mathbf{AA}$$

Nilpotency. A matrix **B** is said to be nilpotent of index m if and only if there exists a real integer $m > 1$ such that

$$\mathbf{B}^m = \mathbf{0}$$

Special notations. A unit matrix, having unity elements along its main diagonal and zeros everywhere else, will often be denoted by the symbol **I**. If it is required explicitly to indicate that the unit matrix is of order n say, it will be written as $\mathbf{I}_n$.

A zero or null matrix, having zero elements in all entries, will usually be denoted by the symbol **0**. If it is required explicitly to show that it is of order m by n say, it will be written as $\mathbf{0}_{m,n}$.

A diagonal matrix, having diagonal elements $\lambda_1, \lambda_2, \ldots, \lambda_n$ say and all other entries zero, will sometimes be represented by the explicit notation

$$\mathbf{A} = \text{diag}[\lambda_1, \lambda_2, \ldots, \lambda_n]$$

The determinant of a matrix **A** will be represented by det(**A**) or by |**A**|. If it is desired to show explicitly that a matrix **A** has an element a_{ij} at the intersection of ith row and jth column the matrix will be written as

$$\mathbf{A} = [a_{ij}]$$

CHAPTER THREE

← Transform Models

In many cases the most convenient method of analysing the behaviour of a dynamical system is in terms of the Laplace transforms of a suitable set of equations. The dynamical system model then used consists of a linear graph representation of the transformed equations. In this chapter the basic Laplace transform theory required is first developed, and then the rules governing the use of block diagrams and signal-flow graphs are derived. Nyquist's stability theory and the root-locus theory for this class of model is then presented, and the chapter concludes with a discussion of z-transform methods for discrete models.

3.1 Laplace Transform

An integral transform sets up a correspondence between a function of an independent variable and a family of functions in the same independent variable which have useful properties for the solution of a particular class of problems. For linear differential equation solution, the family of exponential functions has the valuable property that its derivatives and integrals with respect to the independent variable are also exponential functions. Thus, when α and s are complex constants,

$$\frac{d}{dt}\alpha \exp(st) = s\alpha \exp(st)$$

$$\int \alpha \exp(st)\,dt = \frac{\alpha}{s}\exp(st) + \text{constant}$$

It is therefore natural to seek to express the solution of linear differential equations in terms of exponential functions.

Laplace transform. Let $f(t)$ be a known function of the independent variable t for $t \geq 0$ and let $f(t)$ be identically zero for $t < 0$. The Laplace transform of $f(t)$ is then defined as that function of a complex variable s given by

$$F(s) = \int_0^\infty f(t) \exp(-st)\,dt \qquad (3\text{–}1)$$

It is often convenient to use capital letters for transforms, but the transform will usually be indicated by an explicit statement of the independent variable in the brackets following the function symbol. We could thus have indicated the transform of $f(t)$ by $f(s)$; this convention will often be adopted later when no confusion is likely to arise. In certain cases, as when impulsive forces or discontinuous inputs are concerned, it is necessary to distinguish from which side a limit is taken. The following conventions are adopted for this purpose when required:

$f(0+) = \lim f(t)$ as $t \to 0$ through positive values of t

$f(0-) = \lim f(t)$ as $t \to 0$ through negative values of t

If a similar distinction is required at any other value k of t, or in terms of the running variable t itself, we can put

$$f(k+), \quad f(k-), \quad f(t+) \quad \text{or} \quad f(t-)$$

to indicate the direction of proceeding to the limit concerned. For the single-sided Laplace transform defined by equation (3–1) we normally wish to include in the transform the effects of any peculiarities of $f(t)$ at $t = 0$; the integral will therefore normally be taken as

$$F(s) = \int_{0-}^{\infty} f(t) \exp(-st)\, dt \qquad (3\text{–}1a)$$

Arising from the general properties of the exponential function noted above, the Laplace transform has the extremely useful property of transforming a constant coefficient differential equation relationship between functions of time into an *algebraic* relationship between their Laplace transforms. This enables one to solve such differential equations by simple algebraic manipulations which may be carried out in a routine manner.

Notation for Laplace transforms. The operation of taking a Laplace transform will be denoted by $\mathscr{L}$, i.e.

$$\mathscr{L}f(t) = \int_{0-}^{\infty} f(t) \exp(-st)\, dt \equiv F(s)$$

The inverse transformation operation is denoted by $\mathscr{L}^{-1}$, that is

$$\mathscr{L}^{-1} F(s) = f(t)$$

It follows directly from the definition that $\mathscr{L}$ is a *linear operator* so that

$$\mathscr{L}[f(t) + g(t)] = \mathscr{L}f(t) + \mathscr{L}g(t)$$

and

$$\mathscr{L}\beta f(t) = \beta \mathscr{L}f(t)$$

where β is any complex constant.

Laplace transforms of derivatives. The transforms of derivatives are obtained from an integration by parts using the formula

$$\int_a^b u(t)\frac{dv(t)}{dt}dt = [u(t)v(t)]_a^b - \int_a^b v(t)\frac{du(t)}{dt}dt$$

We thus have

$$\mathscr{L}\frac{df(t)}{dt} = \int_{0-}^{\infty} [\exp(-st)]\frac{df(t)}{dt}dt$$

$$= [\{\exp(-st)\}f(t)]_{0-}^{\infty} - \int_{0-}^{\infty}[-s\exp(-st)]f(t)\,dt$$

$$= s\mathscr{L}f(t) - f(0-) \qquad (3\text{-}2)$$

Again, for the second derivative,

$$\mathscr{L}\frac{d^2f(t)}{dt^2} = \int_{0-}^{\infty}[\exp(-st)]\frac{d^2f(t)}{dt^2}dt$$

$$= \left[\{\exp(-st)\}\frac{df(t)}{dt}\right]_{0-}^{\infty} - \int_{0-}^{\infty}[-s\exp(-st)]\frac{df(t)}{dt}dt$$

$$= s\mathscr{L}\frac{df(t)}{dt} - \frac{df(0-)}{dt}$$

$$= s[s\mathscr{L}f(t) - f(0-)] - \frac{df(0-)}{dt}$$

$$= s^2\mathscr{L}f(t) - sf(0-) - \frac{df(0-)}{dt}$$

giving that

$$\mathscr{L}\frac{d^2f(t)}{dt^2} = s^2\mathscr{L}f(t) - sf(0-) - \frac{df(0-)}{dt}$$

An extension of this form of argument to the case of the *n*th-order derivative gives

$$\mathscr{L}\frac{d^n f(t)}{dt^n} = s^n\mathscr{L}f(t) - s^{n-1}f(0-) - s^{n-2}\frac{df(0-)}{dt} - \cdots - \frac{d^{n-1}f(0-)}{dt^{n-1}} \qquad (3\text{-}3)$$

Laplace transforms of integrals. The Laplace transforms of integrals are obtained using the same formula for integration by parts. We thus have

$$\mathcal{L}\left[\int_{0-}^{t} f(t)\,dt + C\right] = \int_{0-}^{\infty} [\exp(-st)]\left[\int_{0-}^{t} f(t)\,dt + C\right] dt$$

$$= \left[\left\{\int_{0-}^{t} f(t)\,dt + C\right\}\left\{-\frac{1}{s}\exp(-st)\right\}\right]_{0-}^{\infty}$$

$$- \int_{0-}^{\infty} f(t)\left[-\frac{1}{s}\exp(-st)\right] dt$$

$$= \frac{\mathcal{L}f(t)}{s} + \frac{C}{s} \tag{3-4}$$

where C is an arbitrary constant.

Special types of excitation function. The excitation functions most used in the study of dynamical system response are the Heaviside unit step function and the Dirac impulse function.

Heaviside unit step function. The Heaviside unit step function is denoted by $H(t)$ and defined by

$$H(t) = 0 \quad \text{for} \quad t < 0+$$
$$H(t) = 1 \quad \text{for} \quad t \geq 0+$$

A step function of amplitude A is defined as $AH(t)$ and has the Laplace transform

$$\mathcal{L}AH(t) = \int_{0-}^{\infty} A\exp(-st)\,dt = \frac{A}{s}$$

Time-shift theorem. The function $f(t-a)$ is obtained from $f(t)$ by an ideal time delay of amount a. Its Laplace transform may be obtained as follows:

$$\mathcal{L}f(t-a) = \int_{0-}^{\infty} f(t-a)\exp(-st)\,dt$$

Let $\quad t-a = \tau \quad$ then $\quad t = \tau+a \quad$ and $\quad dt = d\tau$

Thus $\quad \mathcal{L}f(t-a) = \int_{0-}^{\infty} f(\tau)\exp[-s(\tau+a)]\,d\tau$

$$= \exp(-sa)\int_{0-}^{\infty} f(\tau)\exp(-s\tau)\,d\tau$$

$$= \exp(-sa)f(s) \tag{3-5}$$

Rectangular pulse function. The rectangular pulse function of amplitude A and duration τ is denoted by $P(t;\tau)$ and defined by

$$P(t;\tau) = AH(t) - AH(t-\tau)$$

It therefore has the Laplace transform

$$\mathscr{L}P(t;\tau) = \frac{A}{s}[1 - \exp(-s\tau)] \qquad (3\text{-}6)$$

Dirac impulse function. It is frequently useful to consider the response of a dynamical system to an idealized form of excitation function having a very large magnitude and acting for a very short time. In mechanical systems, for example, excitations of this type can be used to approximate the effect of impacts and hammer blows. Such an idealized excitation function is termed an *impulse*. Consider a rectangular pulse function of amplitude I/ϵ and duration ϵ. The Laplace transform of such a pulse will be

$$\frac{I}{\epsilon} \cdot \frac{1}{s}[1 - \exp(-s\epsilon)] = \frac{I}{\epsilon s}\left[s\epsilon - \frac{s^2\epsilon^2}{2!} + \frac{s^3\epsilon^3}{3!} - \cdots\right]$$

$$= I - \frac{Is\epsilon}{2!} + \frac{Is^2\epsilon^2}{3!} - \cdots$$

If ϵ is made sufficiently small, this function will have the desired features of an impulse function and will have a Laplace transform which may be closely approximated by I. The quantity I, which is equal to the amplitude of the pulse multiplied by its duration, is defined as the impulse of the rectangular pulse function. To proceed to the limit as $\epsilon \to 0$ in an attempt to define an idealized impulse function raises many mathematical difficulties, since one cannot meaningfully define a function which exists at a single point and yet has a finite value of integral with respect to an independent variable. However the properties of limiting cases of functions of this sort are so valuable and interesting, that the concept of a function has been extended by L. Schwarz that such entities may be rigorously discussed using his Theory of Distributions [S10], [L3]. Since the difficulties involved in proceeding to the limit are known to be surmountable, the simplest procedure, from the practical point of view for dynamical system work, is to *define* an idealized unit impulse function as having a Laplace transform of unity, and an idealized impulse function of impulse I as having a Laplace transform of value I. The unit impulse function is known as the Dirac impulse function, denoted by $\delta(t)$ and defined by

$$\mathscr{L}\delta(t) \triangleq 1$$

The response of dynamical system models may now be formally

investigated by the use of such idealized functions which may be approximated in practice by *any* impulsive function, whose Laplace transform may be sufficiently well approximated by the value of the impulse. If we use step, pulse and impulse functions in dynamical systems analysis we must be able to interpret physically the derivatives of such idealized functions. Again the simplest procedure is to accept the existence of a rigorous theory, and formally proceed to use the Laplace transforms of the various excitation function derivatives, which may then be interpreted in terms of wave forms whose transforms closely approximate the transforms obtained for the derivatives. Since

$$\mathscr{L}\frac{df(t)}{dt} = sf(s) - f(0-)$$

we have that

$$\mathscr{L}\frac{dH(t)}{dt} = s\frac{1}{s} - 0 = 1 = \mathscr{L}\delta(t)$$

This relationship enables us to *formally* treat $\delta(t)$ as though it were the 'derivative' of $H(t)$. What this means from a physical point of view is that if we form *any* sequence of functions which approximate $H(t)$ more and more closely then the corresponding sequence of time derivatives of these functions will approximate more and more closely to a Dirac impulse function *in the sense that* their Laplace transforms will more and more closely approximate to unity. Arguing in this way we may define a sequence of idealized functions which are the formal derivatives of the Dirac impulse function. These are denoted by $\delta^{(1)}(t), \delta^{(2)}(t), \ldots, \delta^{(n)}(t)$, and defined by

$$\delta(t) = \mathscr{L}^{-1}1$$
$$\delta^{(1)}(t) = \mathscr{L}^{-1}s$$
$$\delta^{(2)}(t) = \mathscr{L}^{-1}s^2$$
$$\vdots$$
$$\delta^{(n)}(t) = \mathscr{L}^{-1}s^n$$

In each case we may obtain a physical interpretation of these higher impulse functions by considering the limiting form of an appropriate combination of rectangular pulses. For example consider the double impulse combination of rectangular pulse functions shown in Figure 3–1. Its Laplace transform is

$$\frac{I}{\epsilon^2} \cdot \frac{1}{s} - \frac{2I}{\epsilon^2} \cdot \frac{1}{s} \cdot \exp(-s\epsilon) + \frac{I}{\epsilon^2} \cdot \frac{1}{s} \cdot \exp(-2s\epsilon)$$

which tends to Is as ϵ tends to zero. Thus, for a sufficiently small value

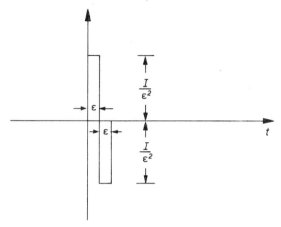

Figure 3–1

of ϵ, this double pulse function is an approximation to the derivative of an ideal impulse function of impulse value I.

Transform pairs. Direct integration together with the use of tables of standard integrals enables one to draw up the Table 3–1 of Laplace transform pairs; this includes most of the transforms required in the routine analysis of linear dynamical systems.

Initial value theorem. Using the relationship derived above for the transform of a derivative, we have

$$\int_{0-}^{\infty} \exp(-st)\frac{df}{dt}\,dt = sf(s) - f(0-)$$

Thus

$$\int_{f(0-)}^{f(\infty)} \exp(-st)\,df = sf(s) - f(0-)$$

so that

$$\int_{f(0-)}^{f(0+)} df + \int_{f(0+)}^{f(\infty)} \exp(-st)\,df = sf(s) - f(0-)$$

giving $\quad f(0+) - f(0-) + \int_{f(0+)}^{f(\infty)} \exp(-st)\,df = sf(s) - f(0-)$

Now let $s \to \infty$ so that $\exp(-st) \to 0$ and we get

$$\lim_{s \to \infty} [sf(s)] = f(0+) \qquad (3\text{--}7)$$

Final value theorem. As before we have

$$sf(s) - f(0-) = \int_{f(0-)}^{f(\infty)} \exp(-st)\,df$$

Table 3–1 Laplace Transform Pairs

A, a, ω are real constants

$f(t)$	$f(s)$	
$A\delta(t)$	A	
$A\delta^{(n)}(t)$	As^n	$n = 1, 2, 3, \ldots$
$AH(t)$	$\dfrac{A}{s}$	
$AtH(t)$	$\dfrac{A}{s^2}$	
$\dfrac{At^{n-1}}{(n-1)!}H(t)$	$\dfrac{A}{s^n}$	$n = 1, 2, 3, \ldots$
$A\exp(-at)H(t)$	$\dfrac{A}{(s+a)}$	
$\dfrac{At^{n-1}}{(n-1)!}\exp(-at)H(t)$	$\dfrac{A}{(s+a)^n}$	$n = 1, 2, 3, \ldots$
$A\sin\omega t H(t)$	$\dfrac{A\omega}{s^2+\omega^2}$	
$A\cos\omega t H(t)$	$\dfrac{As}{s^2+\omega^2}$	
$A\exp(-at)\sin\omega t H(t)$	$\dfrac{A\omega}{(s+a)^2+\omega^2}$	
$A\exp(-at)\cos\omega t H(t)$	$\dfrac{A(s+a)}{(s+a)^2+\omega^2}$	
$A\sinh\omega t H(t)$	$\dfrac{A\omega}{s^2-\omega^2}$	
$A\cosh\omega t H(t)$	$\dfrac{As}{s^2-\omega^2}$	
$A\exp(-at)\sinh\omega t H(t)$	$\dfrac{A\omega}{(s+a)^2-\omega^2}$	
$A\exp(-at)\cosh\omega t H(t)$	$\dfrac{A(s+a)}{(s+a)^2-\omega^2}$	

Now let $s \to 0$ so that $\exp(-st) \to 1$ and we get

$$\lim_{s \to 0} [sf(s)] - f(0-) = f(\infty) - f(0-)$$

giving $$\lim_{s \to 0} [sf(s)] = f(\infty) \qquad (3\text{-}8)$$

Note that, if $f(s)$ has poles in the right half-plane or on the imaginary axis, $f(\infty)$ will not be defined and equation (3-8) is not then applicable.
Region of absolute convergence. For any function defined, as in equation (3-1), by an integral we must establish a set of conditions on the integrand which ensure that the function exists. The existence of the exponential factor in the integral definition suggests the use of a dominating exponential function for establishing convergence. Suppose $f(t)$ satisfies the following pair of conditions

$$\int_0^T f(t)\,dt < K \qquad (3\text{-}9)$$

where K is finite and T is positive and finite, and

$$|f(t)| \leq M \exp(ct) \quad \text{for} \quad t > T \qquad (3\text{-}10)$$

where M and c are finite constants. The theory of integration states that the defining integral will converge absolutely if and only if

$$\int_0^\infty |f(t)\exp(-st)|\,dt < \infty \qquad (3\text{-}11)$$

Let σ be the real part of the complex variable s. We then have that

$$\int_0^\infty |f(t)\exp(-st)|\,dt = \int_0^\infty |f(t)|\exp(-\sigma t)\,dt$$

$$= \int_0^T |f(t)|\exp(-\sigma t)\,dt$$

$$+ \int_T^\infty |f(t)|\exp(-\sigma t)\,dt$$

Now we have that

$$\int_0^T |f(t)|\exp(-\sigma t)\,dt \leq \exp\left(\frac{|c|}{T}\right) \int_0^T |f(t)|\,dt \leq K \exp\left(\frac{|c|}{T}\right)$$

after using equation (3-9). Suppose $\sigma > c$; we then have that, using equation (3-10)

$$\int_T^\infty |f(t)|\exp(-\sigma t)\,dt \leq M \int_T^\infty \exp[-(\sigma - c)t]\,dt$$

so that

$$\int_T^\infty |f(t)| \exp(-\sigma t)\, dt \leq \frac{M}{(\sigma-c)} \exp[-(\sigma-c)T]$$

Combining the two parts of the integral together again, we get that

$$\int_0^\infty |f(t) \exp(-st)|\, dt \leq K \exp\left(\frac{|c|}{T}\right) + \frac{M}{(\sigma-c)} \exp[-(\sigma-c)T]$$
$$< \infty \quad \text{for} \quad \sigma > c$$

This therefore gives a criterion for testing $f(t)$ to ensure absolute convergence of the defining integral for the Laplace transform. The smallest value of σ for which the defining integral converges absolutely is called its *abscissa of absolute convergence*. The half-plane to the right of the abscissa of convergence in the complex plane is called the region of absolute convergence of $F(s)$. The abscissa of absolute convergence is usually denoted by σ_a. Thus $F(s)$ is defined for $\sigma < \sigma_a$.
Inversion of Laplace transforms. Let an operation $\mathscr{L}_{c,k}^{-1}$ on a function of a complex variable s be defined in the following way.

$$\mathscr{L}_{c,k}^{-1} F(s) = \frac{1}{2\pi j} \int_{c-jk}^{c+jk} F(s) \exp(st)\, ds \qquad (3\text{–}12)$$

where c and k are real constants and k is positive. For a given c, k and t this maps $F(s)$ into a number. Now let $F(s)$ be the Laplace transform of $f(t)$ as defined by equation (3–1). Then if $c > \sigma_a$ the fundamental relationship between $\mathscr{L}$ and $\mathscr{L}_{c,k}^{-1}$ is that

$$\lim_{k \to \infty} \mathscr{L}_{c,k}^{-1} \mathscr{L} f(t) = \begin{cases} 0 & \text{for } t < 0 \\ \dfrac{f(0+)}{2} & \text{for } t = 0 \\ f(t) & \text{for } t > 0 \quad \text{when } f(t) \text{ continuous} \\ \tfrac{1}{2}[f(t+)+f(t-)] & \text{for } t > 0 \quad \text{when } f(t) \\ & \qquad\qquad\qquad\qquad \text{discontinuous} \end{cases} \qquad (3\text{–}13)$$

In the proof which follows it is assumed that $f(t)$ is of bounded variation for all t. This is sufficient to legitimatize the various interchanges of limiting and integrating operations performed below, and admits all functions of normal dynamical interest, as may be seen from the following definition of bounded variation.
Bounded variation. Consider an arbitrary selection of values of t such that

$$a = t_0 < t_1 < t_2 \cdots < t_n = b$$

and form the sum

$$g = \sum_{i=1}^{n} |f(t_i) - f(t_{i-1})|$$

The variation of $f(t)$ in the interval $[a, b]$ is the least upper bound of all such sums using all possible methods of subdivision. If the variation is finite then $f(t)$ is said to be of bounded variation in the interval.

Inversion integral. The proof of relationships (3–13) depends on the properties of the following double integral

$$\mathscr{L}_{c,k}^{-1} \mathscr{L} f(t) = I = \frac{1}{2\pi j} \int_{c-jk}^{c+jk} \exp(st)\, ds \int_{0}^{\infty} \exp(-sx) f(x)\, dx \quad (3\text{–}14)$$

in which the dummy variable x has been used in place of t in the second integral to clarify the manipulations below. For $c > \sigma_a$, $\mathscr{L} f(x)$ will converge uniformly and the order of integration may be interchanged to give

$$I = \frac{1}{2\pi j} \int_{0}^{\infty} f(x)\, dx \int_{c-jk}^{c+jk} \exp[s(t-x)]\, ds \quad (3\text{–}15)$$

Now consider the second integral in expression (3–15). We have

$$\int_{c-jk}^{c+jk} \exp[s(t-x)]\, ds = \frac{1}{(t-x)} \{\exp[(c+jk)(t-x)] - \exp[(c-jk)(t-x)]\}$$

$$= \frac{2j}{(t-x)} \sin k(t-x) \exp[c(t-x)] \quad (3\text{–}16)$$

Substituting from (3–16) into (3–15) then gives

$$I = \frac{1}{\pi} \int_{0}^{\infty} f(x) \exp[c(t-x)] \frac{\sin k(t-x)}{(t-x)}\, dx \quad (3\text{–}17)$$

Now consider the following three cases in turn:

Case (a): $t < 0$ Case (b): $t = 0$ Case (c): $t > 0$

Case (a) If we make the substitution

$$\lambda = k(t-x)$$

then we get

$$I = \frac{1}{\pi} \int_{-\infty}^{kt} f\left(t - \frac{\lambda}{k}\right) \exp\left(\frac{c\lambda}{k}\right) \frac{\sin \lambda}{\lambda}\, d\lambda \quad (3\text{–}18)$$

Now $k \to +\infty$ ensures that $kt \to -\infty$ if $t < 0$. Thus in the limit, the upper and lower limits of integration in (3–18) coincide and I vanishes.

Therefore
$$\lim_{k\to\infty} I = 0 \quad \text{for} \quad t < 0 \quad \text{and} \quad c > \sigma_a$$

Case (b) Splitting the limits of integration, we can write the integral (3–18) as

$$I = \frac{1}{\pi}\int_{-\infty}^{0} f\left(t-\frac{\lambda}{k}\right)\exp\left(\frac{c\lambda}{k}\right)\frac{\sin\lambda}{\lambda}d\lambda + \frac{1}{\pi}\int_{0}^{kt} f\left(t-\frac{\lambda}{k}\right)\exp\left(\frac{c\lambda}{k}\right)\frac{\sin\lambda}{\lambda}d\lambda \quad (3\text{–}19)$$

For $t = 0$, the upper and lower limits of the second integral in (3–19) coincide and it thus vanishes. This leaves the first integral for which we have after making the further substitution $\xi = -\lambda$

$$\lim_{k\to\infty} I = \lim_{k\to\infty}\frac{1}{\pi}\int_{0}^{\infty} f\left(\frac{\xi}{k}\right)\exp\left(\frac{-c\xi}{k}\right)\frac{\sin\xi}{\xi}d\xi \quad (3\text{–}20)$$

Interchanging the operations of limiting and integrating then gives

$$\lim_{k\to\infty} I = \frac{1}{\pi}\int_{0}^{\infty} \lim_{k\to\infty} f\left(\frac{\xi}{k}\right)\exp\left(\frac{-c\xi}{k}\right)\frac{\sin\xi}{\xi}d\xi$$

$$= \frac{1}{\pi}f(0+)\int_{0}^{\infty}\frac{\sin\xi}{\xi}d\xi$$

$$= \tfrac{1}{2}f(0+)$$

since
$$\int_{0}^{\infty}\frac{\sin\xi}{\xi}d\xi = \frac{\pi}{2}$$

Case (c) For the final case of $t > 0$, interchange the operations of limiting and integrating in (3–19). This gives, after again making the substitution $\xi = -\lambda$ in the first integral,

$$\lim_{k\to\infty} I = \frac{1}{\pi}\int_{0}^{\infty} \lim_{k\to\infty} f\left(t+\frac{\xi}{k}\right)\exp\left(\frac{-c\xi}{k}\right)\frac{\sin\xi}{\xi}d\xi$$

$$+ \frac{1}{\pi}\int_{0}^{\infty} \lim_{k\to\infty} f\left(t-\frac{\lambda}{k}\right)\exp\left(\frac{c\lambda}{k}\right)\frac{\sin\lambda}{\lambda}d\lambda$$

$$= \frac{1}{\pi}f(t+)\frac{\pi}{2} + \frac{1}{\pi}f(t-)\frac{\pi}{2}$$

$$= \tfrac{1}{2}[f(t+)+f(t-)]$$

which, of course, is simply $f(t)$ if the function $f(t)$ is continuous at the point in question.

The results of equation (3–13) follow immediately from the arguments presented above.

Use of contour integrals. The open contour consisting of a straight line from $c-jk$ to $c+jk$ is often called the Bromwich contour after T.J.I'A. Bromwich who introduced it in an important paper justifying the Heaviside operational calculus [B12]. Closed contours are much more useful for the evaluation of inversion integrals since use can then be made of Cauchy's integral theorem. In Figure 3–2, the Bromwich contour is shown as the straight line AB. We now consider the evaluation of the integral

$$I_\Gamma = \int_\Gamma F(s) \exp(st)\, ds \qquad (3\text{–}21)$$

along various *other* specified contours Γ. We first assume that the following restrictions are imposed on $F(s)$.

(a) The only singularities possessed by $F(s)$ in the left-half plane are poles of finite order. ($F(s)$ is said to be meromorphic in the left-half plane.)

(b) $F(s)$ tends uniformly to zero as $|s| \to \infty$. The contours to be considered are the sides of the rectangle ABCD in Figure 3–2. In view of the above restrictions on $F(s)$ we can make k such that, inside and on the rectangle ABCD,

$$|F(s)| < K \qquad (3\text{–}22)$$

where K is some positive finite constant.

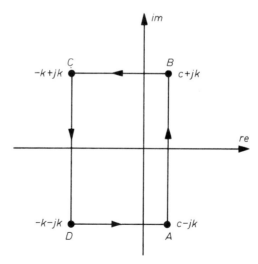

Figure 3–2

Contour BC. Consider the integral I_Γ of equation (3–21) when Γ is the straight line BC. Along this path we may put

$$s = x+jk \qquad x = s-jk \qquad dx = ds$$

so that

$$I_\Gamma = \int_c^{-k} F(s)\exp[(x+jk)t]\,dx = -\int_{-k}^c F(s)\exp(jkt)\exp(xt)\,dx \tag{3-23}$$

and therefore

$$|I_\Gamma| = \left|\int_{-k}^c \exp(xt)\exp(jkt)F(s)\,dx\right| < K\int_{-k}^c \exp(xt)\,dx \tag{3-24}$$

as
$$|F(s)| < K \quad \text{and} \quad |\exp jkt| = 1$$

Evaluating the right-hand integral in equation (3–24), we therefore have that

$$|I_\Gamma| < \frac{K}{t}[\exp(ct)-\exp(-kt)] \quad \text{along BC} \tag{3-25}$$

Contour DA. An exactly similar argument gives

$$|I_\Gamma| < \frac{K}{t}[\exp(ct)-\exp(-kt)] \quad \text{along DA} \tag{3-26}$$

Contour CD. Along the path CD we may put

$$s = -k+jy \qquad ds = j\,dy \qquad y = -j(s+k)$$

so that
$$I_\Gamma = \int_k^{-k} \exp[t(-k+jy)]F(s)j\,dy$$

and therefore

$$|I_\Gamma| = \left|\int_{-k}^k \exp(-kt)\exp(jyt)F(s)j\,dy\right|$$

$$= \left|\exp(-kt)\int_k^{-k} \exp(jyt)F(s)j\,dy\right| \tag{3-27}$$

$$< 2Kk\exp(-kt) \quad \text{along CD}$$

after again using inequality (3–22).

Contour BC+CD+DA. Taking Γ as the sum of the above three paths, we have that

$$|I_\Gamma| < 2K\left[\frac{1}{t}\exp(ct) + k\exp(-kt)\right] \quad \text{along BC+CD+DA} \quad (3\text{-}28)$$

Now, for any given ϵ, because of the stipulated conditions on $F(s)$, we can choose our rectangle in such a way that we have

(a) a K such that $K < \frac{1}{4}t\epsilon \exp(-ct)$
(b) a k such that $2Kk\exp(-kt) < \frac{1}{2}\epsilon$

It follows from this that we can choose a rectangle such that

$$|\Gamma_\Gamma| < \epsilon \quad \text{along BC+CD+DA}$$

This implies that, for a sufficiently large rectangle ABCD, the integral I_Γ along the Bromwich contour AB may be replaced by the *closed* contour integral around the complete rectangle ABCD.

Cauchy's integral formula. [P4] This basic result of the theory of functions of a complex variable states that the integral of a function whose only singularities in the finite part of the complex plane are poles (meromorphic functions), taken in an anticlockwise direction around a closed contour containing a number of simple poles, is equal to $2\pi j$ times the sum of the residues at these poles.

Evaluation of inverse transforms by closed contour integrals. We have shown above that

$$\mathcal{L}^{-1}F(s) = \frac{1}{2\pi j}\int_{c-j\infty}^{c+j\infty} F(s)\exp(st)\,ds \quad (3\text{-}29a)$$

$$= \frac{1}{2\pi j}\int_\Gamma F(s)\exp(st)\,ds \quad (3\text{-}29b)$$

where Γ is a sufficiently large rectangle of the form shown in Figure 3–1 with $c > \sigma_a$. Use of the Cauchy integral formula immediately allows us to replace Γ for equation (3–39b) by *any* closed contour containing all the poles of a meromorphic function $F(s)$. Furthermore it enables us to give an explicit formula for the inverse Laplace transform in this case.

Evaluation of inverse transforms using residues at poles. Let

$$F(s) = \frac{H(s)}{(s-s_0)^n}$$

where $H(s)$ is an analytic function which is nonzero in the neighbourhood of $s = s_0$. Then $F(s)$ is said to have a *pole of n-th order* at $s = s_0$. The *residue* of the function $F(s)$ at this nth order pole is *defined* to be

$$\frac{1}{(n-1)!}\left[\frac{d^{n-1}(s-s_0)^n F(s)}{ds^{n-1}}\right]_{s=s_0}$$

Let $F(s)$ have poles at $s_1, s_2, \ldots, s_n$ of order $r_1, r_2, \ldots, r_n$ respectively so that

$$F(s) = \frac{H(s)}{\prod_{j=1}^{n}(s-s_j)^{r_j}} \qquad (3\text{--}30)$$

where $H(s)$ is analytic over all the finite part of the complex plane (say a polynomial in s). Then, using the Cauchy integral formula with equations (3–29b) we have

$$\mathscr{L}^{-1}F(s) = \sum_{t=1}^{n} \frac{1}{(r_t-1)!} \left[\frac{d^{r_t-1}}{ds^{r_t-1}}(s-s_t)^{r_t}F(s)\exp(st) \right]_{s=s_t} \qquad (3\text{--}31)$$

Examples 3.1

Let
$$F(s) = \frac{1}{(s+a)}$$

Then the inversion integral formula of equation (3–29a) gives

$$f(t) = \frac{1}{2\pi j} \int_{c-j\infty}^{c+j\infty} \frac{\exp(st)}{(s+a)} ds$$

Using the form of equation (3–29b) gives

$$f(t) = \frac{1}{2\pi j} \int_{\Gamma} \frac{\exp(st)}{(s+a)} ds$$

where Γ is any closed contour enclosing the point $s = -a$. Now $\exp(st)/(s+a)$ has a simple pole at $s = -a$ with residue $\exp(-at)$. Thus

$$f(t) = \exp(-at)$$

Example 3.2

Let
$$F(s) = \frac{s+b}{(s+a)^2}$$

then
$$f(t) = \frac{1}{2\pi j} \int_{\Gamma} \frac{[\exp(st)](s+b)}{(s+a)^2} ds$$

where Γ is any closed contour enclosing $s = -a$. At this double pole the integrand has a residue

$$\frac{d}{ds}[(s+b)\{\exp(st)\}]_{s=-a} = \exp(-at) + t(b-a)\exp(-at)$$

so that
$$f(t) = \exp(-at)[1+(b-a)t]$$

Heaviside's expansion formula. Again let

$$F(s) = \frac{H(s)}{\prod_{j=1}^{n}(s-s_j)^{r_j}}$$

where $H(s)$ is analytic over the whole complex plane. Then, expanding $F(s)$ in partial fractions we have

$$F(s) = \sum_{j=1}^{n} \sum_{i=1}^{r_j} \frac{A_{ij}}{(s-s_j)^i} \tag{3-32}$$

Now consider the Laplace transform of a function $g(t)\exp(s_j t)$. We have

$$\mathscr{L}g(t)\exp(s_j t) = \int_0^\infty g(t)\exp(-st)\exp(s_j t)\,dt$$

$$= \int_0^\infty g(t)\exp[-(s-s_j)]t\,dt \tag{3-33}$$

$$= g(s-s_j)$$

so that
$$\mathscr{L}^{-1} g(s-s_j) = g(t)\exp(s_j t) \tag{3-34}$$

We also have, from the standard integrals of Table 3-1, that

$$\mathscr{L}^{-1}\frac{A_{ij}}{s^i} = \frac{A_{ij}}{(i-1)!}t^{i-1} H(t) \tag{3-35}$$

Using equations (3-34) and (3-35) together with equation (3-32) we have

$$f(t) = \sum_{j=1}^{n} \exp(s_j t) \sum_{i=1}^{r_j} A_{ij}\frac{t^{i-1}}{(i-1)!} \tag{3-36}$$

The coefficients A_{ij} may be obtained by multiplying both sides of equation (3-32) by $(s-s_j)^{r_j}$ and differentiating $(r_j - i)$ times. This gives

$$A_{ij} = \frac{1}{(r_j-i)!}\left[\frac{d^{r_j-1}}{ds^{r_j-1}}(s-s_j)^{r_j}F(s)\right]_{s=s_j} \tag{3-37}$$

Combining equations (3-36) and (3-37) we obtain the following extended form of the expansion formula originally due to Heaviside.

$$f(t) = \sum_{j=1}^{n}\exp(s_j t)\sum_{i=1}^{n}\frac{t^{(i-1)}}{(i-1)!(r_j-i)!}\left[\frac{d^{r_j-1}}{ds^{r_j-1}}(s-s_j)^{r_j}F(s)\right]_{s=s_j} \tag{3-38}$$

Solution of Linear, Constant Coefficient Differential Equations by use of Laplace Transforms

Consider the linear nth order constant coefficient differential equation

$$a_0 \frac{d^n y}{dt^n} + a_1 \frac{d^{n-1} y}{dt^{n-1}} + a_2 \frac{d^{n-2} y}{dt^{n-2}} + \cdots + a_n y = f(t)$$

Taking the Laplace transform of both sides of the equation, collecting terms and rearranging, gives the Laplace transform of the solution as

$$y(s) = \frac{1}{\Delta} \{ f(s) + [a_0 s^{n-1} + \cdots + a_{n-1}] y(0-)$$

$$+ [a_0 s^{n-2} + \cdots + a_{n-2}] y^{(1)}(0-) + \cdots + a_0 y^{(n-1)}(0-) \}$$

where $\Delta = a_0 s^n + a_1 s^{n-1} + a_2 s^{n-2} + \cdots + a_n$

and where
$$y^{(r)}(0-) \triangleq \left[\frac{d^r y}{dt^r} \right]_{t=0-} \quad (3\text{-}39)$$

The way in which the initial conditions explicitly enter this expression give the Laplace transform solution a great advantage over the classical method of solution in which the appropriate constants must be obtained by special argument in each case.

Example 3.3

Solve the differential equation

$$\frac{d^2 x}{dt^2} + \omega^2 x = \cos \omega t H(t) \quad \text{for} \quad t > 0$$

with initial conditions $x(0) = x_0$ $(dx/dt)(0) = x_1$. Taking Laplace transforms of both sides of the given equation, we have

$$(s^2 + \omega^2) X(s) = \frac{s}{s^2 + \omega^2} + s x_0 + x_1$$

giving
$$X(s) = \frac{s}{(s^2 + \omega^2)^2} + \frac{s x_0 + x_1}{(s^2 + \omega^2)}$$

and taking inverse transforms then gives

$$x(t) = \frac{t}{2\omega} \sin \omega t + x_0 \cos \omega t + \frac{x_1}{\omega} \sin \omega t$$

3.2 Transfer Function Relationships

Transfer function. Let a pair of input and output variables $s_i(t)$ and $s_0(t)$ respectively, for some dynamical system model be related by the

differential equation

$$a_0 \frac{d^n s_0(t)}{dt^n} + a_1 \frac{d^{n-1} s_0(t)}{dt^{n-1}} + a_2 \frac{d^{n-2} s_0(t)}{dt^{n-2}} + \cdots + a_{n-1} \frac{ds_0(t)}{dt} + a_n s_0(t)$$

$$= b_0 \frac{d^m s_i(t)}{dt^m} + b_1 \frac{d^{m-1} s_i(t)}{dt^{m-1}} + \cdots + b_{m-1} \frac{ds_i(t)}{dt} + b_m s_i(t) \qquad (3\text{-}40)$$

Suppose that at time $t = 0-$ the dynamical system is relaxed, that is there is, relative to some equilibrium condition of the system, zero stored energy in the system. Then the value of $s_0(t)$ and all its derivatives are zero at $t = 0-$ if $s_0(t)$ describes a departure from the equilibrium condition. Let $s_i(t)$ be applied at $t = 0$ under these conditions and take the Laplace transform of both sides of equation (3-40). Then

$$\frac{s_0(s)}{s_i(s)} \triangleq \Phi(s) = \frac{b_0 s^m + b_1 s^{m-1} + \cdots + b_m}{a_0 s^n + a_1 s^{n-1} + \cdots + a_n} \qquad (3\text{-}41)$$

and $\Phi(s)$ is defined as the *transfer function* relating the input and output variable pair. A study of the way in which such differential equations are obtained from physically meaningful dynamical system models shows that $n \geqslant m$, as discussed in Section 3.2.3. below.

We have

$$s_0(t) = \mathscr{L}^{-1} \Phi(s) s_i(s)$$

Given any general transfer function, $\psi(s)$ say, which is the ratio of two polynomials, its inverse transform may be found in the following way

(1) If the order of the numerator is greater than or equal to the order of the denominator (that is, if $\psi(s)$ is an improper fraction), divide the numerator by the denominator to obtain a polynomial in s, $p(s)$ say, plus a proper fraction, $\Theta(s)$ say. That is put

$$\psi(s) = p(s) + \Theta(s)$$

(2) We then have

$$\mathscr{L}^{-1} \psi(s) = \mathscr{L}^{-1} p(s) + \mathscr{L}^{-1} \Theta(s)$$

(3) $\mathscr{L}^{-1} p(s)$ is an appropriate set of impulse functions.
(4) $\mathscr{L}^{-1} \Theta(s)$ is a set of time-weighted exponential functions which are determined using the expansion formula of equation (3-28).

Weighting function. If the input applied to the dynamical system model is a Dirac impulse function $\delta(t)$, the corresponding output

$$w(t) = \mathscr{L}^{-1} \Phi(s) \qquad (3\text{-}42)$$

Transfer Function Relationships

is defined as the *weighting function* for signal transmission between the specified input and output points.

3.2.1 Convolution

The convolution of two functions $w(t)$ and $x(t)$ is defined by

$$y(t) = \int_0^t w(t-\tau)x(\tau)\,d\tau \qquad (3\text{-}43)$$

Consider the Laplace transform of the convolution. We have

$$\begin{aligned}\mathscr{L}y(t) &= \int_0^\infty y(t)\exp(-st)\,dt \\ &= \int_0^\infty \exp(-st)\,dt \int_0^t w(t-\tau)x(\tau)\,d\tau\end{aligned} \qquad (3\text{-}44)$$

Suppose $w(t)$ is identically zero for $t < 0$. Then $w(t-\tau)$ will be identically zero for $\tau > t$, and we may then write equation (3–44) as

$$\mathscr{L}y(t) = \int_0^\infty \exp(-st)\,dt \int_0^\infty w(t-\tau)x(\tau)\,d\tau$$

Changing the order of integration then gives

$$\mathscr{L}y(t) = \int_0^\infty x(\tau)\,d\tau \int_0^\infty w(t-\tau)\exp(-st)\,dt$$

and introducing the change of variable $u = t - \tau$ we get

$$\mathscr{L}y(t) = \int_0^\infty x(\tau)\,d\tau \int_{-\tau}^\infty w(u)\exp(-su)\exp(-s\tau)\,du$$

which, since $w(u)$ vanishes for $u < 0$ may be written as

$$\begin{aligned}\mathscr{L}y(t) &= \int_0^\infty x(\tau)\exp(-s\tau)\,d\tau \int_0^\infty w(u)\exp(-su)\,du \\ &= [\mathscr{L}x(t)][\mathscr{L}w(t)]\end{aligned} \qquad (3\text{-}45)$$

Thus the Laplace transform of the convolution of two functions is the product of their individual Laplace transforms. The integral in equation (3–43) is termed a *convolution* integral. It is often convenient to denote the convolution symbolically by ∗, writing equation (3–43) as

$$y(t) = w(t) * x(t) \qquad (3\text{-}46)$$

Suppose an input signal $s_i(t)$ is applied to a linear, autonomous system from an ideal source. Let $s_0(t)$ be a particular output response; let

$\Phi(s)$ be the transfer function for signal transmission between the input and output points considered; and let $w(t)$ be the corresponding weighting function. Then

$$s_0(s) = \Phi(s)s_i(s) \tag{3-47}$$

and therefore we have that

$$\mathscr{L}[s_0(t)] = [\mathscr{L}w(t)][\mathscr{L}s_i(t)]$$

We may thus express the output as the convolution integral

$$s_0(t) = \int_0^t s_i(\tau)w(t-\tau)\,d\tau \tag{3-48}$$

The system output may also be expressed in terms of a convolution integral involving the step response. Let $a(t)$ be the output response to a unit Heaviside step function applied at the input. Then, using equation (3-47), we have that

$$a(t) = \mathscr{L}^{-1}\frac{1}{s}\Phi(s)$$

If we write equation (3-47) in the form

$$s_0(s) = [s_i(s)]\left[\frac{s\Phi(s)}{s}\right]$$

then an application of the convolution theorem gives

$$s_0(t) = [\mathscr{L}^{-1}s_i(s)] * \left[\mathscr{L}^{-1}\frac{s\Phi(s)}{s}\right] \tag{3-49}$$

Using the relationship for the transform of a derivative we have

$$\frac{s\Phi(s)}{s} = \mathscr{L}\frac{da(t)}{dt} + a(0-)$$

Thus
$$\mathscr{L}^{-1}\left[\frac{s\Phi(s)}{s}\right] = \frac{da(t)}{dt} + a(0-)\delta(t)$$

and the convolution integral corresponding to equation (3-49) is

$$s_0(t) = \int_0^t s_i(\tau)\left[\frac{da(t-\tau)}{dt} + a(0-)\delta(t-\tau)\right]d\tau$$

$$= \int_0^t s_i(\tau)\frac{da(t-\tau)}{dt}\,d\tau + \int_0^t a(0-)s_i(\tau)\delta(t-\tau)\,d\tau \tag{3-50}$$

$$= \int_0^t s_i(\tau)\frac{da(t-\tau)}{dt}\,d\tau + a(0-)s_i(t)$$

Transfer Function Relationships

Since $a(0-)$ is zero for a physically realizable system

$$\frac{da(t-\tau)}{dt} = w(t-\tau)$$

then the convolution integral of equation (3–50) is equivalent to that of (3–48).

Another form of convolution integral may be obtained by integrating equation (3–50) by parts. This gives, remembering for the allocation of signs that the integration is with respect to τ

$$s_0(t) = a(0-)s_i(t) - [s_i(\tau)a(t-\tau)]_{0-}^{t} + \int_0^t \frac{ds_i(\tau)}{d\tau} a(t-\tau) \, d\tau$$

$$= s_i(0-)a(t) + \int_0^t \frac{ds_i(\tau)}{d\tau} a(t-\tau) \, dt \quad (3\text{–}51)$$

It is important to have a physical interpretation of these convolution integrals. To interpret the integral of equation (3–48), suppose the input signal $s_i(\tau)$ is approximated by a sequence of rectangular pulses of height $s_i(\tau)$ and width $\Delta\tau$ as shown in Figure 3–3. If the time interval $\Delta\tau$ is made sufficiently small, the Laplace transforms of the rectangular pulses may be well approximated by $s_i(\tau)\Delta\tau$, and the system response to the pulse may then be well approximated by the response to an

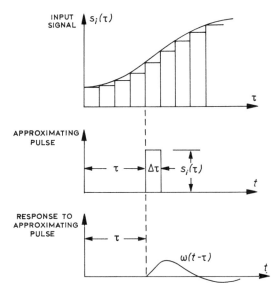

Figure 3–3

impulse of this value. The response at the instant t will be approximately equal to the sum of the responses of all the rectangular pulses from $\tau = 0$ to $\tau = t$ and hence, if $\Delta\tau$ is small, is well approximated by

$$s_0(t) = \sum_{\tau=0}^{\tau=t} s_i(\tau)\,\Delta\tau w(t-\tau)$$

which, in the limit as $\Delta\tau \to 0$ gives

$$s_0(t) = \int_0^t s_i(\tau) w(t-\tau)\, d\tau$$

To interpret the integral of equation (3–51), suppose the input signal $s_i(\tau)$ is approximated by a sequence of step functions of amount Δs_i applied at intervals of time $\Delta\tau$ as shown in Figure 3–4. If the response

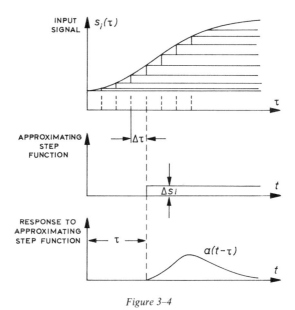

Figure 3–4

to a unit Heaviside step function is $a(t)$ then the output signal may thus be approximated by

$$s_0(t) = s_i(0-)a(t) + \sum_{\tau=\Delta t}^{\tau=t} \Delta s_i a(t-\tau)$$

$$= s_i(0-)a(t) + \sum_{\tau=\Delta t}^{\tau=t} \frac{\Delta s_i}{\Delta\tau} a(t-\tau)\,\Delta\tau$$

Transfer Function Relationships

which gives, in the limit as $\Delta\tau \to 0$

$$s_0(t) = s_i(0-)a(t) + \int_0^t \frac{ds_i(\tau)}{d\tau} a(t-\tau)\, d\tau$$

3.2.2 Asymptotic Relations between Weighting Function and Transfer Function

Suppose a transfer function for a stable, linear, autonomous dynamical system has a Taylor series expansion

$$\Phi(s) = \sum_{k=1}^{\infty} \frac{a_k}{s^k} \quad \text{giving} \quad w(t) = H(t) \sum_{k=1}^{\infty} \frac{a_k}{(k-1)!} t^{k-1} \quad (3\text{-}52)$$

where $|s| > R$ and R is a positive constant. Therefore as $t \to 0$, $w(t)$ will be well approximated by the first few terms of this series. This gives a useful way of checking inverse transforms by computing the response for a short initial period of time.

Again suppose that a transfer function for such a system is a rational function of s and that $|\Phi(s)| \to 0$ as $|s| \to \infty$. Let $\Phi(s)$ be analytic for Re $(s) > \alpha$, where α is a real constant, and let $p_1, p_2, \ldots, p_n$ be the only poles of $\Phi(s)$ in a strip of the complex s-plane of width δ lying to the left of the line $s = \alpha$. We may then expand $\Phi(s)$ in partial fractions as

$$\Phi(s) = \sum_{k=1}^{n} \left[\frac{b_{1k}}{(s-p_k)} + \frac{b_{2k}}{(s-p_k)^2} + \cdots + \frac{b_{m_k k}}{(s-p_k)^{m_k}} \right] + \psi(s)$$

where $\psi(s)$ represents the contribution of those poles of $\Phi(s)$ not lying in the nominated strip. Taking inverse transforms then gives

$$w(t) = H(t) \left[\sum_{k=1}^{n} \left\{ b_{1k} + b_{2k}t + \cdots + \frac{b_{m_k k} t^{k-1}}{(k-1)!} \right\} \exp(p_k t) \right] + \mathscr{L}^{-1}\psi(s)$$

where $\mathscr{L}^{-1}\psi(s)$ will contain terms of the form

$$t^l \exp(v_l t)$$

where
$$\text{Re}(v_l) < (\alpha - \delta)$$

Thus since

$$a \exp(p_k t) + b \exp(v_l t) \to a \exp(p_k t) \quad \text{as} \quad t \to \infty$$

we have that, for $t \to \infty$

$$w(t) \to H(t) \left[\sum_{k=1}^{n} \left\{ b_{1k} + \cdots + \frac{b_{m_k k} t^{k-1}}{(k-1)!} \right\} \exp(p_k t) \right] \quad (3\text{-}53)$$

This shows that the contributions due those transfer function poles which lie furthest to the right in the s-plane dominate the impulse response for large values of t; for this reason they are usually called the *dominant* poles of the transfer function.

3.2.3 Response to Sinusoidal Input

Let $\Phi(s)$ be a transfer function of a stable system, and let $s_i(s)$, $s_0(s)$ be the Laplace transforms of a pair of input and output signals respectively, so that

$$s_0(s) = \Phi(s)s_i(s)$$

Suppose the input signal is

$$s_i(t) = \sin \omega t H(t)$$

and is applied at time $t = 0$ to a system in a relaxed equilibrium condition. Then

$$s_i(s) = \frac{\omega}{s^2 + \omega^2}$$

and

$$s_0(s) = \Phi(s) \frac{\omega}{(s - j\omega)(s + j\omega)}$$

If the poles of the transfer function $\Phi(s)$ are assumed distinct and denoted by $p_1, p_2, \ldots, p_n$, then expansion by partial fractions gives

$$s_0(s) = \sum_{j=1}^{n} \frac{L_j}{(s - p_j)} + \frac{L_{n+1}}{(s - j\omega)} + \frac{L_{n+2}}{(s + j\omega)}$$

where $L_1, L_2, \ldots, L_n, L_{n+1}, L_{n+2}$ are real or complex constants and

$$L_{n+1} = \left[\frac{\Phi(s)\omega(s - j\omega)}{(s - j\omega)(s + j\omega)} \right]_{s = j\omega} = \frac{1}{2j} \Phi(+j\omega)$$

$$L_{n+2} = \left[\frac{\Phi(s)\omega(s + j\omega)}{(s - j\omega)(s + j\omega)} \right]_{s = -j\omega} = -\frac{1}{2j} \Phi(-j\omega)$$

We thus have, on taking inverse transforms

$$s_0(t) = \sum_{j=1}^{n} L_j \exp(p_j t) + \frac{1}{2j}\Phi(j\omega)\exp(j\omega t) - \frac{1}{2j}\Phi(-j\omega)\exp(-j\omega t)$$

For a stable system, all the complex numbers $p_1, p_2, \ldots, p_n$ will have negative real parts and so the first part of this expression will become vanishingly small as t becomes arbitrarily large. Thus the steady state response to an input sinusoid, which we will denote by $s_0(\infty)$, is

$$s_0(\infty) = \frac{1}{2j}\Phi(j\omega)\exp(j\omega t) - \frac{1}{2j}\Phi(-j\omega)\exp(-j\omega t) \qquad (3\text{-}54)$$

Transfer Function Relationships

If we put

$$\Phi(j\omega) = |\Phi(j\omega)| \exp j[\angle \Phi(j\omega)] \qquad (3\text{-}55)$$

where $|\Phi(j\omega)|$ and $\angle \Phi(j\omega)$ denote the modulus and phase of $\Phi(j\omega)$ respectively, then

$$\Phi(-j\omega) = |\Phi(j\omega)| \exp j[-\angle \Phi(j\omega)] \qquad (3\text{-}56)$$

Putting expressions (3-55) and (3-56) in equation (3-54) then gives

$$s_0(\infty) = |\Phi(j\omega)| \left[\left\{ \frac{1}{2j} \exp j[\angle \Phi(j\omega)] \right\} \exp (j\omega t) \right.$$
$$\left. - \left\{ \frac{1}{2j} \exp j[-\angle \Phi(j\omega)] \right\} \exp (-j\omega t) \right] \qquad (3\text{-}57)$$
$$= |\Phi(j\omega)| \sin [\omega t + \angle \Phi(j\omega)]$$

This shows that if a sinusoidal signal is applied to a linear, stable, autonomous system having a transfer function $\Phi(s)$ between a pair of specified input and output points then:

(a) The amplitude of the resulting steady state output sinusoid is $|\Phi(j\omega)|$ times the amplitude of input sinusoid.
(b) The phase shift between input and output sinusoids is $\angle \Phi(j\omega)$.

For any practical system we will have $|\Phi(j\omega)| \to 0$ as $\omega \to \infty$. It follows that for a practical system having a transfer function which is a rational fraction in s, the order of the numerator polynomial must be less than the order of the denominator polynomial.

3.2.4 Determination of System Response from Transfer Function Pole and Zero Distribution

Suppose the transfer function relating a specified input and output variable pair of a linear, autonomous dynamical system is given by

$$\Phi(s) = \frac{K(s-z_1)(s-z_2)\cdots(s-z_m)}{(s-p_1)(s-p_2)\cdots(s-p_n)} \qquad (3\text{-}58)$$

where:

(a) K is a real constant.
(b) $z_1, z_2, \ldots, z_m$ are a set of real or complex constant called the *zeros* of the transfer function. They are those values of s for which $\Phi(s)$ vanishes.
(c) $p_1, p_2, \ldots, p_n$ are a set of real or complex constants called the *poles* of the transfer function. They are those values of s for which the modulus of the transfer function becomes infinite.

This type of transfer function is completely specified by the location of its poles and zeros in the complex plane together with the value of the multiplier constant K. It is conventional to mark the complex plane with small circles to denote the location of the zeros, and small crosses to mark the location of the poles; such a set of markings in the complex plane is usually termed the pole-zero constellation for the transfer function. If an accurate plot is made of the pole-zero constellation for a transfer function, the corresponding complex number $\Phi(s)$ may be evaluated at any point in the complex plane by simple graphical measurements on this plot.

Let the zeros of the transfer function be located at points $Z_1, Z_2, \ldots, Z_m$ in the complex plane and let the poles be located at the points $P_1, P_2, \ldots, P_n$. Consider some arbitrary point M of the complex plane corresponding to the complex number s as shown in Figure 3–5. Then we have that:

(a) The line $Z_1 M$, joining Z_1 to M, represents the complex number $(s - z_1)$, the line $Z_2 M$ represents the complex number $(s - z_2), \ldots$, the line $Z_m M$ represents the complex number $(s - z_m)$.

(b) The line $P_1 M$ represents the complex number $(s - p_1)$, the line $P_2 M$ represents the complex number $(s - p_2), \ldots$, the line $P_n M$ represents the complex number $(s - p_n)$.

For any complex number s we may locate the corresponding point M and then measure the contributions of the zeros to the modulus and phase of $\Phi(s)$ as

$|s - z_j|$ = length of $Z_j M$
$\angle(s - z_j)$ = angle made by $Z_j M$ with the positive direction of the real axis, measured in an anticlockwise direction.

In a similar way the contributions of the poles may be measured as

$|s - p_i|$ = length of $P_i M$
$\angle(s - p_i)$ = angle made by $P_i M$ with the positive direction of the real axis, measured in an anticlockwise direction.

We thus have that the modulus and phase of the overall transfer function $\Phi(s)$ are

$$|\Phi(s)| = \frac{K(Z_1 M)(Z_2 M)\ldots(Z_m M)}{(P_1 M)(P_2 M)\ldots(P_n M)} \qquad (3\text{-}59)$$

$$\angle \Phi(s) = \sum_{j=1}^{m} (\text{angles made by } Z_j M) - \sum_{i=1}^{n} (\text{angles made by } P_i M) \qquad (3\text{-}60)$$

This procedure is particularly useful for graphically evaluating the frequency response corresponding to a given pole-zero constellation.

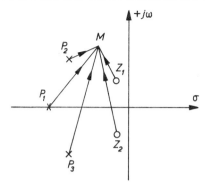

Figure 3-5

This is done by taking for the point M considered a sequence of values along the imaginary axis in the complex plane.

It may also be used for a graphical evaluation of those residues at the transfer function poles which are required in the determination of the system response to an impulse or step input. The unit impulse response corresponding to the transfer function is given by

$$w(t) = \mathscr{L}^{-1}\Phi(s) = \sum_{i=1}^{n} k_i \exp(p_i t)$$

where, for simple distinct poles, the residues k_i are given by

$$k_i = [(s - p_i)\Phi(s)]_{s=p_i} \quad i = 1, 2, \ldots, n$$

The residues at each pole are thus measured by ignoring the contribution of the pole being considered, and measuring $\Phi(s)$ at the point where the pole is located using the method given above. The modulus of the residue at each pole is then the product of the distances measured from the pole to all the other zeros, divided by the distance measured from the pole to all the other poles. If the pole is complex, the residue will be complex and is obtained in polar form by measuring the modulus as above and the phase angle as in the general case. The unit step response for the transfer function is given by

$$a(t) = \mathscr{L}^{-1} \frac{1}{s} \Phi(s)$$

The residues at the poles for the evaluation of the response to a unit step are therefore obtained by adding an extra pole at the origin to the pole zero constellation in the s-plane, and then proceeding as for the impulse response case.

By a simultaneous use of both of the techniques outlined above, one may correlate the frequency and time responses corresponding to any given transfer function.

Potential Analogy

If we divide both sides of equation (3-58) by K and take the natural logarithm we have:

$$\ln\left[\frac{\Phi(s)}{K}\right] = \ln\left|\frac{\Phi(s)}{K}\right| + j \measuredangle \Phi(s) = \sum_{j=1}^{m} \ln(s-z_j) - \sum_{i=1}^{n} \ln(s-p_i)$$

The term on the right-hand side of this expression may be considered as the electric potential due to an appropriate set of uniform *line charges* (of strength 1 c/m) perpendicular to the s-plane such that positive line charges pass through the zeros and negative line charges pass through the poles. The associated electrostatic field has equipotentials which are contours of constant logarithmic magnitude, and fieldlines which are contours of constant phase angle. This potential analogy is very useful for purposes of visualization, and forms the basis of several analogue devices for computing transfer function behaviour.

3.2.5 Relationships between Real and Imaginary Parts of Transfer Function

Let $\Phi(s)$ be a transfer function for a stable, linear, autonomous dynamical system, so that $\Phi(s)$ is analytic in the right-half of the s-plane and

$$\Phi(s) = \text{complex conjugate of } \Phi(s^*)$$

where $\qquad s^* = \text{complex conjugate of } s$

In addition let $\Phi(s)$ have no more than a finite number of imaginary axis singularities, and let $\Phi(s)$ be such that

$$\left|\frac{\Phi(s)}{s}\right| \to 0 \quad \text{as} \quad |s| \to \infty$$

Imaginary Part of $\Phi(j\omega)$ as a Functional of Real Part

Let

$$\text{Im } \Phi(j\omega) = \text{imaginary part of } \Phi(j\omega)$$

$$\text{Re } \Phi(j\omega) = \text{real part of } \Phi(j\omega)$$

$$R_1 = [\text{Re } \Phi(j\omega)]_{\omega=\omega_1}$$

Then $\qquad [\text{Im } \Phi(j\omega)]_{\omega=\omega_1} = \dfrac{2\omega_1}{\pi} \int_0^\infty \dfrac{\text{Re } \Phi(j\omega) - R_1}{\omega^2 - \omega_1^2} d\omega \qquad$ (3-61)

To prove this we consider the contour integral

$$I = \int_\Gamma \psi(s)\, ds$$

where
$$\psi(s) = \frac{2j\omega_1[\Phi(s)-R_1]}{s^2+\omega_1^2}$$
$$= \frac{[\Phi(s)-R_1]}{s-j\omega_1} - \frac{[\Phi(s)-R_1]}{s+j\omega_1}$$

and Γ is the contour shown in Figure 3-6.

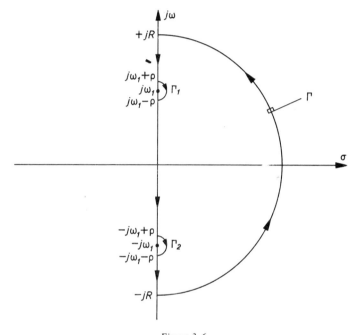

Figure 3-6.

For the moment assume $\Phi(s)$ has no imaginary axis singularities. In this case $\psi(s)$ will be analytic in the right-half plane and on the imaginary axis, except for the two simple poles at $s = \pm j\omega_1$ and thus

$$I = 0$$

Now as
$$|s| \to \infty$$
$$s\psi(s) \approx \frac{\Phi(s)}{s}$$

and thus the assumption that $\Phi(s)/s$ vanishes for $|s| \to \infty$ ensures that the contribution to I due to the large semi-circular arc of the contour Γ vanishes as the radius becomes arbitrarily large. For an arbitrarily large semi-circular arc we may therefore write

$$I = I_1 + I_2 + I_3 + I_4 + I_5 + I_6 + I_7 = 0 \qquad (3\text{-}62)$$

where
$$I_1 = \int_{-\infty}^{j\omega_1 - \rho} \frac{2\omega_1[\Phi(j\omega) - R_1]}{\omega^2 - \omega_1^2} d\omega$$

$$I_2 = \int_{-j\omega_1 + \rho}^{j\omega_1 - \rho} \frac{2\omega_1[\Phi(j\omega) - R_1]}{\omega^2 - \omega_1^2} d\omega$$

$$I_3 = \int_{j\omega_1 + \rho}^{\infty} \frac{2\omega_1[\Phi(j\omega) - R_1]}{\omega^2 - \omega_1^2} d\omega$$

$$I_4 = \int_{\Gamma_1} \frac{[\Phi(s) - R_1]}{(s - j\omega_1)} ds$$

$$I_5 = -\int_{\Gamma_1} \frac{[\Phi(s) - R_1]}{(s + j\omega_1)} ds$$

$$I_6 = \int_{\Gamma_2} \frac{[\Phi(s) - R_1]}{(s - j\omega_1)} ds$$

$$I_7 = -\int_{\Gamma_2} \frac{[\Phi(s) - R_1]}{(s + j\omega_1)} ds$$

As the radius, ρ, of the small semi-circular indentations bypassing the imaginary axis is made arbitrarily small, $I_5 \to 0$ and $I_6 \to 0$ since their integrands are analytic in the neighbourhood of $+j\omega_1$ and $-j\omega_1$ respectively. If the integrands of I_4 and I_7 are expanded in a Laurent series about $s = j\omega_1$ and $s = -j\omega_1$ respectively we have:

$$\frac{[\Phi(s) - R_1]}{s - j\omega_1} = \frac{j[\operatorname{Im} \Phi(j\omega)]_{\omega = \omega_1}}{s - j\omega_1} + \left\{ \frac{d}{ds}[\Phi(s) - R_1] \right\}_{s = j\omega_1} + \cdots$$

$$\frac{[\Phi(s) - R_1]}{s + j\omega_1} = \frac{j[\operatorname{Im} \Phi(j\omega)]_{\omega = \omega_1}}{s + j\omega_1} + \left\{ \frac{d}{ds}[\Phi(s) - R_1] \right\}_{s = -j\omega_1} + \cdots$$

As the radius of the semi-circular indentations bypassing the imaginary axis singularities is made arbitrarily small, only the first terms of these expansions will contribute to the integrals I_4 and I_7 and we will have

$$I_4 \to j[\operatorname{Im} \Phi(j\omega)]_{\omega = \omega_1} \int_{\Gamma_1} \frac{ds}{s - j\omega_1} = -\pi[\operatorname{Im} \Phi(j\omega)]_{\omega = \omega_1}$$

$$I_7 \to -j[\text{Im }\Phi(j\omega)]_{\omega=\omega_1}\int_{\Gamma_2}\frac{ds}{s+j\omega_1} = -\pi[\text{Im }\Phi(j\omega)]_{\omega=\omega_1}$$

The final result of making the semi-circular indentations arbitrarily small will be that

$$I_1+I_2+I_3 \to \int_{-\infty}^{\infty}\frac{2\omega_1[\Phi(j\omega)-R_1]}{\omega-\omega_1^2}d\omega$$

We therefore have, in the limit as $\rho \to 0$ and $R \to \infty$

$$I_1+I_2+I_3 \to -I_4-I_7$$

that is

$$\begin{aligned}
2\pi[\text{Im }\Phi(j\omega)]_{\omega=\omega_1} &= \int_{-\infty}^{+\infty}\frac{2\omega_1[\Phi(j\omega)-R_1]}{\omega^2-\omega_1^2}d\omega \\
&= \int_{-\infty}^{+\infty}\frac{2\omega_1[\text{Re }\Phi(j\omega)-R_1]}{\omega^2-\omega_1^2}d\omega \\
&\quad - \int_{-\infty}^{+\infty}\frac{2\omega_1[\text{Im }\Phi(j\omega)-R_1]}{\omega^2-\omega_1^2}d\omega \\
&= 4\omega_1\int_0^{\infty}\frac{[\text{Re }\Phi(j\omega)-R_1]}{\omega^2-\omega_1^2}d\omega
\end{aligned} \quad (3\text{-}63)$$

since $\text{Re }\Phi(j\omega)$ is an even function of ω and $\text{Im }\Phi(j\omega)$ is an odd function of ω. The result of equation (3-61) then follows immediately.

If $\Phi(s)$ has any singularities on the imaginary axis then these singularities may be bypassed by suitably small detours of Γ. A detailed consideration of the resulting contour integral then shows that the integration along these additional detours does not contribute, in the limit, to the final value of I so that the result holds in this more general case.

Real Part of $\Phi(j\omega)$ as a Functional of Imaginary Part

If $\Phi(s)$ satisfies the assumptions above, and if it is analytic for arbitrarily large values of $|s|$ with a corresponding Taylor series expansion

$$\Phi(s) = R_\infty + \frac{I_\infty}{s} + \frac{R_2}{s^2} + \cdots$$

then

$$[\text{Re }\Phi(j\omega)]_{\omega=\omega_1} = R_\infty - \frac{2}{\pi}\int_0^\infty \frac{\omega\,\text{Im }\Phi(j\omega) - \omega_1[\text{Im }\Phi(j\omega)]_{\omega=\omega_1}}{\omega^2-\omega_1^2}d\omega$$

(3-64)

This result may be proved by an exactly similar argument to that used for the imaginary part case.

If we make the change of variable

$$u = \ln\left(\frac{\omega}{\omega_1}\right)$$

then we get

$$[\operatorname{Im} \Phi(j\omega)]_{\omega=\omega_1} = \frac{1}{\pi} \int_0^\infty \frac{[\operatorname{Re} \Phi(j\omega) - R_1]}{\sinh u} du$$

$$= \frac{1}{\pi} \int_{-\infty}^\infty \frac{d\operatorname{Re} \Phi(j\omega)}{du} \ln\left|\coth \frac{u}{2}\right| du \quad (3\text{–}65)$$

which shows that the value of the imaginary part of $\Phi(j\omega)$ at $\omega = \omega_1$ may be expressed as a weighted average of the slope of the real part of $\Phi(j\omega)$ when plotted on a logarithmic frequency scale of u.

3.2.6 Minimum-phase Transfer Functions

Let $p_1, p_2, \ldots, p_n$ and $z_1, z_2, \ldots, z_m$ be two sets of complex constants with negative real parts with $m \leq n$. Consider the set of transfer functions

$$\Phi_1(s) = \frac{K(s-z_1)(s-z_2)(s-z_3)\ldots(s-z_m)}{(s-p_1)(s-p_2)(s-p_3)\ldots(s-p_n)}$$

$$\Phi_2(s) = \frac{K(s+z_1)(s-z_2)(s-z_3)\ldots(s-z_m)}{(s-p_1)(s-p_2)(s-p_3)\ldots(s-p_n)}$$

$$\Phi_3(s) = \frac{K(s+z_1)(s+z_2)(s-z_3)\ldots(s-z_m)}{(s-p_1)(s-p_2)(s-p_3)\ldots(s-p_n)}$$

$$\Phi_4(s) = \frac{K(s+z_1)(s+z_2)(s+z_3)\ldots(s-z_m)}{(s-p_1)(s-p_2)(s-p_3)\ldots(s-p_n)}$$

$$\vdots$$

$$\Phi_{m+1}(s) = \frac{K(s+z_1)(s+z_2)(s+z_3)\ldots(s+z_m)}{(s-p_1)(s-p_2)(s-p_3)\ldots(s-p_n)}$$

Using the relationships developed in Section 3.2.4 we then see that

$$|\Phi_1(j\omega)| = |\Phi_2(j\omega)| = |\Phi_3(j\omega)| = \cdots = |\Phi_{m+1}(j\omega)|$$

and for $\omega > 0$

$$\angle \Phi_1(j\omega) < \angle \Phi_2(j\omega) < \angle \Phi_3(j\omega) < \cdots < \angle \Phi_{m+1}(j\omega)$$

We thus see that of this set of transfer functions having the same amplitude frequency response characteristic, that with all its zeros in the left-half plane will have the least phase shift for each value of frequency. A stable rational transfer function $\Phi(s)$ is said to be *minimum phase* if and only if it has no finite zeros or poles in the right-half s-plane. A stable transfer function which has one or more finite zeros or poles in the right-half s-plane is said to be non-minimum phase.

3.3 Block Diagrams

Block diagram. A block diagram is defined as an oriented linear graph whose line segments represent variables and whose vertices represent relationships between these variables.

The name block diagram is used because it is conventional to draw the vertices as blocks (or circles) inside which one writes symbols denoting the relationships involved. The orientation arrows on the line segments of a block diagram are associated with causal sense, that is, they designate for each relationship which is the dependent and which the independent variable. To represent any possible set of relationships between variables, we require a complete, independent set of operations on the variables. The analytical theory of approximation shows us that any required relationship between dynamical system variables may be represented to any required degree of accuracy in terms of the five basic operations of summation, multiplication, functional relationship, integration and ideal delay. In block diagram representations however we do not normally limit ourselves to a basic set of operations and synthesize all operations from them but introduce special types of vertex as required, for example, for division.

3.3.1 Block Diagram Conventions

The principal conventions for representing relationships between variables by block diagram vertices are as follows:

(a) *Summation vertex.* The summation vertex has m input line segments, associated with variables $x_1, x_2, \ldots, x_m$ say, and a single output line segment, associated with an output variable y say, such that

$$y = \sum_{i=1}^{m} s_i x_i$$

where $s_i = \pm 1$ according to the algebraic sign required in the summation. It is conventionally represented by a capital sigma enclosed in a circle, as shown in Figure 3–7. The orientation arrows point towards the vertex for the input line segments and away from the vertex for the output line segment. The sign required in the summation is indicated by appending to each input line segment a plus or minus sign as required.

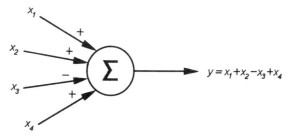

Figure 3-7

(b) *Multiplication vertex.* The multiplication vertex has m input line segments, associated with variables $x_1, x_2, \ldots, x_m$ say, and a single output line segment, associated with an output variable y say, such that

$$y = \prod_{i=1}^{m} s_i x_i$$

where $s_i = \pm 1$ according to the sign required in the multiplication. It is conventionally represented by a capital Π enclosed in a circle, as shown in Figure 3-8. The sign and orientation conventions are the same as for the summation vertex.

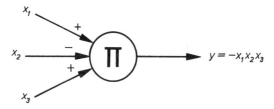

Figure 3-8

(c) *Functional relationship vertex.* A functional relationship vertex has one input line segment, associated with a variable x say, and one output line segment, associated with a variable y say, such that

$$y = \Phi(x)$$

where Φ is some specified functional relationship for which a unique value of output may be determined for every possible value of input variable. The orientation arrow points towards the vertex for the independent variable and away from the vertex for the dependent variable in the functional relationship. It is represented by the vertex symbol shown in Figure 3-9.

Figure 3-9

(d) *Integrator vertex.* The temporal relationships between dynamical system variables are usually represented in their integral form on block diagrams by an integrator vertex. This has one input line segment associated with a function of time, $x(t)$ say, and one output line segment associated with a variable, $y(t)$, say, such that

$$y(t) = \int_{t_0}^{t} x(t)\,dt$$

where t_0 is a reference value of time at which integration commences. It is represented by the vertex symbol shown in Figure 3-10(a). If the value of the output variable is non-zero at the time t_0 corresponding

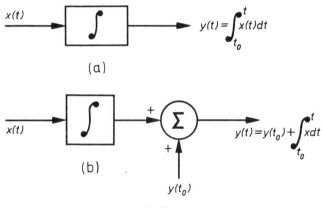

Figure 3-10

to the lower limit of integration, a summation vertex is added as shown in Figure 3-10(b) to represent

$$y(t) = y(t_0) + \int_{t_0}^{t} x(t)\,dt$$

(e) *Ideal delay vertex.* An ideal delay vertex has one input line segment associated with a function of time, $x(t)$ say, and one output line segment associated with a function of time, $y(t)$ say, such that

$$y(t) = x(t-T)$$

where T is the amount of time delay represented. It is represented by the vertex symbol shown in Figure 3-11.

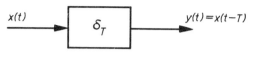

Figure 3-11

(f) *Identity vertex.* An identity vertex (or splitting, pick-off or take-off vertex) has one line segment associated with an input variable incident upon it and any required number of output segments associated with a set of output variables which are all identically equal to the input variable. It is represented by the vertex symbols shown in Figures 3-12(a) and 3-12(b).

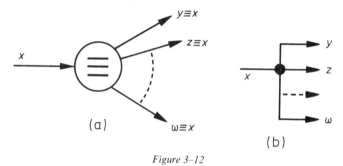

Figure 3-12

(g) *Division vertex.* The convention adopted to represent division is shown in Figure 3-13.

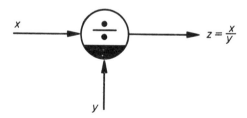

Figure 3-13

(h) *Scaling vertex.* A scaling vertex is simply a special form of functional relationship vertex for which

$$y = kx$$

where k is a specified constant which may be positive or negative. It is represented by the vertex symbol shown in Figure 3–14.

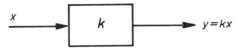

Figure 3–14

3.3.2 Block Diagram Manipulations

Block diagrams are frequently used to represent sets of algebraic relationships between the variables associated with a dynamical system model. Such diagrams may often be manipulated into simpler forms by using a set of rules which follow directly from the defining vertex relationships. The most important of these manipulations are as follows:

(a) *Simple cascade combination rule.* Consider the simple cascaded arrangement of scaling vertices of Figure 3–15(a). We have

$$x_{j+1} = a_j x_j \qquad j = 1, 2, \ldots, n$$

so that

$$x_{n+1} = \prod_{i=1}^{n} a_i x_i$$

and thus the cascade of scaling vertices of Figure 3–15(a) may be replaced by the single scaling vertex of Figure 3–15(b).

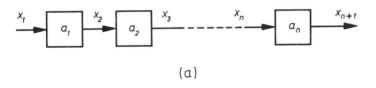

(a)

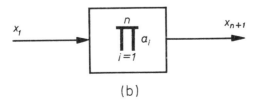

(b)

Figure 3–15

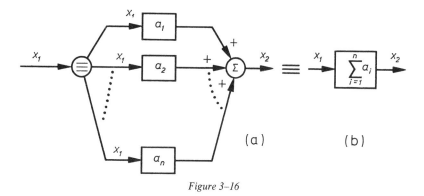

Figure 3–16

(b) *Simple parallel combination rule.* Consider the simple parallel arrangement of Figure 3–16. We have

$$x_2 = a_1 x_1 + a_2 x_1 + \cdots + a_n x_1 = \sum_{i=1}^{n} a_i x_1$$

so that the parallel arrangements of scaling vertices of Figure 3–16(a) may be replaced by the single scaling vertex of Figure 3–16(b).

Cascade block diagram. A cascade block diagram is one in which no consistently-oriented circuits exist. An example of a cascade block diagram is given in Figure 3–17. A consistently-oriented circuit is one in which all the arrows are in the same direction when the circuit is traced out. (Definition on page 166.)

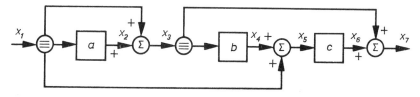

Figure 3–17

(c) *Cascade rule.* The transmittance between any pair of line segments on a cascade block diagram is the sum of all possible separate transmittances between them.

To prove this we first note that if it were true for all variables in the block diagram preceding some particular line segment variable x_i in the cascade, it would also be true for x_i since x_i would be expressed in terms of the preceding variables in the cascade by a relation of the form:

$$x_i = \sum_{j=1}^{i-1} a_{ij} x_j$$

and thus related to the input variables by the extended sum of transmittances through all possible paths leading to x_i. Since the cascade rule is obviously true for the first line segment variable in the cascade, it follows by a simple induction that it must be true for all the block diagram variables.

Applying the rule to Figure 3–17 we have

$$x_7 = [abc + a + c + bc]x_1$$

(d) *Shifting scaling vertex back through summation vertex.* Since

$$a(x_1 + x_2) = ax_1 + ax_2$$

the two block diagrams of Figures 3–18(a) and 3–18(b) are equivalent.

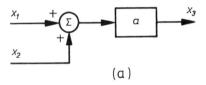

(a)

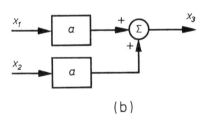

(b)

Figure 3–18

(e) *Shifting scaling vertex forward through summation vertex.* Since

$$ax_1 + x_2 = a\left[x_1 + \frac{1}{a}x_2\right]$$

the two block diagrams of Figure 3–19(a) and 3–19(b) are equivalent.

(f) *Shifting scaling vertex forward through identity vertex.* Inspection shows that the two block diagrams of Figure 3–20(a) and 3–20(b) are equivalent.

(g) *Shifting scaling vertex back through identity vertex.* Inspection shows that the two block diagrams of Figure 3–21(a) and 3–21(b) are equivalent.

(h) *Removal of simple consistently-oriented circuit.* Consider the block diagram of Figure 3–22(a) which has a simple consistently-oriented

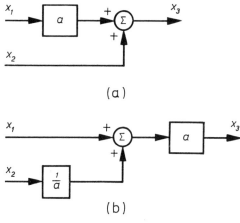

Figure 3–19

circuit, that is one in which all the orientation arrows point in the same direction when the circuit is traced out.

We have
$$e = x + bae$$
so that
$$e(1 - ab) = x$$
and
$$y = ae = \left[\frac{a}{1-ab}\right]x$$

so that the simple consistently-oriented circuit of Figure 3–22(a) can be replaced by the equivalent scaling vertex of Figure 3–22(b).

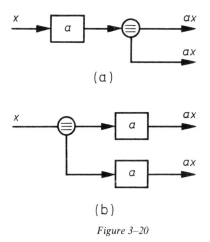

Figure 3–20

Block Diagrams

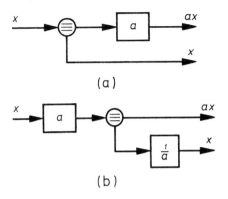

Figure 3–21

Example 3.4

Consider the simple electrical network of Figure 3–23(a), with the positive reference directions for currents and voltages as shown. Denote the Laplace transforms i_1, i_2, etc. by $i_1(s)$, $i_2(s)$, etc. Then we have

$$i_c(s) = i_1(s) - i_2(s)$$

$$v_0(s) = \frac{1}{sC} i_c(s)$$

$$i_2(s) = \frac{v_0(s)}{R_2}$$

$$i_1(s) = \frac{v_i(s) - v_0(s)}{R_1}$$

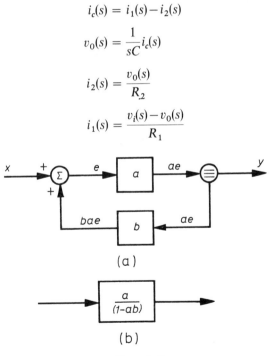

Figure 3–22

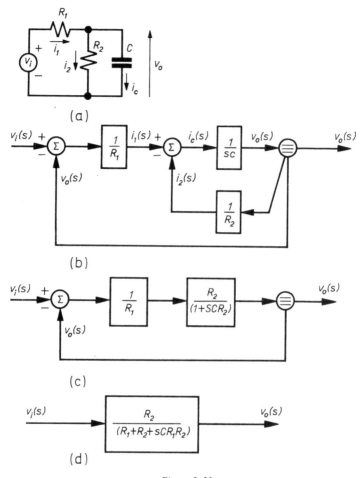

Figure 3–23

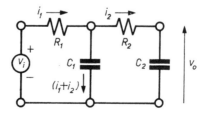

Figure 3–24

and these algebraic relationships may be represented by the block diagram of Figure 3–23(b). An application of the simple block diagram manipulation rules developed above gives, via Figure 3–23(c), the simple transfer function vertex representation of Figure 3–23(d).

A further example of this type of representation and reduction is shown in Figure 3–25 for the network of Figure 3–24.

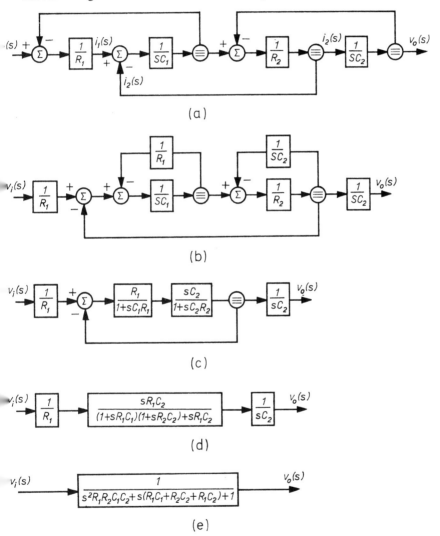

Figure 3–25

3.4 Signal-flow Graphs

Signal-flow graphs. A signal-flow graph is defined as an oriented linear graph on which vertices represent variables and line segments represent certain specific relationships between the vertex variables.

In one sense a signal-flow graph is the dual representation to a block diagram, since the roles played by vertex and line segment are interchanged. There is one extremely important difference between them however, in that on a block diagram every kind of relationship is *explicitly* represented by a vertex while on a signal-flow graph *the relationships of summation and identity are implicit in the diagram by definition*. For this reason the block diagram form of representation is more flexible and general. On the other hand signal-flow graphs are neater and, for linear systems, are associated with a much more powerful set of manipulation and reduction rules. While it is important to realize that the use of signal-flow graphs is in no way prohibited for nonlinear systems, their greatest utility stems from their powerful reduction techniques for linear systems, and so we will present the properties of signal-flow graphs in terms of linear system representation.

3.4.1 Signal-flow Graph Conventions

The line segments of a signal-flow graph will represent coefficients or operators if we are dealing with relationships between Laplace transforms. The coefficients associated with line segments will be called transmittances.

The vertices of a signal-flow graph represent variables.

Summation conventions. A vertex variable is equal to the summation of all terms associated with oriented line segments which are incident on it. For example in the signal-flow graph of Figure 3–26

$$x_4 = ax_1 + bx_2 + cx_3 + dx_4$$

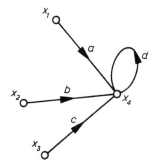

Figure 3–26

Signal-flow Graphs

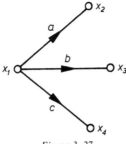

Figure 3-27

Note that the signs required for algebraic combination are taken care of by the signs of the line segment transmittances.

Self-circuit. The line-segment, d, originating from and incident upon vertex x_4 in Figure 3-26 is called a self-circuit (or self-loop).

Identity convention. In a signal-flow graph such as that shown in Figure 3-27 the identity convention is such that

$$x_2 = ax_1 \qquad x_3 = bx_1 \qquad x_4 = cx_1$$

that is if

$$a = b = c = 1$$

then

$$x_1 \equiv x_2 \equiv x_3 \equiv x_4$$

With these conventions a signal-flow graph may be used to represent a set of linear equations. For example the signal-flow graph of Figure 3-28 represents the set of equations

$$x_1 = a_{11}x_1 + a_{12}x_2 + c_1$$
$$x_2 = a_{21}x_1 + a_{22}x_2 + c_2$$

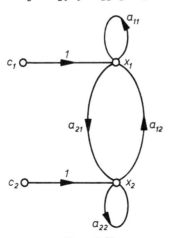

Figure 3-28

3.4.2 Signal-flow Graph Manipulations

Certain simple rules for the manipulation of signal-flow graphs follow directly from the above definition and conventions.

(a) Simple cascade rule. Consider the simple cascade arrangement of Figure 3–29(a). We have

$$x_2 = a_1 x_1$$
$$x_3 = a_2 x_2$$
$$x_4 = a_3 x_3$$
$$\vdots$$
$$x_{n+1} = a_n x_n$$

so that

$$x_{n+1} = \prod_{i=1}^{n} a_i x_1$$

and thus the cascade of line segments in Figure 3–29(a) may be replaced by the single equivalent line segment of Figure 3–29(b).

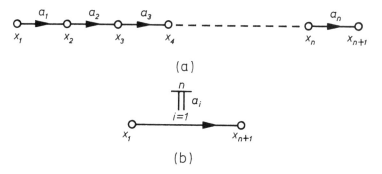

Figure 3–29

(b) Simple parallel rule. Consider the simple parallel arrangement of Figure 3–30(a). We have

$$x_2 = a_1 x_1 + a_2 x_1 + \cdots + a_n x_1 = \sum_{i=1}^{n} a_i x_i$$

and thus the parallel arrangement of line segments in Figure 3–30(a) may be replaced by the single equivalent line segment of Figure 3–30(b). *Cascade flowgraph.* A cascade signal-flow graph is defined as one in which there are no consistently-oriented circuits containing more than one vertex. A circuit is any route traced out through the graph which

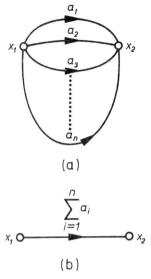

Figure 3-30

returns to its starting point and in which no vertex is encountered more than once. A consistently-oriented circuit is one for which all the orientation arrows point in the same direction when the circuit is traced out. Alternatively, it is one in which no variable is encountered more than once in tracing out any set of oriented line segments in the arrow directions from any one variable to any other. An example of a cascade graph is shown in Figure 3–31(a) and of a noncascade graph in Figure 3–31(b).

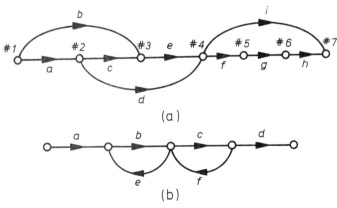

Figure 3-31

(c) *Cascade reduction rule.* The transmittance through a cascade signal-flow graph from an input vertex #1 to an output vertex #n, T_{1n} say, is given by

$$T_{1n} = \sum L_{1n}$$

where ΣL_{1n} is the sum of the transmittances for all possible paths between vertex #1 and vertex #n. For example, for the signal-flow graph of Figure 3–32 we have

$$\begin{aligned} x_4 &= ex_3 + dx_2 \\ &= e(cx_2 + bx_1) + d(ax_1) \\ &= e[c(ax_1) + bx_1] + dax_1 \\ &= (eca + eb + da)x_1 \end{aligned}$$

and for the signal-flow graph of Figure 3–31(a) the transmittance between vertices #1 and #7 is

$$acefgh + befgh + adi + adfgh + bei$$

This cascade rule may be proved by induction in a manner similar to that for the corresponding rule in block diagram reduction.

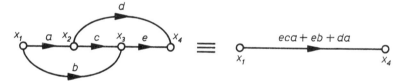

Figure 3–32

(d) *Removal of intermediate vertex.* We have seen above how to manipulate and reduce simple cascade signal-flow graphs. Any given signal-flow graph may be reduced to a cascade graph by a suitable removal of intermediate vertices, from which reverse coefficients originate or which have self-circuits. This process, which is analogous to the removal of unwanted variables in a Gaussian reduction of linear algebraic equations, is best carried out in two stages. We first eliminate self-circuits from any vertex to be removed and then remove the vertex in one further step.

(d)(i) *Elimination of self-circuit on a given vertex.* First consider as an example the removal of the self-circuit on the x_1 vertex of the signal-flow graph of Figure 3–28. For the graph of Figure 3–28 we have

$$x_1 = a_{11}x_1 + a_{12}x_2 + c_1 \qquad (3\text{-}66)$$

$$x_2 = a_{21}x_1 + a_{22}x_2 + c_2 \qquad (3\text{-}67)$$

From equation (3–66) we get

$$x_1 = \frac{a_{12}}{(1-a_{11})}x_2 + \frac{1}{(1-a_{11})}c_1 \qquad (3\text{–}68)$$

Now equations (3–67) and (3–68) correspond to the graph of Figure 3–33, and we see that removal of the self-circuit from vertex x_1 has

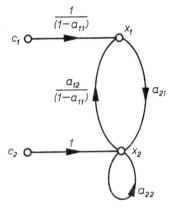

Figure 3–33

resulted in a replacement of all the transmittances of line segments incident on vertex x_1 by their original value divided by one minus the self-circuit transmittance for vertex x_1. To show that this procedure holds in general situation for some vertex x_j for which

$$\sum_{i=1}^{n} a_{ij}x_i = x_j$$

write this out in the form

$$\sum_{i=1}^{j-1} a_{ij}x_i + a_{jj}x_j + \sum_{i=j+1}^{n} a_{ij}x_i = x_j$$

so that

$$\sum_{\substack{i=1 \\ i \neq j}}^{n} a_{ij}x_i = (1-a_{jj})x_j \qquad (3\text{–}69)$$

giving

$$\sum_{\substack{i=1 \\ i \neq j}}^{n} \frac{a_{ij}}{(1-a_{jj})}x_i = x_j$$

We thus have the following:

Rule for removal of self-circuit from a vertex. Remove the self-circuit from the vertex, and replace all transmittances incident on the vertex by their original value divided by one minus the value of the self-circuit transmittance removed.

(d)(ii) Elimination of intermediate vertex. Consider further the example above for the signal-flow graph of Figure 3-33. Multiply equation (3–68) above by a_{21}. This gives

$$a_{21}x_1 = \frac{a_{21}a_{12}}{(1-a_{11})}x_2 + \frac{a_{21}}{(1-a_{11})}c_1 \qquad (3\text{--}70)$$

Subtract equation (3–70) from equation (3–67). This gives

$$a_{22}x_2 + c_2 = x_2 - \frac{a_{21}a_{12}}{(1-a_{11})}x_2 - \frac{a_{21}}{(1-a_{11})}c_1$$

so that

$$\left[a_{22} + \frac{a_{21}a_{12}}{(1-a_{11})}\right]x_2 + c_2 + \frac{a_{21}}{(1-a_{11})}c_1 = x_2 \qquad (3\text{--}71)$$

corresponding to the signal-flow graph of Figure 3-34.

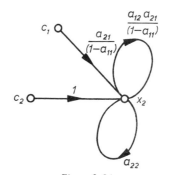

Figure 3-34

Comparing Figure 3-34 with Figure 3-33 shows that, in removing the intermediate vertex x_1, we have replaced all paths through the vertex removed by direct paths between vertices concerned which have the same product transmittances. To show that this procedure holds in the general case we proceed as follows: in the set of relationships for any node x_k

$$\sum_{i=1}^{n} b_{ik}x_i = x_k \qquad (3\text{--}72)$$

we may substitute for x_j from equation (3–69) to get

$$\sum_{\substack{i=1 \\ i \neq j}}^{n} \left[b_{ik} + \frac{a_{ij}b_{jk}}{(1-a_{jj})}\right]x_i = x_k \qquad (3\text{--}73)$$

Signal-flow Graphs

We thus have:
Rule for removal of an intermediate vertex. Remove the vertex and replace every path between vertices through the vertex removed by direct paths, whose transmittances are the product transmittances of the line segments removed, divided by one minus the self-circuit transmittance for the vertex removed.

A simple example of the rule is shown in Figure 3–35.

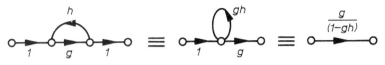

Figure 3–35

3.4.3 Mason's Circuit Rule [L4]

If one is solely concerned with an evaluation of the transmittances between stipulated vertices of a signal-flow graph, the above processes of reduction may be avoided and the required transmittance written down by inspection. This procedure uses the extremely useful result known as Mason's Rule. Before introducing this result we require some further definitions.

Determinant of a signal-flow graph. The determinant, Δ, of a signal-flow graph is defined in terms of the consistently oriented circuits of the graph. *In what follows circuit is to be taken to mean consistently-oriented circuit.*

Self-circuit. A self-circuit is a line segment starting and finishing on the same vertex.

Circuit of m variables. A circuit of m variables is a circuit which passes once through each of m vertices corresponding to m distinct variables.

Disjoint circuits. Two circuits are said to be disjoint when they have no vertices in common.

Circuit transmittance. The circuit transmittance is the product of all the transmittances of the line segments in the circuit.

Circuit product. The circuit product of a set of disjoint circuits is the product of their individual circuit transmittances.

Determinant. In terms of the quantities defined above, the determinant of a signal-flow graph is *defined* to be

$$\Delta = S_1 - \sum_i L_i^1 S_i + \sum_j L_j^2 S_j - \sum_k L_k^3 S_k + \cdots \quad (3\text{–}74)$$

where:

(a) $\{L_i^1\}$ is the set of all circuit transmittances of disjoint circuits of two or more variables.

(b) $\{L_j^2\}$ is the set of all circuit products of pairs of disjoint circuits of two or more variables.
(c) $\{L_k^3\}$ is the set of all circuit products of trios of disjoint circuits of two or more variables.
(d) In general $\{L_m^r\}$ is defined as a straightforward continuation of the definitions (a), (b) and (c) above.
(e) S_1 denotes the product of one minus the graph self-circuit transmittances. That is

$$S_1 = \prod_{v=1}^{l} (1 - a_v)$$

where $a_1, a_2, \ldots, a_l$ are the self-circuit transmittances of the graph.
(f) S_2 denotes the product of one minus the self-circuit transmittances of the self-circuits *disjoint* from the circuits involved in the set $\{L_j^2\}$.
(g) S_3 denotes the product of one minus the self-circuit transmittances of the self-circuits *disjoint* from the circuits involved in the set $\{L_k^3\}$.
(h) In general S_r is defined as a straightforward continuation of the definitions (e), (f) and (g) above.
(i) In the expression (3–74) above for Δ, the summations in each case are over all possible combinations of the circuits involved.

Co-factor. The co-factor of a path through a signal-flow graph is defined as the determinant of the signal-flow graph obtained by removing the path.

Mason's circuit rule. This states that the transmittance T_{ij} between any pair of vertices $^\#i$ and $^\#j$ is given by

$$T_{ij} = \frac{\sum_k t_{ij_k} \Delta_{ij_k}}{\Delta} \qquad (3\text{–}75)$$

where:

(a) t_{ij_k} is the transmittance of the kth direct path (that is not traversing any circuits) from vertex $^\#i$ to vertex $^\#j$.
(b) Δ is the determinant of the signal-flow graph.
(c) Δ_{ij_k} is the co-factor of the kth direct path from vertex $^\#i$ to vertex $^\#j$.

Preliminary considerations relating to proof of circuit rule. Inspection of equation (3–75) shows that the circuit rule gives the correct value of transmittance when applied to the simple case of a single segment. It is shown below that the transmittance between vertices given by the circuit rule is invariant under the removal of an intermediate vertex, that is it gives the same answer for the transmittance when applied to the original signal-flow graph and to the reduced graph obtained by removal of an intermediate vertex. Now the value of the transmittance between a stipulated pair of vertices is unique. Thus any quantity associated with a pair of vertices of the graph which is invariant under the removal of intervening vertices, and which assumes the correct value for the case of the single line segment obtained at the end of the reduction process, must be the transmittance between vertices of the graph. This is the basis of the proof of the circuit rule which follows.

Change in Determinant Consequent on Removal of an Intermediate Vertex

We first examine what effect the removal of an intermediate vertex has on the value of a signal-flow graph determinant. Consider some internal portion of the signal-flow graph from which some intermediate vertex, x say, is to be removed, as illustrated in Figure 3–36. From the rules developed above for signal-flow graph reduction we know that the removal of a self-circuit of transmittance a from the internal vertex x results in the transmittances of all line segments incident upon vertex x being multiplied by $1/(1-a)$ and leaves the transmittances of all line segments directed away from the vertex x unaltered. An inspection of Figure 3–36 thus shows that the effect of removing an internal vertex of self-circuit transmittance a is that all circuits of m variables through the vertex x are replaced by circuits of $(m-1)$ variables, and the corresponding circuit transmittances are multiplied by $1/(1-a)$. Let the determinant of the original signal-flow graph be

$$\Delta = S_1 - \sum_i L_i^1 S_i + \sum_j L_j^2 S_j - \sum_k L_k^3 S_k + \cdots$$

and the determinant of the reduced signal-flow graph from which an internal vertex of self-circuit transmittance a has been removed

$$\Delta' = S_1' - \sum_i (L_i^1)' S_i' + \sum_j (L_j^2)' S_j' - \sum_k (L_k^3)' S_k' + \cdots$$

Considering the terms of Δ' in turn we have that

(a) $$S_1' = \frac{S_1}{(1-a)}$$

since the term contributed by the self-circuit of transmittance a has been removed from the product of one minus the self-circuits.

(b) $$\sum_i (L_i^1)' S_i' = \frac{1}{(1-a)} \sum_i L_i^1 S_i$$

Each circuit involved in the summation either did or did not originally pass through the vertex x. If it did pass through it, then the appropriate member of the set $\{(L_i^1)'\}$ is the corresponding member of

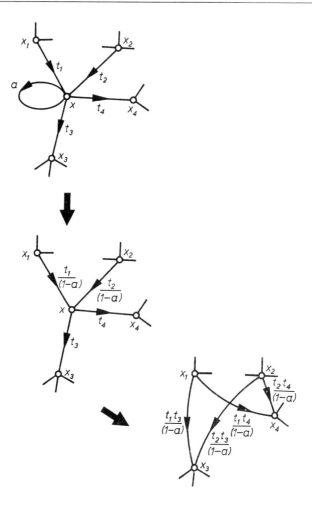

Figure 3–36

the set $\{L_i^1\}$ multiplied by $1/(1-a)$ since circuits of m variables through vertex x are replaced by circuits of $(m-1)$ variables with the corresponding circuit transmittances multiplied by $1/(1-a)$.

If it did not pass through vertex x, then the appropriate member of the set $\{S_i'\}$ is the corresponding member of the set $\{S_i\}$ multiplied by $1/(1-a)$ since a disjoint self-circuit of loop transmittance a has been removed from the product of one minus the self-circuit transmittances of the self-circuits disjoint from the circuits involved in the set $\{L_i^1\}$.

(c)
$$\sum_j (L_j^2)' S_j = \frac{1}{(1-a)} \sum_j L_j^2 S_j$$

$$\sum_k (L_k^3)' S_k = \frac{1}{(1-a)} \sum_k L_k^3 S_k$$

by an argument exactly similar to that in (b) above. This shows that

$$\Delta' = \frac{\Delta}{(1-a)} \qquad (3\text{-}76)$$

so that the removal of an internal vertex having a self-circuit transmittance a from a given signal-flow graph of determinant Δ gives a reduced signal flow graph whose determinant is $\Delta/(1-a)$.

In order to use this result to establish the circuit rule by the induction outlined above our next step is to construct a signal-flow graph whose determinant, Δ_n, is equal to the numerator term in the circuit rule, that is such that

$$\Delta_n = \sum_k t_{ij_k} \Delta_{ij_k}$$

To do this we first note that any signal-flow graph determinant may be expanded in terms of the circuit transmittances and the co-factors of a set of circuits as

$$\Delta = \Delta_0 - \sum_k L_k \Delta_k \qquad (3\text{-}77)$$

where:

(a) Δ_0 is the determinant of the signal-flow graph obtained by removing a set of k circuits of circuit transmittances.

(b) $\Delta_1, \Delta_2, \ldots, \Delta_k$ are the co-factors of the corresponding circuits, that is the values of the determinants for the signal-flow graphs obtained by removing these circuits one at a time.

Suppose the signal-flow graph under consideration is of the form shown in Figure 3–37(a). Now consider the signal-flow graph shown in Figure 3–37(b) obtained by adding to that of Figure 3–37(a) a line segment of transmittance -1 from output vertex to input vertex, and a self-circuit of transmittance 1 on the output vertex. The line segment of transmittance -1 turns the k forward paths from vertex $^\#i$ to vertex $^\#j$ having transmittances t_{ij_k} into consistently-oriented circuits of circuit transmittance $-t_{ij_k}$. The self-circuit of transmittance 1 on the output vertex ensures that only circuits passing through the output vertex contribute to the value of the determinant of the signal-flow graph of Figure 3–37(b), since all those terms in the determinant which correspond

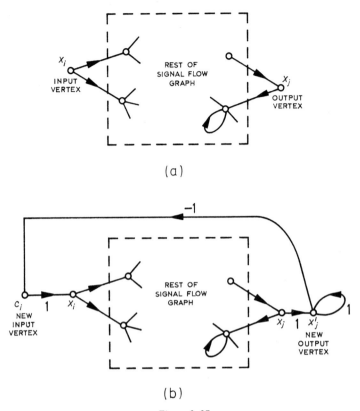

Figure 3-37

to all other circuits will be multiplied by the zero factor of one minus a unity value of self-circuit transmittance. From these two facts, it immediately follows that if we expand the determinant of the graph of Figure 3-37(b) in terms of the circuits formed by the forward paths from input to output vertex and the line segment of transmittance -1 from output to input vertex, that

$$\Delta_n = -\sum_k (-t_{ij_k})\Delta_{ij_k} = \sum_k t_{ij_k}\Delta_{ij_k}$$

since

(a) the product of one minus the self-circuits vanishes,
(b) no other circuits contributes to Δ_n,
(c) the co-factors of the circuits involved in the graph of Figure 3-37(b) are the same as those of the forward paths from input vertex to output vertex on Figure 3-37(a).

Signal-flow Graphs

We may thus write the transmittance between a pair of input and output vertices, $^\#i$ and $^\#j$, of some given signal-flow graph as

$$T_{ij} = \frac{\Delta_n}{\Delta}$$

where:

(a) Δ is the determinant of the original signal-flow graph,
(b) Δ_n is the determinant of the signal-flow graph obtained by standing off the input and output vertices through line segments of unity gain and adding a self-circuit of unity transmittance to the new output vertex and a line segment of transmittance -1 from the new output vertex to the new input vertex. Having thus prepared the ground we may present the proof of the circuit rule.

Proof of circuit rule. For any given signal-flow graph express the circuit rule transmittance for a specified vertex pair i and j, by the means described above, as the ratio of two signal-flow graph determinants

$$T_{ij} = \frac{\Delta_n}{\Delta}$$

Now remove the same intermediate node from both signal-flow graphs, and re-compute the circuit rule transmittance. This will give

$$T'_{ij} = \frac{\Delta'_n}{\Delta'}$$

where Δ' and Δ'_n are the determinants of the reduced signal-flow graphs obtained. Now

$$\Delta'_n = \frac{\Delta_n}{(1-a)}$$

and

$$\Delta' = \frac{\Delta}{(1-a)}$$

so that

$$T'_{ij} = \frac{\Delta_n/(1-a)}{\Delta/(1-a)} = \frac{\Delta_n}{\Delta} = T_{ij}$$

and thus the value of transmittance between a pair of vertices obtained by application of the circuit rule is invariant under the removal of an intermediate vertex, that is, gives the same value when applied to the reduced graph as would be obtained if it were applied to the original graph.

Now the value of the transmittance between vertices of a signal-flow graph is unique, and the circuit rule manifestly gives the correct value for the case of a single line segment connecting input and output vertices. Since the circuit rule transmittance is also invariant under the removal of an intermediate vertex, and since we can always reduce any given graph to a single line segment between any stipulated pair of vertices, it follows that the circuit rule must give the correct value for the transmittance between vertices in all cases.

3.5 Nyquist's Stability Criterion

Nyquist's criterion, first published in 1932, has a unique position among stability criteria by virtue of its direct interpretation in terms of practical measurements. Its proof rests on certain properties of functions of a complex variable [N1].

3.5.1 Complex Plane Mappings

A single-valued function $\Phi(s)$ of a complex variable s is said to be analytic at a point $s = a$ if and only if it is differentiable throughout a neighbourhood of $s = a$. The function $\Phi(s)$ maps points of the s-plane into the corresponding points of a $\Phi(s)$-plane. At every point where $\Phi(s)$ is analytic and does not vanish, the mapping is *conformal*, that is the angle between two curves through such a point in the s-plane is reproduced in magnitude and sense between the mapped curves in the $\Phi(s)$-plane. Suppose C is a simple closed curve in the s-plane, and let $\Phi(s)$ be a rational function of s which has neither zeros nor poles on C. Let $\Phi(s)$ map C into the curve Γ on the $\Phi(s)$-plane. Then as the point s is traced clockwise round the curve C in the s-plane, the corresponding point $\Phi(s)$ in the $\Phi(s)$-plane follows a closed curve Γ and encircles the origin of the $\Phi(s)$ plane $(Z - P)$ times in the clockwise direction, where Z and P are the numbers of zeros and poles of $\Phi(s)$ inside C, taking proper account of multiplicities.

The practical application of Nyquist's criterion depends on the fact that the mapping of the positive imaginary axis of the s-plane under a transfer function $\Phi(s)$ may be directly measured (as shown in Section 3.2.3 above) by means of steady state sinusoidal response determination.

Thus the positive imaginary axis mapping $\Phi(j\omega)$ may be determined by direct measurements on a system, and plotted out. The complex function $\Phi(j\omega)$ will be termed the *frequency response characteristic* for the specified input and output points. A plot of its modulus versus ω will be termed the amplitude response characteristic and a plot of its phase versus ω will be termed the phase shift characteristic.

For practical systems

$$|\Phi(j\omega)| \to 0 \quad \text{as} \quad \omega \to \infty$$

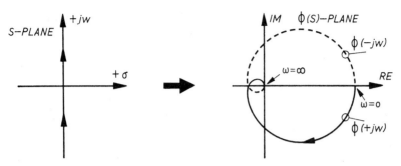

Figure 3-38

and $\Phi(-j\omega)$ = complex conjugate of $\Phi(j\omega)$

and so the whole imaginary axis is mapped into a symmetrical closed curve in the $\Phi(s)$-plane if there are no singularities on the imaginary axis or at the origin, as shown in Figure 3-38. The closed curve in the s-plane shown in Figure 3-39 consisting of that part of the imaginary axis corresponding to all frequencies of interest in an investigation plus a large semi-circle in the right-half of the s-plane is called *Nyquist's contour*.

For the transfer function of a practical system, it will map into a closed curve in the $\Phi(s)$-plane as shown in Figure 3-39. It immediately follows from the conformal nature of the mapping that the region of the right-half of the s-plane enclosed by Nyquist's contour will map into the *inside* of the corresponding closed curve in the $\Phi(s)$-plane. By letting the semi-circular part of the Nyquist contour tend to be of infinite radius we see that for a stable system transfer function:

(a) the whole imaginary axis of the s-plane maps into a closed curve in the $\Phi(s)$-plane,

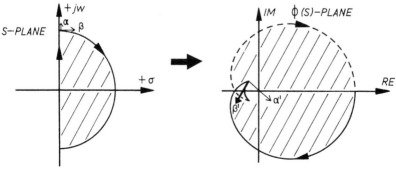

Figure 3-39

(b) as large a portion as required of the right-half of the s-plane may be mapped into the inside of this closed curve.

In considering the mapping of Nyquist's contour we must give special consideration to those cases where they are singularities of $\Phi(s)$ at the origin, or on the imaginary axis of the s-plane. If there is a singularity at the origin then the Nyquist contour may be indented to pass it round an arbitrarily small semi-circle into the right-half plane. Simple conformal mapping arguments then show that the $\Phi(s)$-plane map of $\Phi(j\omega)$ is then closed by an arbitrarily large semi-circle in the right-half $\Phi(s)$-plane as shown in Figure 3–40. If there is a singularity on the

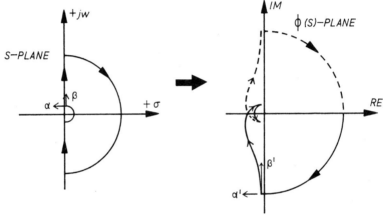

Figure 3–40

imaginary axis similar arguments show that it should again be avoided by a small semi-circular indentation into the right-half of the s-plane. Such small semi-circles map into approximately circular arcs in the $\Phi(s)$-plane as shown in Figure 3–41.

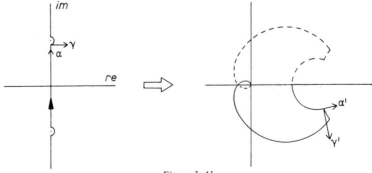

Figure 3–41

3.5.2 Open- and Closed-loop Transfer Function Relationships

A large class of linear autonomous feedback-controlled systems may be represented by the form of block diagram shown in Figure 3–42. If the reference input signal and the transduced output signal variables

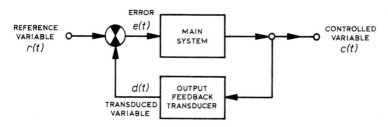

Figure 3–42. The left-hand vertex symbol is a differencing unit.

are denoted by $r(t)$ and $d(t)$ respectively then the input to the main system is

$$e(t) = r(t) - d(t)$$

which is defined as the system error signal. Let

$R(s)$ = Laplace transform of reference input signal
$C(s)$ = Laplace transform of controlled output signal
$E(s)$ = Laplace transform of error signal
$G(s)$ = Transfer function of main system
$H(s)$ = Transfer function of output feedback transducer

The relationships between the signal transforms and transfer functions is then as shown by the signal-flow graph of Figure 3–43.

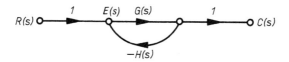

Figure 3–43

Thus
$$C(s) = G(s)E(s)$$
$$E(s) = R(s) - H(s)C(s)$$
whence
$$C(s) = G(s)[R(s) - H(s)C(s)]$$
$$= G(s)R(s) - G(s)H(s)C(s)$$

which gives the *closed-loop transfer function* as

$$\frac{C(s)}{R(s)} = \frac{G(s)}{[1+G(s)H(s)]} \tag{3-78}$$

Also

$$C(s) = \frac{G(s)}{[1+G(s)H(s)]} R(s) \tag{3-79}$$

and

$$E(s) = \frac{C(s)}{G(s)} = \frac{1}{[1+G(s)H(s)]} R(s) \tag{3-80}$$

The function $G(s)H(s)$ will be called the system *open-loop transfer function*; it corresponds to the input and output response pair shown in Figure 3–44. It is sometimes convenient to call $G(s)$ the *opened-loop* transfer function.

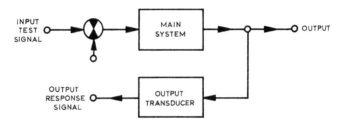

Figure 3–44

3.5.3 Basic Closed-loop Stability Theorem

Let $G(s)$ and $H(s)$ both be rational functions of s, and let the product $G(s)H(s)$ be such that the only cancellations of factors, if any, involve poles and zeros in the left-half of the s-plane. Then the closed-loop transfer function of (3–78) above will be stable if and only if $[1+G(s)H(s)]$ has no zeros in the right-half s-splane.

The proof is straightforward.

Suppose there was an s_1 in the right-half s-plane such that

$$1 + G(s_1)H(s_1) = 0$$

Since no zero of $G(s)$ in the right-half plane has been removed by cancellation, this implies that

$$G(s_1) \neq 0$$

and thus the closed-loop transfer function

$$\frac{G(s)}{1+G(s)H(s)}$$

has a pole at $s = s_1$ in the right-half s-plane and the closed-loop system is therefore unstable. It is therefore sufficient that $[1 + G(s)H(s)]$ has no zeros in the right-half plane. The closed-loop transfer function could only have a pole at $s = s_1$

if   (a) $1 + G(s_1)H(s_1) = 0$ with $G(s_1) \neq 0$

or if (b) $G(s)$ had a pole of order m at $s = s_1$ and $1 + G(s)H(s)$

had a pole of order $< m$ at $s = s_1$.

Possibility (a) has already been considered above. Possibility (b) would imply a cancellation of factors of the form $(s - s_1)$ in the product $G(s)H(s)$ which has been excluded by the assumptions of the theorem. It is thus necessary for stability that $[1 + G(s)H(s)]$ have no zeros in the right-half plane.

Hence the closed loop transfer function (CLTF) is stable if and only if $[1 + G(s)H(s)]$ has no zeros in the right-half of the s-plane.

It is most important to realise that we have *not* assumed either $G(s)$ or $H(s)$ to be stable in the above argument. To show that a system can be opened-loop unstable and closed-loop stable we have only to put

$$G(s) = \frac{K}{s-1} \text{ (which is unstable)}$$

and $H(s) = 1$

for which $\dfrac{C(s)}{R(s)} = \dfrac{K}{s + (K - 1)}$

which is stable for $K > 1$.

3.5.4 Simple Form of Nyquist's Criterion

Consider the mapping of Nyquist's contour traversed clockwise, with suitable indentations where necessary, under

$$\Phi(s) = G(s)H(s)$$

This will give a closed curve encircling the origin of the $\Phi(s)$-plane $(Z - P)$ times in the clockwise direction, where Z and P are the number of zeros and poles of $G(s)H(s)$ in the right-half plane enclosed by the contour. In what follows we assume that the radius of the semi-circular portion of the contour is taken sufficiently large to include all the right-half plane zeros and poles of $G(s)H(s)$.

First consider the case where $G(s)H(s)$ has neither zeros nor poles in the right-half plane. Then we get the Nyquist contour mapped into a closed curve which does not encircle the origin of the $\Phi(s)$-plane.

190   *Transform Models*

Simple consideration arising from the conformal nature of the mapping show that the region of the right-half s-plane enclosed by the Nyquist contour is unambiguously mapped into the interior of the resulting mapped curve in the $\Phi(s)$-plane. By letting the radius of the semicircular part of the Nyquist contour grow arbitrarily large we can say that the whole right-half s-plane maps into the inside of the closed-curve in the $\Phi(s)$-plane. A simple way of looking at this is to imagine two people walking clockwise around the Nyquist contour in the s-plane and its map in the $\Phi(s)$-plane, with their right hands outstretched normal to the direction of travel. Then the region indicated by the first walker's right hand is being mapped into that indicated by the second walker's right hand.

On further considering the case when $G(s)H(s)$ has right-half plane zeros but no right-half plane poles it will be seen that an unambiguous mapping of the right-half s-plane into the interior of a closed curve in the $\Phi(s)$-plane is still achieved. However, when $G(s)H(s)$ has right-half plane poles it will be found that this is no longer the case, due to the reversal of direction of origin encirclement. We therefore consider two special cases of Nyquist's criterion:

(a)   the simple case where $G(s)H(s)$ has no right-half plane poles,
(b)   the general case.

Simple Form of Nyquist's Criterion

If $G(s)H(s)$ has no right-half plane poles the function $[1+G(s)H(s)]$ will have no right-half plane zeros if the map of Nyquist's contour (with suitably large semi-circular radius) under $G(s)H(s)$ does not enclose the point $(-1, 0)$. Thus the closed-loop system of Figure 3–43 will be stable if this map does not enclose $(-1, 0)$.

The proof is immediate, given the basic closed-loop stability theorem, since any s_1 in the right-half s-plane which is a zero of $[1+G(s)H(s)]$ must map into $(-1, 0)$ under $G(s)H(s)$. We can thus only have a right-half plane zero of $[1+G(s)H(s)]$ if the Nyquist contour mapped under $G(s)H(s)$ encloses the point $(-1, 0)$. The point $(-1, 0)$ is called the *critical point*.

The great importance of the Nyquist criterion arises from the fact that $G(j\omega)H(j\omega)$ can be directly plotted by steady-state sinusoidal measurements to give the imaginary axis map known as a *Nyquist diagram*. We can therefore infer closed-loop stability from open-loop experimental measurements. More important still we have a clear indication of how to modify the frequency response characteristics of the open-loop in order to achieve or improve stability. In practice it is only necessary to construct the sinusoidal vector response locus $G(j\omega)H(j\omega)$ over a relatively limited range of frequencies. The Nyquist stability criterion for a feedback control system which is open-loop

Nyquist's Stability Criterion

stable may then be stated in the simplified form of the following Left Hand Rule.

The left-hand rule. If the plot of open-loop sinusoidal vector response is traced out from low to high frequencies the closed-loop system of Figure 3–43 will be stable if the critical point $(-1, 0)$ lies to the left of all points of $G(j\omega)H(j\omega)$. If the plot passes through the critical point, or if the critical point lies to the right of $G(j\omega)H(j\omega)$, the closed-loop system will be unstable.

3.5.5 General Form of Nyquist's Criterion

If $G(s)H(s)$ has any right-half plane poles the open-loop system is unstable and practical measurements of open-loop steady state sinusoidal response become impossible. It is of interest to note however that the general form of Nyquist's criterion admits the case where $G(s)H(s)$ has right-half plane poles. To establish the general form we use the following theorem:

If $\Phi(s)$ has poles as its only singularities inside a closed contour C, and has no poles or zeros on C, then

$$\frac{1}{2\pi j}\int_C \frac{\Phi'(s)}{\Phi(s)}\,ds = P - Z \qquad (3\text{–}81)$$

where Z is the number of zeros and P is the number of poles of $\Phi(s)$ within C.

Here $\Phi'(s) \equiv d\Phi(s)/ds$ and C is traversed in the clockwise direction.

Let C be the Nyquist contour and let

$$\Phi(s) = 1 + G(s)H(s) = r\exp(j\theta)$$

Then

$$\log \Phi = \log r + j\theta$$

$$d(\log \Phi) = d(\log r) + j\,d\theta$$

Now

$$\frac{\Phi'(s)}{\Phi(s)} = \frac{d(\log \Phi)}{ds}$$

Hence equation (3–81) gives

$$\frac{1}{2\pi j}\int_C d(\log \Phi) = \frac{1}{2\pi j}\int_C d(\log r) + \frac{1}{2\pi}\int_C d\theta = P - Z$$

Since C is a closed contour, the first of the component integrals vanishes giving

$$N = \frac{1}{2\pi} \int_C d\theta = P - Z$$

where N is the number of times the map of the contour C by $\Phi(s)$ goes round the origin of the $\Phi(s)$-plane in the *clockwise* direction.

Thus the number of zeros of $1 + G(s)H(s)$ in the right-half plane is

$$Z = P - N$$

where P is the number of poles of $[1 + G(s)H(s)]$ in the right-half plane and N is the number above.

Alternatively since $[1 + G(s)H(s)]$ must have the same poles as $G(s)H(s)$ and as the map of $G(s)H(s)$ is the map of $[1 + G(s)H(s)]$ with the origin shifted to the critical point we have:

General Form of Nyquist Stability Criterion

The function $[1 + G(s)H(s)]$ will have no right-half plane zeros and the closed-loop system of Figure 3–43 will thus be stable if the map of a suitable Nyquist contour by $G(s)H(s)$ encircles the critical point P times in the clockwise direction where P is the number of right-half plane poles of $G(s)H(s)$.

3.5.6 Relative Stability Criteria

A consideration of the mapping of lines parallel to the imaginary axis of the s-plane under $G(s)H(s)$ will show that the closed-loop performance of a feedback control system will be critically dependent on the nearness of approach of the open-loop frequency response locus to the critical point. Two parameters have been introduced to specify the behaviour of the open-loop frequency response locus in the vicinity of the critical point, and thus to give a measure of the *relative stability* of the closed-loop system

(a) *Phase margin.* The phase margin is the angle which the line joining the origin to the point on the open-loop response locus corresponding to unit modulus of gain makes with the negative real axis. It is thus a measure of the additional phase lag allowable in the open-loop transmission path before the closed-loop system becomes unstable.

(b) *Gain margin.* If the modulus of the sinusoidal steady-state open-loop gain of the system is X for a phase shift of 180°, the gain margin is defined as

$$\text{gain margin} = \frac{1}{X}$$

The gain margin is usually specified in decibels where

$$\text{decibel gain margin} = 20 \log_{10} \frac{1}{X}$$

The decibel gain margin is thus positive for a closed-loop-stable system and negative for a closed-loop-unstable system.

The definitions of gain and phase margin are illustrated by Figure 3-45. As a rough working rule, a reasonable value of closed-loop damping is given by a phase margin of about 40° and a gain margin of at least 6 dB.

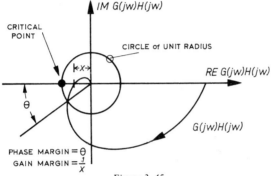

Figure 3-45

Conditionally Stable Systems

A feedback control system is said to be unconditionally stable over a specified region of variation of system gain if it remains stable for all increased and reduced values of gain in the specified range. If a system is closed-loop stable for a given value of open-loop gain, but unstable for closed-loop operation when the value of the opened-loop gain is reduced, it is said to be conditionally stable. The form of the vector plot of $\Phi(+j\omega)$ for a conditionally stable system is shown in Figure 3-46.

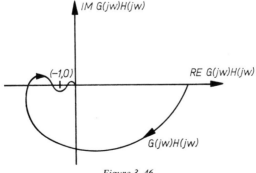

Figure 3-46

The system is stable as the Left Hand Rule is satisfied, but will become unstable if the static value of opened-loop gain is either increased or reduced without changing the shape of the frequency response characteristic.

3.6 Root Locus Method

The root locus method is essentially a means of determining graphically the loci in the complex plane of the roots of the characteristic equation of a linear feedback control system as the system gain is varied. Consider the class of autonomous, linear, feedback control systems for which the relationship between the Laplace transforms of input and output variables may be represented by the transform signal-flow graph of Figure 3–43. Let $R(s)$ and $C(s)$ be the Laplace transforms of the reference input and controlled output variables respectively. Let $G(s)$ be the transfer function of the main system, and let $H(s)$ be the transfer function of the feedback path. The relationships among the transformed variables then gives that

$$[1 + G(s)H(s)]C(s) = G(s)R(s) \tag{3-82}$$

and the characteristic equation of the closed-loop system is therefore

$$1 + G(s)H(s) = 0 \tag{3-83}$$

Let
$$G(s)H(s) = \frac{kN(s)}{D(s)} \tag{3-84}$$

where k is a real constant and $N(s)$ and $D(s)$ are the monic polynomials

$$N(s) = s^m + n_1 s^{m-1} + n_2 s^{m-2} + \cdots + n_{m-1} s + n_m$$
$$D(s) = s^n + d_1 s^{n-1} + d_2 s^{n-2} + \cdots + d_{n-1} s + d_n$$

(A monic polynomial in s is one having unity coefficient for the highest power in s.) We can now write the closed-loop system characteristic equation as

$$1 + \frac{kN(s)}{D(s)} = 0$$

If this is put in the form

$$\frac{N(s)}{D(s)} = \frac{-1}{k} \tag{3-85}$$

then it follows that for a complex number s_r to be a root of the characteristic equation, the complex number $N(s_r)/(D(s_r))$ must be such that

(a) $$\text{modulus} \frac{N(s_r)}{D(s_r)} = \frac{1}{k} \tag{3-86}$$

(b) $$\text{phase} \frac{N(s_r)}{D(s_r)} = 180° + l360° \tag{3-87}$$

where l is zero or integral.

Let $z_1, z_2, \ldots, z_m$ be the roots of the equation

$$N(s) = 0$$

and let $p_1, p_2, \ldots, p_n$ be the roots of the equation

$$D(s) = 0$$

The polynomials $N(s)$ and $D(s)$ may therefore be written in factored form as

$$N(s) = \prod_{j=1}^{m} (s - z_j) \qquad D(s) = \prod_{i=1}^{n} (s - p_i)$$

The numbers $z_1, z_2, \ldots, z_m$ are the zeros of the open-loop transfer function, and the numbers $p_1, p_2, \ldots, p_n$ are the poles of the open-loop transfer function. Thus, if the complex number s_r is a root of the closed-loop system characteristic equation we must have a *modulus* condition

$$\frac{\prod_{j=1}^{m} |s_r - z_j|}{\prod_{i=1}^{n} |s_r - p_i|} = \frac{1}{k} \tag{3-88}$$

and an *angle condition*

$$\sum_{j=1}^{m} \text{phase}(s_r - z_j) - \sum_{i=1}^{n} \text{phase}(s_r - p_i) = 180° + l360° \tag{3-89}$$

For any given complex number s_k, the determination of the quantities

$$\text{modulus}(s - s_k) \quad \text{and} \quad \text{phase}(s - s_k)$$

for any other complex number s may be carried out by means of the simple measurements on the complex plane shown in Figure 3–47. We may therefore use simple graphical measurements in the complex plane to determine the totality of points s_r in the complex plane which satisfy the modulus and angle conditions given above for a given set of open-loop system poles and zeros. The complete set of such points constitutes the root locus plot.

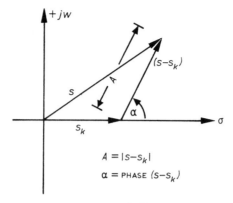

Figure 3–47

The root locus is constructed by determining all the points in the complex plane which satisfy the angle condition. It is then calibrated in k by using the modulus condition. As k is directly proportional to system static gain, the root locus procedure is an efficient method for plotting the variation of the roots of the closed-loop system characteristic equation with variation in gain. After a few initial points have been determined by a trial-and-error procedure, the locus may be quickly constructed on a drawing board with sufficient accuracy for most practical purposes. In addition, once experience of the method has been gained, it is invaluable for sketching out root variations quickly and thus gaining a rapid qualitative insight into the behaviour of a projected feedback control system. Before deriving a set of rules for the construction of root locus diagrams it will be helpful to consider a simple example.

Example 3.5

Consider the root locus for

$$H(s)G(s) = \frac{k}{(s+a)(s+b)(s+c)}$$

Let

$(s+a)$ have modulus A and phase α

$(s+b)$ have modulus B and phase β

$(s+c)$ have modulus C and phase γ

Then the points on the root locus must be such that

$$\alpha + \beta + \gamma = 180° + l360°$$

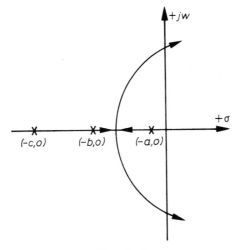

Figure 3-48

The set of points which satisfies this angle condition is shown in Figure 3-48. After construction, the locus may be calibrated in k using the modulus condition

$$k = ABC$$

3.6.1 General Rules for Construction of Root Loci

In order to construct and sketch root locus diagrams satisfactorily, we need to know the behaviour of the root locus for the limiting conditions $k = 0$ and $k = \infty$. Taken together with the modulus and angle conditions, and some further rules for the real axis, this gives a set of rules which enables the complete form of the root locus diagram to be drawn.

Rule 1. For $k = 0$ the root loci start at the poles of the opened-loop transfer function.

If the closed-loop system characteristic equation is written in the form

$$D(s) + kN(s) = 0$$

this follows immediately, since the roots of

$$D(s) = 0$$

are the poles of the opened-loop system transfer function. If the opened-loop system transfer function has n distinct poles, it follows that the root locus has n distinct branches.

Rule 2. The set of points s_r in the complex plane which constitutes the root locus plot is such that

$$\sum_{j=1}^{m} \text{phase}\,(s_r - z_j) - \sum_{i=1}^{n} \text{phase}\,(s_r - p_i) = 180° + l360°$$

where l is any integer.

Rule 3. After construction of the root locus in a region of the complex plane, the locus may be calibrated in k by use of the modulus relationship

$$k = \frac{\prod_{i=1}^{n} |(s_r - p_i)|}{\prod_{j=1}^{m} |(s_r - z_j)|}$$

Rule 4. As $k \to \infty$, m branches of the root locus terminate on the m zeros of the opened-loop system transfer function, and the remaining $(n-m)$ branches tend to infinity.

If the closed-loop system characteristic equation is written in the form

$$N(s) + \frac{D(s)}{k} = 0$$

then we see that, as $k \to \infty$ for s remaining finite, the characteristic equation becomes

$$N(s) = 0$$

having the m opened-loop system zeros as roots. Hence m branches of the root locus tend to the m opened-loop zeros as $k \to \infty$. If we write the closed-loop system characteristic equation in the form

$$\frac{D(s)}{N(s)} + k = 0$$

we see that, as $s \to \infty$ the characteristic equation tends to the form

$$Q(s) + k = 0$$

where $Q(s)$ is a polynomial of order $(n-m)$. Thus, as $s \to \infty$ and $k \to \infty$, we see that $(n-m)$ of the roots must tend to infinity.

Rule 5. As $s \to \infty$ $(n-m)$ of the branches approach asymptotically a set of straight lines making angles with the positive real axis of

$$\psi = \frac{(1+2\lambda)\pi}{(m-n)} \quad \text{where } \lambda \text{ is any integer.}$$

Root Locus Method

As s_r approaches a point on a circle of very large radius, the angle of each of the vectors $(s-z_j)$ and $(s-p_i)$ must approach a common value ψ. Using Rule 2, we therefore have that

$$(m-n)\psi = (1+2\lambda)\pi$$

where λ is integer; giving

$$\psi = \frac{(1+2\lambda)\pi}{(m-n)} \qquad (3\text{-}90)$$

Rule 6. The asymptotes of Rule 5 all emanate from the common point

$$s_g = \frac{\sum_{i=1}^{n} p_i - \sum_{j=1}^{m} z_j}{(n-m)} \qquad \begin{array}{l} n \ne m \\ n \ne m+1 \end{array} \qquad (3\text{-}91)$$

Suppose $n \ne m$ and $n \ne (m+1)$. Then two or more asymptotes will be involved since $n \ge m$ for a physically realizable system. If one asymptote is involved it must be the real axis (see also Rule 7 below). Considerations of symmetry arising from the fact that the roots are always complex conjugate show that the asymptotes must be symmetrically disposed above and below the real axis and intersect in a common point on the real axis. Let this common point be $s = s_g$. Then for a point s_r sufficiently far from the origin in the complex plane we will have, using Rule 2,

$$(n-m)\,\text{phase}\,(s_r-s_g) = \sum_{j=1}^{m} \text{phase}\,(s_r-z_j) - \sum_{i=1}^{n} \text{phase}\,(s_r-p_i)$$

Furthermore, since from an infinite distance all the poles and zeros will appear to be clustered arbitrarily close together, all the moduli of the numbers (s_r-s_g), (s_r-z_j) and (s_r-p_i) will be the same. Thus we will have

$$(n-m)(s_r-s_g) = \sum_{j=1}^{m}(s_r-z_j) - \sum_{i=1}^{n}(s_r-p_i)$$

from which we obtain

$$s_g = \frac{\sum_{i=1}^{n} p_i - \sum_{j=1}^{m} z_j}{(n-m)}$$

Rule 7. For $k > 0$, those points on the real axis which lie to the left of an odd number of poles and zeros are on the locus of roots.
This follows directly from Rule 2.

Rule 8. A point γ at which the root locus breaks away from the real axis is such that

$$\sum_{i=1}^{n} \frac{\gamma + \operatorname{Re}(p_i)}{[\gamma + \operatorname{Re}(p_i)]^2 + [\operatorname{Im}(p_i)]^2} - \sum_{j=1}^{m} \frac{\gamma + \operatorname{Re}(z_j)}{[\gamma + \operatorname{Re}(z_j)]^2 + [\operatorname{Im}(z_j)]^2} \quad (3\text{-}92)$$

where $\operatorname{Re}(s)$ and $\operatorname{Im}(s)$ denote the real and imaginary parts of a complex number s.

Suppose $s_r = -\gamma + j\delta$ is just above a breakaway point on the real axis as shown in Figure 3-49. At s_r we have

$$\text{phase } G(s_r)H(s_r) = \sum_{j=1}^{m} \arctan \left[\frac{\delta - \operatorname{Im}(z_j)}{-[\gamma + \operatorname{Re}(z_j)]} \right]$$

$$- \sum_{i=1}^{n} \arctan \left[\frac{\delta - \operatorname{Im}(p_i)}{-[\gamma + \operatorname{Re}(p_i)]} \right] \quad (3\text{-}92a)$$

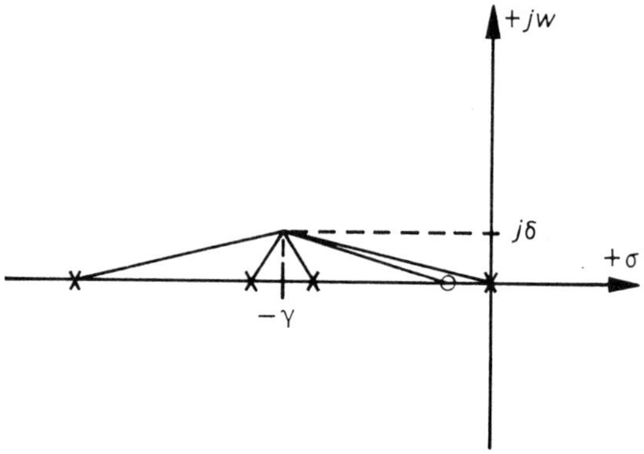

Figure 3-49

The two branches of the root locus which meet at the breakaway point must be orthogonal to the real axis at this point. It follows that

$$\lim_{\delta \to 0} \frac{d}{d\delta} \text{phase } G(s_r)H(s_r) = 0$$

Differentiating (3-92a) gives

$$\frac{d}{d\delta} \text{phase } G(s_r)H(s_r) = \sum_{i=1}^{n} \frac{\gamma + \operatorname{Re}(p_i)}{[\gamma + \operatorname{Re}(p_i)]^2 + [\delta - \operatorname{Im}(p_i)]^2}$$

$$- \sum_{j=1}^{m} \frac{\gamma + \operatorname{Re}(z_j)}{[\gamma + \operatorname{Re}(z_j)]^2 + [\delta - \operatorname{Im}(z_j)]^2}$$

and taking the limit as $\delta \to 0$ then gives Rule 8.

Rule 9. The angle of departure of the root locus from a complex pole or the angle of arrival of the root locus at a complex zero is determined by applying the angle criterion of Rule 2 to any point in the immediate vicinity of the complex pole or complex zero concerned.

3.7 Optimal Linearization and the Describing Function [W3]

Suppose a block diagram model of a dynamical system contains an instantaneous (that is non-dynamic) nonlinearity of the form

$$y = \phi(x) \tag{3-95}$$

where x is an input variable, y an output variable, and ϕ denotes a single-valued functional dependence of y on x. In certain circumstances, as discussed below, it is useful to approximate the nonlinear relationship of equation (3–93) by a linear relationship

$$\bar{y} = Nx \tag{3-94}$$

where N is a real constant. If an error functional is defined such a choice of linear approximation may be carried out in an optimal way. If x, and thus y, are functions of time, t, then we may choose N so as to minimize

$$E = \int_{t_0}^{t_1} (\bar{y} - y)^2 \, dt = \int_{t_0}^{t_1} [Nx - y]^2 \, dt \tag{3-95}$$

where t_0 and t_1 are specified values of time. In this case necessary and sufficient conditions for a minimum of the error functional E are

$$\frac{\partial E}{\partial N} = 0 \qquad \frac{\partial^2 E}{\partial N^2} > 0$$

We then have

$$\frac{\partial}{\partial N} \int_{t_0}^{t_1} [N^2 x^2 - 2Nxy + y^2] \, dt = 0$$

which gives

$$2N \int_{t_0}^{t_1} x^2 \, dt = 2 \int_{t_0}^{t_1} xy \, dt$$

so that

$$N = \frac{\int_{t_0}^{t_1} xy \, dt}{\int_{t_0}^{t_1} x^2 \, dt} \tag{3-96}$$

and this will give a minimum value of error functional since

$$\frac{\partial^2 E}{\partial N^2} = 2 \int_{t_0}^{t_1} x^2 \, dt > 0$$

Now consider the particular input

$$x = a \sin \omega t$$

and put
$$t_0 = 0 \qquad t_1 = \frac{2\pi}{\omega}$$

We may thus determine, according to the chosen error functional criterion, an optimal linear approximation averaged over one cycle of a sinusoidal input. This gives

$$N = \frac{\int_0^{2\pi/\omega} (a \sin \omega t) y \, dt}{\int_0^{2\pi/\omega} a^2 \sin^2 \omega t \, dt} = \frac{a \int_0^{2\pi/\omega} y \sin \omega t \, dt}{a^2 \pi / \omega}$$

$$= \frac{\omega}{a\pi} \int_0^{2\pi/\omega} y \sin \omega t \, dt$$

The output of the nonlinearity will be a periodic waveform whose first harmonic amplitude will be

$$a_1 = \frac{\omega}{\pi} \int_0^{2\pi/\omega} y \sin \omega t \, dt$$

so that we have

$$N = \frac{\text{amplitude of 1st harmonic in output}}{\text{amplitude of input}} \tag{3-97}$$

In general the amplitude of 1st harmonic in the output will depend on the amplitude of the input sinusoid, that is we have

$$N = N(a)$$

The function $N(a)$ corresponding to an optimal linear approximation averaged over one cycle of sinusoidal input is called the *describing function* of the nonlinearity. For any given nonlinearity it is usually computed by a direct Fourier analysis of the output waveform resulting from an input sinusoid of known amplitude. The describing function definition may be extended to include non-single-valued nonlinearities such as a hysteresis characteristic; in this case it becomes a complex number and specifies both a phase and an amplitude for the resulting first harmonic output. Describing function techniques are used when the nonlinearity approximated in this way is followed by energy-storing

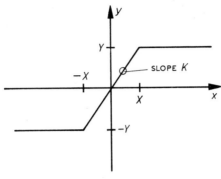

Figure 3–50

elements which heavily attenuate the higher harmonies produced by the nonlinearity, for example when a saturating amplifier feeds the highly inductive field of a servomotor. In such circumstances the final output is well approximated by a sinusoid and the describing function technique gives an acceptably accurate prediction of overall response to sinusoidal forcing.

For an example of the describing function consider the nonlinearity shown in Figure 3–50. The corresponding input and output waveforms

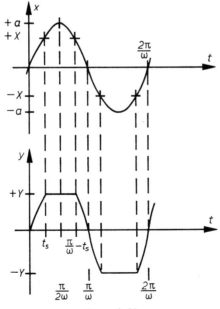

Figure 3–51

are in Figure 3–51. Because of the symmetry of the resulting waveforms we may compute the values of the integral involved over one cycle as four times its value over a quarter-cycle. This gives

$$N(a) = \frac{4\omega}{\pi a} \int_0^{\pi/2\omega} y \sin \omega t \, dt$$

$$= \frac{4\omega}{\pi a} \int_0^{t_s} Ka \sin \omega t \sin \omega t \, dt + \frac{4\omega}{\pi a} \int_{t_s}^{\pi/2\omega} KaX \sin \omega t \, dt$$

where the output waveform saturates at

$$t_s = \frac{1}{\omega} \sin^{-1} \frac{X}{a}$$

Let

$$\sin^{-1} \frac{X}{a} = \alpha$$

Then

$$N(a) = \frac{4Ka\omega}{\pi a} \int_0^{\alpha/\omega} \sin^2 \omega t \, dt + \frac{4Ka\omega}{\pi a} \int_0^{\pi/2\omega} X \sin \omega t \, dt$$

$$= \frac{2KX}{\pi a} \left[\frac{a\alpha}{X} - \frac{a \sin 2\alpha}{2X} + 2 \cos \alpha \right]$$

$$= \frac{2KX}{\pi a} \left[\frac{a}{X} \sin^{-1}\left(\frac{X}{a}\right) + \cos \alpha \right]$$

$$= \frac{2KX}{\pi a} \left[\frac{a}{X} \sin^{-1}\left(\frac{X}{a}\right) + \sqrt{\left(1 - \frac{X^2}{a^2}\right)} \right] \quad \text{for } a > X$$

and has the form sketched in Figure 3–52.

Describing function may be used in some simple cases to extend the Nyquist stability criterion to systems which include a nonlinearity.

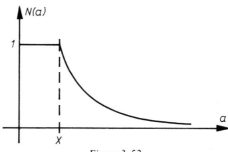

Figure 3–52

Optimal Linearization and the Describing Function

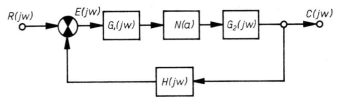

Figure 3-53

Consider the system shown in the block diagram of Figure 3-53, where $N(a)$ is the describing function of the saturation-type nonlinearity shown in Figure 3-52. The closed-loop frequency response characteristic will be

$$\frac{C(j\omega)}{R(j\omega)} = \frac{G_1(j\omega)N(a)G_2(j\omega)}{1 + G_1(j\omega)N(a)G_2(j\omega)H(j\omega)}$$

The possibility of self-oscillation obviously exists when

$$G_1(j\omega)N(a)G_2(j\omega)H(j\omega) = -1$$

that is when

$$G_1(j\omega)G_2(j\omega)H(j\omega) = -\frac{1}{N(a)} \tag{3-99}$$

The complex-plane locus of $-1/N(a)$ for varying a is called the *critical locus* and is shown in Figure 3-54 for the saturation-type nonlinearity. If the critical locus does *not* intersect the open-loop frequency response locus $G_1(j\omega)G_2(j\omega)H(j\omega)$ then the system will be completely stable.

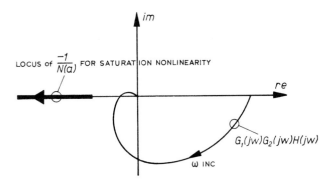

Figure 3-54

3.8 z-transforms for Discrete Systems

In digital devices, continuously varying functions are replaced by discrete data sequences. The dynamical behaviour of such devices may be considered in terms of discrete models in which the dynamical variables are sequences of data values and whose elements are represented in terms of discrete operators on such sequences.

3.8.1 Basis Sequences

In what follows we are only concerned with one-sided sequences which start at some given reference in time and may be considered to be equally spaced in time. Any such sequence

$$f_0 f_1 f_2 \ldots f_{k-1} f_k f_{k+1} \ldots$$

has a general term which will be alternatively denoted by f_k or $f(k)$. The suffix k plays the role of the independent variable for a continuous function. In any application k would correspond to the kth multiple of some conveniently chosen unit of time, T say, and the kth member of the sequence would appear kT seconds after the sequence started. Any such sequence can be expressed as a linear combination of a set of standard *basis sequences*

$$\delta_0: 1\ 0\ 0\ 0\ 0\ \ldots$$
$$\delta_1: 0\ 1\ 0\ 0\ 0\ \ldots$$
$$\delta_2: 0\ 0\ 1\ 0\ 0\ \ldots$$
$$\delta_3: 0\ 0\ 0\ 1\ 0\ \ldots$$
$$\ldots$$

In terms of these basis sequences the sequence above can obviously be expressed as $\sum_{k=0}^{\infty} f(k)\delta_k$. These standard basis sequences play a role analogous to the Dirac impulses $\delta(t-kT)$ in continuous system theory. By analogy with the Heaviside step function of continuous systems it is also convenient to introduce the unit step sequence H_k such that $H_k = 1$ for $k = 1, 2, 3, \ldots$, that is

$$H_k: 1\ 1\ 1\ 1\ 1\ \ldots$$

3.8.2 Discrete Operators

The fundamental operations on sequences are shifting and differencing.

Forward shift. The forward shift operator E is defined by

$$E[f(k)] = f(k+1) \qquad (3\text{--}100)$$

Backward shift. The backward shift operator is conveniently denoted by E^{-1} and is defined by

$$E^{-1}[f(k)] = f(k-1) \qquad (3\text{-}101)$$

Forward difference. A sequence of first differences may be obtained from any given sequence by subtracting each member of the sequence from the next to get a sequence whose general term is $f(k+1)-f(k)$. The corresponding forward difference operator is denoted by Δ and defined by

$$\Delta f(k) = f(k+1) - f(k) \qquad (3\text{-}102)$$

Backward difference. Alternatively we can form a sequence of first differences by taking as the general term $f(k)-f(k-1)$. The use of the previous member of a sequence rather than the next is obviously more likely to be used in certain practical devices, and the corresponding backward difference operator is denoted by ∇ and defined by

$$\nabla f(k) = f(k) - f(k-1) \qquad (3\text{-}103)$$

Simple manipulations readily show that the algebraic equivalences among these operators may be expressed as

$$\begin{aligned} E &= (1+\Delta) \\ E^{-1} &= (1-\nabla) \end{aligned} \qquad (3\text{-}104)$$

3.8.3 z-transform

The z-transform sets up a reciprocal correspondence between the set of standard basis sequences introduced above and the integer powers of a complex variable z, such that

$$\delta_k \leftrightarrow z^{-k}$$

Since any sequence $f_0 f_1 \ldots f_k \ldots$ may be expressed as the sum

$$\sum_{k=0}^{\infty} f_k \delta_k$$

we *define* the z-transform of the sequence to be

$$F(z) = \sum_{k=0}^{\infty} f_k z^{-k} \qquad (3\text{-}105)$$

The utility of the z-transform stems from exactly the same considerations as apply for the Laplace transform; it replaces operational relationships between sequences by algebraic relationships between their z-transforms. The basic properties of the z-transform are now considered; they are exactly analogous to the corresponding relationships for the Laplace transform discussed in the previous sections of this chapter.

It is convenient to denote the operation of taking a z-transform by $\mathscr{Z}$ and the corresponding inverse operation by $\mathscr{Z}^{-1}$. In what follows we will frequently use $f(k)$ to denote *the whole sequence* whose general term is $f(k)$ where no possibility of confusion can arise, and $F(z)$ the corresponding z-transform of the sequence. Thus we put

$$\mathscr{Z}[f(k)] = \mathscr{Z}\left[\sum_{k=0}^{\infty} f_k \delta_k\right] = \sum_{k=0}^{\infty} f_k z^{-k} = F(z)$$

Linearity. $\mathscr{Z}$ is a linear operator since

$$\mathscr{Z}[\alpha f(k)] = \sum_{k=0}^{\infty} \alpha f_k z^{-k} = \alpha \sum_{k=0}^{\infty} f_k z^{-k} = \alpha \mathscr{Z}[f(k)]$$

$$\mathscr{Z}[f(k) + g(k)] = \sum_{k=0}^{\infty} (f_k + g_k) z^{-k} = \sum_{k=0}^{\infty} f_k z^{-k} + \sum_{k=0}^{\infty} g_k z^{-k}$$
$$= \mathscr{Z}[f(k)] + \mathscr{Z}[g(k)]$$

Shifting. If the sequence is shifted, we have

$$\mathscr{Z}[f(k+m)] = \sum_{k=0}^{\infty} f_{k+m} z^{-k}$$
$$= \sum_{j=m}^{\infty} f_j z^{-j+m} \quad \text{on putting } k+m = j$$
$$= z^m \sum_{j=m}^{\infty} f_j z^{-j}$$
$$= z^m \left[\sum_{j=0}^{\infty} f_j z^{-j} - \sum_{j=0}^{m-1} f_j z^{-j}\right]$$
$$= z^m \left[F(z) - \sum_{j=0}^{m-1} f_j z^{-j}\right] \qquad (3\text{--}106)$$

In particular, when $m = 1$ we have

$$\mathscr{Z}[f(k+1)] = zF(z) - zf(0) \qquad (3\text{--}107)$$

that is

$$\mathscr{Z}[Ef(k)] = zF(z) - zf(0) \qquad (3\text{--}108)$$

Differencing

$$\mathscr{L}[f(k)] = \sum_{k=0}^{\infty} f_k z^{-k} = f(0) + f(1)z^{-1} + f(2)z^{-2} + \cdots$$

$$\mathscr{L}[f(k+1)] = f(1) + f(2)z^{-1} + f(3)z^{-2} + \cdots$$
$$= z[F(z) - f(0)]$$

Thus

$$\mathscr{L}[\Delta f(k)] = \mathscr{L}[f(k+1)] - \mathscr{L}[f(k)] = (z-1)F(z) - zf(0) \quad (3\text{-}109)$$

Again we have

$$\mathscr{L}[f(k-1)] = f(-1) + f(0)z^{-1} + f(1)z^{-2} + \cdots$$
$$= z^{-1}F(z) + f(-1)$$

Thus

$$\mathscr{L}[\nabla f(k)] = \mathscr{L}[f(k)] - \mathscr{L}[f(k-1)] = (1 - z^{-1})F(z) - f(-1) \quad (3\text{-}110)$$

Initial-value theorem.

$$\mathscr{L}[f(k)] = f(0) + z^{-1}f(1) + z^{-2}f(2) + \cdots$$

so that we immediately obtain

$$\lim_{z \to \infty} \mathscr{L}[f(k)] = f(0) \quad (3\text{-}111)$$

Final-value theorem. If a sequence converges to a finite limit, then the corresponding z-transform will be analytic outside the circle $|z| < 1$, as shown in Section 3.8.5 below. For any integer N we have, for $|z| > 1$

$$\sum_{k=0}^{\infty} f_{N+k} z^{-(N+k)} = z^{-N} \sum_{k=0}^{\infty} f_{N+k} z^{-k}$$
$$= z^{-N} \left[\sum_{k=0}^{\infty} (f_{N+k} - f_\infty) z^{-k} + \sum_{k=0}^{\infty} f_\infty z^{-k} \right]$$
$$= z^{-N} \left[\sum_{k=0}^{\infty} (f_{N+k} - f_\infty) z^{-k} + f_\infty (1 - z^{-1})^{-1} \right]$$

Now we may write

$$F(z) = \sum_{k=0}^{N-1} f_k z^{-k} + \sum_{k=0}^{\infty} f_{N+k} z^{-(N+k)}$$

so we may put

$$F(z) = \sum_{k=0}^{N-1} f_k z^{-k} + \frac{z^{-N} f_\infty}{(1 - z^{-1})} + z^{-N} \left[\sum_{k=0}^{\infty} (f_{N+k} - f_\infty) z^{-k} \right]$$

from which we have

$$(z-1)F(z) = (z-1)\sum_{k=0}^{\infty} f_k z^{-k} + z^{-N+1}f_{\infty}$$
$$+ z^{-N}(z-1)\left[\sum_{k=0}^{\infty}(f_{N+k}-f_{\infty})z^{-k}\right]$$

If we now let $z \to 1$ and $N \to \infty$ the first and third items vanish and we get, when $f(\infty)$ exists

$$\lim_{z \to 1}(z-1)F(z) = f(\infty) \qquad (3\text{-}112)$$

Convergence. If for a one-sided sequence $f(k)$ we have that
(a) $|f(k)| < \infty$ for all $k < \infty$, and
(b) there exists a set of positive constants N, r and β such that

$$|f(k)| \leq \beta r^k \quad \text{for} \quad k \geq N$$

then the corresponding z-transform

$$F(z) = \sum_{k=0}^{\infty} f_k z^{-k}$$

converges absolutely for

$$\rho = |z| > r$$

The smallest value of $|z|$ for which absolute convergence is achieved is called the radius of convergence of the z-transform.

To see how these two conditions guarantee convergence, write the z-transform as

$$F(z) = \sum_{k=N}^{\infty} f_k z^{-k} + \sum_{k=0}^{N-1} f_k z^{-k}$$

so that we have

$$|F(z)| \leq \sum_{k=0}^{N-1} |f_k|\rho^{-k} + \sum_{k=N}^{\infty} |f_k|\rho^{-k} \qquad (3\text{-}113)$$

where

$$\rho = |z|$$

Condition (a) ensures that there exists some number M (possibly dependent on N) such that

$$\sum_{k=0}^{N-1} |f_k|\rho^{-k} \leq M \sum_{k=0}^{N-1} \rho^{-k} \qquad (3\text{-}114)$$

Using condition (b) for $k \geq N$ gives that

$$\sum_{k=N}^{\infty} |f_k| \rho^{-k} \leq \beta \sum_{k=N}^{\infty} \left(\frac{r}{\rho}\right)^k \qquad (3\text{-}115)$$

If we now use inequalities (3–114) and (3–115) in inequality (3–113) we have

$$|F(z)| \leq M \sum_{k=0}^{N-1} \rho^{-k} + \beta \sum_{k=N}^{\infty} \left(\frac{r}{\rho}\right)^k \qquad (3\text{-}116)$$

The first term on the right-hand side of inequality (3–116) is finite for all finite N and ρ and the second term is a geometric series which converges for all $\rho > r$. This establishes the required result.

Table of z-transforms

A number of useful z-transform pairs can be obtained from the basic definition in a very direct way. For the unit step sequence we have

$$\mathscr{L}[H_k] = \sum_{k=0}^{N-1} z^{-k} = \frac{1}{1-z^{-1}} = \frac{z}{z-1} \qquad (3\text{-}117)$$

If $f_k = kH_k$, then

$$\Delta f(k) = H_k$$

so that

$$\mathscr{L}[\Delta f(k)] = \frac{z}{z-1} = (z-1)F(z) - zf(0)$$

$$= (z-1)F(z)$$

since $f(0)$ is zero for sequence kH_k.

This gives

$$\mathscr{L}[kH_k] = \frac{z}{(z-1)^2} \qquad (3\text{-}118)$$

Proceeding in this way we obtain

$$\mathscr{L}\left[\binom{k}{m} H_k\right] = \frac{z}{(z-1)^{m+1}} \qquad (3\text{-}119)$$

where

$$\binom{k}{m} = \frac{k!}{(k-m)!m!}$$

If α is a complex constant then

$$\mathscr{L}[\alpha^k f(k)] = \sum_{k=0}^{\infty} f_k \alpha^k z^{-k} = \sum_{k=0}^{\infty} f_k \left(\frac{z}{\alpha}\right)^{-k} = F\left(\frac{z}{\alpha}\right) \qquad (3\text{-}120)$$

Using this result in equation (3–119) gives

$$\mathscr{L}\left[\binom{k}{m}\alpha^k H_k\right] = \frac{z\alpha^m}{(z-\alpha)^{m+1}} \qquad (3\text{–}121)$$

and in particular for $m = 0$ we have

$$\mathscr{L}[\alpha^k H_k] = \frac{z}{z-\alpha} \qquad (3\text{–}122)$$

In equation (3–122), express the complex constant α in polar form as

$$\alpha = \rho \exp(j\gamma)$$

We then have:

$$\mathscr{L}[\rho^k \exp(jk\gamma)H_k] = \frac{z}{z - \rho \exp(j\gamma)}$$

so that

$$\mathscr{L}[\rho^k(\cos k\gamma + j \sin k\gamma)H_k] = \frac{z}{z - \rho \cos \gamma - j\rho \sin \gamma}$$

$$= \frac{z(z - \rho \cos \gamma + j\rho \sin \gamma)}{z^2 - 2\rho z \cos \gamma + \rho^2}$$

and equating real and imaginary parts gives

$$\mathscr{L}[\rho^k \cos k\gamma H_k] = \frac{z(z - \rho \cos \gamma)}{z^2 - 2\rho z \cos \gamma + \rho^2} \qquad (3\text{–}123)$$

$$\mathscr{L}[\rho^k \sin k\gamma H_k] = \frac{\rho z \sin \gamma}{z^2 - 2\rho z \cos \gamma + \rho^2} \qquad (3\text{–}124)$$

A collection of useful z-transform pairs is given in Table 3–2.

Evaluation of Inverse z-transform
An inverse z-transform may be evaluated in three different ways.

(a) In simple cases the z-transform may be expressed as a sum of partial fractions and the inversion performed using a table of transform pairs.
(b) The transform may be expressed as a power series expansion.

$$F(z) = \sum_{k=0}^{\infty} f_k z^{-k} = f(0) + f(1)z^{-1} + f(2)z^{-2} + \cdots$$

and the coefficient of z^{-k} in the expansion is then the value of the kth member of the sequence.

Table 3-2. z-transform pairs

$(k, m, \rho, \gamma, a, \omega$ are real constants)

Sequence $f(k)$	z-transform $F(z)$
δ_k	z^{-k}
H_k	$\dfrac{z}{z-1}$
kH_k	$\dfrac{z}{(z-1)^2}$
$\binom{k}{m} H_k$	$\dfrac{z}{(z-1)^{m+1}}$
$\binom{k}{m} \alpha^r H_k$	$\dfrac{z\alpha^m}{(z-\alpha)^{m+1}}$
$\rho^k \sin k\gamma \, H_k$	$\dfrac{\rho z \sin \gamma}{z^2 - 2\rho z \cos \gamma + \rho^2}$
$\rho^k \cos k\gamma \, H_k$	$\dfrac{z(z - \rho \cos \gamma)}{z^2 - 2\rho z \cos \gamma + \rho^2}$
$\exp(-ak) H_k$	$\dfrac{z}{z - \exp(-a)}$
$k \exp(-ak) H_k$	$\dfrac{z \exp(-a)}{[z - \exp(-a)]^2}$
$\tfrac{1}{2} k^2 \exp(-ak) H_k$	$\dfrac{z \exp(-a)[z + \exp(-a)]}{2[z - \exp(-a)]^3}$
$\exp(-ak) \sin \omega k \, H_k$	$\dfrac{z \sin \omega \exp(-a)}{z^2 - 2z \exp(-a) \cos \omega + \exp(-2a)}$
$\exp(-ak) \cos \omega k \, H_k$	$\dfrac{z^2 - z \exp(-a) \cos \omega}{z^2 - 2z \exp(-a) \cos \omega + \exp(-2a)}$
$\sinh \omega k \, H_k$	$\dfrac{z \sinh \omega}{z^2 - 2z \cosh \omega + 1}$
$\cosh \omega k \, H_k$	$\dfrac{z(z - \cosh \omega)}{z^2 - 2z \cosh \omega + 1}$
$\dfrac{1}{k!} H_k$	$\exp\left(\dfrac{1}{z}\right)$

214   *Transform Models*

(c) The inverse transform may be obtained in terms of the contour integral [F1]

$$f_k = \frac{1}{2\pi j} \int_\Gamma F(z) z^{k-1} \, dz \qquad (3\text{–}125)$$

where Γ is any simple closed rectifiable curve enclosing the origin of the complex plane and lying outside the radius of convergence of $F(z)$.

3.8.4 Convolution of Sequences

Suppose we have a device which on being fed with the standard basis sequence δ_0 produces an output sequence

$$w_0 \quad w_1 \quad w_2 \ldots w_m$$

This sequence will be called the *weighting sequence* of the device by analogy with the unit impulse response of a continuous system. Now let this device have applied to its input a sequence

$$f_0 \quad f_1 \quad f_2 \ldots f_n$$

Further suppose that the device is linear, that is such that:

(a) its output to a combined set of discrete input sequences is the sum of the individual outputs when the inputs are applied separately, and

(b) increasing any input by amount K produces K times the resulting output.

We then have that:

(i) the input f_0 will produce an output sequence

$$f_0 w_0 \quad f_0 w_1 \quad f_0 w_2 \quad \ldots \quad f_0 w_m \quad 0 \quad 0 \quad \ldots$$

(ii) the input f_1 will produce an output sequence

$$0 \quad f_1 w_0 \quad f_1 w_1 \quad \ldots \quad f_1 w_{m-1} \quad f_1 w_m \quad 0 \quad 0 \quad \ldots$$

(iii) the input f_2 will produce an output sequence

$$0 \quad 0 \quad f_2 w_0 \quad \ldots \quad f_2 w_{m-2} \quad f_2 w_{m-1} \quad f_2 w_m \quad 0 \quad \ldots$$

and so on.

We therefore conclude that the output sequence resulting from the input sequence is obtained by adding the above sequences together and so is

$$r_0 \quad r_1 \quad r_2 \quad \ldots$$

where
$$r_0 = f_0 w_0$$
$$r_1 = f_0 w_1 + f_1 w_0$$
$$r_2 = f_0 w_2 + f_1 w_1 + f_2 w_0$$
$$r_3 = f_0 w_3 + f_1 w_2 + f_2 w_1 + f_3 w_0$$
and so on, with the general term given by

$$r_k = \sum_{j=0}^{k} f_j w_{k-j} \qquad (3\text{-}126)$$

The output sequence is said to be the *convolution* of the input sequence and the weighting sequence. The convolution sum of equation (3-126) is exactly analogous to the convolution integral introduced in Section 3.2.1 for continuous functions of time.

3.8.5 z-transfer Functions

Consider again the device introduced above in the discussion of convolution. Let the z-transform of the device's weighting sequence be $G(z)$, and again suppose we apply an input sequence

$$f_0 \quad f_1 \quad \cdots \quad f_m$$

whose z-transform is $F(z)$. We then have that:

(a) the input f_0 produces an output whose z-transform is
$$f_0 w_0 + f_0 w_1 z^{-1} + f_0 w_2 z^{-2} + \cdots = f_0 G(z)$$

(b) the input f_1 produces an output whose z-transform is
$$f_1 w_0 z^{-1} + f_1 w_1 z^{-2} + \cdots = f_1 z^{-1} G(z)$$

(c) the input f_2 produces an output whose z-transform is
$$f_2 w_0 z^{-2} + f_2 w_1 z^{-3} + \cdots = f_2 z^{-2} G(z)$$

and so on.

The z-transform of the response sequence is obviously the sum of these individual response z-transforms and so is

$$R(z) = f_0 G(z) + f_1 z^{-1} G(z) + f_2 z^{-2} G(z) + \cdots$$
$$= G(z)[f_0 + f_1 z^{-1} + f_2 z^{-2} + \cdots]$$
$$= G(z)F(z) \qquad (3\text{-}127)$$

where $G(z)$ is called the z-transfer function of the device. Thus the z-transform of the response sequence of a linear device may be obtained as the product of the z-transfer function of the device and the z-transform of the input sequence.

In general, for devices consisting of elements performing a finite number of discrete operations, the associated z-transfer function will be of the form

$$G(z) = \frac{b_0 + b_1 z^{-1} + b_2 z^{-2} + \cdots + b_m z^{-m}}{a_0 + a_1 z^{-1} + a_2 z^{-2} + \cdots + a_n z^{-n}}$$

$$= \frac{K \prod_{i=1}^{m} (z - \alpha_i)}{\prod_{j=1}^{n} (z - \beta_j)}$$

where $\beta_1, \beta_2, \ldots, \beta_n$ are the poles of the z-transfer function and K is a real constant. For $G(z)$ to be the z-transfer function of a stable device we must have that

$$\sum_{k=0}^{\infty} |w_k||z^{-k}| < \infty \quad \text{for} \quad |z| > 1 \qquad (3\text{--}128)$$

For any weighting sequence we will have

$$\sum_{k=0}^{\infty} w_k z^{-k} \leq \sum_{k=0}^{\infty} |w_k||z^{-k}| \qquad (3\text{--}129)$$

So that, combining inequalities (3–128) and (3–129) we have the necessary condition

$$\sum_{k=0}^{\infty} w_k z^{-k} < \infty \quad \text{for} \quad |z| > 1$$

That is we must have

$$G(z) < \infty \quad \text{for} \quad |z| > 1 \qquad (3\text{--}130)$$

if the device is to be stable. Thus $G(z)$ must be analytic everywhere outside the unit circle centred on the origin in the complex plane; that is $G(z)$ must have no poles inside this unit circle in the complex z-plane if the device is to be stable.

CHAPTER FOUR

Hamiltonian Models

In general, a dynamical system model is specified in terms of a set of component relationships together with a set of constraint relationships which define the way in which the components are assembled together to form the system. Network analysis handles these constraint relationships explicitly in terms of sets of specified relationships between component variables. In Chapter 1 we have associated with the various types of dynamical system component certain functions such as energy, co-energy, content and co-content. These functions are given there in terms of the individual component variables, but they may equally well be expressed in terms of any complete, independent set of system variables which uniquely define the component variable set. Any such complete, independent set of system variables in terms of which all component energies, co-energies, contents and co-contents may be expressed, is termed a set of system *generalized coordinates*. In Hamiltonian model analysis, the *complete system* scalar function set is expressed in terms of *any* appropriate set of generalized coordinates, and the system constraint relationships are then handled in an *implicit* form via this generalized coordinate set. Hamiltonian model analysis is based on Hamilton's postulates; these involve the stationary value of an integral, and we therefore first consider the relevant processes of the Calculus of Variations. [P1], [L1]

4.1 Fundamental Processes of the Calculus of Variations

The definite integral

$$I = \int_{\sigma_1}^{\sigma_2} L\left(x, \frac{dx}{d\sigma}, \sigma\right) d\sigma \qquad (4-1)$$

sets up a correspondence between a function $x(\sigma)$ and a number I; I is called a *functional* of x. Consider stationary values of I obtained by investigating changes in the value of I which arise from changes in the function $x(\sigma)$. These stationary values of a functional can be dealt with in terms of simple function extremization, if we consider the consequences of evaluating the integral for a slightly modified function

$$\bar{x}(\sigma) = x(\sigma) + \epsilon\eta(\sigma) \qquad (4-2)$$

where ϵ is small and $\eta(\sigma)$ is arbitrary. The difference between $\bar{x}(\sigma)$

and $x(\sigma)$ is called the *variation* of x, and denoted by δx. It is obviously a function of σ

$$\delta x(\sigma) = \bar{x} - x = \epsilon\eta(\sigma) \tag{4-3}$$

It is most important to be clear about the fundamental difference between a *variation* δx and a *differential* dx. Both represent small changes in a function x: however dx refers to a small change in x caused by a small change $d\sigma$ in the independent variable σ, while δx is a small change in x which produces a *new function* of σ, namely $\{x(\sigma) + \delta x(\sigma)\}$. This point is illustrated in Figure 4–1.

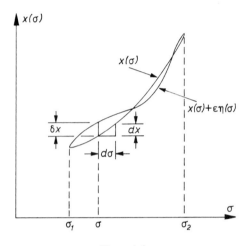

Figure 4–1

Laws of Variation

It is easily verified directly, from the definition above, that the laws governing the variation of sums, powers, ratios, products, etc. are completely analogous to the corresponding laws of differentiation. In particular

$$\delta(x_1 + x_2) = \delta x_1 + \delta x_2 \tag{4-4}$$

$$\delta(x_1 x_2) = x_1 \delta x_2 + x_2 \delta x_1 \tag{4-5}$$

$$\delta\left(\frac{x_1}{x_2}\right) = \frac{x_2 \delta x_1 - x_1 \delta x_2}{x_2^2} \tag{4-6}$$

Commutative Properties of Variation

Since δx is a function of σ, we may take its derivative with respect to σ. We can also take the difference between the derivative of $\bar{x}$ and the derivative of x; this difference will naturally be called the variation

Fundamental Processes of the Calculus of Variations   219

of the derivative $dx/d\sigma$. In this way we can compare the derivative of the variation with the variation of the derivative

$$\frac{d}{d\sigma}(\delta x) = \frac{d}{d\sigma}(\bar{x}-x) = \frac{d}{d\sigma}[\epsilon\eta(\sigma)] = \epsilon\frac{d\eta}{d\sigma}$$

$$\delta\left(\frac{dx}{d\sigma}\right) = \frac{d\bar{x}}{d\sigma} - \frac{dx}{d\sigma} = \frac{d}{d\sigma}[x+\epsilon\eta] - \frac{dx}{d\sigma} = \epsilon\frac{d\eta}{d\sigma}$$

So that
$$\frac{d}{d\sigma}(\delta x) = \delta\left(\frac{dx}{d\sigma}\right) \qquad (4\text{--}7)$$

showing that the operations of variation and differentiation are commutative, that is the variation of the derivative is the derivative of the variation.

In a similar way we may consider the variation of a definite integral. By this we mean that we take the difference between the definite integral evaluated for a modified integrand $\overline{L(\sigma)}$, say, and the definite integral evaluated for the original integrand $L(\sigma)$

$$\delta\left[\int_{\sigma_1}^{\sigma_2} L(\sigma)\,d\sigma\right] = \int_{\sigma_1}^{\sigma_2} \overline{L(\sigma)}\,d\sigma - \int_{\sigma_1}^{\sigma_2} L(\sigma)\,d\sigma$$

$$= \int_{\sigma_1}^{\sigma_2} [\overline{L(\sigma)} - L(\sigma)]\,d\sigma$$

$$= \int_{\sigma_1}^{\sigma_2} \delta[L(\sigma)]\,d\sigma \qquad (4\text{--}8)$$

This shows that the operations of variation and definite integration are commutative, that is, the variation of a definite integral is the definite integral of the variation.

Fundamental Lemma of the Calculus of Variations

The derivations of some of the fundamental results required in the Calculus of Variations are less cumbersome if we first establish a result known as the Fundamental Lemma of the Calculus of Variations:

If $G(\sigma)$ is continuous on the closed interval $[\sigma_1, \sigma_2]$, and if

$$\int_{\sigma_1}^{\sigma_2} G(\sigma)\eta(\sigma)\,d\sigma = 0 \qquad (4\text{--}9)$$

for arbitrary $\eta(\sigma)$, then $G(\sigma)$ must vanish identically over $\sigma_1 \leq \sigma \leq \sigma_2$. To prove this suppose $G(\sigma)$ does not vanish over some interval $[\sigma_3, \sigma_4]$ with $\sigma_1 \leq \sigma_3 \leq \sigma_4 \leq \sigma_2$. Then the sign of $G(\sigma)$ must be constant over this interval. Choose the arbitrary $\eta(\sigma)$ to be zero on the closed intervals $[\sigma_1, \sigma_3]$ and $[\sigma_4, \sigma_2]$, and to be nonzero and have the same sign as $G(\sigma)$ in the interval $[\sigma_3, \sigma_4]$. Then

$$\int_{\sigma_1}^{\sigma_2} G(\sigma)\eta(\sigma)\, d\sigma = \int_{\sigma_3}^{\sigma_4} G(\sigma)\eta(\sigma)\, d\sigma > 0$$

since $\eta(\sigma)$ and $G(\sigma)$ are nonzero and of the same sign in $[\sigma_3, \sigma_4]$. It follows that the assumption of non-vanishing $G(\sigma)$ in the interval $[\sigma_1, \sigma_2]$ is inadmissible, and we therefore conclude that $G(\sigma)$ must vanish identically in the interval $[\sigma_1, \sigma_2]$.

4.1.1 Conditions for Stationary Values of Definite Integral

We first consider the simple case of one independent variable. Let I, as defined in equation (4–1), be stationary for $x = s(\sigma)$. That is, $s(\sigma)$ is such that variations of I about $s(\sigma)$ become arbitrarily small as the amount of variation is continually reduced. Denote the stationary value of I by I_s. Then

$$I_s + \delta I_s = \int_{\sigma_1}^{\sigma_2} L\{\sigma, s + \epsilon\eta(\sigma), s' + \epsilon\eta'(\sigma)\}\, d\sigma \qquad (4\text{–}10)$$

where the primes denote differentiation with respect to σ. Assume that $L(s, s', \sigma)$ possesses continuous partial derivatives; this allows us to expand it in a Taylor series. If the derivatives are continuous up to at least the third order we have

$$L\{\sigma, s + \epsilon\eta(\sigma), s' + \epsilon\eta'(\sigma)\} = L(\sigma, s, s') + \epsilon\left[\eta\frac{\partial L}{\partial s} + \eta'\frac{\partial L}{\partial s'}\right]$$
$$+ \frac{\epsilon^2}{2!}\left[\eta^2\frac{\partial^2 L}{\partial s^2} + 2\eta\eta'\frac{\partial^2 L}{\partial s\, \partial s'} + (\eta')^2\frac{\partial^2 L}{\partial s'^2}\right] + 0(\epsilon^3)$$

where $0(\epsilon^3)$ denotes a term of at least third order in ϵ. Thus

$$\delta I_s = \epsilon\int_{\sigma_1}^{\sigma_2}\left[\eta\frac{\partial L}{\partial s} + \eta'\frac{\partial L}{\partial s'}\right] d\sigma$$
$$+ \frac{\epsilon^2}{2!}\int_{\sigma_1}^{\sigma_2}\left[\eta^2\frac{\partial^2 L}{\partial s^2} + 2\eta\eta'\frac{\partial^2 L}{\partial s\, \partial s'} + (\eta')^2\frac{\partial^2 L}{\partial s'^2}\right] d\sigma + 0(\epsilon^3) \qquad (4\text{–}11)$$

$$\delta I_s = \epsilon I_1 + \frac{\epsilon^2}{2!} I_2 + 0(\epsilon^3)$$

where I_1 and I_2 are called the first and second variations of I.

For I_s to be a maximum, δI_s must be negative for both positive and negative values of ϵ and must vanish for $\epsilon = 0$ as shown in Figure 4–2.

Fundamental Processes of the Calculus of Variations 221

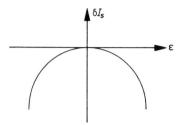

Figure 4-2 Variation of I_s versus ϵ will have this form for a maximum value of I_s.

Since
$$\delta I_s = \epsilon I_1 + \frac{\epsilon^2}{2!} I_2 + 0(\epsilon^3)$$

$$\frac{d(\delta I_s)}{d\epsilon} = I_1 + 0(\epsilon)$$

and
$$\frac{d^2(\delta I_s)}{d\epsilon} = I_2 + 0(\epsilon)$$

where $0(\epsilon)$ denotes a term of at least first order in ϵ. For I_s to be a maximum we must have

$$\left[\frac{d(\delta I_s)}{d\epsilon}\right]_{\epsilon=0} = 0 \quad \text{and} \quad \left[\frac{d^2(\delta I_s)}{d\epsilon^2}\right]_{\epsilon=0} < 0$$

that is, we must have

$$[I_1]_{\epsilon=0} = 0 \quad \text{and} \quad [I_2]_{\epsilon=0} < 0 \tag{4-12}$$

Similarly for a minimum we must have

$$[I_1]_{\epsilon=0} = 0 \quad \text{and} \quad [I_2]_{\epsilon=0} > 0 \tag{4-13}$$

Substituting the expression for I_1 from equation (4-11) we have that a necessary condition for a stationary value of I is that

$$I_1 = \int_{\sigma_1}^{\sigma_2} \left[\eta \frac{\partial L}{\partial s} + \eta' \frac{\partial L}{\partial s'}\right] d\sigma = 0 \tag{4-14}$$

for arbitrary η. Now η and η' are obviously interdependent; but we cannot express this interdependence in simple algebraic terms. To get over this difficulty we integrate the second term in the integrand of (4-14), namely $\eta'(\partial L/\partial s')$, by parts to get

$$\int_{\sigma_1}^{\sigma_2} \frac{\partial L}{\partial s'} \eta' \, d\sigma = \left[\frac{\partial L}{\partial s'} \eta\right]_{\sigma_1}^{\sigma_2} - \int_{\sigma_1}^{\sigma_2} \left[\frac{d}{d\sigma}\left(\frac{\partial L}{\partial s'}\right)\right] \eta \, d\sigma \tag{4-14a}$$

Variation between Fixed End-points

Restrict $\eta(\sigma)$ to be such that

$$\eta(\sigma_1) = \eta(\sigma_2) = 0$$

Then we have for the case of variation between fixed end-points

$$\int_{\sigma_1}^{\sigma_2} \left[\eta \frac{\partial L}{\partial s} - \eta \left\{ \frac{d}{d\sigma} \left(\frac{\partial L}{\partial s'} \right) \right\} \right] d\sigma = 0$$

so that

$$\int_{\sigma_1}^{\sigma_2} \left[\frac{\partial L}{\partial s} - \frac{d}{d\sigma} \left(\frac{\partial L}{\partial s'} \right) \right] \eta(\sigma) \, d\sigma = 0$$

Since $\eta(\sigma)$ is arbitrary, it follows from the Fundamental Lemma of the Calculus of Variations that a necessary condition for $s(\sigma)$ to make the integral I stationary for small variations is that

$$\frac{d}{d\sigma} \left(\frac{\partial L}{\partial s'} \right) - \frac{\partial L}{\partial s} = 0 \qquad (4\text{--}15)$$

This is the Euler–Lagrange equation of the Calculus of Variations.

Transversality Conditions

Suppose the allowed variation has only one end-point fixed so that

$$\delta s(\sigma_1) = 0 \quad \text{and} \quad \delta s(\sigma_2) \neq 0$$

Again considering equation (4–14a) we see that the Euler–Lagrange equation must still hold since $\delta s(\sigma)$ is not zero over the interval $[\sigma_1, \sigma_2]$. For δI_s to vanish we therefore require the further condition

$$\left[\frac{\partial L}{\partial s'} \right]_{\sigma = \sigma_2} = 0 \qquad (4\text{--}15a)$$

This *transversality condition* is used to furnish the required extra boundary condition

Example 4.1

Find the minimizing curve for

$$I = \int_0^1 \left\{ \left(\frac{dx}{d\sigma} \right)^2 + 12 x \sigma \right\} d\sigma$$

with   (a)   fixed end-points $x(0) = 0$ and $x(1) = 1$;

         (b) $x(0) = 0$ and $x(1)$ free.

We have, using a prime to denote differentiation with respect to σ,

$$L = (x')^2 + 12 x \sigma$$

Fundamental Processes of the Calculus of Variations

so that
$$\frac{\partial L}{\partial x} = 12\sigma \qquad \frac{\partial L}{\partial x'} = 2x'$$

The Euler–Lagrange equation is thus

$$2\frac{d^2 x}{d\sigma^2} - 12\sigma = 0$$

having as solution

$$x = \sigma^3 + \alpha\sigma + \beta$$

where α and β are constants determined by the boundary conditions.
Case (a). For the fixed end-points we have

$$x(0) = \beta = 0 \qquad x(1) = 1 + \alpha = 1 \quad \text{so that} \quad \alpha = 0$$

The minimizing curve with given fixed end-points is thus

$$x = \sigma^3$$

Case (b). The left-hand fixed point again gives $\beta = 0$. To evaluate α we use the transversality condition

$$\left[\frac{\partial L}{\partial x'}\right]_{\sigma=\sigma_2} = [2x']_{\sigma=1} = 0$$

Thus
$$x'(1) = 0$$
where
$$x' = 3\sigma^2 + \alpha$$
This gives
$$3 + \alpha = 0$$
$$\alpha = -3$$

The minimizing curve with a free right-hand end-point is thus

$$x = \sigma^3 - 3\sigma$$

4.1.2 Stationary Value of Integral with Fixed End-points and Several Dependent Variables

We now seek a set of equations which are necessarily satisfied by a set of twice-differentiable functions $\{x_1(\sigma), x_2(\sigma), \ldots, x_n(\sigma)\}$ such that the definite integral

$$I = \int_{\sigma_1}^{\sigma_2} L(x_1, x_2, \ldots, x_n; x'_1, x'_2, \ldots, x'_n; \sigma)\, d\sigma \qquad (4\text{–}16)$$

is stationary with respect to those variations which retain prescribed values of the function set $\{x_1(\sigma), x_2(\sigma), \ldots, x_n(\sigma)\}$ at the fixed limits of integration σ_1, and σ_2 where $\sigma_1 < \sigma_2$. As before, the prime denotes differentiation with respect to σ.

Suppose a set of functions $\{s_1(\sigma), s_2(\sigma), \ldots, s_n(\sigma)\}$ give a stationary value to I. Proceeding as for the single-variable case, we form the one-parameter family of comparison functions

$$\bar{s}_i(\sigma) = s_i(\sigma) + \epsilon \eta_i(\sigma) \qquad i = 1, 2, \ldots, n \qquad (4\text{-}17)$$

where $\{\eta_i(\sigma): i = 1, 2, \ldots, n\}$ is a set of arbitrary differentiable functions such that

$$\eta_1(\sigma_1) = \eta_2(\sigma_1) = \cdots = \eta_n(\sigma_1) = \eta_1(\sigma_2) = \eta_2(\sigma_2) = \cdots = \eta_n(\sigma_2) = 0$$

and ϵ is the parameter of the family. The conditions imposed on the set of functions $\{\eta_i(\sigma): i = 1, 2, \ldots, n\}$ ensure that the end conditions are satisfied by all members of the comparison-function family; also, regardless of the choice of function set, the set of functions $\{s_i(\sigma): i = 1, 2, \ldots, n\}$ is a member of each comparison family for $\epsilon = 0$.

Thus if we form the integral

$$I(\epsilon) = \int_{\sigma_1}^{\sigma_2} L(\bar{s}_1, \ldots, \bar{s}_n; \bar{s}'_1, \ldots, \bar{s}'_n; \sigma) \, d\sigma \qquad (4\text{-}18)$$

by replacing $\{x_i(\sigma): i = 1, 2, \ldots, n\}$ by $\{\bar{s}_i(\sigma): i = 1, 2, \ldots, n\}$ in (4–16), we have that the stationary value is given by

$$[I(\epsilon)]_{\epsilon=0}$$

and we conclude that a necessary condition for the stationary value is

$$\left[\frac{dI(\epsilon)}{d\epsilon}\right]_{\epsilon=0} = 0 \qquad (4\text{-}19)$$

From equations (4–17) we have that

$$\frac{d\bar{s}_j}{d\sigma} = \frac{ds_j}{d\sigma} + \epsilon \frac{d\eta_j}{d\sigma} \qquad j = 1, 2, \ldots, n$$

Using the rule for differentiation of an integral on (4–18) we have

$$\frac{dI}{d\epsilon} = \int_{\sigma_1}^{\sigma_2} \sum_{i=1}^{n} \left\{ \frac{\partial L}{\partial \bar{s}_i} \eta_i + \frac{\partial L}{\partial \bar{s}'_i} \eta'_i \right\} d\sigma \qquad (4\text{-}20)$$

since

$$\frac{\partial \bar{s}_i}{\partial \epsilon} = \eta_i \qquad i = 1, 2, \ldots, n$$

Now setting $\epsilon = 0$ is equivalent to replacing $\{\bar{s}_i(\sigma): i = 1, 2, \ldots, n\}$ by

$$\{s_i(\sigma): i = 1, 2, \ldots, n\}$$

Fundamental Processes of the Calculus of Variations 225

and
$$\{\bar{s}_i'(\sigma) : i = 1, 2, \ldots, n\}$$
by
$$\{s_i'(\sigma) : i = 1, 2, \ldots, n\}$$
so that from (4-20) we obtain, for $\epsilon = 0$

$$\left[\frac{dI}{d\epsilon}\right]_{\epsilon=0} = \int_{\sigma_1}^{\sigma_2} \sum_{i=1}^{n} \left\{\frac{\partial L}{\partial s_i}\eta_i + \frac{\partial L}{\partial s_i'}\eta_i'\right\} d\sigma = 0 \quad (4\text{-}21)$$

This relation must hold for *all* $\{\eta_i(\sigma) : i = 1, 2, \ldots, n\}$. In particular, it holds for the special choice in which $\eta_2(\sigma), \eta_3(\sigma), \ldots, \eta_n(\sigma)$ are all identically zero in the interval $[\sigma_1, \sigma_2]$ but for which $\eta_1(\sigma)$ is arbitrary, consistent with the stipulated end conditions. With this particular choice of $\eta_2(\sigma), \ldots, \eta_n(\sigma)$ we can integrate, as before, the second part of (4-20) by parts to get

$$\int_{\sigma_1}^{\sigma_2} \left[\frac{\partial L}{\partial s_1} - \frac{d}{d\sigma}\left(\frac{\partial L}{\partial s_1'}\right)\right] \eta_1(\sigma) \, d\sigma = 0$$

This must hold for all admissible $\eta_1(\sigma)$ so that the Fundamental Lemma of the Calculus of Variations gives

$$\frac{d}{d\sigma}\left(\frac{\partial L}{\partial s_1'}\right) - \frac{\partial L}{\partial s_1} = 0$$

A similar treatment of all the successive terms of (4-21) gives that a set of necessary conditions which must hold for a set of functions to make the integral I stationary is that the set of simultaneous Euler–Lagrange equations below be satisfied:

$$\frac{d}{d\sigma}\left(\frac{\partial L}{\partial s_i'}\right) - \frac{\partial L}{\partial s_i} = 0 \qquad i = 1, 2, \ldots, n \quad (4\text{-}22)$$

Transversality Conditions

When the second term on the right-hand side of equation (4-20) is integrated by parts we get

$$\int_{\sigma_1}^{\sigma_2} \sum_{i=1}^{n} \frac{\partial L}{\partial s_i'} \eta_i' \, d\sigma = \sum_{i=1}^{n} \left[\frac{\partial L}{\partial s_i'} \eta_i\right]_{\sigma_1}^{\sigma_2} - \sum_{i=1}^{n} \int_{\sigma_1}^{\sigma_2} \frac{d}{d\sigma}\left(\frac{\partial L}{\partial s_i'}\right) \eta_i \, d\sigma \quad (4\text{-}23)$$

If we again take the variations with the right-hand end-point free so that

$$\delta s_i(\sigma_1) = 0 \quad \text{and} \quad \delta s_i(\sigma_2) \neq 0 \qquad i = 1, 2, \ldots, n$$

then, for the first term on the right-hand side of equation (4-23) to vanish, we must have that

$$\left[\frac{\partial L}{\partial s_i'}\right]_{\sigma=\sigma_2} = 0 \qquad i = 1, 2, \ldots, n$$

This set of transversality conditions gives the required extra boundary conditions.

4.2 Generalized Coordinates

Generalized coordinates specify dynamical system behaviour with respect to a reference framework. A set of generalized coordinates is *complete* if their specification uniquely defines the dynamical condition of *every* element in the system. A set of generalized coordinates is *independent* if a range of values exists for any single coordinate which, when all other coordinates are held fixed, corresponds to a range of differing system conditions. The same concepts of completeness and independence apply to small admissible variations of generalized coordinates. A set of admissible variations is complete if every admissible variation in the system-generalized coordinate set can be expressed as a linear combination of the variation set; it is said to be independent if, after fixing all except one member of the variation set, the system condition may still be varied in an admissible way. For mechanical systems it is convenient to call the maximal number of independent admissible variations for a system the number of degrees of freedom of the system. If the number of degrees of freedom coincides with the number of independent variables in some complete set of generalized coordinates, then the independent variations may be taken as variations of the independent generalized coordinates, and the system constraints are said to be *holonomic*. If the number of independent variations is not the same as the number of generalized coordinates, the system constraints are said to be *non-holonomic*. Only holonomic constraints are considered here.

4.2.1 Generalized Velocities

Consider a *mechanical* system with holonomic constraints associated with a set of n generalized coordinates $\{q_1, q_2, \ldots, q_n\}$ and denote the n independent small admissible variations compatible with the system constraints by $\{\delta q_1, \delta q_2, \ldots, \delta q_n\}$. Since the set of generalized coordinates is complete, the displacement x of any particle in the system may be expressed in terms of them, that is,

$$x = x(q_1, q_2, \ldots, q_n)$$

and so the velocity of the particle may be expressed as:

$$v = \frac{dx}{dt} = \sum_{i=1}^{n} \frac{\partial x}{\partial q_i} \frac{dq_i}{dt} = \sum_{i=1}^{n} \frac{\partial x}{\partial q_i} \dot{q}_i(t) \qquad (4\text{--}24)$$

The functions $\dot{q}_i(t)$ are called *generalized velocities*; the actual particle velocities are obviously combinations of the generalized velocities.

4.2.2 Generalized Forces

Suppose that, associated with a variation of the system-generalized coordinates, there is a corresponding amount of virtual work δW. This virtual work may be expressed in terms of the set of actual forces associated with all the components $\{f_i : i = 1, 2, \ldots, n\}$ and the associated set of actual in-line component displacements during the variation $\{\delta x_i : i = 1, 2, \ldots, n\}$ as

$$\delta W = \sum_{i=1}^{n} f_i \delta x_i \qquad (4\text{--}25)$$

This amount of virtual work may equally well be expressed in terms of the variations in the generalized coordinate set as

$$\sum_{i=1}^{n} F_i \delta q_i = \sum_{i=1}^{n} f_i \delta x_i \qquad (4\text{--}26)$$

The quantities $\{F_i : i = 1, 2, \ldots, n\}$ which are defined by equation (4-26) are called *generalized forces*. Their dimensions depend on the dimensions of the corresponding variations $\{\delta q_i : i = 1, 2, \ldots, n\}$. If δq_i has the dimensions of length, then F_i will have the dimension of force; if δq_j is an angle, then F_j will be a torque; and so on. The simplest way to evaluate the individual members of the set $\{F_i\}$ is to consider a variation of generalized coordinates in which only one coordinate is varied at a time so that

$$\delta q_k \neq 0$$

and $\qquad \delta q_i = 0 \quad \text{for} \quad i \neq k$

A study of the system constraints then determines which members of the set $\{\delta x_i\}$ are associated with this variation and how much virtual work is involved. Dividing this amount of virtual work by δq_k then gives the corresponding generalized force.

4.3 Primal Form of Hamilton's Postulate and the Set of Lagrangian Equations [C4]

Lagrangian

Let the total co-kinetic energy and potential energy of a mechanical system be expressed in terms of a complete independent set of generalized coordinates and generalized velocities

$$T^* = T^*(\dot{q}_1, \dot{q}_2, \ldots, \dot{q}_n)$$
$$U = U(q_1, q_2, \ldots, q_n)$$

The system Lagrangian is then defined as

$$L(q_1,\ldots,q_n;\dot{q}_1,\ldots,\dot{q}_n) = T^* - U$$

Primal Form of Hamilton's Postulate [C4]

The primal form of Hamilton's Postulate states that the natural motion of a dissipationless dynamical system, free from external disturbance, from a fixed configuration at time t_1 to another fixed configuration at time t_2 is such that the integral

$$I = \int_{t_1}^{t_2} L\, dt$$

is stationary on the path followed. That is

$$\delta I = \delta \int_{t_1}^{t_2} L\, dt = 0 \qquad (4\text{-}27)$$

for arbitrary admissible variations satisfying the geometrical constraints of the system between the specified fixed end-points of the trajectory. It immediately follows from the fundamental processes of the Calculus of Variations considered above that the generalized coordinates associated with the behaviour of such a dynamical system must necessarily satisfy the set of Lagrangian equations of motion

$$\frac{d}{dt}\left(\frac{\partial L}{\partial \dot{q}_i}\right) - \frac{\partial L}{\partial q_i} = 0 \qquad i = 1, 2,\ldots, n \qquad (4\text{-}28)$$

These equations have the immensely *important property of being true for every set of generalized coordinates*. This is simply because they result from the stationary property of the integral I, which ensures that its first variation δI is zero for any set of admissible variations in *any* set of generalized coordinates $\{q_i\}$.

To extend Hamilton's Postulate to the case where dissipation and external disturbing forces are present, the form of variation of an integral cannot be retained, and the integral of a variation plus extra terms involving the virtual work now associated with source and dissipation variations must be used. This extended form of the primal postulate states that the natural motion of a dynamical system from a fixed configuration at a time t_1 to another fixed configuration at a time t_2 is such that

$$\int_{t_1}^{t_2} (\delta L + \delta W)\, dt = 0 \qquad (4\text{-}29)$$

where δL is the variation in the Lagrangian, and δW is the virtual work done by dissipators and sources, for an arbitrary admissible variation satisfying the imposed geometrical constraints between the

Primal Form of Hamilton's Postulate

fixed end-points of the trajectory. The virtual work is taken as positive†
when work is done *by* an object and expressed as

$$\delta W = \sum_{i=1}^{n} \bar{F}_i \delta q_i$$

where $\bar{F}_i$ is a suitable set of generalized forces. It is convenient to distinguish between the virtual work associated with system dissipation and that associated with the effect of an ideal set of external forces acting on the system.

Let
$$\bar{F}_i = F_i + \Xi_i \qquad i = 1, 2, \ldots, n$$

Where $\{\Xi_i : i = 1, 2, \ldots, n\}$ is the set of contributions to the generalized force set $\{\bar{F}_i\}$ arising from the effect of ideal external forces and $\{F_i : i = 1, 2, \ldots, n\}$ arise from dissipation.

It will be helpful at this point to give some simple examples of the direct application of Hamilton's Postulate in its primal form.

Example 4.2

Consider the linear translational mechanical system of Figure 4–3. Let M, B and K be respectively the mass of the trolley, the viscous damping coefficient of the dissipator, and the stiffness of the spring; and let F be an externally applied force. Take the generalized displacement characterizing the system behaviour as simply the displacement of the trolley. We can then have that, since the kinetic co-energy and kinetic energy will be equal

$$L = \tfrac{1}{2} M \dot{q}^2 - \tfrac{1}{2} K q^2$$

so that
$$\delta L = \delta[\tfrac{1}{2} M \dot{q}^2 - \tfrac{1}{2} K q^2] = \tfrac{1}{2} M \delta(\dot{q}^2) - \tfrac{1}{2} K \delta(q^2)$$

$$= \tfrac{1}{2} M 2\dot{q} \delta(\dot{q}) - \tfrac{1}{2} K 2q \delta q = M\dot{q}\frac{d}{dt}(\delta q) - Kq \delta q$$

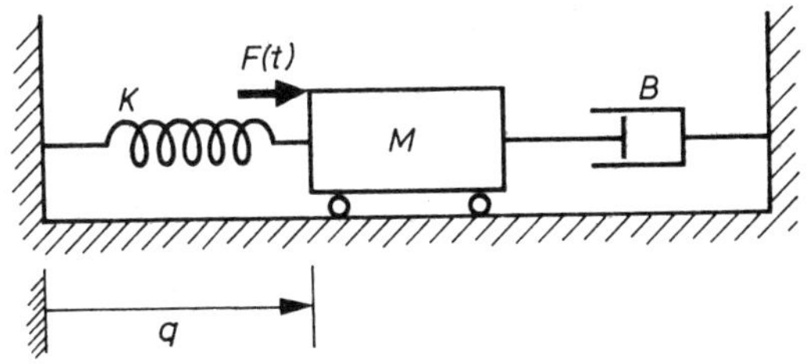

Figure 4–3

† This is the standard convention in analytical mechanics. In network theory, positive work is done on an object.

Taking the virtual work done *by* an object as positive we have
$$\delta W = F\delta q - B\dot{q}\delta q$$
Applying Hamilton's Postulate we thus get
$$\int_{t_1}^{t_2} \left[M\dot{q}\frac{d}{dt}(\delta q) - Kq\delta q + (F - B\dot{q})\delta q \right] dt = 0$$
Integrating the first term in the integral by parts gives
$$[M\dot{q}\delta q]_{t_1}^{t_2} - \int_{t_1}^{t_2} M\ddot{q}\delta q \, dt + \int_{t_1}^{t_2} (-Kq + F - B\dot{q})\delta q \, dt = 0$$
The first term must vanish, since the admissible variations $\delta q(t)$ must vanish at times t_1 and t_2. This gives
$$\int_{t_1}^{t_2} [M\ddot{q} + B\dot{q} + Kq - F]\delta q \, dt = 0$$
and since $\delta q(t)$ is arbitrary, the Fundamental Lemma of the Calculus of Variations gives that the system behaviour is governed by the differential equation
$$M\ddot{q} + B\dot{q} + Kq = F$$

Example 4.3

Now consider the two-degree-of-freedom translational mechanical system shown in Figure 4-4, and take the various parameters involved to be as indicated on the figure. Take as generalized coordinates characterizing the system dynamic behaviour the displacements of the two trollies. We then have
$$U = \tfrac{1}{2}K_1 q_1^2 + \tfrac{1}{2}K_2(q_2 - q_1)^2 + \tfrac{1}{2}K_3 q_2^2$$
$$T^* = \tfrac{1}{2}M_1 \dot{q}_1^2 + \tfrac{1}{2}M_2 \dot{q}_2^2$$
$$\delta L = \delta T^* - \delta U$$
$$= M_1\dot{q}_1\frac{d}{dt}(\delta q_1) + M_2\dot{q}_2\frac{d}{dt}(\delta q_2) - K_1 q_1 \delta q_1 - K_2 q_2 \delta q_2$$
$$+ K_2 q_1 \delta q_2 + K_2 q_2 \delta q_1 - K_2 q_1 \delta q_1 - K_3 q_2 \delta q_2$$
Again taking the virtual work done by an object as positive we have
$$\delta W = F_1 \delta q_1 + F_2 \delta q_2 - B_1(\dot{q}_2 - \dot{q}_1)\delta(q_2 - q_1) - B_2\dot{q}_2 \delta q_2$$
$$= F_1 \delta q_1 + F_2 \delta q_2 - B_1 \dot{q}_2 \delta q_2 + B_1 \dot{q}_2 \delta q_1 + B_1 \dot{q}_1 \delta q_2$$
$$- B_1 \dot{q}_1 \delta q_1 - B_2 \dot{q}_2 \delta q_2$$

Primal Form of Hamilton's Postulate

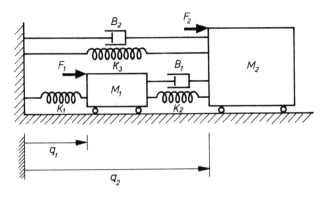

Figure 4-4

An application of Hamilton's Postulate then gives

$$\int_{t_1}^{t_2} \Big[M_1\dot{q}_1 \frac{d}{dt}(\delta q_1) + M_2\dot{q}_2 \frac{d}{dt}(\delta q_2) - K_1 q_1 \delta q_1 - K_2 q_2 \delta q_2 + K_2 q_1 \delta q_2$$
$$+ K_2 q_2 \delta q_1 - K_2 q_1 \delta q_1 - K_3 q_2 \delta q_2 + F_1 \delta q_1 + F_2 \delta q_2 - B_1 \dot{q}_2 \delta q_2$$
$$+ B_1 \dot{q}_2 \delta q_1 + B_1 \dot{q}_1 \delta q_2 - B_1 \dot{q}_1 \delta q_1 - B_2 \dot{q}_2 \delta q_2 \Big] dt = 0$$

On integrating by parts as usual, the first two terms vanish and are replaced under the integral sign by $-M_1\ddot{q}_1\delta q_1$ and $-M_2\ddot{q}_2\delta q_2$ respectively. We thus get

$$\int_{t_1}^{t_2} [-M_1\ddot{q}_1 - K_1 q_1 + K_2 q_2 - K_2 q_1 + F_1 + B_1 \dot{q}_2 - B_1 \dot{q}_1]\delta q_1 \, dt$$
$$+ \int_{t_1}^{t_2} [-M_2\ddot{q}_2 - K_2 q_2 + K_2 q_1 - K_3 q_2 + F_2 - B_1 \dot{q}_2 + B_1 \dot{q}_1$$
$$- B_2 \dot{q}_2]\delta q_2 \, dt = 0$$

Since the variations $\delta q_1(t)$ and $\delta q_2(t)$ are both arbitrary, we may first choose a combined variation in which $\delta q_2(t)$ is identically zero for arbitrary $\delta q_1(t)$. The Fundamental Lemma of the Calculus of Variations then gives that

$$M_1\ddot{q}_1 + B_1(\dot{q}_1 - \dot{q}_2) + K_1 q_1 - K_2(q_2 - q_1) = F_1$$

Similarly by considering a combined variation in which $\delta q_1(t)$ is identically zero for arbitrary $\delta q_2(t)$ we get

$$M_2\ddot{q}_2 + B_1(\dot{q}_2 - \dot{q}_1) + B_2\dot{q}_2 + K_2(q_2 - q_1) + K_3 q_2 = F_2$$

4.3.1 Lagrangian Equation Set

The Lagrangian equations given by (4–28) above only apply in the very restricted case where dissipation and external forces are absent. We now proceed to derive a general form of such equations which includes both of these circumstances. The individual system components representing dissipation are associated with a set of component forces $\{f_i : i = 1, 2, \ldots, n\}$ which may be evaluated in terms of individual dissipator component co-content functions as in Chapter 1:

$$f_i = \frac{\partial J_i}{\partial v_i} \qquad i = 1, 2, \ldots, d$$

where d is the number of dissipator components and J_i are the individual dissipator co-contents, which are assumed to be known functions of the dissipator component velocities $\{v_i : i = 1, 2, \ldots, n\}$. Let the *total* system co-content be expressed in terms of the set of generalized velocities

$$J = J(\dot{q}_1, \ldots, \dot{q}_n)$$

and define the corresponding set of generalized forces by

$$F_i = \frac{\partial J}{\partial \dot{q}_i} \qquad i = 1, 2, \ldots, n \qquad (4\text{–}30)$$

Let the actual in-line displacement variations for the dissipative components be related to the generalized displacement variations by

$$\delta q_i = \sum_{j=1}^{m} t_{ij} \delta x_j$$

or, in matrix and vector form

$$\delta \mathbf{q} = \mathbf{T} \delta \mathbf{x} \qquad (4\text{–}31)$$

where $\delta \mathbf{q}$ is a vector whose components are the set $\{\delta q_i\}$ and $\delta \mathbf{x}$ is a vector whose components are the set $\{\delta x_i\}$, and $\mathbf{T}$ is a suitable transformation matrix. If we also let $\mathbf{f}$ be a vector whose components are the actual dissipator component force set $\{f_i\}$ and $\mathbf{v}$ be a vector whose components are the individual dissipator velocities, then we have, for the virtual work done on the dissipative elements:

$$\delta W_d = \langle \delta \mathbf{x}, \mathbf{f} \rangle$$

$$= \left\langle \delta \mathbf{x}, \left(\frac{\partial J}{\partial \mathbf{v}} \right)^t \right\rangle$$

$$= \left\langle \delta \mathbf{x}, \mathbf{T}^t \left(\frac{\partial J}{\partial \mathbf{T} \mathbf{v}} \right)^t \right\rangle$$

Primal Form of Hamilton's Postulate

$$= \left\langle \delta\mathbf{x}, \mathbf{T}^t \left(\frac{\partial J}{\partial \dot{\mathbf{q}}}\right)^t \right\rangle$$

$$= \left\langle \mathbf{T}\delta\mathbf{x}, \left(\frac{\partial J}{\partial \dot{\mathbf{q}}}\right)^t \right\rangle$$

$$= \langle \delta\mathbf{q}, \mathbf{F} \rangle$$

where $\mathbf{F}$ is a vector having components $F_1, F_2, \ldots, F_n$. Thus we have

$$\delta W_d = \sum_{i=1}^{n} \frac{\partial J}{\partial \dot{q}_i} \delta q_i \qquad (4\text{-}32)$$

where δW_d is the virtual work associated with dissipative elements. This gives

$$\delta W = \sum_{i=1}^{n} \frac{\partial J}{\partial \dot{q}_i} \delta q_i + \sum_{i=1}^{n} \Xi_i \delta q_i \qquad (4\text{-}33)$$

The extended primal form of Hamilton's Postulate thus gives

$$\int_{t_1}^{t_2} (\delta L + \delta W) \, dt$$

$$= \int_{t_1}^{t_2} \left[\sum_{i=1}^{n} \frac{\partial L}{\partial \dot{q}_i} \delta \dot{q}_i + \sum_{i=1}^{n} \frac{\partial L}{\partial q_i} \delta q_i + \sum_{i=1}^{n} \frac{\partial J}{\partial \dot{q}_i} \delta q_i + \sum_{i=1}^{n} \Xi_i \delta q_i \right] dt = 0 \qquad (4\text{-}34)$$

Consider the first term in the integral, namely

$$\int_{t_1}^{t_2} \sum_{i=1}^{n} \frac{\partial L}{\partial \dot{q}_i} \delta \dot{q}_i \, dt$$

Since the variation of a derivative is equal to the derivative of a variation, this first term may be written as

$$\int_{t_1}^{t_2} \sum_{i=1}^{n} \frac{\partial L}{\partial \dot{q}_i} \frac{d}{dt}(\delta q_i) \, dt$$

and then integrated by parts to give

$$\int_{t_1}^{t_2} \sum_{i=1}^{n} \frac{\partial L}{\partial \dot{q}_i} \frac{d}{dt}(\delta q_i) \, dt = \left[\sum_{i=1}^{n} \frac{\partial L}{\partial \dot{q}_i} \delta q_i \right]_{t_1}^{t_2} - \int_{t_1}^{t_2} \sum_{i=1}^{n} \frac{d}{dt}\left(\frac{\partial L}{\partial \dot{q}_i}\right) \delta q_i \, dt$$

Since the system configurations are fixed at times t_1 and t_2 the admissible variations δq_i must vanish at t_1 and t_2 and thus the first term arising from this integration by parts must vanish. Thus we have

$$\int_{t_1}^{t_2} (\delta L + \delta W)\, dt$$

$$= \int_{t_1}^{t_2} \sum_{i=1}^{n} \left[-\frac{d}{dt}\left(\frac{\partial L}{\partial \dot q_i}\right) + \frac{\partial L}{\partial q_i} + \frac{\partial J}{\partial \dot q_i} + \Xi_i \right] \delta q_i\, dt = 0 \qquad (4\text{-}35)$$

Since the variations δq_i are arbitrary, it follows, as above, from the Calculus of Variations that the system must obey the extended set of Lagrangian equations

$$\frac{d}{dt}\left(\frac{\partial L}{\partial \dot q_i}\right) - \frac{\partial J}{\partial \dot q_i} - \frac{\partial L}{\partial q_i} = \Xi_i \qquad i = 1, 2, \ldots, n \qquad (4\text{-}36)$$

Two examples are now considered of the Lagrangian equations of motion for mechanical systems. The chief point illustrated in these examples is the way in which generalized coordinates are used.

Example 4.4   Equations of Motion of Flyball Governor

Lagrange's equations may be used to determine the motion of the flyball governor shown in Figure 4–5. It consists of four light rigid rods pin-jointed at their junctions in such a way, that the mass M may move freely up and down the vertical axis, and the whole arrangement may rotate freely about this vertical axis. As it rotates, the masses m move outwards under the action of centrifugal force, moving M vertically. As a set of generalized coordinates, we may take

$q_1 =$ the angle made by the plane of m, m and M with some initial reference plane.
$q_2 =$ the angle between a light rigid rod and the vertical axis.

Let I be the moment of inertia of the mass M about the vertical axis, and assume that the masses m can be treated as particles for the purposes of computing rotational inertia. We must first express the system potential energy and kinetic co-energy in terms of the adopted generalized coordinates and associated generalized velocities. Take the zero potential energy state to be that in which the two masses m lie in the vertical axis. The system potential energy is then

$$U = 2mgl(1 - \cos q_2) + Mgl2(1 - \cos q_2)$$
$$= 2gl(M + m)(1 - \cos q_2)$$

(Note that M moves twice the vertical distance moved by m.)

The total kinetic co-energy consists of the kinetic co-energy T_1^* associated with rotation about the vertical axis, plus the kinetic co-energy T_2^* associated with motion in the plane of m, m and M. For any

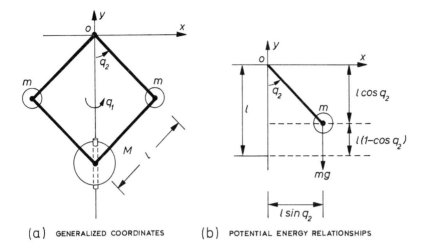

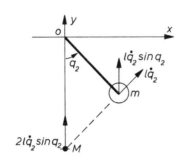

Figure 4-5 Flyball governor.

mechanical system in which the velocities involved are very small compared to the velocity of light we may, as discussed in Chapter 1, equate kinetic co-energy and kinetic energy. The two constituent parts of the kinetic co-energy are therefore as follows. Obviously

$$T_1^* = \tfrac{1}{2}(I + 2ml^2 \sin^2 q_2)\dot{q}_1^2$$

For motion in the plane of the light rods we have

instantaneous velocity of each $m = l\dot{q}_2$
instantaneous velocity of $M \quad = 2l\dot{q}_2 \sin q_2$

so that
$$T_2^* = \tfrac{1}{2}(2ml^2\dot{q}_2^2 + 4Ml^2\dot{q}_2^2 \sin^2 q_2)$$

The system Lagrangian is thus
$$L = T_1^* + T_2^* - U$$
$$= \tfrac{1}{2}I\dot{q}_1^2 + ml^2 \sin^2 q_2 \dot{q}_1^2 + ml^2 \dot{q}_2^2 + 2Ml^2 \dot{q}_2^2 \sin^2 q_2$$
$$- 2gl(m+M) + 2gl(m+M)\cos q_2$$
so that
$$\frac{\partial L}{\partial \dot{q}_1} = (I + 2ml^2 \sin^2 q_2)\dot{q}_1$$

$$\frac{\partial L}{\partial \dot{q}_2} = (2ml^2 + 4Ml^2 \sin^2 q_2)\dot{q}_2$$

$$\frac{\partial L}{\partial q_1} = 0$$

$$\frac{\partial L}{\partial q_2} = 2l^2 \sin q_2 \cos q_2 (m\dot{q}_1^2 + 2M\dot{q}_2^2) - 2gl(m+M)\sin q_2$$

From these the system Lagrangian equations of motion are obtained as
$$\frac{d}{dt}[(I + 2ml^2 \sin^2 q_2)\dot{q}_1] = 0$$
$$\frac{d}{dt}[(2ml^2 + 4Ml^2 \sin^2 q_2)\dot{q}_2] - 2l^2 \sin q_2 \cos q_2 (m\dot{q}_1^2 + 2M\dot{q}_2^2)$$
$$+ 2gl(m+M)\sin q_2 = 0$$

Note that the first of these equations shows that the angular momentum about the vertical axis is conserved. q_1 is said to be a cyclic or ignorable coordinate; this concept is further discussed below.

Example 4.5 Double Pendulum

As a further simple example of the use of the Lagrangian equations, consider the double pendulum shown in Figure 4–6. This consists of two massive particles m_1 and m_2 on light rigid pin-jointed rods and constrained to move in one plane. As a set of generalized coordinates we may take, as shown in the figure, the angles made by the pair of rods with the vertical. If we take the zero potential energy reference state to be that in which the pair of generalized coordinates chosen is zero, then we have for the system total potential energy
$$U = m_1 g l_1 (1 - \cos q_1) + m_2 g[l_1(1 - \cos q_1) + l_2(1 - \cos q_2)]$$
The kinetic co-energy must equal the kinetic energy and so is
$$T^* = \tfrac{1}{2}m_1(l_1\dot{q}_1)^2 + \tfrac{1}{2}m_2[(l_1\dot{q}_1)^2 + (l_2\dot{q}_2)^2 + 2l_1 l_2 \cos(q_2 - q_1)\dot{q}_1\dot{q}_2]$$

Primal Form of Hamilton's Postulate

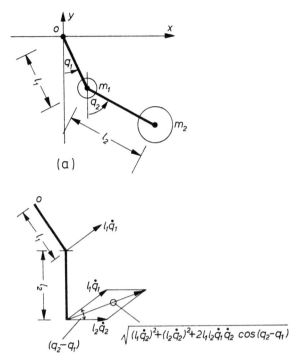

Figure 4-6 Double pendulum.

using the kinetic relationships of Figure 4-6(b). Calculating the various derivatives involved in the Lagrangian equations we have

$$\frac{\partial L}{\partial q_1} = \frac{\partial T^*}{\partial q_1} - \frac{\partial U}{\partial q_1} = -m_2 l_1 l_2 \dot{q}_1 \dot{q}_2 \sin(q_2 - q_1) - (m_1 + m_2) g l_1 \sin q_1$$

$$\frac{\partial L}{\partial q_2} = \frac{\partial T^*}{\partial q_2} - \frac{\partial U}{\partial q_2} = m_2 l_1 l_2 \dot{q}_1 \dot{q}_2 \sin(q_2 - q_1) - m_2 g l_2 \sin q_2$$

$$\frac{\partial L}{\partial \dot{q}_1} = \frac{\partial T^*}{\partial \dot{q}_1} = m_1 l_1^2 \dot{q}_1 + m_2 l_1^2 \dot{q}_1 + m_2 l_1 l_2 \cos(q_2 - q_1) \dot{q}_2$$

$$\frac{\partial L}{\partial \dot{q}_2} = \frac{\partial T^*}{\partial \dot{q}_2} = m_2 l_2^2 \dot{q}_2 + m_2 l_1 l_2 \cos(q_2 - q_1) \dot{q}_1$$

The Lagrangian equations are thus

$$\frac{d}{dt}[(m_1+m_2)l_1^2\dot{q}_1 + m_2 l_1 l_2 \cos(q_2-q_1)\dot{q}_2]$$

$$+ m_2 l_1 l_2 \dot{q}_1 \dot{q}_2 \sin(q_2-q_1) + (m_1+m_2)gl_1 \sin q_1 = 0$$

$$\frac{d}{dt}[m_2 l_2^2 \dot{q}_2 + m_2 l_1 l_2 \cos(q_2-q_1)\dot{q}_1]$$

$$- m_2 l_1 l_2 \dot{q}_1 \dot{q}_2 \sin(q_2-q_1) + m_2 g l_2 \sin q_2 = 0$$

giving

$$(m_1+m_2)l_1^2 \ddot{q}_1 + m_2 l_1 l_2 \cos(q_2-q_1)\ddot{q}_2 + m_2 l_1 l_2 \dot{q}_2(\dot{q}_1-\dot{q}_2)\sin(q_2-q_1)$$

$$+ m_2 l_1 l_2 \dot{q}_1 \dot{q}_2 \sin(q_2-q_1) + (m_1+m_2)gl_1 \sin q_1 = 0$$

$$m_2 l_2^2 \ddot{q}_2 + m_2 l_1 l_2 \cos(q_2-q_1)\ddot{q}_1 + m_2 l_1 l_2 \dot{q}_1(\dot{q}_1-\dot{q}_2)\sin(q_2-q_1)$$

$$- m_2 l_1 l_2 \dot{q}_1 \dot{q}_2 \sin(q_2-q_1) + m_2 g l_2 \sin q_2 = 0$$

4.4 Generalized Momenta

If a mechanical system contains N mass particles, the corresponding system kinetic co-energy may be expressed as

$$T^* = \sum_{j=1}^{3N} \int_0^{v_j} \bar{p}_j \, dv_j \qquad (4\text{-}37)$$

where $\{\bar{p}_j : j = 1, 2, \ldots, 3N\}$ and $\{v_j : j = 1, 2, \ldots, 3N\}$ are respectively the appropriate components of particle momenta and velocity in a set of Cartesian coordinate reference directions. The individual mass component velocities may be expressed in terms of the system generalized coordinate set as

$$\dot{x}_j = v_j = \sum_{i=1}^{3N} \frac{\partial x_j}{\partial q_i} \frac{dq_i}{dt} = \sum_{i=1}^{3N} \frac{\partial x_j}{\partial q_i} \dot{q}_i \qquad j = 1, 2, \ldots, 3N \qquad (4\text{-}38)$$

so that

$$\frac{\partial \dot{x}_j}{\partial \dot{q}_i} = \frac{\partial v_j}{\partial \dot{q}_i} = \frac{\partial x_j}{\partial q_i} \qquad i, j, 1, 2, \ldots, 3N \qquad (4\text{-}39)$$

This now enables us to give a physical interpretation to the terms $\partial L/\partial \dot{q}_i$ which occur in equation (4–36). We have

$$\frac{\partial L}{\partial \dot{q}_i} = \frac{\partial}{\partial \dot{q}_i}(T^* - U) = \frac{\partial T^*}{\partial \dot{q}_i}$$

Now, using (4-37) we have

$$\frac{\partial T^*}{\partial \dot{q}_i} = \sum_{j=1}^{3N} \bar{p}_j \frac{\partial v_j}{\partial \dot{q}_i}$$

Thus, using (4-39) we get

$$\frac{\partial T^*}{\partial \dot{q}_i} = \frac{\partial L}{\partial \dot{q}_i} = \sum_{j=1}^{3N} \bar{p}_j \frac{\partial x_j}{\partial q_i} \qquad (4\text{-}40)$$

The quantity $\partial T^*/\partial \dot{q}_i$ is thus a sum of appropriate components of individual particle momenta and is called a *generalized momentum*, denoted in what follows by

$$p_i \triangleq \frac{\partial L}{\partial \dot{q}_i} \qquad (4\text{-}41)$$

4.5 Dual Form of Hamilton's Postulate and the Set of Co-Lagrangian Equations

Co-Lagrangian

Let the total kinetic energy and potential co-energy of a mechanical system be expressed in terms of a complete independent set of generalized momenta and time rates of change of generalized momenta:

$$T = T(p_1, p_2, \ldots, p_n)$$

$$U^* = U^*(\dot{p}_1, \dot{p}_2, \ldots, \dot{p}_n)$$

In interpreting this, remember that generalized momenta have the dimensions of momentum and that time rate of change of momentum has the dimensions of force. Thus the time rates of change of generalized momenta can be interpreted as a set of generalized forces. The system Co-Lagrangian is then defined as

$$L^*(p_1, \ldots, p_n; \dot{p}_1, \ldots, \dot{p}_n) = U^* - T$$

Dual Form of Hamilton's Postulate

The dual form of Hamilton's Postulate states that the natural motion of a dissipationless dynamical system, free from external disturbances, from a fixed distribution of momenta at time t_1 to another fixed distribution of momenta at time t_2 is such that the integral

$$I^* = \int_{t_1}^{t_2} L^* \, dt$$

is stationary on the path followed. That is

$$\delta I^* = \delta \int_{t_1}^{t_2} L^* \, dt = 0$$

for arbitrary admissible variations satisfying the imposed constraints on forces and momenta. In this dual case the admissible variations must be interpreted in terms of constraints of the form

$$\frac{d\bar{p}_i}{dt} = f_i$$

and

$$\sum_j f_j = 0$$

and not in terms of geometric constraints on displacements and velocities as in the primal case. The fundamental processes of the Calculus of Variations immediately give that the system must satisfy the set of Co-Lagrangian equations of motion

$$\frac{d}{dt}\left(\frac{\partial L^*}{\partial \dot{p}_i}\right) - \frac{\partial L^*}{\partial p_i} = 0 \qquad i = 1, 2, \ldots, n \qquad (4\text{-}42)$$

The dual postulate may be extended to the case where dissipation exists and external agencies do work on the system. In its extended form it states that the motion of a dynamical system from a fixed distribution of momenta at time t_1 to another fixed distribution of momenta at time t_2 is such that the integral

$$\int_{t_1}^{t_2} (\delta L^* + \delta W^*)\, dt = 0 \qquad (4\text{-}43)$$

where δL^* is the variation in the Co-Lagrangian and δW^* is the virtual work done by dissipators and sources for an arbitrary admissible variation of momenta between the fixed end-points of the trajectory. The virtual work is expressed as

$$\delta W^* = \sum_{i=1}^{n} \bar{V}_i \delta p_i \qquad (4\text{-}44)$$

where $\bar{V}_i$ is a suitable set of generalized velocities. It is convenient to distinguish between the virtual work associated with system dissipation and that associated with an ideal set of externally applied velocities acting on the system by putting

$$\bar{V}_i = V_i + \Gamma_i \qquad i = 1, 2, \ldots, n \qquad (4\text{-}45)$$

where $\{\Gamma_i : i = 1, 2, \ldots, n\}$ is the set of contributions to the generalized velocity set $\{\bar{V}_i\}$ arising from the effect of externally applied velocities and $\{V_i : i = 1, 2, \ldots, n\}$ arise from dissipative effects.

Let the *total* system content be expressed in terms of the set of rates of change of generalized momenta as

$$G = G(\dot{p}_1, \dot{p}_2, \ldots, \dot{p}_n) \qquad (4\text{-}46)$$

Conservation of Energy and Momentum 241

and define the corresponding set of generalized velocities by

$$V_i = \frac{\partial G_i}{\partial \dot{p}_i} \qquad i = 1, 2, \ldots, n \qquad (4\text{-}47)$$

Then an argument similar to that given for the primal case gives us that

$$\delta W^* = \sum_{i=1}^n \frac{\partial G_i}{\partial \dot{p}_i} \delta p_i + \sum_{i=1}^n \Gamma_i \delta p_i \qquad (4\text{-}48)$$

and then carrying through a process similar to that given for the primal case gives that the system must obey the extended set of Co-Lagrangian equations:

$$\frac{d}{dt}\left(\frac{\partial L^*}{\partial \dot{p}_i}\right) - \frac{\partial G}{\partial \dot{p}_i} - \frac{\partial L^*}{\partial p_i} = \Gamma_i \qquad i = 1, 2, \ldots, n \qquad (4\text{-}49)$$

4.6 Conservation of Energy and Momentum

For a dissipationless dynamical system free from external disturbances, the set of Lagrangian equations is

$$\frac{d}{dt}\left(\frac{\partial L}{\partial \dot{q}_i}\right) - \frac{\partial L}{\partial q_i} = 0 \qquad i = 1, 2, \ldots, n \qquad (4\text{-}50)$$

Multiplying each equation in this set by $\dot{q}_i$, and adding the resulting equations gives

$$\sum_{i=1}^n \left[\dot{q}_i \frac{d}{dt}\left(\frac{\partial L}{\partial \dot{q}_i}\right) - \dot{q}_i \frac{\partial L}{\partial q_i}\right] = 0 \qquad (4\text{-}51)$$

Now, using the rule for differentiation of a product we have

$$\frac{d}{dt}\left(\dot{q}_i \frac{\partial L}{\partial \dot{q}_i}\right) = \ddot{q}_i \frac{\partial L}{\partial \dot{q}_i} + \dot{q}_i \frac{d}{dt}\left(\frac{\partial L}{\partial \dot{q}_i}\right)$$

so that

$$\dot{q}_i \frac{d}{dt}\left(\frac{\partial L}{\partial \dot{q}_i}\right) = \frac{d}{dt}\left(\dot{q}_i \frac{\partial L}{\partial \dot{q}_i}\right) - \ddot{q}_i \frac{\partial L}{\partial \dot{q}_i} \qquad (4\text{-}52)$$

Inserting the expression for $\dot{q}_i (d/dt)(\partial L/\partial \dot{q}_i)$ given by equation (4-52) into equation (4-51) we have

$$\sum_{i=1}^n \left[\frac{d}{dt}\left(\dot{q}_i \frac{\partial L}{\partial \dot{q}_i}\right) - \ddot{q}_i \frac{\partial L}{\partial \dot{q}_i} - \dot{q}_i \frac{\partial L}{\partial q_i}\right] = 0 \qquad (4\text{-}53)$$

Now

$$L = L(q_1, \ldots, q_n; \dot{q}_1, \ldots, \dot{q}_n)$$

so that
$$\frac{dL}{dt} = \sum_{i=1}^{n} \frac{\partial L}{\partial q_i}\dot{q}_i + \sum_{i=1}^{n} \frac{\partial L}{\partial \dot{q}_i}\ddot{q}_i$$

Substituting this in equation (4-53) gives
$$\frac{d}{dt}\left[\sum_{i=1}^{n} \dot{q}_i \frac{\partial L}{\partial \dot{q}_i}\right] - \frac{dL}{dt} = 0$$

and integrating this gives
$$\sum_{i=1}^{n} \dot{q}_i \frac{\partial L}{\partial \dot{q}_i} - L = \text{constant}$$
(4-54)

so that
$$\sum_{i=1}^{n} \dot{q}_i p_i - L = \text{constant}$$

Now, as shown above in Section 4.4, the individual component momenta and velocities and the generalized momenta and generalized velocities are such that

$$p_i = \sum_{j=1}^{3N} \frac{\partial x_j}{\partial q_i} \bar{p}_j$$

$$v_i = \sum_{j=1}^{3N} \frac{\partial x_i}{\partial q_j} \dot{q}_j$$
(4-55)

Thus if **p** is a vector whose components are the generalized momenta, **p̄** is a vector whose components are the individual component momenta, **v** a vector whose components are the individual component velocities, and **q̇** a vector whose components are the generalized velocities, we have

$$\mathbf{p} = \mathbf{T}\bar{\mathbf{p}}$$
$$\mathbf{v} = \mathbf{T}^t\dot{\mathbf{q}}$$
(4-56)

where the components of the transformation matrix **T** are

$$T_{ij} = \frac{\partial x_j}{\partial q_i}$$

We thus have
$$\langle \bar{\mathbf{p}}, \mathbf{v} \rangle = \langle \bar{\mathbf{p}}, \mathbf{T}^t\dot{\mathbf{q}} \rangle = \langle \mathbf{T}\bar{\mathbf{p}}, \dot{\mathbf{q}} \rangle = \langle \mathbf{p}, \dot{\mathbf{q}} \rangle$$

so that
$$\sum_{i=1}^{n} p_i \dot{q}_i = \sum_{j=1}^{3N} \bar{p}_j v_j = T + T^*$$
(4-57)

Substituting this in equation (4–54) we get

$$\sum_{i=1}^{n} p_i \dot{q}_i - L = T + T^* - L = T + T^* - (T^* - U) = T + U = \text{constant}$$

and thus the principle of the conservation of system energy has been deduced from the Lagrangian equations.

$$\sum_{i=1}^{n} p_i \dot{q}_i - L \triangleq H \qquad (4\text{–}58)$$

is called the system *Hamiltonian*. For a dissipationless system free from external disturbances, it is equal to the total system stored energy.

For a dissipationless system, the set of Lagrangian equations is

$$\frac{d}{dt}\left(\frac{\partial L}{\partial \dot{q}_i}\right) - \frac{\partial L}{\partial q_i} = \Xi_i \qquad i = 1, 2, \ldots, n$$

Suppose that for some particular generalized coordinate, q_k say, L depends only on the derivative $\dot{q}_k$ and Ξ_k is zero. The relevant Lagrangian equation becomes simply

$$\frac{d}{dt}\left(\frac{\partial L}{\partial \dot{q}_k}\right) = 0$$

So that integrating this we get

$$\frac{\partial L}{\partial \dot{q}_k} = p_k = \text{constant}$$

Such a generalized coordinate, for which the corresponding generalized momentum is conserved, is called a *cyclic* or *ignorable* coordinate.

4.7 Hamilton's Equations [W2]

From equation (4–58) we have firstly that

$$L = \sum_{i=1}^{n} p_i \dot{q}_i - H$$

and secondly that

$$\frac{\partial H}{\partial p_i} = \dot{q}_i$$

Thus we obtain the dual set of Legendre transformation relationships:

$$\dot{q}_i = \frac{\partial H}{\partial p_i} \quad (4\text{-}59) \qquad p_i = \frac{\partial L}{\partial \dot{q}_i} \quad (4\text{-}61)$$

$$H = \sum_{i=1}^{n} p_i \dot{q}_i - L \quad (4\text{-}60) \qquad L = \sum_{i=1}^{n} p_i \dot{q}_i - H \quad (4\text{-}62)$$

Now consider the Lagrangian equations for a dissipationless system free from external disturbances

$$\frac{d}{dt}\left(\frac{\partial L}{\partial \dot{q}_i}\right) - \frac{\partial L}{\partial q_i} = 0$$

which gives

$$\frac{dp_i}{dt} = \frac{\partial L}{\partial q_i} = -\frac{\partial H}{\partial q_i} \quad (4\text{-}63)$$

Equations (4-59) and (4-63) are called the Hamiltonian equations of motion for the system

$$\begin{aligned}\frac{dq_i}{dt} &= \frac{\partial H}{\partial p_i} \\ \frac{dp_i}{dt} &= -\frac{\partial H}{\partial q_i}\end{aligned} \qquad i = 1, 2, \ldots, n \quad (4\text{-}64)$$

For a dissipationless system the Hamiltonian is equal to the total stored energy, and the principle of the conservation of energy then follows directly from Hamilton's equations since

$$\frac{dH}{dt} = \sum_{i=1}^{n} \frac{\partial H}{\partial q_i}\frac{dq_i}{dt} + \sum_{i=1}^{n} \frac{\partial H}{\partial p_i}\frac{dp_i}{dt}$$

and, then substituting from equations (4-64), we get

$$\frac{dH}{dt} = \sum_{i=1}^{n} -\dot{p}_i \dot{q}_i + \sum_{i=1}^{n} \dot{q}_i \dot{p}_i = 0 \quad (4\text{-}65)$$

4.7.1 Hamilton–Jacobi Equation [P2]

If we define an *action* variable S by

$$S = \int_0^t L \, dt \quad (4\text{-}66)$$

then, by starting from any given set of release coordinates we can determine a scalar function

$$S = S(q_1, q_2, \ldots, q_n; t)$$

Hamilton's Equations

for all ensuing natural motions of a dissipationless system free from external disturbances. S is thus defined throughout a region of the *configuration* space of the system. Differentiating S along a system trajectory we have

$$\frac{dS}{dt} = L$$

where

$$\frac{dS}{dt} = \frac{\partial S}{\partial t} + \sum_{i=1}^{n} \frac{\partial S}{\partial q_i} \frac{dq_i}{dt}$$

If we now make the identification

$$p_i = \frac{\partial S}{\partial q_i} \qquad i = 1, 2, \ldots, n$$

this gives

$$\frac{\partial S}{\partial t} + \sum_{i=1}^{n} p_i \dot{q}_i = L$$

so that

$$\frac{\partial S}{\partial t} + \sum_{i=1}^{n} p_i \dot{q}_i - L = 0$$

giving the Hamilton–Jacobi equation

$$\frac{\partial S}{\partial t} + H = 0$$

In Section 4.9 below a form of Hamiltonian equations called Pontryagin's equations is derived for an optimal control problem. In this derivation a wavefront approach is used and the variables p_i are interpreted in terms of outward normals to an expanding wavefront. A similar expanding wavefront theory exists in analytical mechanics. In this theory the Hamilton–Jacobi equation plays a central role and the momentum variable **p** is interpreted as the gradient in configuration space of the least-action function S.

4.7.2 Liouville's Theorem

The Hamiltonian equations specify a *velocity field* in the 2n-dimensional space whose coordinates are the generalized coordinates and generalized momenta of the dynamical system. Such a space is called a *state space* for the system; an extensive discussion of state spaces is given in Chapter 6. For any given set of initial conditions of the system generalized coordinates and generalized momenta, the solution of the Hamiltonian equations specifies a unique trajectory in this space. The *totality* of solution trajectories associated with the velocity field may be considered as analogous to the motion of a fluid in the state space. The divergence of the velocity field may be calculated, for a system governed by the Hamiltonian equations (4–64) as

$$\text{divergence} = \sum_{i=1}^{n} \left[\frac{\partial \dot{q}_i}{\partial q_i} + \frac{\partial \dot{p}_i}{\partial p_i} \right]$$

$$= \sum_{i=1}^{n} \left[\frac{\partial}{\partial q_i}\left(\frac{\partial H}{\partial p_i}\right) - \frac{\partial}{\partial p_i}\left(\frac{\partial H}{\partial q_i}\right) \right]$$

$$= 0$$

This result for Hamiltonian systems is known as Liouville's Theorem. It has the interesting physical interpretation that the state space flow is analogous to that of an incompressible fluid.

4.8 Hamiltonian Principles for Electrical Networks

The primal form of Hamilton's Postulate, when applied to electrical networks, takes the form

$$\int_{t_1}^{t_2} \left[\delta(T_e^* - U_e) + \sum_j v_j \delta q_j \right] dt = 0 \qquad (4\text{–}67)$$

where T_e^* is the total inductive co-energy, U_e is the total capacitive energy, and $\sum_j v_j \delta q_j$ represents the virtual work done by dissipative and source elements. The admissible variations in charges and currents must satisfy the constraints imposed by the network, that is conservation of charge and the Kirchhoff Current Law. If these constraints are satisfied, then Hamilton's Postulate automatically yields a set of network equations which satisfy the element constitutive relations and the network voltage constraints. In applying Hamilton's Postulate in its primal or dual form to networks, we work entirely in terms of sets of variables denoted by q, λ, $\dot{q}$ and $\dot{\lambda}$. For a capacitor, q is the charge and $\dot{q}$ the current, and for an inductor, λ is the flux-linkage and $\dot{\lambda}$ the voltage. For a non-conservative element, such as a resistor or a source, associated with a current i_n and a voltage v_n say, we may *define* associated variables:

$$\lambda_n = \int_0^t v_n \, dt \qquad \frac{d\lambda_n}{dt} = v_n$$
$$q_n = \int_0^t i_n \, dt \qquad \frac{dq_n}{dt} = i_n \qquad (4\text{–}68)$$

These defined variables for non-conservative elements are simply integrals of voltage and current and are *not* in any sense flux-linkages of coils or voltages of capacitors. They have been defined in such a way that their variations are subject to the same types of constraint as those of conservative elements. The network generalized coordinates, which will normally be denoted simply by sets $\{q_i\}$ or $\{\lambda_j\}$ will be combinations of the q's and λ's defined above. The appropriate dual form of Hamilton's Postulate is

$$\int_{t_1}^{t_2} \left[\delta(U_e^* - T_e) + \sum_k i_k \delta \lambda_k \right] dt = 0 \qquad (4\text{-}69)$$

where U_e^* is the total capacitative co-energy, T_e is the total inductive energy, and $\sum_k i_k \delta \lambda_k$ represents the work done by dissipative and source elements. The admissible variations in flux-linkages and voltages, must satisfy the constraints imposed by the circuit that is $v_L = d\lambda_L/dt$ and Kirchhoff's Voltage Law. If these constraints are satisfied, then Hamilton's Postulate automatically yields a set of network equations which satisfy the element constitutive relations and the network current constraints.

The Lagrangian and Co-Lagrangian forms of electrical network equations may be derived from the primal and dual forms of Hamilton's Postulate in an exactly similar manner to that adopted for mechanical systems, so there is little point in repeating such a derivation. Furthermore, these equations for networks are derived using a network approach in Chapter 5. It will be useful however to consider an example of the direct derivation of electrical network equations from Hamilton's Postulate.

Example 4.6   Application of Hamilton's Postulate to Electrical Network Analysis

For the electrical network of Figure 4–7, choose as generalized coordinates the pair of generalized charges q_1 and q_2 such that $\dot{q}_1$ and $\dot{q}_2$ are the pair of circuit currents shown. In what follows, the network Lagrangian will be denoted by the symbol $\mathscr{L}$ to avoid confusion with the inductance symbol L. We have

$$\text{charge on capacitor } C = (q_1 - q_2)$$

so that

$$\mathscr{L} = \tfrac{1}{2}L_1 \dot{q}_1^2 + \tfrac{1}{2}L_2 \dot{q}_1^2 - \tfrac{1}{2}S(q_1 - q_2)^2$$
$$= \tfrac{1}{2}L_1 \dot{q}_1^2 + \tfrac{1}{2}L_2 \dot{q}_2^2 - \tfrac{1}{2}Sq_1^2 + Sq_1 q_2 - \tfrac{1}{2}Sq_2^2$$

where $S = 1/C$. Thus

$$\delta \mathscr{L} = L_1 \dot{q}_1 \frac{d}{dt}(\delta q_1) + L_2 \dot{q}_2 \frac{d}{dt}(\delta q_2) - Sq_1 \delta q_1 + Sq_1 \delta q_2 + Sq_2 \delta q_1 - Sq_2 \delta q_2$$

The virtual work term is

$$\delta W = E \delta q_1 - R_1 \dot{q}_1 \delta q_1 - R_2 \dot{q}_2 \delta q_2$$

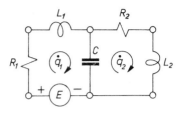

Figure 4-7

An application of Hamilton's Postulate thus gives

$$\int_{t_1}^{t_2} \left[L_1\dot{q}_1 \frac{d}{dt}(\delta q_1) + L_2\dot{q}_2 \frac{d}{dt}(\delta q_2) - Sq_1\delta q_1 + Sq_1\delta q_2 + Sq_2\delta q_1 \right.$$
$$\left. - Sq_2\delta q_2 + E\delta q_1 - R_1\dot{q}_1\delta q_1 - R_2\dot{q}_2\delta q_2 \right] dt = 0$$

The first two terms may be integrated by parts to give

$$[L_1\dot{q}_1\delta q_1]_{t_1}^{t_2} - \int_{t_1}^{t_2} L_1\ddot{q}_1\delta q_1 \, dt$$

and $\quad [L_2\dot{q}_2\delta q_2]_{t_1}^{t_2} - \int_{t_1}^{t_2} L_2\ddot{q}_2\delta q_2 \, dt \quad$ respectively.

Since δq_1 and δq_2 vanish at t_1 and t_2, the first terms in each of these expressions disappear and the whole integral expression becomes

$$\int_{t_1}^{t_2} [-L_1\ddot{q}_1 - Sq_1 + Sq_2 + E - R_1\dot{q}_1]\delta q_1 \, dt$$
$$+ \int_{t_1}^{t_2} [-L_2\ddot{q}_2 + Sq_1 - Sq_2 - R_2\dot{q}_2]\delta q_2 \, dt = 0$$

An application of the Fundamental Lemma of the Calculus of Variations in the usual way then gives that the network behaviour must satisfy the pair of equation

$$L_1\ddot{q}_1 + R_1\dot{q}_1 + S(q_1 - q_2) = E$$
$$L_2\ddot{q}_2 + R_2\dot{q}_2 + S(q_2 - q_1) = 0$$

4.9 Pontryagin's Equations [P6], [S11], [G4]

The designer of an automatic control system has three main tasks:

(a) to define the behaviour of his engineering system in mathematical terms;

(b) to construct a quantitative measure of its behaviour and relate this to a performance specification;
(c) to select, according to the criteria developed for the performance specification, the best achievable system subject to the satisfaction of certain constraints. The constraints involved arise because only finite amounts of power and of movement of controlling actuators, are available in any practical system. The design of complex automatic control systems may be approached from the point of view of classical dynamical theory; a close analogy then emerges between optimal control theory and Hamilton's wave-front treatment of mechanics.

In general we may describe the behaviour of any engineering dynamical system as a set of first order equations (see Chapter 6)

$$\frac{dx_i}{dt} = f_i(x_1, x_2, \ldots, x_n; u_1, u_2, \ldots, u_r) \qquad i = 1, 2, \ldots, n \qquad (4\text{--}70)$$

$$y_j = g_j(x_1, x_2, \ldots, x_n; u_1, u_2, \ldots, u_r) \qquad j = 1, 2, \ldots, m$$

where:

(a) $x_1, x_2, \ldots, x_n$ are a set of system state variables,
(b) $u_1, u_2, \ldots, u_r$ are a set of system input variables,
(c) $y_1, y_2, \ldots, y_m$ are a set of system output variables,
(d) $f_1, \ldots, f_n$ are a set of functions defining the rates of change of the state variables in terms of the instantaneous values of the state variables and the input variables,
(e) $g_1, \ldots, g_m$ are a set of functions defining the output variables in terms of the system state variables and input variables.

The solution of the set of differential equations (4–70) will give a set of state functions of time for any given set of initial values of state variables and any given set of input functions of time. In order to use standard analytical techniques to select from the multiply-infinite solution state functions those that are, in some defined sense, the best we must allocate to each function a number which is a measure of its suitability. The establishment of a correspondence between functions and numbers is said to define a *functional*. A quantity F is said to be a functional of a function $x(t)$ in an interval $[a, b]$ when it depends on all the values taken by $x(t)$ as the independent variable t varies in the interval $[a, b]$. In the design problem, the functions considered are the behaviour of certain system variables and the corresponding numbers are quantitative measures of some aspect of system performance. A typical functional would be

$$F = \int_0^T f(t) z^2(t)\, dt$$

where $z(t)$ is the departure of a system state variable from an equilibrium condition and $f(t)$ a weighting function to discriminate between the relative importance of the departure from equilibrium at various instants of time.

The dynamical model, given by equation set (4–70) above, representing a practicable control system, will have input variables $\{u_1, u_2, \ldots, u_r\}$ with a limited range of permissible values. These limitations correspond to the limited rates of energy conversion and movement which may be achieved in practice, and result in corresponding restrictions on the amplitude and speed of variations in the model system variables. The designer's problem is to choose the input variables, subject to a set of imposed constraints, in such a way that some specified measure of performance achieves an extremum value (that is a maximum or minimum). For example, a system controlling the position of a radar aerial or radio telescope could be investigated to determine what variation of system inputs resulted in a desired position change in the minimum time. If the position-control system were housed in a satellite, the careful husbanding of fuel might be more important than the speed of response and the system model might be investigated to discover what variation of system inputs result in a desired change of position for a minimum expenditure of fuel. The appropriate measure of system performance is called a *performance index*. The most general performance indices depend on all the values taken by the system state and input variables in a specified interval of time; they are therefore functionals of appropriate sets of state and input variables.

Since any extremum problem can be turned into an equivalent minimization problem by, if necessary, replacing a maximization by the minimization of the negative of the quantity to be maximized, it is only necessary to consider minimization problems. Minimization problems associated with the set of equations (4–70) are normally classified into the following three types:

(a) *Lagrangian problems*: of the form

$$\text{Minimize} \int_0^T L(x_1, \ldots, x_n; u_1, \ldots, u_r)\, dt$$

where the integrand L in the functional is called the system Lagrangian.

(b) *Mayer problems*: of the form

$$\text{Minimize } \Phi[x_1(T), x_2(T), \ldots, x_n(T)]$$

where Φ is some stipulated function of the terminal values of the system state variables.

(c) *Bolza problems*: of the form

$$\text{Minimize} \left\{ \psi[x_1(T), \ldots, x_n(T)] + \int_0^T L(x_1, \ldots, x_n; u_1, \ldots, u_r) \, dt \right\}$$

where both sorts of condition are involved.

Although at first sight different, these three problems are equivalent; for this reason we may accordingly only consider the Mayer problem in what follows. To transform a Lagrangian problem or a Bolza problem into Mayer form, one introduces the additional state variable

$$x_0 = \int_0^t L \, dt$$

which satisfies the associated differential equation

$$\frac{dx_0}{dt} = L \tag{4-71}$$

If we adjoin equation (4–71) to the set of equations (4–70), then a Lagrangian or Bolza problem for the set of equations (4–70) obviously becomes an equivalent Mayer problem in the augmented set of equations (4–69) plus (4–70). For this reason we normally only consider Mayer problems in what follows.

4.9.1 Optimal Control Problem

Consider the problem of controlling the inputs to a system, whose dynamical behaviour is governed by the set of state space equations (4–70). The inputs are to be manipulated so that, at a time T, the value of a performance index

$$V(T) = \sum_{i=1}^n c_i x_i(T)$$

is minimized, where $\{c_1, c_2, \ldots, c_n\}$ is a set of real constants and the terminal components of the state vector $\{x_1(T), x_2(T), \ldots, x_n(T)\}$ are not stipulated. It is assumed that the choice of possible source output functions of time is to be limited to some stipulated admissible set of source outputs, and this is denoted symbolically by writing

$$\mathbf{u} \in \{U\}$$

4.9.2 Event Vector and Event Space

The set of variables $\{x_1, x_2, \ldots, x_n, t\}$ are defined as the components of an *event vector*

$$\xi = \begin{bmatrix} x_1 \\ x_2 \\ \vdots \\ x_n \\ t \end{bmatrix}$$

in an $(n+1)$-dimensional *event space*.

The use of the event space greatly simplifies discussion of this optimal control problem. It is shown below that optimal system trajectories, that is trajectories which result in an optimum value of the performance index, must lie in a certain surface in the event space; this particular surface is then used to derive a method of defining optimal trajectories in the state space.

4.9.3 The Set of Possible Events [H1]

Equations (4-70) may be regarded as specifying the velocities of the tips of the state vector component projections along the directions of the basis vectors in the state space. If the possible system inputs are restricted, this will naturally result in corresponding restrictions on the state variable component velocities which may be attained. It follows that, if the state vector **x** is at a given point **x**(0) of the state space at the time $t = 0$, the distances in the state space which may be traversed in a finite time T are limited by virtue of the limitations on the attainable velocities. If the solutions of the equation set (4-70) are suitably continuous, then for any given starting point **x**(0) in the state space there will be a closed, bounded region of the state space corresponding to all possible points **x**(T) which may be reached under the action of all possible choices of source output functions from the permissible set $\{U\}$. Consider the components of an event vector defined by:

$$t = \tau$$
$$x_i = \int_0^t f_i(x_1, \ldots, x_n; u_1, \ldots, u_r) \, dt \tag{4-72}$$

If the source outputs are restricted to the permissible set $\{U\}$, the event vector is said to define a *possible event*. For the continuous set of all possible values of the time τ, and for all possible choices of source output functions of time, the equations will define a bounded point set in the event space; this bounded set is defined as the *set of possible events* for the time interval $[0, \tau]$. The components of the set are assumed to be piece-wise continuous and the function set $\{f_1, f_2, \ldots, f_n\}$

is assumed to be continuous and sufficiently differentiable in all arguments to ensure that the boundary surface of the set of possible events is continuous in the event space. The boundary surface of the set of possible events plays a central role in the solution to the optimal control problem presented here.

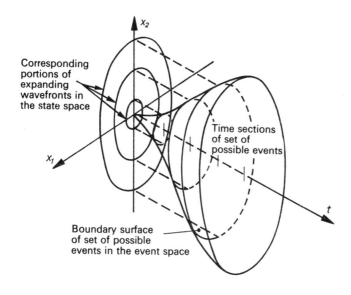

Figure 4-8

4.9.4 Expanding Wave-fronts in the State Space [H1]

Figure 4-8 shows the boundary surface of the set of possible events for a second-order system for which the event space is three-dimensional. An inspection of this diagram will show that successive 'time-sections' of the boundary surface of the set of possible events define an expanding wavefront in the state space when projected on to the state space. The key to the solution of the optimal control problem posed lies in a method of defining the generation of these expanding wavefronts. Suppose we have one of these wavefronts corresponding to time t; denote this wavefront by $W(t)$. The corresponding wavefront for a successive time $(t+dt)$ may be constructed in the following way, illustrated by Figure 4-9. For the small interval of time dt there will be a limited region of the state space attainable by starting a trajectory at the point A on the wavefront $W(t)$. If a sufficient number of trajectories for permissible source outputs are computed, we may construct the boundary of this attainable region of the state space corresponding

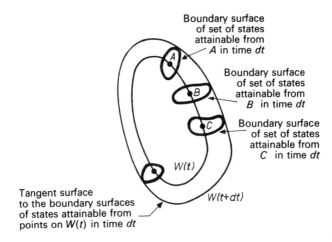

Figure 4–9

to point A and time dt. This boundary surface will be a closed, continuous surface surrounding point A. Similar closed, continuous surfaces may be constructed surrounding the points $B, C, \ldots$, and so on for $W(t)$; suppose these boundary surfaces are constructed for a very large number of points on the wavefront $W(t)$. Now consider a surface $W(t+dt)$ constructed to be tangent to all these closed surfaces and exterior to them. Since $W(t+dt)$ must be the boundary surface of the set of states attainable from the states on $W(t)$ by proceeding under the influence of all permissible source outputs for time dt, it follows that $W(t+dt)$ is the projected time-section of the boundary surface of the set of possible events corresponding to the time $(t+dt)$.

4.9.5 Analogy with Huygens' Principle in Geometrical Optics [M14]

Huygens' Principle in simple geometrical optics postulates that each point on an expanding wavefront acts as a point source emitting spherical wavefronts propagated with constant velocity. The principle can be extended to the propagation of wavefronts through an anisotropic medium through which disturbances propagate with different velocities for different directions of propagation. A strong analogy obviously exists between the latter form of Huygens' Principle and the construction described above for the wavefront $W(t+dt)$. The finite attainable velocities of the tips of the projections of the state vector in the base vector directions are directly analogous to the finite anisotropic propagation velocities in the extended form of Huygens' Principle in geometrical optics.

4.9.6 Pontryagin's Maximum Principle [P6]

In most problems the construction of these boundary surfaces by extensive computation is impracticable, and some other means must be found of defining points on the wavefront $W(t+dt)$, given points on $W(t)$. Pontryagin's method requires the introduction of a *co-state vector*, **p**, which is an outward normal to the surface $W(t)$ at the point corresponding to the state **x**. Figure 4–10 shows the way in which the

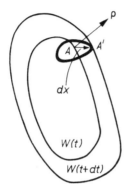

Figure 4–10

co-state vector **p** is used to define points on the required expanding wavefronts in the state space. Suppose that the vector **p** is known for some point A on the wavefront $W(t)$. Let **dx** be the change in state **x** in a time dt for some chosen set of permissible source outputs. Then the point A' shown on the wavefront $W(t+dt)$ corresponds to the maximum value of the scalar product $\langle \mathbf{p}, \mathbf{dx} \rangle$ for all the states attainable from the state for point A in time dt. Thus points on the surface $W(t+dt)$ corresponding to given points on $W(t)$ may be obtained by choosing the source outputs for the given points on $W(t)$ in such a way as to maximize the scalar product $\langle \mathbf{p}, \mathbf{dx} \rangle$. By letting the time interval dt become arbitrarily small, one arrives at one form of the Pontryagin Maximum Principle. This may be stated in terms of the expanding wavefronts being considered as follows: those expanding wavefronts in the state space which correspond to successive time-sections of the boundary surface of the set of possible events are obtained by continuously choosing the system source outputs so that the scalar product $\langle \mathbf{p}, \dot{\mathbf{x}} \rangle$ is continuously maximized.

4.9.7 Generation of Optimal Trajectory

An optimal trajectory for the problem being considered is one which, the source restrictions being complied with throughout the trajectory, results in a *minimum* value of

$$V = \langle \mathbf{c}, \mathbf{x}(T) \rangle$$

where **c** is a vector having components $c_1, c_2, \ldots, c_n$ and $\mathbf{x}(T)$ is the terminal state vector. Alternatively, we may say that an optimal trajectory results in a *maximum* value of

$$-V = \langle -\mathbf{c}, \mathbf{x}(T) \rangle$$

Now consider the terminal wavefront $W(T)$. The optimal terminal point, $\mathbf{x}°(T)$ say, must lie in $W(T)$ since for any point not in $W(T)$, such as A in the simple illustration of Figure 4–11 which corresponds to some terminal point $\mathbf{x}'(T)$, we can always find another point, say B in Figure 4–11, which lies in the boundary $W(T)$ which gives a larger value to $\langle -\mathbf{c}, \mathbf{x} \rangle$. It therefore follows that:

(a) the optimal terminal point $\mathbf{x}°(T)$ must lie in the terminal wavefront $W(T)$;
(b) the optimal trajectory is generated by the Maximum Principle since this generates trajectories which lie in the boundary surface of the set of possible events in the event space.

In order to find the *particular* trajectory in the event space which is the solution to the problem posed, we must obtain a suitable set of boundary conditions which uniquely define it.

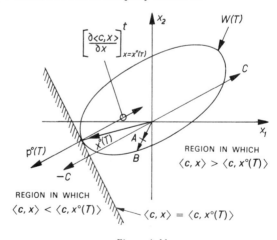

Figure 4–11

The equation

$$\langle [\mathbf{x} - \mathbf{x}(T)], \mathbf{c} \rangle = 0$$

defines a hyperplane in the state space through $\mathbf{x}(T)$ and orthogonal to c. The equation

$$\langle \mathbf{x}, \mathbf{c} \rangle = \langle \mathbf{x}(T), \mathbf{c} \rangle$$

thus defines a hyperplane through $\mathbf{x}(T)$ and orthogonal to $\mathbf{c}$, which splits the state space into two disjoint regions in which

$\langle \mathbf{x}, \mathbf{c} \rangle$ is greater than $\langle \mathbf{x}(T), \mathbf{c} \rangle$

and $\langle \mathbf{x}, \mathbf{c} \rangle$ is less than $\langle \mathbf{x}(T), \mathbf{c} \rangle$

respectively.

Now $W(T)$ must be tangent to the hyperplane

$$\langle \mathbf{x}, \mathbf{c} \rangle = \langle \mathbf{x}^\circ(T), \mathbf{c} \rangle$$

at the point $\mathbf{x}^\circ(T)$, since all other points on, or interior to, $W(T)$ must correspond to larger values of the quantity $\langle \mathbf{c}, \mathbf{x}^\circ(T) \rangle$ which we are trying to minimize. This is illustrated by Figure 4–11. It therefore follows that the outward normal to $W(T)$ at $\mathbf{x}^\circ(T)$, that is the corresponding value of $\mathbf{p}^\circ(T)$, must equal the negative gradient of the scalar function $\langle \mathbf{c}, \mathbf{x} \rangle$ at the point $\mathbf{x}^\circ(T)$. Thus the terminal condition for the co-state vector corresponding to the desired optimal trajectory must be

$$-\left[\frac{\partial V(T)}{\partial \mathbf{x}}\right]^t = \mathbf{p}^\circ(T) = -\left[\frac{\partial \langle \mathbf{c}, \mathbf{x} \rangle}{\partial \mathbf{x}}\right]^t = -\mathbf{c} \qquad (4\text{–}73)$$

To complete the solution of the optimal control problem posed we require a set of differential equations which may be solved with appropriate boundary conditions to define an optimal trajectory. These are obtained by deriving a differential equation set for the components of the co-state vector $\mathbf{p}$ which may be adjoined to those for the state vector $\mathbf{x}$ to give a complete set of differential equations which define an optimal trajectory.

4.9.8 Derivation of Pontryagin's Equations

To obtain the differential equation set for the components of the co-state vector $\mathbf{p}$ we consider a 'bundle' of trajectories about a specific trajectory and define the expanding wavefront $W(t)$ in terms of a set $\{\delta \mathbf{x}^{(1)}, \delta \mathbf{x}^{(2)}, \ldots, \delta \mathbf{x}^{(n)}\}$ of small departures from the specific trajectory. Our first requirement is the differential equation set governing the behaviour of this small-departure set of vectors $\{\delta \mathbf{x}^{(i)} : i = 1, 2, \ldots, n\}$. For some specific set of source output variables $\{u_1(t), u_2(t), \ldots, u_r(t)\}$ let $\{z_1(t), z_2(t), \ldots, z_n(t)\}$ be the corresponding solution trajectory components. Let $\{\delta x_i : i = 1, 2, \ldots, n\}$ be the components of a small departure $\delta \mathbf{x}$ in the state variables from the solution set, and consider the equation set for $\{z_i + \delta x_i : i = 1, 2, \ldots, n\}$. We have

$$\frac{d}{dt}(z_i + \delta x_i) = f_i(z_1 + \delta x_1, z_2 + \delta x_2, \ldots, z_n + \delta x_n; u_1, \ldots, u_r; t)$$

$$i = 1, 2, \ldots, n$$

The multivariable form of Taylor's theorem then gives

$$\frac{dz_i}{dt} + \frac{d}{dt}(\delta x_i) = f_i(z_1, \ldots, z_n; u_1, \ldots, u_r; t)$$

$$+ \sum_{r=1}^{n} \left[\frac{\partial f_i}{\partial x_r}\right]_{x=z} \delta x_r + \frac{1}{2!} \sum_{r=1}^{n} \sum_{s=1}^{n} \left[\frac{\partial^2 f_i}{\partial x_r \partial x_s}\right]_{x=z} \delta x_r \delta x_s$$

$$+ \text{etc.} \qquad (4\text{-}74)$$

Now, since $\{z_1(t), \ldots, z_n(t)\}$ are solutions of equation (4-70) for the specified $\{u_1(t), \ldots, u_r(t)\}$, we have

$$\frac{dz_i}{dt} = f_i(z_1, \ldots, z_n; u_1, \ldots, u_r; t) \qquad i = 1, 2, \ldots, n$$

Equation (4-74) thus becomes

$$\frac{d}{dt}(\delta x_i) = \sum_{r=1}^{n} \left[\frac{\partial f_i}{\partial x_r}\right]_{x=z} \delta x_r + 0(\delta x_i^2) \qquad i = 1, 2, \ldots, n$$

This shows that, for sufficiently small departures from a specific trajectory, the time evolution of small departures is given by

$$\frac{d}{dt}(\delta x_i) = \sum_{r=1}^{n} \frac{\partial f_i}{\partial x_r} \delta x_r \qquad (4\text{-}75)$$

where the partial derivatives $\partial f_i / \partial x_r$ are evaluated along the specific solution trajectory involved.

For the situation illustrated in Figure 4-12 suppose that the source output vector under whose action the point A' is reached from the point A is the vector $\mathbf{u}^*$. Consider the surface $W^*(t+dt)$ obtained from the surface $W(t)$ by carrying out the construction described above for $W(t+dt)$ but using the *fixed* control vector $\mathbf{u}^*$ for all points on the surface $W(t)$. The surfaces $W^*(t+dt)$ and $W(t+dt)$ will meet tangentially at the point A' and have the same outward normal at A'. It follows that we may derive the equations governing the behaviour of the components of $\mathbf{p}$ which correspond to a system trajectory through A by considering the mapping of $W(t)$ into $W^*(t+dt)$. The equations relating the outward normal of $W(t)$ at A to the outward normals of $W^*(t+dt)$ and $W(t+dt)$ at the point A' will be the same; it is therefore

Pontryagin's Equations

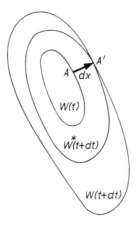

Figure 4–12

sufficient to consider the simpler mapping of $W(t)$ into $W^*(t+dt)$ under the action of a fixed vector **u** in order to determine the required differential equation set in the components of **p**.

Let $\delta\mathbf{x}^{(2)}, \delta\mathbf{x}^{(3)}, \ldots, \delta\mathbf{x}^{(n)}$ be a set of $(n-1)$ independent vectors which are tangent to the surface $W(t)$ in the neighbourhood of the point A, and let $\delta\mathbf{x}^{(1)}$ be a further vector at point A, which is directed towards the *inside* of the closed surface $W(t)$. An *outward* normal to the surface $W(t)$ at the point A may then be defined in terms of this independent vector set by means of the equations

$$\langle \mathbf{p}, \delta\mathbf{x}^{(1)} \rangle = -d$$
$$\langle \mathbf{p}, \delta\mathbf{x}^{(k)} \rangle = 0 \quad \text{for} \quad k = 2, 3, \ldots, n \qquad (4\text{--}76)$$

where d is an arbitrary constant. Figure 4–13 illustrates the situation

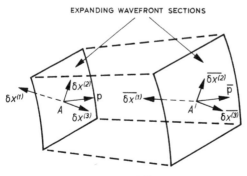

Figure 4–13

for $n = 3$. The differential equations governing the behaviour of the components of the co-state vector **p** follow from the fact that the relations (4–76) are, by definition, invariant along a trajectory. That is

$$\frac{d}{dt}\langle \mathbf{p}, \delta \mathbf{x}^{(i)}\rangle = 0 \qquad i = 1, 2, \ldots, n \qquad (4\text{–}77)$$

This gives

$$\left\langle \frac{d\mathbf{p}}{dt}, \delta \mathbf{x}^{(i)}\right\rangle + \left\langle \mathbf{p}, \frac{d}{dt}\delta \mathbf{x}^{(i)}\right\rangle = 0 \qquad i = 1, 2, \ldots, n \qquad (4\text{–}78)$$

Then since

$$\frac{d}{dt}\delta \mathbf{x}^{(i)} = \boldsymbol{\psi}\delta \mathbf{x}^{(i)} \qquad i = 1, 2, \ldots, n$$

where $\boldsymbol{\psi}$ is a matrix having elements

$$\psi_{rs} = \frac{\partial f_r}{\partial x_s} \qquad r, s = 1, 2, \ldots, n$$

we may write equation (4–78) as

$$\left\langle \frac{d\mathbf{p}}{dt}, \delta \mathbf{x}^{(i)}\right\rangle + \langle \mathbf{p}, \boldsymbol{\psi}\delta \mathbf{x}^{(i)}\rangle = 0 \qquad i = 1, 2, \ldots, n$$

so that

$$\left\langle \frac{d\mathbf{p}}{dt}, \delta \mathbf{x}^{(i)}\right\rangle + \langle \boldsymbol{\psi}^t\mathbf{p}, \delta \mathbf{x}^{(i)}\rangle = 0 \qquad i = 1, 2, \ldots, n$$

which gives us that:

$$\left\langle \left(\frac{d\mathbf{p}}{dt} + \boldsymbol{\psi}^t\mathbf{p}\right), \delta \mathbf{x}^{(i)}\right\rangle = 0 \qquad i = 1, 2, \ldots, n \qquad (4\text{–}79)$$

where $\boldsymbol{\psi}^t$ is the transpose of matrix $\boldsymbol{\psi}$. Since the vector set $\{\delta \mathbf{x}^{(1)}, \delta \mathbf{x}^{(2)}, \ldots, \delta \mathbf{x}^{(n)}\}$ is an arbitrarily chosen independent set, equation (4–79) can only be satisfied if

$$\frac{d\mathbf{p}}{dt} + \boldsymbol{\psi}^t\mathbf{p} \equiv \mathbf{0}$$

Thus for the mapping of one wavefront into another in a sufficiently small neighbourhood of the point A, the differential equation governing the evolution of the co-state vector **p** along a trajectory is

$$\frac{d\mathbf{p}}{dt} = -\boldsymbol{\psi}^t\mathbf{p} \qquad (4\text{–}80)$$

In terms of the components of the vector **p**, we thus have

$$\frac{dp_i}{dt} = -\sum_{j=1}^{n} \frac{\partial f_j}{\partial x_i} p_j \qquad i = 1, 2, \ldots, n \qquad (4\text{-}81)$$

Formulation of Equations in Terms of a Scalar Function
Let a scalar function $\bar{\mathcal{H}}$ be defined as

$$\bar{\mathcal{H}} = \left\langle \mathbf{p}, \frac{d\mathbf{x}}{dt} \right\rangle \qquad (4\text{-}82)$$

In terms of this function the equation governing the behaviour of the state vector components may be immediately written as

$$\frac{dx_i}{dt} = \frac{\partial \bar{\mathcal{H}}}{\partial p_i} \qquad i = 1, 2, \ldots, n \qquad (4\text{-}83)$$

Now as
$$\bar{\mathcal{H}} = \sum_{j=1}^{n} f_j p_j$$

we have
$$\frac{\partial \bar{\mathcal{H}}}{\partial x_i} = \sum_{j=1}^{n} \frac{\partial f_j}{\partial x_i} p_j = -\frac{dp_i}{dt}$$

It follows that in terms of the scalar function $\bar{\mathcal{H}}$ the trajectory equations may be written in the form

$$\begin{aligned}\frac{dx_i}{dt} &= \frac{\partial \bar{\mathcal{H}}}{\partial p_i} \\ \frac{dp_i}{dt} &= -\frac{\partial \bar{\mathcal{H}}}{\partial x_i}\end{aligned} \qquad i = 1, 2, \ldots, n \qquad (4\text{-}84)$$

or in vector form
$$\begin{aligned}\dot{\mathbf{x}} &= \left(\frac{\partial \bar{\mathcal{H}}}{\partial \mathbf{p}}\right)^t \\ \mathbf{p} &= -\left(\frac{\partial \bar{\mathcal{H}}}{\partial \mathbf{x}}\right)^t\end{aligned} \qquad (4\text{-}85)$$

These forms of Hamiltonian equations are usually called Pontryagin's equations for the optimal control problem.

Equation Set and Boundary Conditions Defining an Optimal Trajectory
In order that a trajectory be optimal, we have seen that the source outputs must be adjusted continually so that the scalar product

$\langle \mathbf{p}, d\mathbf{x}/dt \rangle$ is continuously maximized. Let a scalar function $\bar{\mathscr{H}}^{\circ}$ be defined as

$$\bar{\mathscr{H}}^{\circ} = \underset{\mathbf{u} \in \{U\}}{\text{maximum}} \bar{\mathscr{H}} \tag{4-86}$$

that is let $\bar{\mathscr{H}}^{\circ}$ be the scalar function obtained throughout the state space by choosing the system source outputs at all instants of time such that the scalar function

$$\bar{\mathscr{H}} = \langle \mathbf{p}, \dot{\mathbf{x}} \rangle$$

achieves its maximum possible value for the permissible source outputs. Then the performance index

$$V = \langle \mathbf{c}, \mathbf{x}(T) \rangle$$

will be minimized if the system satisfies the equation set

$$\begin{aligned} \frac{dx_i}{dt} &= \frac{\partial \bar{\mathscr{H}}^{\circ}}{\partial p_i} \\ \frac{dp_i}{dt} &= -\frac{\partial \bar{\mathscr{H}}_0}{\partial x_i} \end{aligned} \quad i = 1, 2, \ldots, n \tag{4-87}$$

with the 2-point boundary conditions

$$\begin{aligned} x_i(0) &= \text{given starting values} \\ p_i(T) &= -c_i \end{aligned} \quad i = 1, 2, \ldots, n \tag{4-88}$$

4.9.9 Derivation of Pontryagin Equations from a Variational Principle for the Case When the Control Inputs are Unrestricted

The derivation of Pontryagin's equations given above has assumed that the control variables are selected from a restricted set. If the control variables are not restricted, these equations may be derived using standard variational techniques. For simplicity in presenting the argument we will only consider the single-input case; the extension to multiple inputs readily follows.

We have to minimize $V[\mathbf{x}(T)]$ subject to the set or constraints imposed by the fact that the system state variables must satisfy the differential equation set

$$\dot{x}_i = f_i(x_1, x_2, \ldots, x_n; u_1, \ldots, u_r) \quad i = 1, 2, \ldots, n$$

These differential equations may be rearranged to give the set of constraint equations

$$\phi_i(x_1, \ldots, x_n; u_1, \ldots, u_r; \dot{x}_1, \ldots, \dot{x}_n) = f_i - \dot{x}_i = 0 \quad i = 1, 2, \ldots, n$$

In terms of the Calculus of Variations, this constrained minimization

Pontryagin's Equations

problem is solved by introducing a vector Lagrange multiplier $\lambda(t)$, having n elements $\lambda_1(t), \lambda_2(t), \ldots, \lambda_n(t)$, and then finding a minimum of the function

$$G[\mathbf{x}(T)] = V[\mathbf{x}(T)] + \int_0^T \sum_{i=1}^n \lambda_i(t)\phi_i[\mathbf{x}, \dot{\mathbf{x}}, u]\, dt$$

Consider a small variation $u(t) + \epsilon\eta(t)$ about the optimal $u(t)$. We then have

$$G + \delta G = V[\mathbf{x}(T) + \delta\mathbf{x}(T)] + \int_0^T \sum_{i=1}^n \lambda_i(t)\phi_i[\mathbf{x} + \delta\mathbf{x}, \dot{\mathbf{x}} + \delta\dot{\mathbf{x}}, u + \delta u]\, dt$$

so that

$$\delta G = \sum_{j=1}^n \frac{\partial V}{\partial x_j}\delta x_j(T) + \int_0^T \sum_{i=1}^n \lambda_i(t)\left\{\sum_{j=1}^n \frac{\partial \phi_i}{\partial x_j}\delta x_j + \sum_{j=1}^n \frac{\partial \phi_i}{\partial \dot{x}_j}\delta \dot{x}_j \right.$$
$$\left. + \epsilon\eta\frac{\partial \phi_i}{\partial u}\right\} dt + O(\epsilon^2)$$

Integrating by parts, as in previous variational manipulations, and putting $\delta\mathbf{x}(0) = 0$ gives

$$\delta G = \sum_{j=1}^n \frac{\partial V}{\partial x_j}\delta x_j(T) + \sum_{i=1}^n \lambda_i(T) \sum_{j=1}^n \left(\frac{\partial \phi_i}{\partial \dot{x}_j}\right)_{t=T} \delta x_j(T)$$
$$+ \int_0^T \left\{\sum_{i=1}^n \lambda_i(t)\left[\sum_{j=1}^n \frac{\partial \phi_i}{\partial x_j}\delta x_j + \epsilon\eta\frac{\partial \phi_i}{\partial u}\right]\right. \qquad (4\text{-}89)$$
$$\left. - \sum_{i=1}^n \sum_{j=1}^n \frac{d}{dt}\left[\lambda_i(t)\frac{\partial \phi_i}{\partial \dot{x}_j}\delta x_j\right]\right\} dt + O(\epsilon^2)$$

Now for an optimal u, ∂G must vanish with ϵ; this gives a set of necessary conditions for an optimal solution. The $\lambda_i(t)$ must be such that all the coefficients of $\delta x_j (j = 1, 2, \ldots, n)$ in equation (4-89) are zero. This gives the set of equations

$$\sum_{i=1}^n \left\{\lambda_i(t)\frac{\partial \phi_i}{\partial x_j} - \frac{d}{dt}\left[\lambda_i(t)\frac{\partial \phi_i}{\partial \dot{x}_j}\right]\right\} = 0 \qquad j = 1, 2, \ldots, n \qquad (4\text{-}90)$$

If δG is to vanish for all $\eta(t)$ we must have that

$$\sum_{i=1}^n \lambda_i(t)\frac{\partial \phi_i}{\partial u} = 0 \qquad (4\text{-}91)$$

Finally, on setting the coefficients $\delta x_j(T)$ equal to zero, we must have

that
$$\frac{\partial V}{\partial x_j(T)} + \sum_{i=1}^{n} \lambda_i(T)\left[\frac{\partial \phi_i}{\partial \dot{x}_j}\right]_{t=T} = 0 \quad j = 1, 2, \ldots, n \quad (4\text{-}92)$$

From equation (4-92) we have
$$\frac{\partial V}{\partial x_j(T)} = -\sum_{i=1}^{n} \lambda_i(T)\frac{\partial}{\partial \dot{x}_j}(f_i - \dot{x}_i) = \lambda_j(T)$$

Put
$$\mathcal{H}^* = \sum_{i=1}^{n} \lambda_i f_i$$

then
$$f_i = \frac{\partial \mathcal{H}^*}{\partial \lambda_i}$$

which gives
$$\dot{x}_i = \frac{\partial \mathcal{H}^*}{\partial \lambda_i} \quad i = 1, 2, \ldots, n$$

Equation (4-91) then gives
$$\sum_{i=1}^{n} \lambda_i(t)\frac{\partial}{\partial u}(f_i - \dot{x}_i) = \sum_{i=1}^{n} \lambda_i(t)\frac{\partial f_i}{\partial u} = \frac{\partial}{\partial u}\sum_{i=1}^{n} \lambda_i f_i = \frac{\partial \mathcal{H}^*}{\partial u} = 0$$

From equation (4-90) we have
$$\sum_{i=1}^{n} \frac{d}{dt}\left[\lambda_i \frac{\partial(f_i - \dot{x}_i)}{\partial \dot{x}_j}\right] = \sum_{i=1}^{n} \lambda_i \frac{\partial}{\partial x_j}(f_i - \dot{x}_i) \quad j = 1, 2, \ldots, n$$

so that
$$-\frac{d}{dt}[\lambda_j] = \sum_{i=1}^{n} \lambda_i \frac{\partial f_i}{\partial x_j} = \frac{\partial}{\partial x_j}\sum_{i=1}^{n} \lambda_i f_i = \frac{\partial \mathcal{H}^*}{\partial x_j} \quad j = 1, 2, \ldots, n$$

Collecting all these results together we have the set of equations
$$\frac{dx_i}{dt} = \frac{\partial \mathcal{H}^*}{\partial \lambda_i} \quad i = 1, 2, \ldots, n$$

$$\frac{d\lambda_i}{dt} = -\frac{\partial \mathcal{H}^*}{\partial x_i} \quad i = 1, 2, \ldots, n$$

$$\mathcal{H}^* = \sum_{i=1}^{n} \lambda_i \dot{x}_i$$

$$\frac{\partial \mathcal{H}^*}{\partial u} = 0 \qquad x_i(0) = \text{given set of initial conditions}$$

$$\lambda_i(T) = \frac{\partial V}{\partial x_i(T)} \qquad i = 1, 2, \ldots n$$

This shows that the Pontryagin form of equations holds in the unrestricted input case. A consideration of the boundary condition for $\lambda_i(T)$ shows that the vector λ is the *negative* of the co-state vector p introduced previously. This sign reversal occurs in many treatments of the optimal control problem and will be mentioned again below; care is therefore required in comparing the results given in different texts.

4.9.10 Minimization of an Integral Functional of System Motion

The most usual requirement in an optimal control problem is that some integral functional of system performance is minimized. Suppose that the system source outputs are to be chosen such that the performance index functional

$$V = \tfrac{1}{2}x^t(T)\Omega x(T) + \int_0^T \mathscr{L}(x_1, x_2, \ldots, x_n; u_1, \ldots, u_r)\, dt \quad (4\text{-}93)$$

is minimized. Define a variable x_0 as

$$x_0 = \int_0^t \mathscr{L}\, dt$$

giving the associated differential equation

$$\frac{dx_0}{dt} = \mathscr{L} \quad (4\text{-}94)$$

The variable x_0 may be adjoined to the state vector component set $\{x_1, x_2, \ldots, x_n\}$ to define an *augmented state vector* x^* where

$$x^* \triangleq \begin{bmatrix} x_0 \\ x_1 \\ x_2 \\ \vdots \\ x_n \end{bmatrix}$$

is an $(n+1)$-dimensional augmented state space. If a corresponding augmented co-state vector is defined as:

$$p^* \triangleq \begin{bmatrix} p_0 \\ p_1 \\ p_2 \\ \vdots \\ p_n \end{bmatrix}$$

then this problem of minimizing an integral functional may be treated in exactly the same way as the previous Mayer-type problem, by introducing the scalar functions

$$\bar{\mathscr{H}}^* = \left\langle \mathbf{p}^*, \frac{d\mathbf{x}^*}{dt} \right\rangle$$

and
$$\mathscr{H} = \underset{\mathbf{u}\in\{U\}}{\text{maximum}} \bar{\mathscr{H}}^*$$

The optimal system trajectory is then given by the solution of the equations

$$\frac{dx_i}{dt} = \frac{\partial \mathscr{H}}{\partial p_i}$$
$$\frac{dp_i}{dt} = -\frac{\partial \mathscr{H}}{\partial x_i} \qquad i = 1, 2, \ldots, n \qquad (4\text{-}95)$$

At the terminal time T we have, using an appropriately modified form of equation (4-73)

$$\mathbf{p}^*(T) = -\left[\frac{\partial V(T)}{\partial \mathbf{x}^*(T)}\right]^t$$

Now
$$\left[\frac{\partial V(T)}{\partial \mathbf{x}^*(T)}\right]^t = \left[\frac{1}{\Omega \mathbf{x}(T)}\right]$$

It follows that the required two-point boundary conditions are:

$x_0(0) = 0$

$x_i(0) =$ given set of initial values for $i = 1, 2, \ldots, n$

$p_0(T) = -1$

$p(T) = -\Omega x(T)$

Now
$$\frac{dp_0}{dt} = -\frac{\partial \mathscr{H}}{\partial x_0} = 0$$

and $\quad p_0(T) = -1$

so that $\quad p_0(t) \equiv -1$

along a trajectory. Therefore since

$$\mathscr{H} = \langle \mathbf{p}, \dot{\mathbf{x}} \rangle + p_0 \mathscr{L}$$

it follows that
$$\mathscr{H} = \langle \mathbf{p}, \dot{\mathbf{x}} \rangle - \mathscr{L}$$

that is
$$\mathscr{H} = \sum_{i=1}^{n} p_i \dot{x}_i - \mathscr{L} \qquad (4\text{-}96)$$

Pontryagin's Equations

If we summarize the Pontryagin equation set as follows:

$$\frac{dx_i}{dt} = \frac{\partial \mathcal{H}}{\partial p_i}$$

$$\frac{dp_i}{dt} = -\frac{\partial \mathcal{H}}{\partial x_i} \qquad i = 1, 2, \ldots, n \qquad (4\text{-}97)$$

$$\mathcal{H} = \underset{u \in \{U\}}{\text{maximum}} \left[\sum_{i=1}^{n} p_i \dot{x}_i - \mathcal{L} \right] \qquad (4\text{-}98)$$

$x_i(0) = $ given initial values $\qquad i = 1, 2, \ldots, n$

$p(T) = -\Omega x(T)$

then we see that they are exactly analogous to the Hamiltonian equations of mechanics. The analogy may be pursued most effectively by considering that the Pontryagin equations represent a 'wave-mechanical' solution of the optimal control problem, in which the expanding 'cost-wavefronts' play the part of the action wavefronts in the wavefront treatment of Hamiltonian mechanics. For this reason it is conventional to refer to $\mathcal{L}$ and $\mathcal{H}$ as respectively the Hamiltonian and Lagrangian functions for the optimal control problem. The use of these equations is now illustrated by a set of examples.

Example 4.7

The state $x_1(t)$ of a first-order dynamical system is described in terms of a single control input variable $u(t)$ by the linear differential equation

$$\frac{dx_1}{dt} = ax_1(t) + bu(t) \qquad (4\text{-}99)$$

where a and b are real constants. Suppose $u(t)$ is to be manipulated so that $x_1(t)$ is taken from a given initial state $x_1(0)$ towards a null state $x_1 = 0$ while minimizing the performance criterion function

$$V(T) = x_1^2(T) + \int_0^T u^2 \, dt \qquad (4\text{-}100)$$

over a stipulated, fixed time T. The first term of this performance index represents a cost on deviation from the null state at the stipulated terminal time, and the second term a cost on the use of control. In terms of the general formulation given above we have

$$\mathcal{L} = u^2(t) \qquad \Omega = 2$$

$$\overline{\mathcal{H}}^* = p_1 \dot{x}_1 - u^2 = p_1 a x_1 + p_1 b u - u^2$$

$$\frac{\partial \overline{\mathcal{H}}^*}{\partial u} = p_1 b - 2u$$

Hamiltonian Models

For a maximum of $\bar{\mathcal{H}}^*$ we have

$$\frac{\partial \bar{\mathcal{H}}^*}{\partial u} = 0$$

so that

$$u = \frac{p_1 b}{2} \tag{4-101}$$

giving

$$\mathcal{H} = p_1 a x_1 + \frac{(p_1 b)^2}{2} - \frac{(p_1 b)^2}{4}$$

so that

$$\mathcal{H} = p_1 a x_1 + \frac{p_1^2 b^2}{4}$$

The Pontryagin equations are therefore

$$\frac{dx_1}{dt} = \frac{\partial \mathcal{H}}{\partial p_1} = a x_1 + \tfrac{1}{2} p_1 b^2 \tag{4-102}$$

$$\frac{dp_1}{dt} = -\frac{\partial \mathcal{H}}{\partial x_1} = -a p_1 \tag{4-103}$$

with two-point boundary conditions

$$\begin{aligned} x_1(0) &= \text{given} \\ p_1(T) &= -\Omega x_1(T) = -2 x_1(T) \end{aligned} \tag{4-104}$$

Equations (4–103) and (4–104) give

$$p_1(t) = -2 x_1(T) \exp[a(T-t)] \tag{4-105}$$

which, together with equation (4–101) gives, for an optimal trajectory

$$u(t) = -x_1(T) b \exp[a(T-t)] \tag{4-106}$$

In the above expressions, $x_1(T)$ is the as yet undetermined terminal error deviation. Thus, to solve the problem explicitly for u in terms of x_1 and t, we must find the transformation relating the terminal point $x_1(T)$ to $x_1(t)$ for an optimal trajectory between $x_1(t)$ and $x_1(T)$. To do this, substitute for $p_1(t)$ from (4–105) into (4–102) giving

$$\begin{aligned} \frac{dx_1}{dt} &= a x_1 + \tfrac{1}{2} b^2 \{-2 x_1(T) \exp[a(T-t)]\} \\ &= a x_1 - b^2 x_1(T) \exp[a(T-t)] \end{aligned}$$

This linear equation may be solved by taking Laplace transforms. Thus, denoting the transform of $x_1(t)$ by $X_1(s)$, we have

$$sX_1(s) - x_1(0) = aX_1(s) - \frac{b^2 x_1(T) \exp(aT)}{(s+a)}$$

This gives

$$X_1(s) = \frac{x_1(0)}{(s-a)} - \frac{b^2 x_1(T) \exp(aT)}{(s-a)(s+a)}$$

$$= \frac{x_1(0)}{(s-a)} - \frac{b^2 x_1(T) \exp(aT)}{2a} \left[\frac{1}{(s-a)} - \frac{1}{(s+a)} \right]$$

taking inverse transforms gives

$$x_1(t) = x_1(0) \exp(at) + \frac{b^2 x_1(T)}{2a} \{ \exp[a(T-t)] - \exp[a(T+t)] \} \quad (4\text{-}107)$$

Before finding the relationship between $x_1(t)$ and $x_1(T)$, we must express $x_1(0)$ in terms of $x_1(T)$. Putting $t = T$ in equation (4-107) gives

$$x_1(T) = x_1(0) \exp(aT) + \frac{b^2 x_1(T)}{2a} - \frac{b^2 x_1(T)}{2a} \exp(2aT)$$

from which we obtain

$$x_1(0) = x_1(T) \exp(-aT) \left[1 - \frac{b^2}{2a} + \frac{b^2}{2a} \exp(2aT) \right]$$

Inserting this in equation (4-107) gives

$$x_1(t) = x_1(T) \exp[-a(T-t)] \left\{ 1 - \frac{b^2}{2a} + \frac{b^2}{2a} \exp(2aT) \right\}$$

$$+ \frac{b^2 x_1(T)}{2a} \{ \exp[a(T-t)] - \exp[a(T+t)] \}$$

$$= x_1(T) \exp[-a(T-t)] \left\{ 1 - \frac{b^2}{2a} + \frac{b^2}{2a} \exp[2a(T-t)] \right\}$$

from which

$$x_1(T) = \frac{\exp[a(T-t)] x_1(t)}{\{1 - (b^2/2a) + (b^2/2a) \exp[2a(T-t)]\}}$$

so that the optimal control is given by

$$u^0(t) = -x_1(T) b \exp[a(T-t)]$$

$$= \frac{-b \exp[2a(T-t)] x_1(t)}{\{1 - (b^2/2a) + (b^2/2a) \exp[2a(T-t)]\}}$$

4.10 Maximal-effort or 'Bang-bang' Systems

The form of Pontryagin's optimal control equations shows that, for a large class of systems of practical interest, the form of optimal control action is of a maximal-effort or 'bang-bang' type. The best known examples of such systems are those in which the control variables are separable (there are no cross-products of control variables), the controls occur linearly, and are subject to inequality constraints which are independent of state variables. Such systems will have a state space equation set

$$\frac{dx_i}{dt} = \psi_i(x_1, x_2, \ldots, x_n) + \sum_{k=1}^{r} b_{ik} u_k \qquad i = 1, 2, \ldots, n \qquad (4\text{--}108)$$

where the b_{ik} are real constants and

$$|u_k| \leqslant m_k \qquad k = 1, 2, \ldots, r \qquad (4\text{--}109)$$

where the m_k are positive constants. For systems of this type

$$\bar{\mathscr{H}}^* = \sum_{i=1}^{n} p_i \psi_i + \sum_{i=1}^{n} \sum_{k=1}^{r} p_i b_{ik} u_k - \mathscr{L}$$

If $\mathscr{L}$ is independent of the control variables then $\bar{\mathscr{H}}^*$ is maximized with respect to $u_1, \ldots, u_r$ when

$$u_k = m_k \operatorname{sgn} \sum_{i=1}^{n} p_i b_{ik} \qquad (4\text{--}110)$$

Thus for optimal control, regardless of the form of $\mathscr{L}$, provided that it is independent of control variables, the control variables are always at their maximum permissible modulus. The associated controller is thus of *switching* type, specifying only sign changes in the control variables. (Note: sgn x is $+1$ when x positive and -1 when x negative.)

Example 4.8

Consider a second order system, subject only to inertial forces, which is to be controlled in such a way as to reach a stipulated fixed target in minimum time. Let the governing equation be

$$\frac{d^2 x}{dt^2} = u \quad \text{with} \quad |u| \leqslant 1$$

and suppose the system is to be taken from initial conditions

$$x = 0 \qquad \frac{dx}{dt} = 0$$

to a fixed target

$$x = a \qquad \frac{dx}{dt} = b$$

in minimum time.

The corresponding state space equations are obtained by putting:

$$x_1 = x \qquad x_2 = \frac{dx}{dt}$$

giving

$$\dot{x}_1 = x_2 \qquad \dot{x}_2 = u$$

For minimum time control

$$V(T) = T = \int_0^T dt$$

so that the corresponding Lagrangian and Hamiltonian are given by

$$\mathscr{L} = 1 \qquad \bar{\mathscr{H}}^* = p_1 \dot{x}_1 + p_2 \dot{x}_2 - \mathscr{L}$$
$$= p_1 x_2 + p_2 u - 1$$

$\bar{\mathscr{H}}^*$ is maximized with respect to u when

$$u = \text{sgn } p_2$$

and the equations governing an optimally controlled motion are then found to be

$$\dot{x}_1 = x_2$$
$$\dot{x}_2 = \text{sgn } p_2 \qquad (4\text{-}111)$$

$$\dot{p}_1 = 0$$
$$\dot{p}_2 = -p_1 \qquad (4\text{-}112)$$

Nature of Optimal Control Action

In this simple case we can solve the two-point boundary value problem by considering all the possibilities involved. For an arbitrary set of initial conditions for the co-state vector

$$p_1(0) = c_1 \quad \text{and} \quad p_2(0) = c_2$$

we may integrate equations (4-112) to get

$$p_1(t) = c_1$$
$$p_2(t) = -c_1 t + c_2$$

We may now consider the nature of the optimal control action using

the pair of equations
$$u = \operatorname{sgn} p_2$$
where
$$p_2 = -c_1 t + c_2$$
Consider the various possibilities in turn:

(a) If $c_1 = 0$ and $c_2 < 0$, then $p_2(t) < 0$ for $t > 0$, and the corresponding control signal will always be -1.

(b) If $c_1 = 0$ and $c_2 > 0$, then $p_2(t) > 0$ for $t > 0$, and the corresponding control signal will always be $+1$.

(c) If $c_1 > 0$, then $p_2(t) < 0$ for $t > c_2/c_1$ and otherwise $p_2(t) > 0$. Thus the corresponding control signal will be -1 for $t > c_2/c_1$ and otherwise it will be $+1$.

(d) If $c_1 < 0$, then $p_2(t) > 0$ for $t > c_2/c_1$ and otherwise $p_2(t) < 0$. Thus the corresponding control signal will be $+1$ for $t > c_2/c_1$ and otherwise it will be -1.

This shows that the only possible types of optimal trajectory in the state space are:

(i) A first arc with $u = +1$ followed by a second arc with $u = -1$.
(ii) A first arc with $u = -1$ followed by a second arc with $u = +1$.
(iii) A single arc with $u = +1$ throughout.
(iv) A single arc with $u = -1$ throughout.

The latter cases represent isolated types of solution for special starting conditions and, in general, the optimal control action will be a period of time with the control variable at the maximum permissible value with one sign, then an instantaneous switch to the maximum permissible value of opposite sign with the control remaining at this value till the target is reached.

Construction of Field of Optimal Trajectories

In this simple case, a knowledge of the nature of the optimal control action enables the field of optimal trajectories to be constructed. Integrating equations (4–111) for $u = \operatorname{sgn} p_2 = +1$ gives

$$\left.\begin{array}{l} x_1(t) = \tfrac{1}{2}t^2 + x_2(0)t + x_1(0) \\ x_2(t) = t + x_2(0) \end{array}\right\} \text{Field 1}$$

and for $u = \operatorname{sgn} p_2 = -1$ gives

$$\left.\begin{array}{l} x_1(t) = -\tfrac{1}{2}t^2 + x_2(0)t + x_1(0) \\ x_2(t) = -t + x_2(0) \end{array}\right\} \text{Field 2}$$

If we call the fields of trajectories for positive and negative drives Fields 1 and 2 respectively, an examination of the corresponding trajectories in Figures 4–14, 4–15 and 4–16 shows that the origin starting

Maximal-effort or 'Bang-bang' Systems 273

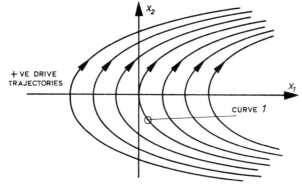

+VE DRIVE TRAJECTORIES

CURVE 1

Figure 4–14

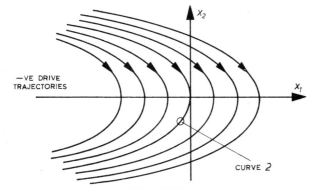

−VE DRIVE TRAJECTORIES

CURVE 2

Figure 4–15

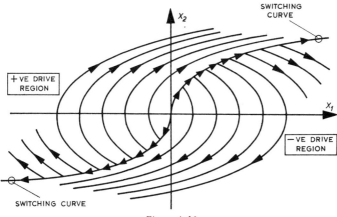

SWITCHING CURVE

+VE DRIVE REGION

−VE DRIVE REGION

SWITCHING CURVE

Figure 4–16

point may only be left along the two curves denoted as curve 1 and 2 for positive and negative drives respectively. It immediately follows from this that the field of optimal trajectories is as shown in Figure 4–16 and that the optimal control is specified by a *switching curve* separating the regions of positive and negative drive. For release from any arbitrary value of initial state, the system state is driven along a parabolic arc under one sign of drive to appropriate place on the switching curve at which point the sign of the drive is reversed, and the state point travels along the appropriate parabolic trajectory through the target.

The equations of curves 1 and 2 are easily determined as

$$\text{curve 1} \quad x_1 - \tfrac{1}{2}x_2^2 = 0$$
$$\text{curve 2} \quad x_1 + \tfrac{1}{2}x_2^2 = 0$$

from which it follows that the equation of the switching curve is

$$x_1 + \tfrac{1}{2}x_2|x_2| = 0$$

Form of Wavefront $W(T)$

It is instructive to construct the wavefront $W(T)$ for this system. Consider an arbitrary trajectory starting from $x_1(0) = 0$, $x_2(0) = 0$ such that a switch in control variable from $u = +1$ to $u = -1$ takes place at $t = \tau$, that is

$$u = +1 \quad \text{for} \quad 0 \leqslant t \leqslant \tau \quad \text{where} \quad 0 \leqslant \tau \leqslant T$$

and $$u = -1 \quad \text{for} \quad \tau < t \leqslant T$$

A simple calculation then gives

$$\begin{aligned} x_1(T) &= \tfrac{1}{2}\tau^2 - \tfrac{1}{2}(T-\tau)^2 + \tau(T-\tau) \\ &= \tfrac{1}{2}T^2 - (T-\tau)^2 \\ x_2(T) &= \tau - (T-\tau) \\ &= 2\tau - T \end{aligned} \qquad (4\text{–}113)$$

Elimination of τ from equations (4–113) gives a segment of $W(T)$ as the parabolic arc ABC in Figure 4–17 given by

$$x_1(T) = \tfrac{1}{2}T^2 - \left[T - \frac{x_2(T) + T}{2} \right]^2$$

A similar procedure for the trajectory switching from $u = -1$ to $u = +1$ gives the other segment of $W(T)$ as the parabolic arc ADC in Figure 4–17 given by

$$-x_1(T) = \tfrac{1}{2}T^2 - \left\{ T - \frac{[-x_2(T) + T]}{2} \right\}^2$$

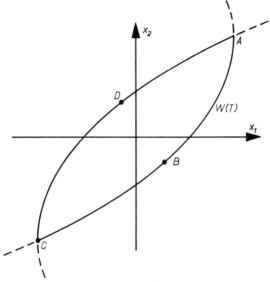

Figure 4–17

The form of expanding wavefront $W(T)$ associated with the optimal trajectories may therefore be constructed for a series of values of T; this is illustrated in Figure 4–18. Examination of the form of these wavefronts shows that the co-state vector $\mathbf{p}(T)$ is not uniquely defined along the switching curve.

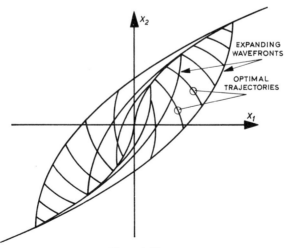

Figure 4–18

Example 4.9

As a final example of the use of Pontryagin's equations consider the following linear regulator problem. Suppose a system is governed by the set of linear state-space equations†

$$\dot{\mathbf{x}} = \mathbf{Ax} + \mathbf{Bu}$$

where the (unbounded) control vector **u** is to be manipulated so that the system is driven from an arbitrary initial state in such a way as to minimize the performance index

$$V(T) = \frac{1}{2}\mathbf{x}^t(T)\mathbf{\Omega}\mathbf{x}(T) + \frac{1}{2}\int_0^T [\mathbf{x}^t\mathbf{Q}\mathbf{x} + \mathbf{u}^t\mathbf{R}\mathbf{u}]\, dt$$

where $\mathbf{\Omega}$, $\mathbf{Q}$ and $\mathbf{R}$ are square symmetric positive-definite matrices. The system Hamiltonian is obtained from

$$\max_{\mathbf{u}} \bar{\mathcal{H}}^* = \mathcal{H} = \max_{\mathbf{u}} \{\mathbf{p}^t\mathbf{Ax} + \mathbf{p}^t\mathbf{Bu} - \tfrac{1}{2}\mathbf{x}^t\mathbf{Q}\mathbf{x} - \tfrac{1}{2}\mathbf{u}^t\mathbf{R}\mathbf{u}\}$$

For a maximum $\bar{\mathcal{H}}^*$ we have

$$\left(\frac{\partial \bar{\mathcal{H}}^*}{\partial \mathbf{u}}\right)^t = \mathbf{B}^t\mathbf{p} - \mathbf{R}\mathbf{u} = 0$$

from which
$$\mathbf{u} = \mathbf{R}^{-1}\mathbf{B}^t\mathbf{p}$$

so that

$$\mathcal{H} = \mathbf{p}^t\mathbf{Ax} + \mathbf{p}^t\mathbf{BR}^{-1}\mathbf{B}^t\mathbf{p} - \tfrac{1}{2}\mathbf{x}^t\mathbf{Q}\mathbf{x} - \tfrac{1}{2}[\mathbf{R}^{-1}\mathbf{B}^t\mathbf{p}]^t\mathbf{R}[\mathbf{R}^{-1}\mathbf{B}^t\mathbf{p}]$$
$$= \mathbf{p}^t\mathbf{Ax} - \tfrac{1}{2}\mathbf{x}^t\mathbf{Q}\mathbf{x} + \tfrac{1}{2}\mathbf{p}^t\mathbf{BR}^{-1}\mathbf{B}^t\mathbf{p}$$

giving
$$\left(\frac{\partial \mathcal{H}}{\partial \mathbf{p}}\right)^t = \mathbf{Ax} + \mathbf{BR}^{-1}\mathbf{B}^t\mathbf{p}$$

$$\left(\frac{\partial \mathcal{H}}{\partial \mathbf{x}}\right)^t = -\mathbf{A}^t\mathbf{p} + \mathbf{Q}\mathbf{x}$$

The equations governing an optimal trajectory are

$$\dot{\mathbf{x}} = \mathbf{Ax} + \mathbf{BR}^{-1}\mathbf{B}^t\mathbf{p}$$
$$\dot{\mathbf{p}} = \mathbf{Q}\mathbf{x} - \mathbf{A}^t\mathbf{p}$$

(4-114)

with two-point boundary conditions

$$\mathbf{x}(0) = \text{given vector} \qquad \mathbf{p}(T) = -\mathbf{\Omega}\mathbf{x}(T)$$

† The properties of this set of equations are fully considered in Chapter 6.

The form of equations (4–114) and the nature of the boundary condition on $\mathbf{p}(T)$ suggest that it is reasonable to infer that the co-state $\mathbf{p}$ will be a function of the state $\mathbf{x}$. Assume that

$$\mathbf{p}(t) = \mathbf{P}(t)\mathbf{x}(t)$$

where the matrix $\mathbf{P}(t)$ represents a time-varying transformation of $\mathbf{x}(t)$ into $\mathbf{p}(t)$. Substituting this into the equation for $\dot{\mathbf{p}}$ gives

$$\dot{\mathbf{p}} = \dot{\mathbf{P}}\mathbf{x} + \mathbf{P}\dot{\mathbf{x}} = \mathbf{Q}\mathbf{x} - \mathbf{A}^t\mathbf{P}\mathbf{x}$$

and using the equation for $\dot{\mathbf{x}}$ then gives

$$\dot{\mathbf{P}}\mathbf{x} + \mathbf{P}(\mathbf{A}\mathbf{x} + \mathbf{B}\mathbf{R}^{-1}\mathbf{B}^t\mathbf{p}) = \mathbf{Q}\mathbf{x} - \mathbf{A}^t\mathbf{P}\mathbf{x}$$

so that, again substituting $\mathbf{P}\mathbf{x}$ for $\mathbf{p}$, we get

$$[\dot{\mathbf{P}} + \mathbf{P}\mathbf{A} + \mathbf{P}\mathbf{B}\mathbf{R}^{-1}\mathbf{B}^t\mathbf{P}]\mathbf{x} = [\mathbf{Q} - \mathbf{A}^t\mathbf{P}]\mathbf{x}$$

If this equation is to hold for all values of $\mathbf{x}$ then the negative-definite matrix $\mathbf{P}$ must satisfy the differential equation

$$\dot{\mathbf{P}} + \mathbf{A}^t\mathbf{P} + \mathbf{P}\mathbf{A} + \mathbf{P}\mathbf{B}\mathbf{R}^{-1}\mathbf{B}^t\mathbf{P} = \mathbf{Q} \qquad (4\text{–}115)$$

with boundary condition

$$\mathbf{P}(T) = -\Omega$$

If the so-called matrix Riccati equation (4–115) is solved to determine the transformation matrix $\mathbf{P}$ then the optimal control is specified by a time-varying feedback controller with

$$\mathbf{u} = \mathbf{R}^{-1}\mathbf{B}^t\mathbf{p} = \mathbf{R}^{-1}\mathbf{B}^t\mathbf{P}\mathbf{x}$$

and the optimal response is found by integrating

$$\dot{\mathbf{x}} = [\mathbf{A} + \mathbf{B}\mathbf{R}^{-1}\mathbf{B}^t\mathbf{P}]\mathbf{x}$$

forward from the given initial conditions.

Note on signs. In many treatments of optimal control the Maximum Principle used here is replaced by an equivalent Minimum Principle. In terms of the equations developed here this leads to a change in the sign of the co-state. The form of the Pontryagin equations remains the same, but the sign of the co-state terminal boundary conditions is changed, and the sign of matrix $\mathbf{P}$ changes together with the sign of its end-point boundary value. Since the sign of $\mathbf{P}$ is changed the sign of all the terms in the matrix Riccati equation is changed except that of the quadratic term in $\mathbf{P}$.

CHAPTER FIVE

Network Models

The essential feature of network analysis is the prediction of the behaviour of an interconnected set of components of the type discussed in Chapter 1 from a knowledge of:

(a) the isolated behaviour of the constituent components, and
(b) the way in which the components are combined to form the network.

The fundamental principle on which the analysis is based is the conservation of energy. For the purposes of network analysis this may be expressed as:

net power summed over all network components is identically zero.

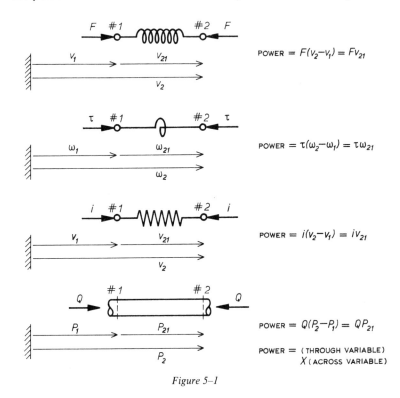

Figure 5–1

Network Models

For each component, the power may be conceptually determined by the measurement of a pair of conjugate variables whose product is component power. This is illustrated for simple translational mechanical, rotational mechanical, electrical and fluid components in Figure 5–1. In each case one of the power-defining variables has the same value at each of the component terminals and the other is the difference between the values of a variable at each terminal. These variables are said to be *through* variables and *across* variables respectively, since one is, in a sense, propagated through the component and the other is defined across the component terminal pair. This classification of variables is listed in Table 5–1.

Table 5–1

System type	Translational mechanical	Rotational mechanical	Fluid	Electrical
Through variable	Force	Torque	Flow	Current
Across variable	Relative velocity	Relative angular velocity	Pressure difference	Voltage difference

Integrating the through and across variables for the relevant kinds of components involved gives the other variables used in dynamical systems analysis, for example

$$\int (\text{force})\, dt = \text{momentum}$$

$$\int (\text{velocity})\, dt = \text{displacement}$$

$$\int (\text{current})\, dt = \text{charge}$$

and so on.

These variables may therefore be classified as integrated across-variables and integrated through-variables as summarized in Table 5–2.

It is convenient, for the discussion of thermal networks, to classify thermal system variables in a related way. In this case we have that the heat-flow itself equals the transmitted power and is defined as the thermal system through-variable. The integrated through-variable is then heat energy and the thermal across-variable is temperature. At

Table 5-2

System type	Translational mechanical	Rotational mechanical	Fluid	Electrical
Integrated through-variable	Translational momentum	Angular momentum	Volume	Charge
Integrated across-variable	Relative displacement	Relative angular displacement	Pressure momentum	Flux linkage

first sight it may seem that temperature is not an across-variable since measurement of the temperature at a specific point in space usually involves placing a thermometer there, and may not seem to involve another spatial reference point. However, the thermometer must be calibrated, and this involves creating a situation (beaker of melting ice, say) at some other point in space to give a reference point for temperature measurement. Alternatively one could consider the measurement of temperature to be carried out using a thermocouple and regard the thermocouple set as connected between two points in space.

The relationship between the various dynamical system variables and the above scheme of variable classification is illustrated in Figures 5-2 to 5-7 inclusive. Dynamical system components may be classified into

(a) Stores, and
(b) Converters

and the stores may be further classified into

(c) Across-stores
(d) Through-stores

according to whether the store constitutive relationship involves an across-variable or a through-variable respectively (as distinct from an integrated across- or integrated through-variable—see Figure 5-7). This gives the component classification shown in Table 5-3.

Table 5-3

Type of system	Translational mechanical	Rotational mechanical	Fluid	Electrical	Thermal
Across-store	Translational mass	Rotational mass	Fluid capacitor	Capacitor	Thermal capacitor
Through-store	Translational spring	Rotational spring	Fluid inertor	Inductor	None

Network Models

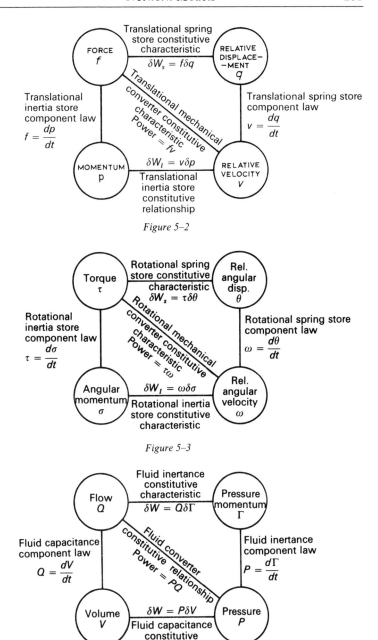

Figure 5–2

Figure 5–3

Figure 5–4

282   Network Models

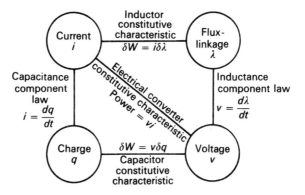

Figure 5-5

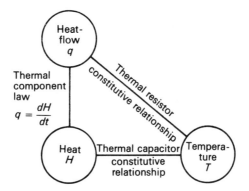

Figure 5-6

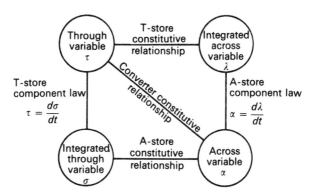

Figure 5-7

Network Models 283

If a consistent set of reference orientation conventions is adopted, as discussed in Chapter 1, each component of a network may be represented by an oriented line segment, and the whole network associated with an oriented linear graph. The most convenient way to construct the linear graph for any given system is to imagine that a complete set of power-measuring instrument pairs with a consistent orientation convention has been connected to every system component to obtain a complete power balance. Every measurement pair is then represented by an arbitrarily-oriented line segment and the complete set of oriented line segments is the system oriented linear graph. A set of examples of simple systems and corresponding oriented linear graphs is given in Figures 5–8 to 5–12. In each case the identifier labels used are letters for spatial points and numbers for system components.

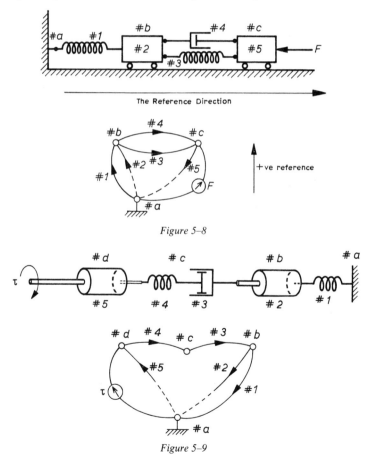

Figure 5–8

Figure 5–9

284   Network Models

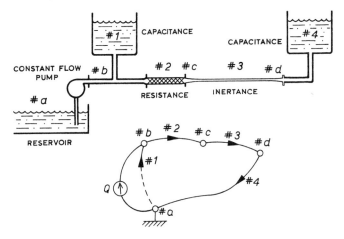

Figure 5–10

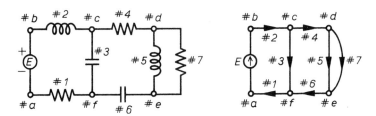

Figure 5–11

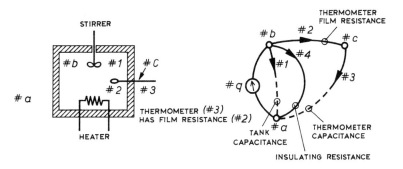

Figure 5–12

5.1 Basic Definitions for Linear Graphs

The network analysis methods used here make considerable use of linear graph theory. The terms used are now defined.
Linear graph. A linear graph is defined as an interconnected set of line segments.
   In what follows, where there is no possibility of confusion, line segment will normally be abbreviated to line. In linear graph theory, the straightness or otherwise of the line segments is irrelevant; we are only concerned with certain topological features of the graph, that is with those features which are invariant under deformations which exclude the tearing or removal of the line segments.
Vertex. An end-point of a line segment, or any single isolated point on a linear graph is called a vertex.
   Figure 5–13 shows an example linear graph on which the line segments have been labelled a, b, c, d, e and f, and the vertices have been labelled A, B, C, D and E. In electrical work the word node is often used for vertex.

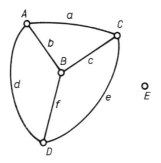

Figure 5–13

Arc. An arc is any route traced out through a linear graph for which no vertex is encountered more than once.
Path. A route traced out through a linear graph which goes through no line segment more than once is called a path.
Subgraph. Any constituent part of a linear graph is called a subgraph.
Connected graph. A connected graph is one for which every vertex is joined to every other vertex by some arc.
Connectivity. The connectivity of a linear graph is its number of separate parts. The example graph of Figure 5–13 has a connectivity of 2.
Circuit. Any arc which returns to its starting point is called a circuit. Alternatively, a circuit may be defined as a connected subgraph having only two lines incident on each vertex. In Figure 5–13, circuits are

formed by line segments a, b, c; b, f, d; a, e, d and so on. In electrical work the words loop and mesh are often used for circuit.

Circuits play a crucial role in the dynamical applications of linear graph theory. Their role in dynamical work is closely related to the concept of a tree of the graph.

Tree. A tree of a connected linear graph is a connected subgraph which contains all the vertices but no circuits. The constituent line segments of the tree are called tree-branches.

Chords. Those line segments of a connected graph which are not in a selected tree are called the chords of that tree.

Co-tree. The complete set of chords for a given tree is called the co-tree of that tree.

In Figure 5–13, trees of the connected left-hand subgraph are formed by line segments b, c, f; d, b, a; e, f, d and so on. The corresponding co-trees are a, e, d; c, f, e; a, b, c and so on.

Forest. If a graph consists of a number of separate parts, it cannot have a single tree connecting all its vertices. We may however form a tree for each separate part of the graph. This set of trees is called a forest of the graph.

Cut-set. A cut-set of a graph is a set of line segments whose removal increases the connectivity of the graph by one provided that, if one or more line segments of the set is omitted, the remaining line segments in the set do not have this property. In Figure 5–13 cut-sets of the left-hand subgraph are formed by line segments d, b, a; b, c, f; d, f, e; d, b, c, e and so on.

Completely connected graph. A completely connected graph is one in which every pair of vertices is connected by a line segment. If a completely connected graph has n vertices, it will have $\frac{1}{2}n(n-1)$ line segments.

Null graph. A graph having only vertices and devoid of line segments will be called a null graph.

Oriented line segment or oriented element. A line segment on which an arrow has been placed to indicate an orientation is called an oriented line segment or oriented element.

Oriented linear graph. A linear graph on which every line segment is oriented is called an oriented linear graph, directed graph or digraph. An example of an oriented linear graph is shown in Figure 5–14.

Consistently-oriented circuit. If all the orientation arrows point in the same direction when a circuit is traced out on an oriented linear graph, the circuit will be called a consistently-oriented circuit. On Figure 5–14, a, b, i; i, j, h; b, d, c are consistently-oriented circuits.

In using linear graphs, one is frequently concerned with the relationships between line segments, vertices and circuits. We first note that a tree of a given graph may be formed by starting with the vertices of the graph and adding on the line segments one at a time. On placing

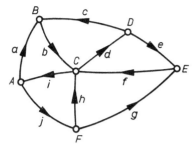

Figure 5-14

in the first line segment we see that it is incident on two vertices. In forming the tree by adding in further line segments, we see that each time a further line segment is added it brings in one new vertex to the growing set of vertices on which line segments are incident. From this we conclude that:

A tree with n vertices contains $(n-1)$ line segments.

By a simple extension of the argument we have that:

A forest with n vertices and k separate parts contains $(n-k)$ line segments.

The most important elementary number associated with a connected graph is the least number of line segments which must be removed to give a subgraph which contains no circuits. This number is called the *circuit rank* or *cyclomatic number* of the graph. Let a given connected graph have e line segments and n vertices. Suppose line segments are removed until a tree remains. The tree must, by definition, have the same number of vertices as the original graph and therefore consists of $(n-1)$ line segments. The number of line segments removed, and thus the number of chords for the tree, is

This gives:
$$c = e - n + 1$$

The least number of line segments which must be removed from a connected graph having e line segments and n vertices to give a tree is $(e-n+1)$.

5.2 Interconnective Constraints on Power Variables

When a set of components are connected together to form a network, the component power variables are no longer independent since the net power must sum to be identically zero. The following *postulates* are necessary and sufficient for the conservation of energy, and are therefore taken as the basis of systematic methods of network analysis. The necessity follows from the consideration of simple series and parallel combinations of components, and the sufficiency follows from Tellegen's theorem in Section 5.4 below.

Vertex postulate for through-variables. The algebraic sum of all through-variables incident on any vertex of a network or oriented linear graph is identically zero.

This postulate may be illustrated by a few examples. For translational mechanical networks, the algebraic sum of all forces incident at a point of connection is identically zero. This is usually called d'Alembert's Principle when, as here, inertial forces are included. For the system of Figure 5-8 we thus have, where f_i is the force for component $^\#i$

Vertex $^\#b$ $\qquad f_1+f_2-f_3-f_4 = 0$

Vertex $^\#c$ $\qquad f_3+f_4-f_5+F = 0$

For rotational mechanical networks, the algebraic sum of all torques, including inertia torques, at any connection point is identically zero. For the system of Figure 5-9 we thus have, where τ_i is the torque for component $^\#i$

Vertex $^\#b$ $\qquad \tau_3-\tau_2-\tau_1 = 0$

Vertex $^\#c$ $\qquad \tau_4-\tau_3 = 0$

Vertex $^\#d$ $\qquad \tau+\tau_5-\tau_4 = 0$

For fluid systems the algebraic sum of the flows at a connection point is identically zero. Considering the system of Figure 5-10 we have, where Q_i denotes the flow for component $^\#i$

Vertex $^\#b$ $\qquad Q+Q_1-Q_2 = 0$

Vertex $^\#c$ $\qquad Q_2-Q_3 = 0$

Vertex $^\#d$ $\qquad Q_3-Q_4 = 0$

In electrical systems Kirchhoff's Current Law (conveniently abbreviated to KCL) states that the algebraic sum of all currents incident on any vertex of the network is identically zero. Considering the network of Figure 5-11, we have, where i_i is the current for component $^\#i$

Vertex $^\#b$ $\qquad i_E-i_2 = 0$

Vertex $^\#c$ $\qquad i_2-i_3-i_4 = 0$

Vertex $^\#e$ $\qquad i_5+i_7-i_6 = 0$

and so on.

Finally, for thermal systems the conservation of energy requires that the algebraic sum of the heat flows incident on any connection point of the thermal network must be identically zero. For the system of Figure 5–12 we have, where q_i is the heat flow through component $^\#i$

Vertex $^\#b$ $\qquad q - q_1 - q_2 - q_4 = 0$

Vertex $^\#c$ $\qquad q_2 - q_3 = 0$

Circuit postulate for across-variables. In any network or linear graph, closed paths may be traced out through connected elements of the network; any such closed path which includes more than one network element is termed a *circuit* of the network. The second fundamental postulate of network analysis states that the algebraic sum of all across-variables taken round any circuit of the network is identically zero.

For mechanical networks, the algebraic sum of the relative velocities between component junction points is identically zero when taken round any closed circuit. For the translational network of Figure 5–8 we have, where v_i is the relative velocity between terminals of component $^\#i$

$$v_1 - v_2 = 0$$
$$v_1 + v_4 + v_5 = 0$$
$$v_2 + v_3 + v_5 = 0$$

and so on. For the rotational network of Figure 5–9 we have, where ω_i is the relative angular velocity across terminals of component $^\#i$

$$\omega_5 + \omega_4 + \omega_3 + \omega_2 = 0$$
$$\omega_1 - \omega_2 = 0$$
$$\omega_5 + \omega_4 + \omega_3 + \omega_1 = 0$$

and so on.

For fluid systems the algebraic sum of the pressure differences taken round a circuit is zero. In the system of Figure 5–10 we have, where P_i is the pressure difference across component $^\#i$

$$P_Q - P_1 = 0$$
$$P_1 + P_2 + P_3 + P_4 = 0$$
$$P_Q + P_2 + P_3 + P_4 = 0$$

In electrical systems Kirchhoff's Voltage Law (conveniently abbreviated to KVL) states that the algebraic sum of the voltages across the component terminal pairs is identically zero when taken round any circuit. For the network of Figure 5-11 we have, where v_i is the voltage across component $^\#i$

$$v_E + v_2 + v_3 + v_1 = 0$$
$$v_6 - v_3 + v_4 + v_5 = 0$$
$$v_7 - v_5 = 0$$
$$v_E + v_2 + v_4 + v_7 + v_6 + v_1 = 0$$

and so on.

Finally, for thermal systems the algebraic sum of the temperature differences round any circuit will be identically zero. For the system of Figure 5-13 we have, where T_i is the temperature across component $^\#i$

$$T_q + T_1 = 0$$
$$-T_1 + T_2 + T_3 = 0$$
$$-T_4 + T_2 + T_3 = 0$$

and so on.

5.3 Topological Relationships between Network Variables [B10]

In any method of network analysis we must select an *independent* set of vertex equations and an *independent* set of circuit equations, since the total sets of these respective equations are obviously not independent. Let the system linear graph have e elements and n vertices.

Independent set of vertex equations. Any set of $(n-1)$ vertex equations corresponding to $(n-1)$ distinct vertices is an independent set of vertex equations.

Independent set of circuit equations. Any set of $(e-n+1)$ circuit equations corresponding to $(e-n+1)$ distinct circuits is an independent set of circuit equations.

To establish these fundamentally important relationships we use the concept of a tree of the linear graph. A tree of the linear graph is a connected subgraph which contains all the vertices but no circuits; the constituent elements of the tree are termed branches and the remaining elements, not belonging to the tree, are termed chords. As shown in Section 5.1 above there are
(a) $(n-1)$ branches, and
(b) $(e-n+1)$ chords.

For each of the $(n-1)$ branches we may select a different vertex and write down a vertex equation which expresses that branch through-variable as a sum of all other element through-variables. These $(n-1)$

vertex equations will be independent since each equation will include at least one through-variable not contained in the other equations. If we write each of this set of equations in the form

$$\sum_i \tau_i = 0$$

using τ as a general symbol for a through-variable, then the equation obtained for the nth vertex will be the negative sum of the equations for these $(n-1)$ vertices. To see this imagine the $(n-1)$ vertices enclosed by a surface cutting all the elements incident upon the nth vertex once. Now apply the vertex law to the $(n-1)$ vertices enclosed in this way. In the equations obtained each through-variable will enter twice, once for each vertex of its element vertex pair. Furthermore it will enter with opposite sign in each case since if it is directed towards one vertex it must be directed away from the other. Thus adding together all the equations for the $(n-1)$ vertices internal to the enclosing surface will result in cancelling all the through-variables associated with branches internal to the surface. This will leave an equation which will involve only through-variables associated with branches crossing the surface, that is with branches incident on the nth vertex. This equation is obviously the negative of the vertex equation for the nth vertex since if a through-variable is directed into the closed surface and thus appears in the sum of equations with a positive sign, it will be directed away from the nth vertex and thus appear in the nth vertex equation with a negative sign. This establishes that *any* $(n-1)$ vertex equations are independent. If the network has m separate parts then each separate part will satisfy the above argument and these will therefore be $(n-m)$ linearly independent vertex equations.

To establish the number of independent circuit equations, consider the insertion of a single chord element into a tree of the linear graph. The insertion of each chord will create a circuit and thus $(e-n+1)$ circuits can be generated by inserting each chord into the tree *one at a time*. Applying the circuit law for across-variables to each circuit generated in this way will give a set of $(e-n+1)$ independent circuit equations, since each equation will contain an across-variable which does not appear in any of the other equations. In addition all other circuits which may be formed can be constructed as combinations of the $(e-n+1)$ fundamental circuits formed in this way. Thus there are $(e-n+1)$ independent circuit equations and these may be formed by inserting the chords into a tree one at a time.

Suppose a network contains p passive components and s sources so that the number of oriented line segments in the oriented linear graph is

$$e = p+s$$

The number of unknowns is $(2p+s)$ since both the across and through variables must be found for the passive components and *either* an across- *or* a through-variable must be found for each of the sources. The number of equations available is therefore:

$(n-1)$ vertex equations
$(e-n+1) = (p+s-n+1)$ circuit equations
p constitutive equations

Thus the total number of equations is

$$(n-1)+(p+s-n+1)+p = 2p+s$$

which is equal to the total number of unknowns and the network problem may therefore be solved using such a set of equations.

The Incidence Matrix and the Reduced Incidence Matrix [B10]

The incidence of the vertices and lines of a linear graph may be described algebraically by means of an n by e matrix called the *incidence matrix* and denoted by $\mathbf{A}_a$. The rows of $\mathbf{A}_a$ correspond to the vertices of the linear graph and the columns correspond to the lines. Its elements A_{jk} are defined as follows:

$A_{jk} = +1$, if element k is incident with vertex j and oriented away from it.
$A_{jk} = -1$, if element k is incident with vertex j and oriented towards it.
$A_{jk} = 0$, if element k is not incident with vertex j.

For the example graph of Figure 5–15 we have

$$\mathbf{A}_a = \begin{bmatrix} -1 & +1 & 0 & 0 & -1 & 0 \\ 0 & 0 & +1 & +1 & +1 & 0 \\ 0 & -1 & -1 & 0 & 0 & -1 \\ +1 & 0 & 0 & -1 & 0 & +1 \end{bmatrix}$$

If τ is a vector whose components are the set of through-variables corresponding to the elements of the linear graph, the set of n vertex equations for the graph may be written as

$$\mathbf{A}_a \tau = \mathbf{0} \tag{5-1}$$

The consideration above of the number of independent vertex equations for a network with n vertices and m separate parts shows that $(n-m)$, and no more than $(n-m)$, of the rows of $\mathbf{A}_a$ will be independent. Thus the rank of $\mathbf{A}_a$ is $(n-m)$. For a connected network the rank will be

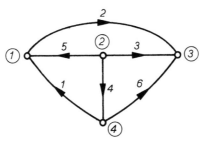

Figure 5-15

$(n-1)$. In what follows we will only consider connected networks, since we can make any linear graph having several separate parts into a connected graph by a suitable superposition of vertices. Since only $(n-1)$ of the vertex equations of a connected network are independent we may delete any row from $\mathbf{A}_a$ to give a *reduced incidence matrix* $\mathbf{A}$ having $(n-1)$ rows and e columns and of rank $(n-1)$.† For the graph of Figure 5-15 we may designate vertex #4 as the reference vertex and thus delete the fourth row from $\mathbf{A}_a$ to get the corresponding reduced incidence matrix $\mathbf{A}$ as:

$$\mathbf{A} = \begin{bmatrix} -1 & +1 & 0 & 0 & -1 & 0 \\ 0 & 0 & +1 & +1 & +1 & 0 \\ 0 & -1 & -1 & 0 & 0 & -1 \end{bmatrix}$$

The Circuit Matrix [B10]

In general a linear graph will contain many circuits. Each circuit may be arbitrarily oriented by defining a positive direction in which the circuit may be traced out, and the circuits may be arbitrarily numbered. If some particular circuit contains a stipulated line of the linear graph, the circuit and line are said to be *incident*. The incidence of circuits and lines may be described algebraically by means of a matrix containing e columns and one row for every circuit of the linear graph. This matrix may be denoted by $\mathbf{B}_a$ and has elements defined as follows:

$B_{jk} = +1$, if element k is incident with circuit j and they are oriented in the same direction.

$B_{jk} = -1$, if element k is incident with circuit j and they are oriented in opposite directions.

$B_{jk} = 0$, if element k and circuit j are not incident.

All the possible circuits of the graph of Figure 5-15 are shown numbered and oriented in Figure 5-16. The corresponding matrix $\mathbf{B}_a$

† The number of trees for a connected graph is equal to the determinant of $\mathbf{AA}^t$.

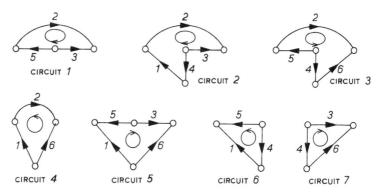

Figure 5-16

is given by

$$\mathbf{B}_a = \begin{bmatrix} 0 & 1 & -1 & 0 & 1 & 0 \\ 1 & 1 & -1 & -1 & 0 & 0 \\ 0 & -1 & 0 & 1 & -1 & 1 \\ -1 & -1 & 0 & 0 & 0 & 1 \\ 1 & 0 & 1 & 0 & -1 & -1 \\ -1 & 0 & 0 & -1 & 1 & 0 \\ 0 & 0 & 1 & -1 & 0 & -1 \end{bmatrix}$$

The matrices $\mathbf{A}_a$ and $\mathbf{B}_a$ have the fundamentally important property that

$$\mathbf{A}_a \mathbf{B}_a^t = \mathbf{0}$$
$$\mathbf{B}_a \mathbf{A}_a^t = \mathbf{0}$$
(5-2)

To prove these relationships, consider the inner product of the ith row of $\mathbf{A}_a$ and the jth row of $\mathbf{B}_a$ (that is, the jth column of $\mathbf{B}_a^t$). The ith row of $\mathbf{A}_a$ and the jth row of $\mathbf{B}_a$ will both contain nonzero elements if, and only if, an element of the linear graph is incident on vertex i and is also included in circuit j. If this is not the case, their inner product is zero. If vertex i is included in circuit j then it follows from the definition of a circuit that there will be two elements incident on vertex i and that both of them are included in circuit j. Thus the inner product of the ith row of $\mathbf{A}_a$ and the jth column of $\mathbf{B}_a^t$ will contain one $+1$ term and one -1 term, so that the inner product is again zero. (To see why this is so, consider the specific examples available in Figure 5-16.) The results of equation (5-2) then follow immediately.

The complete set of circuit equations for a network may be written as

$$\mathbf{B}_a \boldsymbol{\alpha} = 0 \tag{5-3}$$

where $\boldsymbol{\alpha}$ is used as a general symbol for a vector whose components are the set of across-variables corresponding to the elements of the linear graph. Only $(e-n+1)$ of these circuit equations, as shown above, is independent and so we may form a *circuit matrix*, denoted by $\mathbf{B}$, deleting from $\mathbf{B}_a$ all but $(e-n+1)$ rows. Thus the circuit matrix

$$\mathbf{B} = \begin{bmatrix} 1 & 1 & -1 & 1 & 0 & 0 \\ 0 & 1 & -1 & 0 & 1 & 0 \\ 0 & -1 & 0 & 1 & -1 & 1 \end{bmatrix}$$

for the linear graph of Figure 5–15 corresponds to the set of independent loops shown in Figure 5–17.

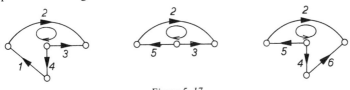

Figure 5–17

The particular set of $(e-n+1)$ independent circuits formed by inserting a set of chords one at a time into a chosen tree is called *a set of fundamental circuits* (where each circuit orientation is defined by the inserted chord orientation) and the corresponding *fundamental circuit matrix* is denoted by $\mathbf{B}_f$. If for the linear graph of Figure 5–15 we select a tree consisting of lines #4, #5 and #6 the corresponding set of fundamental circuits is shown in Figure 5–18 and we have

$$\mathbf{B}_f = \begin{bmatrix} 1 & 0 & 0 & 1 & -1 & 0 \\ 0 & 1 & 0 & -1 & 1 & -1 \\ 0 & 0 & 1 & -1 & 0 & -1 \end{bmatrix}$$

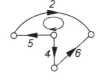

Figure 5–18

Since **A** and **B** are submatrices of $\mathbf{A}_a$ and $\mathbf{B}_a$ we have that

$$\mathbf{AB}^t = \mathbf{0}$$
$$\mathbf{B}^t\mathbf{A} = \mathbf{0} \tag{5-4}$$

5.4 Tellegen's Theorem

Tellegen's Theorem, in simple language, asserts that Kirchhoff's Laws are sufficient for the conservation of energy in a network. Since it is mostly used in electrical network analysis we will discuss it in specifically electrical terms. As its physical content is obvious, this section can be omitted on a first reading. Consider a connected electrical network of e components and n vertices and let the set of Kirchhoff Law constraints be written in the form

$$\mathbf{Ai} = \mathbf{0} \tag{5-5}$$
$$\mathbf{Bv} = \mathbf{0} \tag{5-6}$$

where **A** is an appropriate reduced incidence matrix, **B** is an appropriate circuit matrix, **i** is a vector whose elements are the component currents, and **v** is a vector whose elements are the component voltages. Consider the set of component currents $\{i_1, \ldots, i_e\}$ and the set of component voltages $\{v_1, \ldots, v_e\}$ to be vectors in an e-dimensional Euclidean vector space $\mathscr{E}_e$. The inner product for two vectors $\mathbf{i}, \mathbf{v} \in \mathscr{E}_e$ is

$$\langle \mathbf{i}, \mathbf{v} \rangle = \sum_{\mu=1}^{e} i_\mu v_\mu$$

Let $\mathscr{J}$ be the set of all vectors **i** such that $\mathbf{i} \in \mathscr{J}$ if and only if **i** satisfies equation (5-5). Let $\mathscr{V}$ be the set of all vectors **v** such that $\mathbf{v} \in \mathscr{V}$ if and only if **v** satisfies equation (5-6). Then we may state the important theorem due to Tellegen in the following way.

Tellegen's Theorem. If $\mathbf{i} \in \mathscr{J}$ and $\mathbf{v} \in \mathscr{V}$ then $\langle \mathbf{i}, \mathbf{v} \rangle = 0$. That is to say $\mathscr{J}$ and $\mathscr{V}$ are orthogonal subspaces of $\mathscr{E}_e$. Furthermore $\mathscr{J}$ and $\mathscr{V}$ together span $\mathscr{E}_e$.

To prove Tellegen's theorem we argue as follows. As further discussed in Section 5.7.2 below we may introduce a vector of vertex voltages **V** such that

$$\mathbf{v} = \mathbf{A}^t \mathbf{V}$$

The individual component voltages $\{v_\mu : \mu = 1, 2, \ldots, e\}$ are then each expressible in terms of the difference between a pair of vertex voltages of the set $\{V_1, V_2, \ldots, V_n\}$. Denote the *total* current flowing from vertex k to vertex l by i_{kl}, and take i_{kl} as zero if no components lie between the the vertices involved. Then if the ρth to the νth components connect vertices k and l we have

$$v_\rho i_\rho + \cdots + v_\nu i_\nu = (V_k - V_l)i_{kl} = (V_l - V_k)i_{lk}$$

and, arising from this symmetry in the indices k and l, we may express the inner product $\langle \mathbf{i}, \mathbf{v} \rangle$ as a sum in the form

$$\langle \mathbf{i}, \mathbf{v} \rangle = \sum_{\mu=1}^{e} i_\mu v_\mu = \frac{1}{2} \sum_{k,l} (V_k - V_l) i_{kl}$$

so that
$$\langle \mathbf{i}, \mathbf{v} \rangle = \frac{1}{2} \left[\sum_k V_k \left(\sum_l i_{kl} \right) - \sum_l V_l \left(\sum_k i_{kl} \right) \right] \quad (5-7)$$

If we consider the current summations involved we have that

$$\sum_l i_{kl} = \text{sum of currents incident on vertex } k = 0$$

and
$$\sum_k i_{lk} = \text{sum of currents incident on vertex } l = 0$$

So that, inserting these values in equation (5–7) we have

$$\langle \mathbf{i}, \mathbf{v} \rangle = 0 \quad (5-8)$$

Now consider equation (5–5). If we let

$$\mathbf{a}_\nu^t \quad \text{for} \quad \nu = 1, 2, \ldots, (n-1)$$

denote the rows of $\mathbf{A}$, then equation (5–5) gives

$$\langle \mathbf{a}_\nu, \mathbf{i} \rangle = 0 \quad \nu = 1, 2, \ldots, (n-1)$$

for all vectors $\mathbf{i} \in \mathscr{I}$. This shows that the $(n-1)$ vectors $\mathbf{a}_\nu$ span the orthogonal complement to $\mathscr{I}$. Since $\mathscr{V}$ is of dimension $(n-1)$, it follows that $\mathscr{V}$ is the entire orthogonal complement to $\mathscr{I}$ and that the set of vectors $\{\mathbf{a}_\nu : \nu = 1, 2, \ldots, (n-1)\}$ is a basis for $\mathscr{V}$.

We may consider equation (5–6) in a similar way. Let

$$\mathbf{b}_\lambda^t \quad \text{for} \quad \lambda = 1, 2, \ldots, (e-n+1)$$

denote the rows of $\mathbf{B}$. Then equation (5–6) gives

$$\langle \mathbf{b}_\lambda, \mathbf{v} \rangle = 0 \quad \lambda = 1, 2, \ldots, (e-n+1)$$

for all vectors $\mathbf{v} \in \mathscr{V}$. This shows that the $(e-n+1)$ vectors $\mathbf{b}_\lambda$ span the orthogonal complement to $\mathscr{V}$. Since $\mathscr{I}$ is of dimension $(e-n+1)$, it follows that $\mathscr{I}$ is the entire orthogonal complement to $\mathscr{V}$ and that the set of vectors $\{\mathbf{b}_\lambda : \lambda = 1, 2, \ldots, (e-n+1)\}$ is a basis for $\mathscr{I}$.

5.5 The Dynamical Transformation Matrix

From any network and its associated linear graph we may obtain a set of $(n-1)$ linearly independent vertex equations which may be written in the form

$$\mathbf{A}\tau = 0 \qquad (5\text{--}9)$$

where $\mathbf{A}$ is an appropriate reduced incidence matrix, and τ is a vector whose elements are the component through-variables. We may also, after choosing a tree of the linear graph, obtain a set of $(e-n+1)$ linearly independent circuit equations, associated with a set of fundamental circuits, which may be written in the form

$$\mathbf{B}_f \alpha = 0 \qquad (5\text{--}10)$$

where $\mathbf{B}_f$ is an appropriate fundamental circuit matrix, and α is a vector whose elements are the component across-variables. *Note that we may define the orientation of the fundamental circuits to coincide with the orientations of the defining chords.*

Group the components into branches and chords and partition equation (5–10) accordingly. This gives, when orientations coincide

$$[\mathbf{I}_{(e-n+1)} \mid \mathbf{F}] \begin{bmatrix} \alpha_c \\ \alpha_t \end{bmatrix} = 0 \qquad (5\text{--}11)$$

where

(a) $\mathbf{I}_{(e-n+1)}$ is a unit matrix of order $(e-n+1)$ which results from the partitioning *after such a grouping.*
(b) $\mathbf{F}$ is an $(e-n+1) \times (n-1)$ matrix resulting from the partitioning of $\mathbf{B}_f$.
(c) α_c is a vector whose components are the across-variables of the chord (or co-tree) elements.
(d) α_t is a vector whose components are the across-variables of the tree elements.

Multiplying out equation (5–11) we get

$$\alpha_c + \mathbf{F}\alpha_t = 0 \qquad (5\text{--}12)$$

For the same grouping of components, suppose the reduced incidence matrix $\mathbf{A}$ partitions such that equation (5–9) becomes

$$[\mathbf{X} \mid \mathbf{K}] \begin{bmatrix} \tau_c \\ \tau_t \end{bmatrix} = 0 \qquad (5\text{--}13)$$

where

(a) $\mathbf{X}$ and $\mathbf{K}$ are matrices obtained by suitably partitioning matrix $\mathbf{A}$.

The Dynamical Transformation Matrix

(b) τ_c is a vector whose components are the through-variables of the chord (or co-tree) elements.

(c) τ_t is a vector whose components are the through-variables of the tree elements.

Multiplying out equation (5–13) we get

$$X\tau_c + K\tau_t = 0 \qquad (5\text{–}14)$$

Now the through-variables of the tree components can always be expressed in terms of the through-variables of a set of chords, since the set of $(e-n+1)$ linearly independent equations obtained by creating a set of fundamental circuits by inserting the chords one at a time enable us to express each tree element through-variable as a linear combination of chord element through-variables. It follows from this that the inverse of K must exist and so equation (5–14) gives

$$\tau_t + K^{-1}X\tau_c = 0 \qquad (5\text{–}15)$$

Now multiply equation (5–12) by τ_c^t to get

$$\tau_c^t \alpha_c + \tau_c^t F \alpha_t = 0 \qquad (5\text{–}16)$$

and multiply equation (5–15) by α_t^t to get

$$\alpha_t^t \tau_t + \alpha_t^t K^{-1} X \tau_c = 0 \qquad (5\text{–}17)$$

Adding equations (5–16) and (5–17) together we have

$$-\alpha_t^t [F^t + K^{-1}X]\tau_c = \tau_c^t \alpha_c + \alpha_t^t \tau_t$$
$$= \tau_c^t \alpha_c + \tau_t^t \alpha_t$$
$$= \langle \tau, \alpha \rangle \qquad (5\text{–}18)$$

Now Tellegen's theorem, established in Section 5.4 above, shows that the right-hand side of equation (5–18) is identically zero. (Alternatively one could simply appeal to the principle of conservation of energy and say that $\langle \tau, \alpha \rangle$ is the net power summed over all components which must vanish identically.) Thus an application of Tellegen's theorem to equation (5–18) gives that

$$K^{-1}X = -F^t \qquad (5\text{–}19)$$

Thus
$$\tau_t = F^t \tau_c$$
$$\alpha_c = -F \alpha_t \qquad (5\text{–}20)$$

If we introduce a *dynamical transformation matrix* D such that

$$D = F^t \qquad (5\text{–}21)$$

then we get the pair of equations

$$\tau_t = \mathbf{D}\tau_c \quad (5\text{-}22)$$

$$\alpha_c = -\mathbf{D}^t\alpha_t \quad (5\text{-}23)$$

Note that $\quad \mathbf{D} = -\mathbf{K}^{-1}\mathbf{X}$

These equations will be used a great deal in certain systematic methods of network analysis given below. Their particular advantages stem from the simple fact that we are defining independent sets of *both* vertex *and* circuit equations in terms of a *single* matrix $\mathbf{D}$. To write down the dynamical transformation matrix $\mathbf{D}$ for a given network, we select a tree of the linear graph and form a set of fundamental circuits by inserting the chords into the chosen tree one at a time. The matrix $\mathbf{D}$ then has elements:

$d_{jk} = +1$ if the orientation of branch j and the orientation of chord k are in the same direction in the circuit formed by insertion of chord k into the chosen tree.

$d_{jk} = -1$ if the orientation of branch j and the orientation of chord k are in opposite directions in the circuit formed by insertion of chord k into the chosen tree.

$d_{jk} = 0$ if branch j does not lie in the circuit formed by insertion of chord k into the chosen tree.

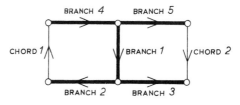

Figure 5–19

For example, the linear graph shown in Figure 5–19 would have a dynamical transformation matrix formed by drawing up a table as follows:

	Chord 1	Chord 2
Branch 1	+1	−1
Branch 2	+1	0
Branch 3	0	−1
Branch 4	+1	0
Branch 5	0	+1

giving

$$D = \begin{bmatrix} +1 & -1 \\ +1 & 0 \\ 0 & -1 \\ +1 & 0 \\ 0 & +1 \end{bmatrix}$$

5.6 Analogues, Duals and Dualogues

Two networks having the same linear graph and the same types of sources (that is across- and through-sources corresponding in both cases) distributed in the same relative positions on the linear graphs will be called *structurally analogous*. The *same* vertex and circuit equations will hold for both networks. If in addition the corresponding lines in the two linear graphs relate to elements of the *same* type (that is across-stores, through-stores, dissipators, across-sources, through-sources) the two networks are called *analogous*. One network is then said to be the analogue of the other. Some examples of analogous systems are shown in Figure 5–20. Network A is the electrical analogue of the mechanical network B, and the fluid network C. Network B is the mechanical analogue of A and C, and so on. Thermal analogues can only exist when no through-stores are involved, since there is no thermal analogue of electrical inductance. Mechanical and fluid analogues of electrical networks will only exist in a direct sense when all the electrical network capacitors have a single terminal in common which may be taken as the reference terminal. This particular difficulty may be overcome by adding ideal unity-ratio electrical transformers in the electrical network to isolate the electrical capacitors and break the direct connections between them. These transformers may then be replaced by their analogues (levers, gear ratios, piston transformers, etc.) in the analogous networks formed.

If two linear graphs are such that the dynamical transformation matrix for one is the negative transpose of the dynamical transformation matrix for the other for every choice of tree, the two graphs are said to be *dual*. As shown by equations (5–22) and (5–23) this implies that the vertex equations for one network are identical with the circuit equations for the other network. The particular circuits considered are the 'windows' or meshes of a *planar graph*, that is a graph which can be contained in a plane without any of its lines intersecting at other than a vertex of the graph. *A nonplanar graph does not have a dual.* To form the dual graph for a given planar graph we proceed in the following way (which cannot be extended to deal with nonplanar graphs).

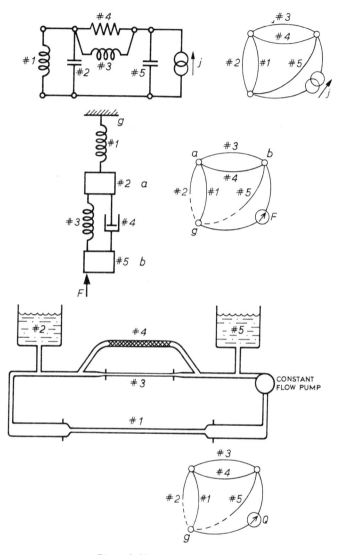

Figure 5–20 Analogous systems.

(a) Place a vertex inside each 'window' of the original graph.
(b) Add a further vertex at some arbitrary point exterior to the original graph.
(c) Connect these vertices by the same number of lines as on the original graph in such a way that every vertex of the original graph

appears in a window of the dual graph, except for one vertex which is external to the dual graph.

This procedure is illustrated in Figure 5-21. An inspection of this figure will show that the lines radiating from a vertex of one graph 'cut through' in a one-to-one relationship the lines forming a 'window' of the dual graph. Thus for any orientation of the primal graph we can so orient the dual graph that the vertex equation set for the primal graph is the 'window-circuit' equation set for the dual graph and vice versa.

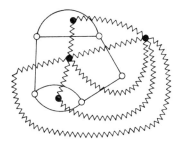

Figure 5-21

If two networks have dual linear graphs and if in addition the roles played by the through- and across-variables for the *corresponding* elements in the dual linear graphs are interchanged (that is across-store → through-store; across-source → through-source; etc.), then the two networks are said to be *dual*. In an electrical system for example all inductors in the primal network would be replaced by capacitors in the dual network and so on.

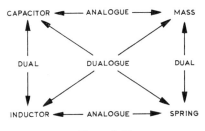

Figure 5-22

Dualogues

An analogue of a dual network is called a *dualogue*. The various possibilities for related electrical and mechanical systems are summarized in Figure 5-22. This shows that we can use either:

(a) The mass-capacitance analogue (or mobility analogue) in which analogous quantities are:

$$\text{force} \leftrightarrow \text{current}$$
$$\text{velocity} \leftrightarrow \text{voltage}$$
$$\text{mass} \leftrightarrow \text{capacitance}$$
$$\text{spring} \leftrightarrow \text{inductance}$$

or
(b) The mass-inductance dualogue (or classical analogue) in which analogous quantities are:

$$\text{force} \leftrightarrow \text{voltage}$$
$$\text{velocity} \leftrightarrow \text{current}$$
$$\text{mass} \leftrightarrow \text{inductance}$$
$$\text{spring} \leftrightarrow \text{capacitance}$$

In historical terms the mass-inductance analogy (used by Kelvin and Maxwell) [M15] preceded the mass-capacitance analogy (introduced by Nickle and Firestone [F2] and given a definitive treatment by Trent [T1]). While it is pointless to discuss which analogy is 'correct', as both are equally valid when they exist, the mass-capacitance analogy has considerable advantages. These chiefly stem from the fact that while we can always construct the mass-capacitance analogue of a *nonplanar* electrical or mechanical system, the mass-inductance analogue of a nonplanar electrical or mechanical system will not exist. This is because we cannot construct a dualogue when the dual linear graph does not exist.

5.7 Circuit, Vertex and Mixed Transform Analysis Methods for Linear Electrical Networks [B10]

In the detailed discussions of network analysis methods which follow, we will normally consider electrical networks since other networks can then be analysed using analogues.

In this section we consider linear connected networks composed of general electrical elements of the type shown in Figure 5-23. If any given network is not connected, say because it contains separate parts coupled by mutual inductance, then it may be made into a connected network by identifying a set of individual parts' reference vertices into a common reference vertex. The outputs of the network sources are assumed known as functions of time and our problem is to determine the voltage and currents of the passive elements as functions of time.

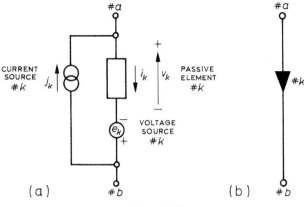

Figure 5-23

The discussions of source distributions given in Section 5.12 below show that no loss of generality is involved in considering only elements as shown in Figure 5-23.

We will use Laplace transforms of all variables involved. The basic circuit relationship, Kirchhoff's Voltage Law, applied to a set of independent circuits gives a set of m independent equations which may be written in matrix form as

$$\mathbf{Bv}(s) - \mathbf{Be}(s) = \mathbf{0} \qquad (5\text{-}24)$$

where $\mathbf{B}$ is the appropriate circuit matrix, and $\mathbf{v}(s)$ and $\mathbf{e}(s)$ are vectors whose elements are the Laplace transforms of the voltages of the passive network components, and the Laplace transforms of the voltage source outputs respectively. The basic vertex relationship, Kirchhoff's Current Law, applied to a set of $(n-1)$ vertices gives a set of $(n-1)$ independent equations which may be written in matrix form as

$$\mathbf{Ai}(s) - \mathbf{Aj}(s) = \mathbf{0} \qquad (5\text{-}25)$$

where $\mathbf{A}$ is a reduced incidence matrix, and $\mathbf{i}(s)$ and $\mathbf{j}(s)$ are vectors whose components are the Laplace transforms of the currents of the passive network components and the Laplace transforms of the outputs of the current sources respectively.

Now consider the relationships between the Laplace transforms of the component variables for the various types of component involved. For a resistive component, $^\#k$ say,

$$v_k(s) = R_k i_k(s) \qquad (5\text{-}26)$$

where R_k is the resistance of the resistor. For a capacitive component

$$i_k(s) = sC_k v_k(s) \tag{5-27}$$

where C_k is the capacitance of the capacitor. The ratio of voltage transform to current transform

$$\frac{v_k(s)}{i_k(s)} = \frac{1}{sC_k} \tag{5-28}$$

is called the *generalized impedance* of the capacitor. For an inductive element not coupled inductively to any other inductors we have

$$v_k(s) = sL_k i_k(s) \tag{5-29}$$

giving the corresponding generalized impedance as

$$\frac{v_k(s)}{i_k(s)} = sL_k \tag{5-30}$$

In the general inductive case

$$v_k(s) = L_{k1} si_1(s) + L_{k2} si_2(s) + \cdots + L_{kl} si_l(s) \tag{5-31}$$

where $\{L_{k1}, \ldots, L_{kl}\}$ are the appropriate elements of a mutual inductance matrix and l is the number of inductive elements in the network.

If the element numbering in a network is taken to be all inductors, followed by all resistors, and finally all capacitors, then the component power variable transforms are related by

$$\mathbf{v}(s) = \mathbf{Z}(s)\mathbf{i}(s) \tag{5-32}$$

where $\mathbf{v}(s)$ and $\mathbf{i}(s)$ are vectors whose elements are the component voltage transforms and current transforms respectively. The *generalized impedance matrix* $\mathbf{Z}(s)$ will, for this grouping of components, be the direct sum of three generalized impedance matrices

$$\mathbf{Z}(s) = s\mathbf{L} \oplus \mathbf{R} \oplus \frac{1}{s}\mathbf{S} \tag{5-33}$$

where $\oplus$ denotes the direct sum and $\mathbf{L}$ is a non-singular, positive-definite matrix of self and mutual-inductance values, $\mathbf{R}$ is a non-singular diagonal matrix of resistance values, and $\mathbf{S} = \mathbf{C}^{-1}$ where $\mathbf{C}$ is a non-singular diagonal matrix of capacitance values.

Since we have stipulated non-zero values of resistances, inductances and capacitances, equation (5-32) may be inverted to give

$$\mathbf{i}(s) = \mathbf{Y}(s)\mathbf{v}(s) \tag{5-34}$$

where

$$\mathbf{Y}(s) = \mathbf{Z}^{-1}(s) \tag{5-35}$$

Circuit, Vertex and Mixed Transform Analysis Methods

and
$$Y(s) = \frac{1}{s}\boldsymbol{\Gamma} \oplus \mathbf{G} \oplus s\mathbf{C} \tag{5-36}$$

where $\boldsymbol{\Gamma} = \mathbf{L}^{-1}$, $\mathbf{G} = \mathbf{R}^{-1}$ and $\mathbf{C} = \mathbf{S}^{-1}$

The matrix $Y(s)$ is called the *generalized admittance matrix* of the network.

Replacement of Initial Conditions by Equivalent Source Excitation Functions

In the systematic treatment of transform methods of analysis given below it is convenient to ignore initial conditions and deal only in terms of source outputs. We must therefore treat any initial conditions which may be present in terms of equivalent source excitation functions. To see how this is done consider the terminal behaviour of the storage elements involved. For the capacitive element of Figure 5–24(a) we have

$$i(t) = i_k(t) - j_k(t)$$
$$v(t) = v_k(t) - e_k(t)$$
$$i_k(t) = C_k \frac{dv_k(t)}{dt}$$

Taking Laplace transforms gives

$$i_k(s) = C_k[sv_k(s) - v_k(0-)]$$
$$i(s) = sC_kv_k(s) - C_kv_k(0-) - j_k(s)$$

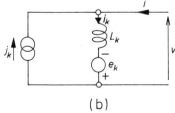

Figure 5–24

This shows that, as seen by the remainder of the network, an initial voltage of $v_k(0-)$ on the capacitor is equivalent to a current source output whose transform is

$$j_k(s) = C_k v_k(0-)$$

that is, a current source whose time variation is

$$j_k(t) = C_k v_k(0-)\delta(t) \tag{5-37}$$

Thus, for the purpose of *linear* network analysis, an initial condition associated with a capacitor may be replaced by an equivalent impulsive current source output whose impulse value is given by equation (5-37).

For an inductive component of Figure 5-24(b) we have

$$i(t) = i_k(t) - j_k(t)$$
$$v(t) = v_k(t) - e_k(t)$$
$$v_k(t) = L_k \frac{di_k}{dt}$$

Thus
$$v_k(s) = L_k[si_k(s) - i_k(0-)]$$
$$v(s) = sL_k i_k(s) - L_k i_k(0-) - e_k(s)$$

This shows that, as seen by the rest of the network, an initial current of $i_k(0-)$ in the inductor is equivalent to a voltage source output whose transform is

$$e_k(s) = L_k i_k(0-)$$

and whose time variation is therefore

$$e_k(t) = L_k i_k(0-)\delta(t) \tag{5-38}$$

Thus, for the purposes of linear network analysis, an initial condition associated with an inductor may be replaced by an equivalent impulsive voltage source output whose impulse value is given by equation (5-38).

5.7.1 Circuit Method [B10]

We first consider the case in which only voltage sources are present in the network; the general case in which both current and voltage sources are present is treated in Section 5.7.3 below. For this case equations (5-24), (5-25) and (5-32) give

$$\mathbf{Bv}(s) = \mathbf{Be}(s)$$
$$\mathbf{Ai}(s) = \mathbf{0} \tag{5-39}$$
$$\mathbf{v}(s) = \mathbf{Z}(s)\mathbf{i}(s)$$

Circuit, Vertex and Mixed Transform Analysis Methods

Now $\mathbf{Be}(s)$ represents a column vector having m rows. Each of its elements, corresponding to one of the m independent circuits chosen in the network, is the algebraic sum of the transforms of the voltage sources contained in that circuit. Thus if we put

$$\mathbf{Be}(s) = \mathbf{E}(s) \quad (5\text{--}40)$$

then $\mathbf{E}(s)$ is an m by 1 column vector representing the circuit voltage source transforms.

Now consider the equation

$$\mathbf{Ai}(s) = \mathbf{0} \quad (5\text{--}41)$$

We have that, as shown in Section 5.3 above

$$\mathbf{AB}^t = \mathbf{0} \quad (5\text{--}42)$$

where $\mathbf{A}$ is of order $(n-1)$ by e and rank $(n-1)$, and $\mathbf{B}^t$ is of order e by m and rank $m = e - n + 1$, where e is the number of elements in the network and n is the number of vertices. Equation (5–42) gives that

$$\mathbf{Ab}_i^t = \mathbf{0} \quad i = 1, 2, \ldots, m \quad (5\text{--}43)$$

where $\mathbf{b}_1^t, \mathbf{b}_2^t, \ldots, \mathbf{b}_m^t$ are the m linearly independent columns of $\mathbf{B}^t$. Since the $\mathbf{b}_i^t$ are independent, and since equation (5–43) holds, it follows that any solution of

$$\mathbf{Ai} = \mathbf{0}$$

must be expressible as a linear combination of the m vectors $\mathbf{b}_i^t$. That is we must have

$$\mathbf{i}(s) = \sum_{i=1}^{m} \mathbf{b}_i^t I_i(s) \quad (5\text{--}44)$$

where $I_i(s)$ denotes a suitable set of coefficient functions of s. Thus the solution must be of the form

$$\mathbf{i}(s) = \mathbf{B}^t \mathbf{I}(s) \quad (5\text{--}45)$$

for some m by 1 column vector $\mathbf{I}(s)$. In physical terms the components of this vector $\mathbf{I}(s)$ are the transforms of a set of independent circuit currents (also often called loop currents or mesh currents) and equation (5–45) expresses the transforms of the component currents in terms of the transforms of these circulating currents. Using equations (5–40) and (5–45) in the set of equations (5–39) we have

$$\mathbf{E}(s) = \mathbf{Be}(s)$$
$$= \mathbf{Bv}(s)$$
$$= \mathbf{BZ}(s)\mathbf{i}(s)$$
$$= \mathbf{BZ}(s)\mathbf{B}^t \mathbf{I}(s)$$

Thus $\mathbf{I}(s) = [\mathbf{BZ}(s)\mathbf{B}^t]^{-1}\mathbf{E}(s)$

$\mathbf{B}^t\mathbf{I}(s) = \mathbf{B}^t[\mathbf{BZ}(s)\mathbf{B}^t]^{-1}\mathbf{Be}(s)$

so that $\mathbf{i}(s) = \mathbf{B}^t[\mathbf{BZ}(s)\mathbf{B}^t]^{-1}\mathbf{Be}(s)$ (5-46)

giving $\mathbf{i}(t) = \mathscr{L}^{-1}\{\mathbf{B}^t[\mathbf{BZ}(s)\mathbf{B}^t]^{-1}\mathbf{Be}(s)\}$ (5-47)

Equation (5-47) gives a direct solution for the component currents in terms of an appropriate circuit matrix and impedance matrix, provided that $\mathbf{BZB}^t$ is non-singular. That this is always the case may be shown as follows. The matrix $\mathbf{Z}$ is always non-singular, since it is stipulated that only nonzero impedances are considered. Now $\mathbf{B}$ and $\mathbf{B}^t$ are of rank m, so that the m rows of $\mathbf{B}$ and the m columns of $\mathbf{B}^t$ are linearly independent. Because of this independence the determinant of $\mathbf{BZB}^t$ will not vanish if the determinant of $\mathbf{Z}$ does not vanish. Thus $\mathbf{BZB}^t$ is always non-singular.

This form of direct solution was first obtained by Kron [K9].

It is often more convenient however to work in terms of circuit voltages and circuit currents. In this case we proceed as above to obtain

$$\mathbf{E}(s) = \mathbf{BZ}(s)\mathbf{B}^t\mathbf{I}(s)$$

and put $\mathbf{E}(s) = \mathbf{Q}(s)\mathbf{I}(s)$ (5-48)

where the matrix $\mathbf{Q}(s)$ is defined by

$$\mathbf{Q}(s) \triangleq \mathbf{BZ}(s)\mathbf{B}^t \qquad (5\text{-}49)$$

Explicit expressions for the elements of the matrix $\mathbf{Q}(s)$ may be obtained from equation (5-49) when $\mathbf{Z}(s)$ is diagonal that is when there is no mutual inductive coupling. For this case we have

$$Q_{ij}(s) = \sum_{k=1}^{e} B_{ik} Z_{kk}(s) B_{jk} \qquad i,j = 1, 2, \ldots, m \qquad (5\text{-}50)$$

Thus, if $i \neq j$, $Q_{ij}(s)$ is made up of those component generalized impedances *common to both circuits* i and j, each element being included with a positive sign if the two circuits have the same orientation along the common component, and being included with a negative sign if they have opposite orientation along the common component. If $i = j$, then

$$Q_{ii}(s) = \sum_{k=1}^{e} (B_{ik})^2 Z_{kk}(s) \qquad i = 1, 2, \ldots, m \qquad (5\text{-}51)$$

showing that $Q_{ii}(s)$ is composed of the positive sums of all the component generalized impedances contained in the circuit i.

$\mathbf{Q}(s)$ is called the *circuit generalized impedance matrix* of the network, and equation (5-48) is called the set of transformed circuit equations. It has been shown above that $\mathbf{Q}(s)$ is non-singular, so that equation

(5-48) may be inverted to give

$$\mathbf{I}(s) = \mathbf{Q}^{-1}(s)\mathbf{E}(s) \quad (5\text{-}52)$$

so that
$$\mathbf{I}(t) = \mathcal{L}^{-1}[\mathbf{Q}^{-1}(s)\mathbf{E}(s)] \quad (5\text{-}53)$$

and thus
$$\mathbf{i}(t) = \mathbf{B}^t \mathcal{L}^{-1}[\mathbf{Q}^{-1}(s)\mathbf{B}\mathbf{e}(s)] \quad (5\text{-}54)$$

Example 5.1

For a simple example of these relationships, consider the network of Figure 5–25(a), with the corresponding oriented linear graph shown in Figure 5–25(b). In drawing the oriented linear graph, the voltage source e_1 has been associated with inductor L_1 and the voltage source e_2 with resistor R_2. Figure 5–25(b) shows the element numbering and orientation, and also the circuit numbering and orientation adopted. For this numbering of components, the network generalized impedance matrix is

$$\mathbf{Z}(s) = \begin{bmatrix} sL_1 & 0 & 0 & 0 & 0 & 0 \\ 0 & sL_2 & 0 & 0 & 0 & 0 \\ 0 & 0 & R_1 & 0 & 0 & 0 \\ 0 & 0 & 0 & R_2 & 0 & 0 \\ 0 & 0 & 0 & 0 & \dfrac{1}{sC_1} & 0 \\ 0 & 0 & 0 & 0 & 0 & \dfrac{1}{sC_2} \end{bmatrix}$$

To form the circuit matrix we draw up a table with entries

Circuit number	Component number →					
	1	2	3	4	5	6
1	+1	0	+1	0	+1	−1
2	0	+1	0	+1	0	+1

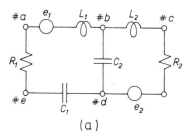
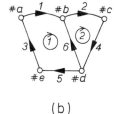

(a)   (b)

Figure 5–25

giving
$$\mathbf{B} = \begin{bmatrix} 1 & 0 & 1 & 0 & 1 & -1 \\ 0 & 1 & 0 & 1 & 0 & 1 \end{bmatrix}$$

The component voltage source vector is

$$\mathbf{e}(s) = \begin{bmatrix} e_1(s) \\ 0 \\ 0 \\ e_2(s) \\ 0 \\ 0 \end{bmatrix}$$

so that the circuit voltage source vector is

$$\mathbf{E}(s) = \mathbf{Be}(s) = \begin{bmatrix} 1 & 0 & 1 & 0 & 1 & -1 \\ 0 & 1 & 0 & 1 & 0 & 1 \end{bmatrix} \begin{bmatrix} e_1(s) \\ 0 \\ 0 \\ e_2(s) \\ 0 \\ 0 \end{bmatrix} = \begin{bmatrix} e_1(s) \\ e_2(s) \end{bmatrix}$$

The circuit generalized impedance matrix is

$$\mathbf{Q}(s) = \mathbf{BZ}(s)\mathbf{B}^t = \begin{bmatrix} 1 & 0 & 1 & 0 & 1 & -1 \\ 0 & 1 & 0 & 1 & 0 & 1 \end{bmatrix}$$

$$\cdot \begin{bmatrix} sL_1 & 0 & 0 & 0 & 0 & 0 \\ 0 & sL_2 & 0 & 0 & 0 & 0 \\ 0 & 0 & R_1 & 0 & 0 & 0 \\ 0 & 0 & 0 & R_2 & 0 & 0 \\ 0 & 0 & 0 & 0 & \dfrac{1}{sC_1} & 0 \\ 0 & 0 & 0 & 0 & 0 & \dfrac{1}{sC_2} \end{bmatrix} \begin{bmatrix} 1 & 0 \\ 0 & 1 \\ 1 & 0 \\ 0 & 1 \\ 1 & 0 \\ -1 & 1 \end{bmatrix}$$

$$= \begin{bmatrix} \left(sL_1 + R_1 + \dfrac{1}{sC_1} + \dfrac{1}{sC_2} \right) & -\dfrac{1}{sC_2} \\ -\dfrac{1}{sC_2} & \left(R_2 + \dfrac{1}{sC_2} + sL_2 \right) \end{bmatrix}$$

Circuit, Vertex and Mixed Transform Analysis Methods   313

The circuit current transforms are thus given by

$$\begin{bmatrix} I_1(s) \\ I_2(s) \end{bmatrix} = \begin{bmatrix} \left(sL_1 + R_1 + \dfrac{1}{sC_1} + \dfrac{1}{sC_2}\right) & -\dfrac{1}{sC_2} \\ -\dfrac{1}{sC_2} & \left(R_2 + \dfrac{1}{sC_2} + sL_2\right) \end{bmatrix}^{-1} \begin{bmatrix} e_1(s) \\ e_2(s) \end{bmatrix}$$

5.7.2 Vertex Method [B10]

We now consider the case in which only current sources are present; the general case is considered in Section 5.7.3 below. For this case equations (5–24), (5–25) and (5–34) give:

$$\mathbf{B}\mathbf{v}(s) = \mathbf{0}$$

$$\mathbf{A}\mathbf{i}(s) = \mathbf{A}\mathbf{j}(s) \qquad (5\text{–}55)$$

$$\mathbf{i}(s) = \mathbf{Y}(s)\mathbf{v}(s)$$

Consider first the term $\mathbf{A}\mathbf{j}(s)$. This represents a column vector of $(n-1)$ rows, each element corresponding to one of the labelled vertices $1, 2, \ldots, (n-1)$ of the network, and whose elements are the algebraic sum of the transforms of the current source transforms entering that vertex. If we put

$$\mathbf{A}\mathbf{j}(s) = \mathbf{J}(s) \qquad (5\text{–}56)$$

we may therefore take $\mathbf{J}(s)$ as an $(n-1)$ by 1 column vector representing what we may call the vertex current source transforms. Now consider the equation

$$\mathbf{B}\mathbf{v}(s) = \mathbf{0} \qquad (5\text{–}57)$$

We have that $\qquad \mathbf{B}\mathbf{A}^t = \mathbf{0} \qquad (5\text{–}58)$

where $\mathbf{B}$ is of order m by e and rank m, while $\mathbf{A}^t$ is of order b by $(n-1)$ and rank $(n-1) = e - m$. Equation (5–58) gives that

$$\mathbf{B}\mathbf{a}^t_j = \mathbf{0} \qquad j = 1, 2, \ldots, (n-1) \qquad (5\text{–}59)$$

where $\mathbf{a}^t_1, \mathbf{a}^t_2, \ldots, \mathbf{a}^t_{(n-1)}$ are the $(n-1)$ columns of $\mathbf{A}^t$. The set of columns of $\mathbf{A}^t$ therefore form a basis for the solutions of equation (5–57) since they are linearly independent and since equation (5–59) holds. It follows that any solution of

$$\mathbf{B}\mathbf{v} = \mathbf{0}$$

must be expressible as a linear combination of the $(n-1)$ vectors $\mathbf{a}^t_j$. That is we must have

$$\mathbf{v}(s) = \sum_{j=1}^{n-1} \mathbf{a}^t_j V_j(s) \qquad (5\text{–}60)$$

where $V_j(s)$ denotes a suitable set of coefficient functions of s. Thus the solution must be of the form

$$\mathbf{v}(s) = \mathbf{A}^t \mathbf{V}(s) \qquad (5\text{-}61)$$

for some $(n-1)$ by 1 column vector $\mathbf{V}(s)$. In physical terms, the components of this vector $\mathbf{V}(s)$ are the transforms of the voltages of vertices $1, 2, 3, \ldots, (n-1)$ relative to that of node n, and equation (5-61) expresses the transforms of the component voltages in terms of these *vertex voltages*. Using equations (5-56) and (5-61) in the set of equations (5-55) we have

$$\mathbf{J}(s) = \mathbf{A}\mathbf{j}(s)$$
$$= \mathbf{A}\mathbf{i}(s)$$
$$= \mathbf{A}\mathbf{Y}(s)\mathbf{v}(s)$$
$$= \mathbf{A}\mathbf{Y}(s)\mathbf{A}^t\mathbf{V}(s)$$

Thus $\qquad \mathbf{V}(s) = [\mathbf{A}\mathbf{Y}(s)\mathbf{A}^t]^{-1}\mathbf{J}(s)$

$$\mathbf{A}^t\mathbf{V}(s) = \mathbf{A}^t[\mathbf{A}\mathbf{Y}(s)\mathbf{A}^t]^{-1}\mathbf{A}\mathbf{j}(s)$$

so that $\qquad \mathbf{v}(s) = \mathbf{A}^t[\mathbf{A}\mathbf{Y}(s)\mathbf{A}^t]^{-1}\mathbf{A}\mathbf{j}(s)$

giving $\qquad \mathbf{v}(t) = \mathscr{L}^{-1}\{\mathbf{A}^t[\mathbf{A}\mathbf{Y}(s)\mathbf{A}^t]^{-1}\mathbf{A}\mathbf{j}(s)\} \qquad (5\text{-}62)$

Equation (5-62) gives a direct solution for the component voltages in terms of an appropriate reduced incidence matrix and an admittance matrix, provided that $\mathbf{A}\mathbf{Y}\mathbf{A}^t$ is non-singular. This will always be so for the following reasons. The matrix $\mathbf{Y}$ is always non-singular since only finite impedances are considered. Now $\mathbf{A}$ and $\mathbf{A}^t$ are of rank $(n-1)$, so that the $(n-1)$ rows of $\mathbf{A}$ and the $(n-1)$ columns of $\mathbf{A}^t$ are linearly independent. Because of this independence, the determinant of $\mathbf{A}\mathbf{Y}\mathbf{A}^t$ will not vanish if the determinant of $\mathbf{Y}$ does not vanish. Thus $\mathbf{A}\mathbf{Y}\mathbf{A}^t$ is non-singular.

This form of solution was first given by Salzer [S12] and is a particular case of an earlier solution due to Kron.

It is often more convenient however to work in terms of vertex voltages and vertex currents. In this case we proceed as above to obtain

$$\mathbf{J}(s) = \mathbf{A}\mathbf{Y}(s)\mathbf{A}^t\mathbf{V}(s)$$

and put $\qquad \mathbf{J}(s) = \mathbf{P}(s)\mathbf{V}(s) \qquad (5\text{-}63)$

where $\qquad \mathbf{P}(s) = \mathbf{A}\mathbf{Y}(s)\mathbf{A}^t \qquad (5\text{-}64)$

When $\mathbf{Y}(s)$ is diagonal, that is when no mutual inductive coupling is present, we can obtain explicit expressions for the elements of $\mathbf{P}(s)$

Circuit, Vertex and Mixed Transform Analysis Methods 315

from equation (5–64). For this case we have

$$P_{ij}(s) = \sum_{k=1}^{e} A_{ik} Y_{kk}(s) A_{jk} \qquad i, j, \ldots, (n-1)$$

Thus, if $i \neq j$, $P_{ij}(s)$ is the *negative* sum of the component generalized admittances (that is the reciprocals of the component generalized impedances) which are incident on both of the vertices i and j.
If $i = j$, then

$$P_{ii}(s) = \sum_{k=1}^{e} (A_{ik})^2 Y_{kk} \qquad i = 1, 2, \ldots, (n-1)$$

showing that $P_{ii}(s)$ is the positive sum of all the branch admittances incident on vertex i. $\mathbf{P}(s)$ is called the *vertex generalized admittance matrix* of the network with respect to the designated reference node n, and equation (5–63) is known as the set of vertex equations of the network. It has been shown above that $\mathbf{P}(s)$ is non-singular, so that equation (5–63) may be inverted to give

$$\mathbf{V}(s) = \mathbf{P}^{-1}(s)\mathbf{J}(s) \qquad (5\text{–}65)$$

so that
$$\mathbf{V}(t) = \mathscr{L}^{-1}\{\mathbf{P}^{-1}(s)\mathbf{J}(s)\} \qquad (5\text{–}66)$$

and thus
$$\mathbf{v}(t) = \mathbf{A}^t \mathscr{L}^{-1}\{\mathbf{P}^{-1}(s)\mathbf{A}\mathbf{j}(s)\} \qquad (5\text{–}67)$$

Example 5.2

For a simple example of these relationships, consider the network of Figure 5–26(a), with the corresponding oriented linear graph shown in Figure 5–26(b). In drawing the oriented linear graph, the current source j_1 has been associated with the resistor R_1, and the current source j_2 with the resistor R_3. Figure 5–26(b) shows the element numbering and orientation and the vertex numbering adopted. For this numbering of components, the network generalized admittance matrix is

$$\mathbf{Y}(s) = \begin{bmatrix} \dfrac{1}{sL_1} & 0 & 0 & 0 & 0 & 0 & 0 \\ 0 & \dfrac{1}{sL_2} & 0 & 0 & 0 & 0 & 0 \\ 0 & 0 & \dfrac{1}{R_1} & 0 & 0 & 0 & 0 \\ 0 & 0 & 0 & \dfrac{1}{R_2} & 0 & 0 & 0 \\ 0 & 0 & 0 & 0 & \dfrac{1}{R_3} & 0 & 0 \\ 0 & 0 & 0 & 0 & 0 & sC_1 & 0 \\ 0 & 0 & 0 & 0 & 0 & 0 & sC_2 \end{bmatrix}$$

To form the reduced incidence matrix we draw up a table with entries

Component number →

Vertex number	1	2	3	4	5	6	7
1	+1	0	−1	0	0	0	0
2	−1	−1	0	−1	0	−1	−1
3	0	0	0	0	−1	0	+1

giving $\quad \mathbf{A} = \begin{bmatrix} 1 & 0 & -1 & 0 & 0 & 0 & 0 \\ -1 & -1 & 0 & -1 & 0 & -1 & -1 \\ 0 & 0 & 0 & 0 & -1 & 0 & 1 \end{bmatrix}$

The component current source vector is

$$\mathbf{j}(s) = \begin{bmatrix} 0 \\ 0 \\ j_1(s) \\ 0 \\ j_2(s) \\ 0 \\ 0 \end{bmatrix}$$

so that the vertex current source vector is

$\mathbf{J}(s) = \mathbf{A}\mathbf{j}(s)$

$= \begin{bmatrix} 1 & 0 & -1 & 0 & 0 & 0 & 0 \\ -1 & -1 & 0 & -1 & 0 & -1 & -1 \\ 0 & 0 & 0 & 0 & -1 & 0 & 1 \end{bmatrix} \begin{bmatrix} 0 \\ 0 \\ j_1(s) \\ 0 \\ j_2(s) \\ 0 \\ 0 \end{bmatrix} = \begin{bmatrix} -j_1(s) \\ 0 \\ -j_2(s) \end{bmatrix}$

Circuit, Vertex and Mixed Transform Analysis Methods 317

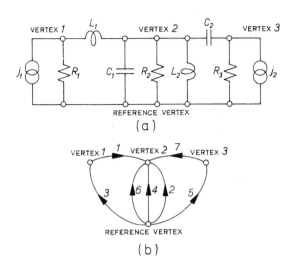

Figure 5-26

The vertex generalized admittance matrix is

$$\mathbf{P}(s) = \mathbf{A}\mathbf{Y}(s)\mathbf{A}^t = \begin{bmatrix} 1 & 0 & -1 & 0 & 0 & 0 & 0 \\ -1 & -1 & 0 & -1 & 0 & -1 & -1 \\ 0 & 0 & 0 & 0 & -1 & 0 & 1 \end{bmatrix}$$

$$\begin{bmatrix} \dfrac{1}{sL_1} & 0 & 0 & 0 & 0 & 0 & 0 \\ 0 & \dfrac{1}{sL_2} & 0 & 0 & 0 & 0 & 0 \\ 0 & 0 & \dfrac{1}{R_1} & 0 & 0 & 0 & 0 \\ 0 & 0 & 0 & \dfrac{1}{R_2} & 0 & 0 & 0 \\ 0 & 0 & 0 & 0 & \dfrac{1}{R_3} & 0 & 0 \\ 0 & 0 & 0 & 0 & 0 & sC_1 & 0 \\ 0 & 0 & 0 & 0 & 0 & 0 & sC_2 \end{bmatrix} \begin{bmatrix} 1 & -1 & 0 \\ 0 & -1 & 0 \\ -1 & 0 & 0 \\ 0 & -1 & 0 \\ 0 & 0 & -1 \\ 0 & -1 & 0 \\ 0 & -1 & 1 \end{bmatrix}$$

$$= \begin{bmatrix} \left(\dfrac{1}{sL_1}+\dfrac{1}{R_1}\right) & -\dfrac{1}{sL_1} & 0 \\ -\dfrac{1}{sL_1} & \left(\dfrac{1}{sL_1}+\dfrac{1}{sL_2}+\dfrac{1}{R_2}+sC_1+sC_2\right) & -sC_2 \\ 0 & -sC_2 & \left(\dfrac{1}{R_3}+sC_2\right) \end{bmatrix}$$

The vertex voltage transforms are therefore given by

$$\begin{bmatrix} V_1(s) \\ V_2(s) \\ V_3(s) \end{bmatrix}$$

$$= \begin{bmatrix} \left(\dfrac{1}{sL_1}+\dfrac{1}{R_1}\right) & -\dfrac{1}{sL_1} & 0 \\ -\dfrac{1}{sL_1} & \left(\dfrac{1}{sL_1}+\dfrac{1}{sL_2}+\dfrac{1}{R_2}+sC_1+sC_2\right) & -sC_2 \\ 0 & -sC_2 & \left(\dfrac{1}{R_3}+sC_2\right) \end{bmatrix}^{-1}$$

$$\cdot \begin{bmatrix} -j_1(s) \\ 0 \\ -j_2(s) \end{bmatrix}$$

5.7.3 Mixed Method [B10]

If both current and voltage sources are present, then a transform network analysis by either the circuit or vertex method requires some modification from the treatment given above. Since we are considering linear networks we may apply the Principle of Superposition, considering the effects of voltage sources and current sources separately and then add the results.

We have

currents through passive elements
= (currents due to voltage sources) + (currents due to current sources)
= (currents due to voltages sources) + (network admittance matrix) ×
  (voltages due to current sources)

Thus, using the results of Sections 5.7.1 and 5.7.2 we have

$$\mathbf{i}(s) = \mathbf{B}^t \mathbf{Q}^{-1}(s)\mathbf{Be}(s) + \mathbf{Y}(s)\mathbf{A}^t \mathbf{P}^{-1}(s)\mathbf{A}\mathbf{j}(s) \qquad (5\text{–}68)$$

Alternatively we have

Voltages across passive elements
 = (voltages due to voltage sources) + (voltages due to current sources)
 = (network impedance matrix) × (currents due to voltage sources) + (voltages due to current sources)

Thus, using the results of Sections 5.7.1 and 5.7.2 we have

$$\mathbf{v}(s) = \mathbf{Z}(s)\mathbf{B}^t\mathbf{Q}^{-1}(s)\mathbf{B}\mathbf{e}(s) + \mathbf{A}^t\mathbf{P}^{-1}(s)\mathbf{A}\mathbf{j}(s) \qquad (5\text{-}69)$$

Now it may be verified by matrix manipulation that

$$\mathbf{Y}(s)\mathbf{A}^t\mathbf{P}^{-1}(s)\mathbf{A} + \mathbf{B}^t\mathbf{Q}^{-1}(s)\mathbf{B}\mathbf{Z}(s) = \mathbf{I}_e \qquad (5\text{-}70)$$

$$\mathbf{Z}(s)\mathbf{B}^t\mathbf{Q}^{-1}(s)\mathbf{B} + \mathbf{A}^t\mathbf{P}^{-1}(s)\mathbf{A}\mathbf{Y}(s) = \mathbf{I}_e \qquad (5\text{-}71)$$

where $\mathbf{I}_e$ is a unit matrix of order e.

Using equation (5-70) we therefore have

$$\mathbf{j}(s) = [\mathbf{Y}(s)\mathbf{A}^t\mathbf{P}^{-1}(s)\mathbf{A} + \mathbf{B}^t\mathbf{Q}^{-1}(s)\mathbf{B}\mathbf{Z}(s)]\mathbf{j}(s) \qquad (5\text{-}72)$$

From equations (5-72) and (5-68) we get

$$\mathbf{i}(s) - \mathbf{j}(s) = \mathbf{B}^t\mathbf{Q}^{-1}(s)\mathbf{B}\mathbf{e}(s) + \mathbf{Y}(s)\mathbf{A}^t\mathbf{P}^{-1}(s)\mathbf{A}\mathbf{j}(s)$$
$$- \mathbf{Y}(s)\mathbf{A}^t\mathbf{P}^{-1}(s)\mathbf{A}\mathbf{j}(s) - \mathbf{B}^t\mathbf{Q}^{-1}(s)\mathbf{B}\mathbf{Z}(s)\mathbf{j}(s)$$

so that
$$\mathbf{i}(s) - \mathbf{j}(s) = \mathbf{B}^t\mathbf{Q}^{-1}(s)\mathbf{B}[\mathbf{e}(s) - \mathbf{Z}(s)\mathbf{j}(s)] \qquad (5\text{-}73)$$

which is the circuit method of solution modified to deal with the presence of current sources.

Again, using equation (5-71) we have

$$\mathbf{e}(s) = [\mathbf{Z}(s)\mathbf{B}^t\mathbf{Q}^{-1}(s)\mathbf{B} + \mathbf{A}^t\mathbf{P}^{-1}(s)\mathbf{A}\mathbf{Y}(s)]\mathbf{e}(s) \qquad (5\text{-}74)$$

From equations (5-74) and (5-69) we get

$$\mathbf{v}(s) - \mathbf{e}(s) = \mathbf{Z}(s)\mathbf{B}^t\mathbf{Q}^{-1}(s)\mathbf{B}\mathbf{e}(s) + \mathbf{A}^t\mathbf{P}^{-1}(s)\mathbf{A}\mathbf{j}(s)$$
$$- \mathbf{Z}(s)\mathbf{B}^t\mathbf{Q}^{-1}(s)\mathbf{B}\mathbf{e}(s) - \mathbf{A}^t\mathbf{P}^{-1}(s)\mathbf{A}\mathbf{Y}(s)\mathbf{e}(s) \qquad (5\text{-}75)$$

so that
$$\mathbf{v}(s) - \mathbf{e}(s) = \mathbf{A}^t\mathbf{P}^{-1}(s)\mathbf{A}[\mathbf{j}(s) - \mathbf{Y}(s)\mathbf{e}(s)] \qquad (5\text{-}76)$$

which is the vertex method of solution modified to deal with the presence of voltage sources.

5.8 Systems Matrix Analysis of Networks [R4]

Consider the set of Laplace-transformed equations:

$$\mathbf{T}(s)\mathbf{z}(s) = \mathbf{U}(s)\mathbf{u}(s)$$
$$\mathbf{y}(s) = \mathbf{V}(s)\mathbf{z}(s) + \mathbf{W}(s)\mathbf{u}(s)$$

where $\mathbf{u}(s)$ is a vector whose components are the Laplace transforms of a set of input functions of time; $\mathbf{z}(s)$ is a vector whose components are the Laplace transforms of a set of dynamical system variables; and $\mathbf{y}(s)$ is a vector whose components are the Laplace transforms of a set of observed or output variables. The transfer function matrix relating the output and input system vectors is $\mathbf{G}(s)$ say, so that

$$\mathbf{y}(s) = \mathbf{G}(s)\mathbf{u}(s)$$

where
$$\mathbf{G}(s) = \mathbf{V}(s)\mathbf{T}^{-1}(s)\mathbf{U}(s) + \mathbf{W}(s) \qquad (5\text{-}77)$$

and $\mathbf{T}(s)$, $\mathbf{U}(s)$, $\mathbf{V}(s)$ and $\mathbf{W}(s)$ are matrices of rational forms. Now consider the determinant

$$\Delta(s) = \begin{vmatrix} \mathbf{T}(s) & \mathbf{U}(s) \\ -\mathbf{V}(s) & \mathbf{W}(s) \end{vmatrix}$$

where $|\mathbf{T}(s)| \not\equiv 0$. We have, on pre-multiplying the first block row by $\mathbf{V}(s)\mathbf{T}^{-1}(s)$ and adding it to the second block row

$$\Delta(s) = \begin{vmatrix} \mathbf{T}(s) & \mathbf{U}(s) \\ 0 & \mathbf{W}(s) + \mathbf{V}(s)\mathbf{T}^{-1}(s)\mathbf{U}(s) \end{vmatrix}$$

so that
$$\Delta(s) = |\mathbf{T}(s)||\mathbf{G}(s)|$$

where $\mathbf{G}(s)$ is the transfer function matrix for the system. We define the *system matrix* corresponding to the system to be [R3], [M5]

$$\mathbf{P}(s) = \begin{bmatrix} \mathbf{T}(s) & \mathbf{U}(s) \\ -\mathbf{V}(s) & \mathbf{W}(s) \end{bmatrix} \qquad (5\text{-}78)$$

so that we have
$$|\mathbf{G}(s)| = \frac{|\mathbf{P}(s)|}{|\mathbf{T}(s)|} \qquad (5\text{-}79)$$

If, in the matrix $\mathbf{P}(s)$, we replace the sub-matrices $\mathbf{U}$, $\mathbf{V}$ and $\mathbf{W}$ by any other matrices $\overline{\mathbf{U}}$, $\overline{\mathbf{V}}$ and $\overline{\mathbf{W}}$ subject only to the conformability required by the partitioning of $\mathbf{P}(s)$ so that

$$\mathbf{U} \to \overline{\mathbf{U}} \qquad \mathbf{V} \to \overline{\mathbf{V}} \qquad \mathbf{W} \to \overline{\mathbf{W}}$$

results in
$$\mathbf{P} \to \overline{\mathbf{P}} \quad \text{and} \quad \mathbf{G} \to \overline{\mathbf{G}}$$

then we will have
$$|\overline{\mathbf{G}}(s)| = \frac{|\overline{\mathbf{P}}(s)|}{|\mathbf{T}(s)|} \qquad (5\text{-}80)$$

Let the minor formed from rows $i_1, i_2, \ldots, i_k$ and columns $j_1, j_2, \ldots, j_k$ of $\mathbf{G}(s)$ be denoted by

$$\mathbf{G}^{i_1 i_2 \ldots i_k}_{j_1 j_2 \ldots j_k}$$

and let the minor of $\mathbf{P}(s)$ obtained by deleting all rows of $\mathbf{V}(s)$ and $\mathbf{W}(s)$ *except* rows $i_1, i_2, \ldots, i_k$ and all columns of $\mathbf{U}(s)$ and $\mathbf{W}(s)$ *except* columns $j_1, j_2, \ldots, j_k$ be denoted by

$$\mathbf{P}^{i_1 i_2 \ldots i_k}_{j_1 j_2 \ldots j_k})$$

Then it is a direct consequence of equations (5–77), (5–79) and (5–80) above that

$$\mathbf{G}^{i_1 i_2 \ldots i_k}_{j_1 j_2 \ldots j_k} = \frac{\mathbf{P}^{i_1 i_2 \ldots i_k}_{j_1 j_2 \ldots j_k})}{\mathbf{P})} \tag{5–81}$$

where $\mathbf{P}) \triangleq |\mathbf{T}(s)|$

In particular, the elements of the transfer function matrix $\mathbf{G}(s)$ are given by

$$g_{ij}(s) = \frac{\mathbf{P}^i_j)}{\mathbf{P})} \tag{5–82}$$

The Laplace transforms of a set of network variables satisfy the set of interconnective constraint equations

$$\begin{aligned} \mathbf{Ai}(s) &= \mathbf{Aj}(s) \\ \mathbf{Bv}(s) &= \mathbf{Be}(s) \end{aligned} \tag{5–83}$$

where

(a) $\mathbf{A}$ is the network reduced incidence matrix.
(b) $\mathbf{B}$ is the network circuit matrix.
(c) $\mathbf{i}(s)$ is a vector whose entries are the transforms of the passive element currents.
(d) $\mathbf{v}$ is a vector whose entries are the transforms of the passive element voltages.
(e) $\mathbf{j}$ is a vector whose entries are the transforms of the outputs of the current sources.
(f) $\mathbf{e}$ is a vector whose entries are the transforms of the outputs of the voltage sources.

Now consider the network component relationships. As discussed above, the Laplace transforms of the component variables will be related such that:
  (i) for a capacitive component

$$i_k(s) = sC_k v_k(s)$$

(ii) for an inductive component
$$v_k(s) = sL_k i_k(s)$$
(iii) for a resistive component
$$v_k(s) = R_k i_k(s)$$
Thus the set of component equations may be combined into one matrix equation of the form
$$\overline{\mathbf{Z}}(s)\mathbf{i}(s) = \overline{\mathbf{Y}}(s)\mathbf{v}(s) \tag{5-84}$$
where:

(a) the matrix $\overline{\mathbf{Z}}(s)$ is (if no mutual inductive coupling is present) a diagonal matrix with terms of the form sL corresponding to inductive components, terms of the form R corresponding to resistive components, and unit entries corresponding to capacitive elements.
(b) the matrix $\overline{\mathbf{Y}}(s)$ is a diagonal matrix with terms of the form sC corresponding to capacitive components and unit entries corresponding to inductive and resistive components.

Treating all the network passive element voltages and currents as output variables and all the source outputs as input variables, we can now write the complete set of network equations in a form suitable for representation by a system matrix. In doing this we must take account of the fact that, by the Principle of Superposition, we must add the individual contributions of the current and voltage sources together to form the net outputs. We thus have the set of equations

$$\begin{bmatrix} \mathbf{A} & \mathbf{0} \\ \mathbf{0} & \mathbf{B} \\ \overline{\mathbf{Z}} & -\overline{\mathbf{Y}} \end{bmatrix} \begin{bmatrix} \mathbf{i} \\ \mathbf{v} \end{bmatrix} = \begin{bmatrix} \mathbf{A} & \mathbf{0} \\ \mathbf{0} & \mathbf{B} \\ \mathbf{0} & \mathbf{0} \end{bmatrix} \begin{bmatrix} \mathbf{j} \\ \mathbf{e} \end{bmatrix}$$

$$\mathbf{y} = \begin{bmatrix} \mathbf{I} & \mathbf{0} \\ \mathbf{0} & \mathbf{I} \end{bmatrix} \begin{bmatrix} \mathbf{i} \\ \mathbf{v} \end{bmatrix} + \begin{bmatrix} \mathbf{0} & \mathbf{0} \\ \mathbf{0} & \mathbf{0} \end{bmatrix} \begin{bmatrix} \mathbf{j} \\ \mathbf{e} \end{bmatrix}$$

where $\mathbf{y}$ is the vector of observed or output quantities and $\mathbf{I}$ denotes a unit matrix of appropriate order.

This gives the system matrix for the general electrical network as

$$\mathbf{N}(s) = \begin{bmatrix} \mathbf{A} & \mathbf{0} & \mathbf{A} & \mathbf{0} \\ \mathbf{0} & \mathbf{B} & \mathbf{0} & \mathbf{B} \\ \overline{\mathbf{Z}} & -\overline{\mathbf{Y}} & \mathbf{0} & \mathbf{0} \\ -\mathbf{I} & \mathbf{0} & \mathbf{0} & \mathbf{0} \\ \mathbf{0} & -\mathbf{I} & \mathbf{0} & \mathbf{0} \end{bmatrix} \tag{5-85}$$

Using this system matrix with the transfer function formulae of equation (5–82) we can, by a suitable selection of input and output variables, calculate any required aspect of network behaviour.

5.9 Lagrangian Equations for Networks

Consider the class of non-linear networks which may be regarded as composed of networks of general elements of the type shown in Figure 5–23, for which the constitutive characteristics of the passive elements may be non-linearities of the types discussed in Chapter 1. Suppose there are only voltage sources present. Then, as in the circuit method of Section 5.7.1, but using functions of time and not transforms, we have

$$\mathbf{Bv} = \mathbf{Be} \qquad (5\text{–}86)$$

where

(a) $\mathbf{B}$ is the network circuit matrix.
(b) $\mathbf{v}$ is a vector whose components are the passive element voltages.
(c) $\mathbf{e}$ is a vector whose components are the active element voltages.

Group the elements together into three sets associated with inductors, resistors and capacitors, and partition equation (5–86) as

$$[\mathbf{B}_L \mathrel{\vdots} \mathbf{B}_R \mathrel{\vdots} \mathbf{B}_C] \begin{bmatrix} \mathbf{v}_L \\ \mathbf{v}_R \\ \mathbf{v}_C \end{bmatrix} = \mathbf{Be} \triangleq \mathbf{E} \qquad (5\text{–}87)$$

where

(a) $\mathbf{v}_L$, $\mathbf{v}_R$, $\mathbf{v}_C$ are vectors whose components are the inductors', resistors' and capacitors' voltages respectively.
(b) $\mathbf{E}$ is a vector of circuit voltage sources.
(c) $\mathbf{B}_L, \mathbf{B}_R, \mathbf{B}_C$ are matrices obtained by a suitable partitioning of matrix $\mathbf{B}$.

Let $\mathbf{I}$ be a vector whose components are a set of independent circuit currents, and let $\mathbf{i}$ be a vector whose components are the passive element currents. Then, as shown in Section 5.7.1, we have

$$\mathbf{i} = \mathbf{B}^t \mathbf{I} \qquad (5\text{–}88)$$

For the grouping of components and corresponding partitioning of $\mathbf{B}$ adopted, we therefore have

$$\mathbf{i} = \begin{bmatrix} \mathbf{i}_L \\ \mathbf{i}_R \\ \mathbf{i}_C \end{bmatrix} = [\mathbf{B}_L \mathrel{\vdots} \mathbf{B}_R \mathrel{\vdots} \mathbf{B}_C]^t \mathbf{I} = \begin{bmatrix} \mathbf{B}_L^t \\ \mathbf{B}_R^t \\ \mathbf{B}_C^t \end{bmatrix} \mathbf{I} = \begin{bmatrix} \mathbf{B}_L^t \mathbf{I} \\ \mathbf{B}_R^t \mathbf{I} \\ \mathbf{B}_C^t \mathbf{I} \end{bmatrix}$$

324   Network Models

so that
$$\mathbf{i}_L = \mathbf{B}_L^t \mathbf{I}$$
$$\mathbf{i}_R = \mathbf{B}_R^t \mathbf{I} \qquad (5\text{-}89)$$
$$\mathbf{i}_C = \mathbf{B}_C^t \mathbf{I}$$

In terms of the component variables we will have

$$\mathbf{v}_L = \frac{d}{dt}\left(\frac{\partial T_e^*}{\partial \mathbf{i}_L}\right)^t$$
$$\mathbf{v}_R = \left(\frac{\partial G}{\partial \mathbf{i}_R}\right)^t \qquad \mathbf{v}_C = \left(\frac{\partial U_e}{\partial \mathbf{q}_c}\right)^t \qquad (5\text{-}90)$$

where

(a) T_e^* is the total inductive co-energy.
(b) G is the total resistor content.
(c) U_e is the total capacitive energy.
(d) $\mathbf{q}_c$ is a vector whose components are the capacitor charges.

Equation (5-87) thus gives

$$\mathbf{B}_L \frac{d}{dt}\left(\frac{\partial T_e^*}{\partial \mathbf{i}_L}\right)^t + \mathbf{B}_R \left(\frac{\partial G}{\partial \mathbf{i}_R}\right)^t + \mathbf{B}_c \left(\frac{\partial U_e}{\partial \mathbf{q}_c}\right)^t = \mathbf{E} \qquad (5\text{-}91)$$

Now define a *circuit charge* variable $\mathbf{Q}$ by

$$\mathbf{Q} = \int_0^t \mathbf{I}\, dt \qquad (5\text{-}92)$$

Then
$$\mathbf{q}_c = \mathbf{B}^t \mathbf{Q} \qquad (5\text{-}93)$$

Substituting from equations (5-89) and (5-93) into equation (5-91), we therefore have

$$\mathbf{B}_L \frac{d}{dt}\left(\frac{\partial T_e^*}{\partial \mathbf{B}_L^t \mathbf{I}}\right)^t + \mathbf{B}_R \left(\frac{\partial G}{\partial \mathbf{B}_R^t \mathbf{I}}\right)^t + \mathbf{B}_C \left(\frac{\partial U_e}{\partial \mathbf{B}_C^t \mathbf{Q}}\right)^t = \mathbf{E}$$

so that using (2-36)
$$\frac{d}{dt}\left(\frac{\partial T_e^*}{\partial \mathbf{I}}\right)^t + \left(\frac{\partial G}{\partial \mathbf{I}}\right)^t + \left(\frac{\partial U_e}{\partial \mathbf{Q}}\right)^t = \mathbf{E}$$

and thus we obtain the corresponding Lagrangian equation

$$\frac{d}{dt}\left(\frac{\partial T_e^*}{\partial \dot{\mathbf{Q}}}\right)^t + \left(\frac{\partial G}{\partial \dot{\mathbf{Q}}}\right)^t + \left(\frac{\partial U_e}{\partial \mathbf{Q}}\right)^t = \mathbf{E} \qquad (5\text{-}94)$$

Example 5.3 Network Lagrangian equations
For the network of Figure 5–27, take as generalized charge coordinates, the pair q_1 and q_2 such that their time derivatives $\dot{q}_1$ and $\dot{q}_2$

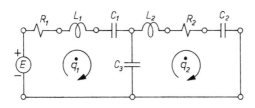

Figure 5-27

are the pair of circuit currents shown. We then have

$$\mathbf{Q} = \begin{bmatrix} q_1 \\ q_2 \end{bmatrix} \quad \dot{\mathbf{Q}} = \begin{bmatrix} \dot{q}_1 \\ \dot{q}_2 \end{bmatrix} \quad \mathbf{E} = \begin{bmatrix} E \\ 0 \end{bmatrix}$$

$$T_e^* = \tfrac{1}{2}L_1(\dot{q}_1)^2 + \tfrac{1}{2}L_2(\dot{q}_2)^2$$

$$U_e = \tfrac{1}{2}S_1 q_1^2 + \tfrac{1}{2}S_2 q_2^2 + \tfrac{1}{2}S_3(q_1 - q_2)^2$$

where
$$S_i = \frac{1}{C_i} \quad i = 1, 2, 3$$

$$G = \tfrac{1}{2}R_1 \dot{q}_1^2 + \tfrac{1}{2}R_2 \dot{q}_2^2$$

Thus

$$\left(\frac{\partial T_e^*}{\partial \dot{\mathbf{Q}}}\right)^t = \begin{bmatrix} L_1 \dot{q}_1 \\ L_2 \dot{q}_2 \end{bmatrix} \quad \frac{d}{dt}\left(\frac{\partial T_e^*}{\partial \dot{\mathbf{Q}}}\right)^t = \begin{bmatrix} L_1 \ddot{q}_1 \\ L_2 \ddot{q}_2 \end{bmatrix}$$

$$\left(\frac{\partial G}{\partial \dot{\mathbf{Q}}}\right)^t = \begin{bmatrix} R_1 \dot{q}_1 \\ R_2 \dot{q}_2 \end{bmatrix} \quad \left(\frac{\partial U_e}{\partial \mathbf{Q}}\right)^t = \begin{bmatrix} S_1 q_1 + S_3(q_1 - q_2) \\ S_2 q_2 + S_3(q_2 - q_1) \end{bmatrix}$$

This gives the network Lagrangian equation set as

$$L_1 \ddot{q}_1 + R_1 \dot{q}_1 + S_1 q_1 + S_3(q_1 - q_2) = E$$
$$L_2 \ddot{q}_2 + R_2 \dot{q}_2 + S_3(q_2 - q_1) = 0$$

5.10 Co-Lagrangian Equations for Networks

If only current sources are present then, as in the vertex method of Section 5.7.2

$$\mathbf{Ai} = \mathbf{Aj} \qquad (5\text{-}95)$$

where

(a) $\mathbf{A}$ is the network reduced incidence matrix.
(b) $\mathbf{i}$ is a vector whose components are the passive element currents.
(c) $\mathbf{j}$ is a vector whose components are the active element voltages.

Again group elements together into three sets associated with inductors, resistors and capacitors, and partition equation (5–95) as

$$[\mathbf{A}_L \mid \mathbf{A}_R \mid \mathbf{A}_C] \begin{bmatrix} \mathbf{i}_L \\ \mathbf{i}_R \\ \mathbf{i}_C \end{bmatrix} = \mathbf{Aj} \triangleq \mathbf{S} \tag{5-96}$$

where

(a) $\mathbf{i}_L$, $\mathbf{i}_R$ and $\mathbf{i}_C$ are vectors whose components are the inductors', resistors' and capacitors' currents respectively.
(b) $\mathbf{S}$ is a vector of vertex current sources. ($\mathbf{S}$ has been used here to avoid confusion with the co-content symbol J.)
(c) $\mathbf{A}_L$, $\mathbf{A}_R$ and $\mathbf{A}_C$ are matrices obtained by a suitable partitioning of matrix $\mathbf{A}$.

Let $\mathbf{V}$ be a vector whose components are a set of independent vertex voltages, and let $\mathbf{v}$ be a vector whose components are the passive element voltages. Then, as shown in Section 5.7.2, we have

$$\mathbf{v} = \mathbf{A}^t \mathbf{V} \tag{5-97}$$

For the grouping of components and corresponding partitioning of $\mathbf{A}$ adopted, we therefore have

$$\mathbf{v} = \begin{bmatrix} \mathbf{v}_L \\ \mathbf{v}_R \\ \mathbf{v}_C \end{bmatrix} = [\mathbf{A}_L \mid \mathbf{A}_R \mid \mathbf{A}_C]^t \mathbf{V} = \begin{bmatrix} \mathbf{A}_L^t \\ \mathbf{A}_R^t \\ \mathbf{A}_C^t \end{bmatrix} \mathbf{V} = \begin{bmatrix} \mathbf{A}_L^t \mathbf{V} \\ \mathbf{A}_R^t \mathbf{V} \\ \mathbf{A}_C^t \mathbf{V} \end{bmatrix}$$

so that
$$\mathbf{v}_L = \mathbf{A}_L^t \mathbf{V}$$
$$\mathbf{v}_R = \mathbf{A}_R^t \mathbf{V} \tag{5-98}$$
$$\mathbf{v}_C = \mathbf{A}_C^t \mathbf{V}$$

In terms of the component variables we will have

$$\mathbf{i}_L = \left(\frac{\partial T_e}{\partial \lambda_L}\right)^t$$
$$\mathbf{i}_R = \left(\frac{\partial J}{\partial \mathbf{v}_R}\right)^t \tag{5-99}$$
$$\mathbf{i}_C = \frac{d}{dt}\left(\frac{\partial U_e^*}{\partial \mathbf{v}_C}\right)^t$$

where

(a) T_e is the total inductive energy.

(b) J is the total resistor co-content.
(c) U_e^* is the total capacitive co-energy.

Equation (5–96) thus gives

$$\mathbf{A}_L\left(\frac{\partial T_e}{\partial \boldsymbol{\lambda}_L}\right)^t + \mathbf{A}_R\left(\frac{\partial J}{\partial \mathbf{v}_R}\right)^t + \mathbf{A}_C\frac{d}{dt}\left(\frac{\partial U_e^*}{\partial \mathbf{v}_C}\right)^t = \mathbf{S} \qquad (5\text{–}100)$$

Now define a *vertex flux-linkage* variable $\boldsymbol{\Lambda}$ by

$$\boldsymbol{\Lambda} = \int_0^t \mathbf{V}\, dt \qquad (5\text{–}101)$$

Then
$$\boldsymbol{\lambda}_L = \mathbf{A}_L^t \boldsymbol{\Lambda} \qquad (5\text{–}102)$$

Substituting from equations (5–98) and (5–102) into equation (5–100) we therefore have

$$\mathbf{A}_L\left(\frac{\partial T_e}{\partial \mathbf{A}_L^t \boldsymbol{\Lambda}}\right)^t + \mathbf{A}_R\left(\frac{\partial J}{\partial \mathbf{A}_R^t \mathbf{V}}\right)^t + \mathbf{A}_C\frac{d}{dt}\left(\frac{\partial U_e^*}{\partial \mathbf{A}_C^t \mathbf{V}}\right)^t = \mathbf{S}$$

so that using (2–36)
$$\left(\frac{\partial T_e}{\partial \boldsymbol{\Lambda}}\right)^t + \left(\frac{\partial J}{\partial \mathbf{V}}\right)^t + \frac{d}{dt}\left(\frac{\partial U_e^*}{\partial \mathbf{V}}\right)^t = \mathbf{S}$$

and thus we obtain the corresponding Co-Langrangian equation

$$\frac{d}{dt}\left(\frac{\partial U_e^*}{\partial \dot{\boldsymbol{\Lambda}}}\right)^t + \left(\frac{\partial J}{\partial \dot{\boldsymbol{\Lambda}}}\right)^t + \left(\frac{\partial T_e}{\partial \boldsymbol{\Lambda}}\right)^t = \mathbf{S} \qquad (5\text{–}103)$$

Example 5.4 *Network Co-Lagrangian equations*

For the network of Figure 5–28 take vertex D as the reference vertex and the voltages of vertices, A, B and C as the generalized voltages V_1, V_2 and V_3 respectively. Define the corresponding generalized vertex flux-linkages as

$$\Lambda_i = \int_0^t V_i\, dt \qquad i = 1, 2, 3$$

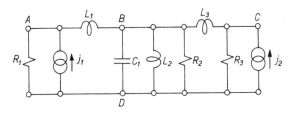

Figure 5–28

and take these to be the network generalized coordinate set. Then

$$\text{flux-linkage of inductor } ^\#1 = \Lambda_1 - \Lambda_2$$
$$\text{flux-linkage of inductor } ^\#2 = \Lambda_2$$
$$\text{flux-linkage of inductor } ^\#3 = \Lambda_2 - \Lambda_3$$

Thus we have
$$T_e = \tfrac{1}{2}\Gamma_1(\Lambda_1 - \Lambda_2)^2 + \tfrac{1}{2}\Gamma_2\Lambda_2^2 + \tfrac{1}{2}\Gamma_3(\Lambda_2 - \Lambda_3)^2$$

where
$$\Gamma_i = \frac{1}{L_i} \quad i = 1, 2, 3$$

$$U_e^* = \tfrac{1}{2}C_1(\dot{\Lambda}_2)^2$$
$$J = \tfrac{1}{2}G_1\dot{\Lambda}_1^2 + \tfrac{1}{2}G_2\dot{\Lambda}_2^2 + \tfrac{1}{2}G_3\dot{\Lambda}_3^2$$

where
$$G_i = \frac{1}{R_i} \quad i = 1, 2, 3$$

$$\mathbf{S} = \begin{bmatrix} j_1 \\ 0 \\ j_2 \end{bmatrix} \quad \mathbf{\Lambda} = \begin{bmatrix} \Lambda_1 \\ \Lambda_2 \\ \Lambda_3 \end{bmatrix} \quad \dot{\mathbf{\Lambda}} = \begin{bmatrix} \dot{\Lambda}_1 \\ \dot{\Lambda}_2 \\ \dot{\Lambda}_3 \end{bmatrix}$$

This gives

$$\left(\frac{\partial U_e^*}{\partial \dot{\mathbf{\Lambda}}}\right)^t = \begin{bmatrix} 0 \\ C_1\dot{\Lambda}_2 \\ 0 \end{bmatrix} \quad \frac{d}{dt}\left(\frac{\partial U_e^*}{\partial \dot{\mathbf{\Lambda}}}\right)^t = \begin{bmatrix} 0 \\ C_1\ddot{\Lambda}_2 \\ 0 \end{bmatrix}$$

$$\left(\frac{\partial J}{\partial \dot{\mathbf{\Lambda}}}\right)^t = \begin{bmatrix} G_1\dot{\Lambda}_1 \\ G_2\dot{\Lambda}_2 \\ G_3\dot{\Lambda}_3 \end{bmatrix}$$

$$\left(\frac{\partial T_e}{\partial \mathbf{\Lambda}}\right)^t = \begin{bmatrix} \Gamma_1\Lambda_1 - \Gamma_1\Lambda_2 \\ -\Gamma_1\Lambda_1 + \Gamma_1\Lambda_2 + \Gamma_2\Lambda_2 + \Gamma_3\Lambda_2 - \Gamma_3\Lambda_3 \\ -\Gamma_3\Lambda_2 + \Gamma_3\Lambda_3 \end{bmatrix}$$

This gives the set of Co-Lagrangian equations as

$$G_1\dot{\Lambda}_1 + \Gamma_1(\Lambda_1 - \Lambda_2) = j_1$$
$$C_1\ddot{\Lambda}_2 + G_2\dot{\Lambda}_2 - \Gamma_1(\Lambda_1 - \Lambda_2) + \Gamma_2\Lambda_2 + \Gamma_3(\Lambda_2 - \Lambda_3) = 0$$
$$G_3\dot{\Lambda}_3 - \Gamma_3(\Lambda_2 - \Lambda_3) = j_2$$

5.11 Special Variational Principles for Networks [C3], [M9]

In addition to Hamilton's Postulate, discussed in Chapter 4, there are a number of specialized variational principles for networks which are of considerable interest and which may be established in a straightforward way from the network constraint relationships.

Stationary Content Principle

For a network composed entirely of sources and dissipative elements, the actual distribution of currents is such that the total content is stationary for current variations subject to Kirchhoff's Current Law.

Suppose there are m elements in the network, and let

$$\{v_j : j = 1, 2, \ldots, m\} \quad \text{and} \quad \{\delta i_j : j = 1, 2, \ldots, m\}$$

be the element voltages and current variations respectively. Let $\{G_j : j = 1, 2, \ldots, m\}$ be the set of element contents and G the total content. Then

$$\delta G = \sum_{j=1}^{m} \frac{\partial G_j}{\partial i_j} \delta i_j = \sum_{j=1}^{m} v_j \delta i_j = \langle \mathbf{v}, \delta \mathbf{i} \rangle \tag{5-104}$$

where $\mathbf{v}$ and $\delta \mathbf{i}$ are vectors whose components are respectively the element voltages and current variations. Express the scalar product $\langle \mathbf{v}, \delta \mathbf{i} \rangle$ in terms of its tree and chord parts, after an arbitrary choice of tree, as:

$$\langle \mathbf{v}, \delta \mathbf{i} \rangle = \mathbf{v}^t \delta \mathbf{i} = [\mathbf{v}_t \mathbf{v}_c]^t \begin{bmatrix} \delta \mathbf{i}_t \\ \delta \mathbf{i}_c \end{bmatrix} = \langle \mathbf{v}_t, \delta \mathbf{i}_t \rangle + \langle \mathbf{v}_c, \delta \mathbf{i}_c \rangle \tag{5-105}$$

Now it is stipulated that the current variations must be subject to Kirchhoff's Current Law. It therefore follows from the network relationships of Section 5.5 that

$$\delta \mathbf{i}_t = \mathbf{D} \delta \mathbf{i}_c \tag{5-106}$$

where $\mathbf{D}$ is the network dynamical transformation matrix. The actual voltages are subject to Kirchhoff's Voltage Law so that

$$\mathbf{v}_c = -\mathbf{D}^t \mathbf{v}_t \tag{5-107}$$

Combining equations (5-104) through (5-107) we have

$$\delta G = \langle \mathbf{v}_t, \delta \mathbf{i}_t \rangle + \langle \mathbf{v}_c, \delta \mathbf{i}_c \rangle$$
$$= \langle \mathbf{v}_t, \delta \mathbf{i}_t \rangle + \langle -\mathbf{D}^t \mathbf{v}_t, \delta \mathbf{i}_c \rangle$$
$$= \langle \mathbf{v}_t, \delta \mathbf{i}_t \rangle - \langle \mathbf{v}_t, \mathbf{D} \delta \mathbf{i}_c \rangle$$
$$= \langle \mathbf{v}_t, \delta \mathbf{i}_t \rangle - \langle \mathbf{v}_t, \delta \mathbf{i}_t \rangle$$
$$= 0$$

which proves the principle.

Stationary Co-content Principle

For a network composed entirely of sources and dissipative elements, the actual distribution of voltages is such that the total co-content is stationary for voltage variations subject to Kirchhoff's Voltage Law.

Suppose there are m elements in the network, and let $\{i_j : j = 1, 2, \ldots, m\}$ and $\{\delta v_j : j = 1, 2, \ldots, m\}$ be the element currents and voltage variations respectively. Let $\{J_j : j = 1, 2, \ldots, m\}$ be the set of element co-contents and J the total co-content. Then

$$\delta J = \sum_{j=1}^{m} \frac{\partial J_j}{\partial v_j} \delta v_j = \sum_{j=1}^{m} i_j \delta v_j = \langle \mathbf{i}, \delta \mathbf{v} \rangle \quad (5\text{-}108)$$

where $\mathbf{i}$ and $\delta \mathbf{v}$ are vectors whose components are respectively the element currents and voltage variations. After an arbitrary choice of tree, express δJ in terms of its tree and chord parts as

$$\delta J = \langle \mathbf{i}_t, \delta \mathbf{v}_t \rangle + \langle \mathbf{i}_c, \delta \mathbf{v}_c \rangle \quad (5\text{-}109)$$

Now the actual currents are such that Kirchhoff's Current Law is satisfied, thus

$$\mathbf{i}_t = \mathbf{D} \mathbf{i}_c \quad (5\text{-}110)$$

and it is stipulated that the voltage variations are subject to Kirchhoff's Voltage Law, so that

$$\delta \mathbf{v}_c = -\mathbf{D}^t \delta \mathbf{v}_t \quad (5\text{-}111)$$

Thus
$$\begin{aligned}\delta J &= \langle \mathbf{D} \mathbf{i}_c, \delta \mathbf{v}_t \rangle + \langle \mathbf{i}_c, \delta \mathbf{v}_c \rangle \\ &= \langle \mathbf{i}_c, \mathbf{D}^t \delta \mathbf{v}_t \rangle + \langle \mathbf{i}_c, \delta \mathbf{v}_c \rangle \\ &= -\langle \mathbf{i}_c, \delta \mathbf{v}_c \rangle + \langle \mathbf{i}_c, \delta \mathbf{v}_c \rangle \\ &= 0\end{aligned}$$

which proves the principle.

Maxwell's Stationary Heat Theorem

If the networks considered contain only linear resistors, then both of the above stationary principles reduce to Maxwell's Stationary Heat Theorem, the words 'content' and 'co-content' being replaced by 'power dissipated as heat'.

Stationary Inductive Energy Principle

For an all-inductor network driven entirely by ideal voltage sources, the actual distribution of passive-element voltages is such that the total inductive energy is stationary for passive-element voltage variations subject to Kirchhoff's Voltage Law.

Special Variational Principles for Networks

Suppose there are m inductors in the network, and let

$$\{\lambda_j : j = 1, 2, \ldots, m\}, \{i_j : j = 1, 2, \ldots, m\}$$

and $\{\delta v_j : j = 1, 2, \ldots, m\}$ be the inductor flux-linkages, currents and voltage variations respectively. Suppose all flux-linkages were initially zero at time $t = 0$. Then, if T is the total inductive energy, we have

$$\delta T = \sum_{j=1}^{m} \frac{\partial T}{\partial \lambda_j} \delta \lambda_j = \sum_{j=1}^{m} i_j \int_0^t \delta v_j \, dt = \left\langle \mathbf{i}, \int_0^t \delta \mathbf{v} \, dt \right\rangle \quad (5\text{-}112)$$

where $\mathbf{i}$ and $\delta \mathbf{v}$ are vectors whose components are respectively the inductor currents and voltage variations. Make an arbitrary choice of tree and group the network components into tree and chord components. Then the actual currents are subject to Kirchhoff's Current Law and there are no current sources, so that

$$\mathbf{i}_t = \mathbf{D}\mathbf{i}_c \quad (5\text{-}113)$$

where $\mathbf{i}$ is partitioned into its chord and tree components $\mathbf{i}_c$ and $\mathbf{i}_t$ respectively.

Thus, using equations (5-111) through (5-113) we have, since the passive-element voltage variations are stipulated to be subject to Kirchhoff's Voltage Law, so that

$$\delta \mathbf{v}_c = -\mathbf{D}^t \delta \mathbf{v}_t \quad (5\text{-}114)$$

$$\delta T = \left\langle \mathbf{i}_t, \int_0^t \delta \mathbf{v}_t \, dt \right\rangle + \left\langle \mathbf{i}_c, \int_0^t \delta \mathbf{v}_c \, dt \right\rangle$$

$$= \left\langle \mathbf{D}\mathbf{i}_c, \int_0^t \delta \mathbf{v}_t \, dt \right\rangle + \left\langle \mathbf{i}_c, \int_0^t \delta \mathbf{v}_c \, dt \right\rangle$$

$$= \left\langle \mathbf{i}_c, \mathbf{D}^t \int_0^t \delta \mathbf{v}_t \, dt \right\rangle + \left\langle \mathbf{i}_c, \int_0^t \delta \mathbf{v}_c \, dt \right\rangle$$

$$= -\left\langle \mathbf{i}_c, \int_0^t \delta \mathbf{v}_c \, dt \right\rangle + \left\langle \mathbf{i}_c, \int_0^t \delta \mathbf{v}_c \, dt \right\rangle$$

$$= 0$$

which proves the principle.

Stationary Inductive Co-energy Principle

For an all-inductor network driven entirely by current sources, the actual distribution of passive-element currents is such that the total inductive co-energy is stationary with respect to passive-element current variations subject to Kirchhoff's Current Law.

If T^* is the total inductive co-energy we have

$$\delta T^* = \sum_{j=1}^{m} \frac{\partial T^*}{\partial i_j} \delta i_j = \sum_{j=1}^{m} \lambda_j \delta i_j$$

$$= \sum_{j=1}^{m} \int_0^t v_j \, dt \, \delta i_j$$

$$= \left\langle \int_0^t \mathbf{v} \, dt, \delta \mathbf{i} \right\rangle \qquad (5\text{-}115)$$

where $\mathbf{v}$ and $\delta \mathbf{i}$ are vectors whose components are the inductor voltages and current variations respectively. For an arbitrary choice of tree, we may again partition the vectors involved into their tree and chord components. The actual passive-element voltages satisfy Kirchhoff's Voltage Law so that

$$\mathbf{v}_t = -\mathbf{D}^t \mathbf{v}_t \qquad (5\text{-}116)$$

and it is stipulated that the passive-element current variations are subject to Kirchhoff's Current Law, so that

$$\delta \mathbf{i}_t = \mathbf{D} \delta \mathbf{i}_c \qquad (5\text{-}117)$$

Thus we have that

$$\delta T^* = \left\langle \int_0^t \mathbf{v}_t \, dt, \delta \mathbf{i}_t \right\rangle + \left\langle \int_0^t \mathbf{v}_c \, dt, \delta \mathbf{i}_c \right\rangle$$

$$= \left\langle \int_0^t \mathbf{v}_t \, dt, \delta \mathbf{i}_t \right\rangle + \left\langle -\mathbf{D}^t \int_0^t \mathbf{v}_t \, dt, \delta \mathbf{i}_c \right\rangle$$

$$= \left\langle \int_0^t \mathbf{v}_t \, dt, \delta \mathbf{i}_t \right\rangle - \left\langle \int_0^t \mathbf{v}_t \, dt, \mathbf{D} \delta \mathbf{i}_c \right\rangle$$

$$= \left\langle \int_0^t \mathbf{v}_t \, dt, \delta \mathbf{i}_t \right\rangle - \left\langle \int_0^t \mathbf{v}_t \, dt, \delta \mathbf{i}_t \right\rangle$$

$$= 0$$

which proves the principle.

Stationary Capacitive Energy Principle

For an all-capacitor network driven entirely by ideal current sources, the actual distribution of passive-element currents is such that the total capacitive energy is stationary for passive-element variations subject to Kirchhoff's Current Law.

Suppose there are m capacitors in the network, and let $\{q_j : j = 1, 2, \ldots, m\}$, $\{v_j : j = 1, 2, \ldots, m\}$ and $\{\delta i_j : j = 1, 2, \ldots, m\}$ be

Special Variational Principles for Networks 333

the capacitor charges, voltages and current variations respectively. Suppose all charges are initially zero at time $t = 0$. Then, if U is the total capacitive energy, we have

$$\delta U = \sum_{j=1}^{m} \frac{\partial U}{\partial q_j} \delta q_j = \sum_{j=1}^{m} v_j \delta q_j = \sum_{j=1}^{m} v_j \int_0^t \delta i_j \, dt = \left\langle \mathbf{v}, \int_0^t \delta \mathbf{i} \, dt \right\rangle$$

(5–118)

where $\mathbf{v}$ and $\delta \mathbf{i}$ are vectors whose components are respectively the capacitor voltages and current variations. For an arbitrary choice of tree, we may again partition the relevant vectors into their tree and chord components. The actual passive-element voltages satisfy Kirchhoff's Voltage Law, so that

$$\mathbf{v}_c = -\mathbf{D}^t \mathbf{v}_t \qquad (5\text{–}119)$$

and it is stipulated that the passive-element current variations must satisfy Kirchhoff's Current Law, so that

$$\delta \mathbf{i}_t = \mathbf{D} \delta \mathbf{i}_c \qquad (5\text{–}120)$$

Equations (5–118) through (5–120) then give

$$\delta U = \left\langle \mathbf{v}_c, \int_0^t \delta \mathbf{i}_c \, dt \right\rangle + \left\langle \mathbf{v}_t, \int_0^t \delta \mathbf{i}_t \, dt \right\rangle$$

$$= \left\langle -\mathbf{D}^t \mathbf{v}_t, \int_0^t \delta \mathbf{i}_c \, dt \right\rangle + \left\langle \mathbf{v}_t, \int_0^t \delta \mathbf{i}_t \, dt \right\rangle$$

$$= -\left\langle \mathbf{v}_t, \mathbf{D} \int_0^t \delta \mathbf{i}_c \, dt \right\rangle + \left\langle \mathbf{v}_t, \int_0^t \delta \mathbf{i}_t \, dt \right\rangle$$

$$= -\left\langle \mathbf{v}_t, \int_0^t \delta \mathbf{i}_t \, dt \right\rangle + \left\langle \mathbf{v}_t, \int_0^t \delta \mathbf{i}_t \, dt \right\rangle$$

$$= 0$$

which proves the principle. Using similar arguments it is straightforward to establish the following principle.

Stationary Capacitive Co-energy Principle
For an all-capacitor network driven entirely by ideal voltage sources, the actual distribution of passive-element voltages is such that the total capacitive co-energy is stationary for passive-element voltage variations which satisfy Kirchhoff's Voltage Law.

Example 5.5 Stationary Co-content Principle
Consider the series connection of ideal voltage source, tunnel diode and linear resistor shown in Figure 5–29. The constitutive characteristics of these elements are shown in Figure 5–30. Denote the co-content

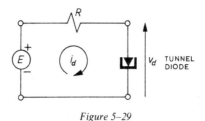

Figure 5-29

of the resistor and the tunnel diode by J_r and J_d respectively. The co-content of the ideal voltage source is zero. Let the tunnel diode constitutive characteristic be

$$i_d = f(v_d)$$

where i_d and v_d are the tunnel diode current and voltage respectively. We thus have

$$J = J_r + J_d = \frac{1}{2}\frac{(E-v_d)^2}{R} + \int_0^{v_d} f(v)\, dv$$

Thus
$$\frac{\partial J}{\partial v_d} = -\frac{(E-v_d)}{R} + f(v_d) = 0$$

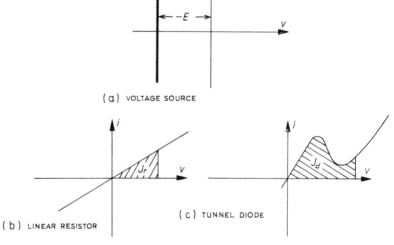

(a) VOLTAGE SOURCE

(b) LINEAR RESISTOR

(c) TUNNEL DIODE

Figure 5-30

Special Variational Principles for Networks 335

The equilibrium tunnel diode voltage v_d must therefore be such that

$$f(v_d) = \frac{(E-v_d)}{R} = i_d$$

This corresponds to the well-known graphical construction shown in Figure 5-31. To investigate the stability of these equilibrium points,

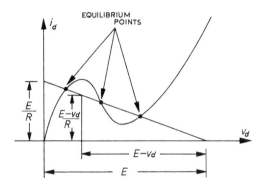

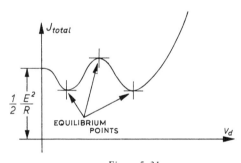

Figure 5-31

let the stray capacitance across the tunnel diode be C, and the current through this stray capacitance be i_c as shown in Figure 5-32. Then we have

$$i_c = C\frac{dv_d}{dt} = \frac{(E-v_d)}{R} - f(v_d)$$

so that for small variations about an operating point

$$C\frac{d}{dt}(\delta v_d) = \delta\left(C\frac{dv_d}{dt}\right) = \delta\left[\frac{(E-v_d)}{R} - f(v_d)\right]$$

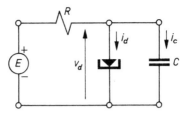

Figure 5-32

so that we have

$$C\frac{d}{dt}(\delta v_d) = \frac{\partial}{\partial v_d}\left[\frac{(E-v_d)}{R} - f(v_d)\right]\delta v_d$$

$$= \left[-\frac{1}{R} - \frac{\partial f}{\partial v_d}\right]\delta v_d$$

$$= -\left[\frac{1}{R} + \frac{\partial f}{\partial v_d}\right]\delta v_d$$

The equilibrium point will be stable if the term in brackets on the right-hand side of this equation is positive, that is, if

(a) $\left[\dfrac{\partial f}{\partial v}\right]_{v=v_d}$ is positive

or (b) $\left[\dfrac{\partial f}{\partial v}\right]_{v=v_d}$ is negative but $\left|\dfrac{\partial f}{\partial v}\right|_{v=v_d} < \dfrac{1}{R}$

The system will be unstable if

(c) $\left[\dfrac{\partial f}{\partial v}\right]_{v=v_d}$ is negative and $\left|\dfrac{\partial f}{\partial v}\right|_{v=v_d} > \dfrac{1}{R}$

These simple stability criteria are illustrated graphically in Figure 5-33.

Lagrange Multiplier Interpretation of Stationary Heat Theorem

It is interesting to briefly examine the stationary theorems for resistive networks in terms of the Lagrange multiplier technique introduced in Chapter 2. This will be done by means of a pair of simple examples.

Example 5.6

Suppose a pair of resistances, of value R_1 and R_2 are to be supplied with currents i_1 and i_2 respectively so that the total power dissipated is minimized subject to the constraint that the sum of the currents is I.

Special Variational Principles for Networks

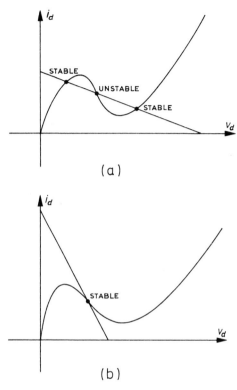

Figure 5-33

That is to say minimize

$$P = i_1^2 R_1 + i_2^2 R_2$$

subject to $\qquad i_1 + i_2 - I = 0$

Introducing a Lagrange multiplier λ, this is equivalent to minimizing

$$P^* = i_1^2 R_1 + i_2^2 R_2 + \lambda(i_1 + i_2 - I)$$

so that the solution is given by the set of equations

$$\frac{\partial P^*}{\partial i_1} = 2i_1 R_1 + \lambda = 0$$

$$\frac{\partial P^*}{\partial i_2} = 2i_2 R_2 + \lambda = 0$$

$$i_1 + i_2 - I = 0$$

from which we obtain

$$i_1 = \frac{IR_2}{(R_1+R_2)} \qquad i_2 = \frac{IR_1}{(R_1+R_2)} \qquad \lambda = -\frac{2IR_1R_2}{(R_1+R_2)}$$

Consideration of this result shows that the currents are apportioned in the way they would be if the resistors were connected in parallel and supplied with the current I. The Lagrange multiplier is proportional to the value of the corresponding voltage across the two resistors.

Example 5.7

Suppose the same pair of resistors as in the previous example are supplied with voltages v_1 and v_2 respectively so that the total power dissipated is minimized subject to the constraint that the sum of the voltages is V. That is to say, if G_1 and G_2 are the conductances of the two resistors,

$$\minimize_{v_1,v_2} [v_1^2 G_1 + v_2^2 G_2]$$

subject to $v_1 + v_2 - V = 0$.

On introducing a Lagrange multiplier λ we put

$$P^* = v_1^2 G_1 + v_2^2 G_2 + \lambda(v_1 + v_2 - V)$$

so that the solution is given by the set of equations

$$\frac{\partial P^*}{\partial v_1} = 2v_1 G_1 + \lambda = 0$$

$$\frac{\partial P^*}{\partial v_2} = 2v_2 G_2 + \lambda = 0$$

$$v_1 + v_2 - V = 0$$

from which

$$v_1 = \frac{VG_2}{(G_1+G_2)} \qquad v_2 = \frac{VG_1}{(G_1+G_2)} \qquad \lambda = -\frac{2VG_1G_2}{(G_1+G_2)}$$

This shows that the voltages are apportioned in the way they would be if the resistors were connected in series and supplied with the voltage V. The Lagrange multiplier is proportional to the value of the current through the two resistors.

5.12 Formulation of Canonical Equation Sets for Linear Networks

We first consider the incorporation of sources in a *linear* network, and show that any given distribution of current and voltage sources may be replaced by an equivalent distribution of sources in which all

Formulation of Canonical Equation Sets for Linear Networks 339

current sources appear connected directly across passive elements of the network, and all voltage sources appear connected in series with passive elements of the network.

Equivalent distribution of current sources. Ideal current sources may be incorporated into a network by connection in series with a set of passive elements or in parallel with a set of passive elements. If an ideal current source is connected to a network of passive elements, an argument of the form given below may always be used to show how the network may be replaced by an equivalent network in which all current sources appear in parallel with passive elements. Consider the portion of network shown in Figure 5–34(a). Since an ideal current source appears directly in series with elements $^\#1$ and $^\#2$, the determination of the currents and voltages for these elements becomes trivial and they may be removed from the network without in any way influencing the remainder of the network; this gives the portion of network shown in Figure 5–34(b). Finally the single current source in the network of Figure 5–34(b) may be replaced by the pair of equivalent current sources shown in Figure 5–34(c), since this pair of current sources gives the same current balance at vertices $^\#a$ and $^\#c$ and does not alter the current balance at vertex $^\#b$. Note that we could equally well have replaced the current generator between vertices $^\#a$ and $^\#c$ by a pair of current generators between $^\#a$ and $^\#d$ and between $^\#d$ and $^\#c$ as shown in Figure 5–34(d).

Obviously arguments of this type may be constructed for any network with any given distribution of current sources. It follows that any network with any given distribution of current sources may be replaced, for the purpose of analysis, by an equivalent network in which all current sources appear in parallel with passive components. The equivalent set of current sources is formed by placing one across each of any set of passive components which form a closed circuit with the original source (after removal of surplus elements). The current source orientations are chosen so as to leave the current balance unaltered at all vertices involved.

Equivalent distribution of voltage sources. Ideal voltage sources may be incorporated into a network by connection in series with a set of passive elements or in parallel with a set of passive elements. If an ideal voltage source is connected to a network of passive elements, an argument of the form given below may always be used to show how the network may be replaced by an equivalent network in which all voltage sources appear in series with passive elements. Consider the portion of network shown in Figure 5–34(e). Since an ideal voltage source appears directly across elements $^\#3$ and $^\#4$, the determination of the voltages and currents for these elements becomes trivial and they may be removed from the network without in any way influencing the behaviour of the

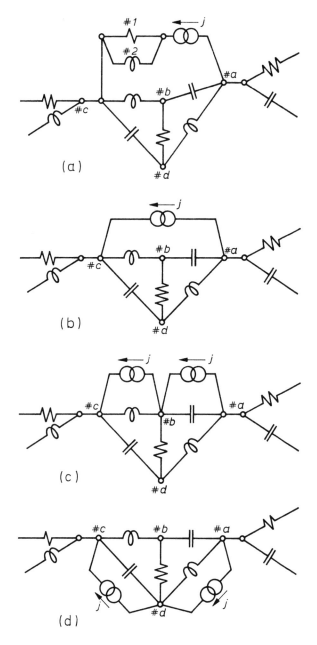

Figure 5-34

Formulation of Canonical Equation Sets for Linear Networks 341

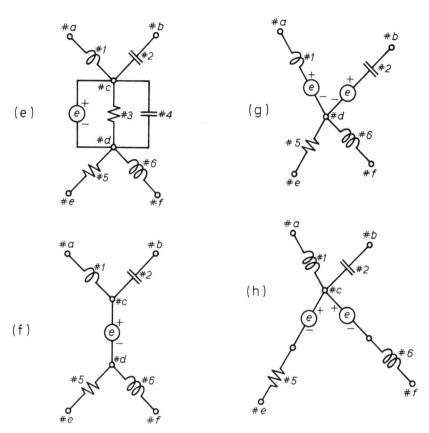

Figure 5–34 continued

remainder of the network; this gives the portion of network shown in Figure 5–34(f). Finally, the single voltage source in the network of Figure 5–34(f) may be replaced by the pair of equivalent voltage sources shown in Figure 5–34(g), since this pair of sources result in the same values of voltage between vertices $^\#a$ and $^\#d$, and between $^\#b$ and $^\#d$ as a result from the single voltage source of Figure 5–34(b). Note that we could equally well have 'shifted the voltage sources down' as shown in Figure 5–34(h).

Obviously arguments of this type may be constructed for any network with any given distribution of voltage sources. It follows that any network with any given distribution of voltage sources may be replaced, for the purposes of analysis, by an equivalent network in which all voltage sources appear in series with network passive elements.

The voltage source may be visualized as 'pushed forward' or 'pulled back' through one of the two vertices which it connects (after removal of surplus elements) and split up into a number of equivalent sources, one for each component incident on the vertex past which the voltage source is 'pushed' or 'pulled' as the case may be.

It follows from the above that for the purposes of analysing the behaviour of *linear* networks it is sufficient to consider interconnected sets of general elements such as that shown in Figure 5–23(a). Such a general branch element with its associated set of reference positive directions may be represented by the oriented line segment of Figure 5–23(b). Associated with any network composed of general elements of this sort is a reduced incident matrix **A** and a fundamental circuit matrix $\mathbf{B}_f$ such that

$$\mathbf{A}(\mathbf{i}-\mathbf{j}) = \mathbf{0} \qquad (5\text{--}121)$$

$$\mathbf{B}_f(\mathbf{v}-\mathbf{e}) = \mathbf{0} \qquad (5\text{--}122)$$

The network's constraint equation set may therefore be written in the form

$$\mathbf{A}\mathbf{i} = \mathbf{A}\mathbf{j} \qquad (5\text{--}123)$$

$$\mathbf{B}_f \mathbf{v} = \mathbf{B}_f \mathbf{e} \qquad (5\text{--}124)$$

where:

(a) **i** is a vector whose components are the passive element currents.
(b) **j** is a vector whose components are the current source outputs.
(c) **v** is a vector whose components are the passive voltages.
(d) **e** is a vector whose components are the voltage source outputs.

For any particular choice of network tree, we may group the network elements into branches and chords. For such a grouping we may associate with a network of e elements and n vertices an $(n-1)$ by $(e-n+1)$ matrix **D** such that

$$\mathbf{B}_f = [\mathbf{I}_{(e-n+1)} \;\vdots\; \mathbf{D}^t] \qquad (5\text{--}125)$$

where **D** is as defined in Section 5.5 above. Equation (5–124) thus gives

$$[\mathbf{I}_{(e-n+1)} \;\vdots\; \mathbf{D}^t] \begin{bmatrix} \mathbf{v}_c \\ \cdots \\ \mathbf{v}_t \end{bmatrix} = [\mathbf{I}_{(e-n+1)} \;\vdots\; \mathbf{D}^t] \begin{bmatrix} \mathbf{e}_c \\ \cdots \\ \mathbf{e}_t \end{bmatrix} \qquad (5\text{--}126)$$

$$\mathbf{v}_c = -\mathbf{D}^t \mathbf{v}_t + \mathbf{e}_c + \mathbf{D}^t \mathbf{e}_t$$

where:

(a) $\mathbf{v}_c$ is a vector whose components are the chord passive element voltages.

Formulation of Canonical Equation Sets for Linear Networks 343

(b) v_t is a vector whose components are the branch passive element voltages.
(c) e_c is a vector whose components are the chord voltage source outputs.
(d) e_t is a vector whose components are the branch voltage source outputs.

For the same grouping of elements into branches and chords we may partition equation (5–123) such that

$$[X \vdots K] \begin{bmatrix} i_c \\ i_t \end{bmatrix} = [X \vdots K] \begin{bmatrix} j_c \\ j_t \end{bmatrix}$$

where
$$D = -K^{-1}X$$

as shown in Section 5.5 above. This gives

$$Xi_c + Ki_t = Xj_c + Kj_t$$
$$K^{-1}Xi_c + i_t = K^{-1}Xj_c + j_t$$
$$-Di_c + i_t = -Dj_c + j_t \qquad (5\text{–}127)$$
$$i_t = Di_c + j_t - Dj_c$$

where:

(a) i_c is a vector whose components are the chord passive element currents.
(b) i_t is a vector whose components are the branch passive element currents.
(c) j_c is a vector whose components are the chord current source outputs.
(d) j_t is a vector whose components are the branch current source outputs.
(e) X and K are matrices obtained by suitably partitioning matrix A.

Taking equations (5–126) and (5–127) together we have

$$\begin{bmatrix} i_t - j_t \\ v_c - e_c \end{bmatrix} = \begin{bmatrix} 0 & \vdots & D \\ -D^t & \vdots & 0 \end{bmatrix} \begin{bmatrix} v_t - e_t \\ i_c - j_c \end{bmatrix} \qquad (5\text{–}128)$$

giving
$$\begin{bmatrix} i_t \\ v_c \end{bmatrix} = \begin{bmatrix} 0 & \vdots & D \\ -D^t & \vdots & 0 \end{bmatrix} \begin{bmatrix} v_t \\ i_c \end{bmatrix} + \begin{bmatrix} I \\ E \end{bmatrix} \qquad (5\text{–}129)$$

where
$$I \triangleq j_t - Dj_c \qquad (5\text{–}130)$$

is a vector representing the effects of the current sources and

$$E \triangleq e_c + D^t e_t \tag{5-131}$$

is a vector representing the effects of the voltage sources.

The general form of linear network relationships may thus be summarized by the block diagram of Figure 5–35. Consideration of this block diagram shows that we may consider the relationships between the passive elements separately from the effects of the network sources. In what follows therefore, we frequently discuss only relationships among passive element in terms of trees and co-trees consisting entirely of passive elements.

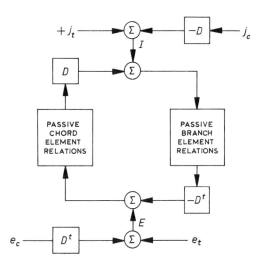

Figure 5–35

Independent Sets of Storage Elements

Independent capacitors. If a set of capacitors form a complete circuit then their voltages, and thus their charges, are dependent. Conversely if a set of capacitors do not form a loop, their voltages and thus their charges, are independent. It is convenient to call a set of capacitors whose voltages are independent an independent set of capacitors.

Independent inductors. If a set of inductors form a cut-set then their currents, and thus their flux-linkages, are independent. Conversely if a set of inductors do not form a cut-set, their currents, and thus flux-linkages are independent. It is convenient to call a set of inductors whose currents are independent an independent set of inductors.

Formulation of Canonical Equation Sets for Linear Networks 345

State, excess and non-state elements. From the complete set of network capacitors and inductors we may select *maximal sets* of independent capacitors and inductors; for a stipulated selected such maximal set, the constituent elements of the set will be termed *state elements*. Those capacitors and inductors which are not part of a stipulated maximal set of independent elements will be termed *excess elements*. The resistors of a network will be termed *non-state elements*.

We now consider the selection of a maximal independent set of capacitors and inductors; the selection process depends on a particular choice of network tree. Consider first the inclusion of capacitors in a tree. The only situations precluding the inclusion of all network capacitors in some network tree arise from the existence of capacitor-only circuits. Now if any such capacitor-only circuits exist, the capacitors comprising them may be split up into a maximal independent set of capacitors which may be included in the tree and a remainder set of excess capacitors which may be included in the chords for the tree. Now consider the inclusion of inductors in the chord set for a network tree. The only situations precluding the inclusion of all network inductors in a chord set for a chosen tree containing a maximal independent set of capacitors arise from the existence of inductor-only cut-sets. If any such cut-sets exist, they may be split up into a maximal independent set of inductors which may be included in the chord set for the chosen tree and a remainder set of excess inductors which may be included in the tree. By treating all capacitor-only circuits and inductor-only cut-sets in this way, a tree may be chosen for any given network whose branches contain a maximal set of independent capacitors and whose chords contain a maximal set of independent inductors. It is convenient to call such a tree a *state tree*. The state tree is completed by an arbitrary selection of resistors.

State tree properties. For the network then, a state tree may be selected having the following properties.

(ST 1) The state tree contains a maximal independent set of capacitors and a minimal set of dependent inductors.

(ST 2) The chords of the state tree (i.e. the state co-tree) contain a maximal independent set of inductors and a minimal dependent set of capacitors.

(ST 3) The only type of circuit which may be formed by the insertion of a chord capacitor into a state tree is a circuit composed entirely of capacitors.

(ST 4) No circuit formed by the insertion of a chord resistor into the state tree may contain a tree inductor. If such a circuit were to contain a tree inductor, then this would imply that the state tree could have been formed using this resistor in place of an inductor. This is impossible, however, since the inductor concerned must have been one of a cut-set and thus the state tree could not have been completed without its inclusion.

Simple Examples of Formulation of State Space Equation Sets

The general procedure given below for the formulation of a set of first order differential equations describing the dynamical behaviour of a general linear electrical network is fairly involved, and too cumbersome to use for very simple networks. For certain simple practical purposes, and also to illuminate the nature of the general procedure below, we give first some examples of the procedure adopted for simple networks.

Example 5.8

Suppose we wish to describe the behaviour of the network of Figure 5–36(a) by means of a set of first order differential equations. Let the components be numbered and oriented as shown in Figure 5–36(b).

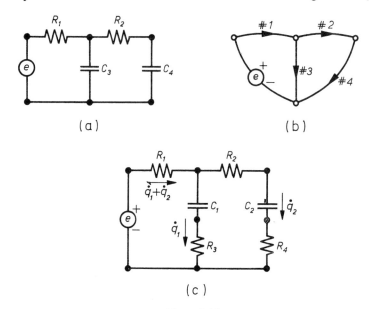

Figure 5–36

The system has two state variables: the capacitor voltages v_3 and v_4; and the system equation set is as follows.

Component constitutive equation

$$i_1 R_1 = v_1 \qquad i_2 R_2 = v_2 \qquad C_3 \frac{dv_3}{dt} = i_3 \qquad C_4 \frac{dv_4}{dt} = i_4$$

Formulation of Canonical Equation Sets for Linear Networks

Circuit equations
$$v_1 + v_3 = e \qquad -v_3 + v_2 + v_4 = 0$$

Vertex equations
$$i_1 - i_3 - i_2 = 0 \qquad i_e - i_1 = 0 \qquad i_2 - i_4 = 0$$

The state equations are then simply obtained by starting with the two capacitor component equations and expressing everything in terms of the capacitor voltages. This gives, where $G_1 = 1/R_1$ and $G_2 = 1/R_2$,

$$C_3 \frac{dv_3}{dt} = i_3 = i_1 - i_2 = G_1 v_1 - G_2 v_2 = G_1(e - v_3) - G_2(v_3 - v_4)$$

$$C_4 \frac{dv_4}{dt} = i_4 = i_2 = G_2 v_2 = G_2(v_3 - v_4)$$

Collecting terms gives, where $S_3 = 1/C_3$ and $S_4 = 1/C_4$

$$\frac{dv_3}{dt} = -S_3(G_1 + G_2)v_3 + S_3 G_2 v_4 + S_3 G_1 e$$

$$\frac{dv_4}{dt} = S_4 G_2 v_3 - S_4 G_2 v_4$$

and, expressing these in matrix form, gives

$$\frac{d}{dt}\begin{bmatrix} v_3 \\ v_4 \end{bmatrix} = \begin{bmatrix} -S_3(G_1 + G_2) & S_3 G_2 \\ S_4 G_2 & -S_4 G_2 \end{bmatrix} \begin{bmatrix} v_3 \\ v_4 \end{bmatrix} + \begin{bmatrix} S_3 G_1 e \\ 0 \end{bmatrix}$$

Example 5.9

An alternative, and often more straightforward, procedure is to write down a set of circuit or vertex equations in terms of a set of charge and/or flux-linkage derivatives and then use simple matrix manipulation techniques to get a set of first-order equations. As an illustration of this approach, consider the network of Figure 5–36(c). The circuit equations are

$$R_1(\dot{q}_1 + \dot{q}_2) + S_1 q_1 + R_3 \dot{q}_1 = e$$

$$-R_3 \dot{q}_1 - S_1 q_1 + R_2 \dot{q}_2 + S_2 q_2 + R_4 \dot{q}_2 = 0$$

Expressing these in matrix terms we have

$$\begin{bmatrix} (R_1 + R_3) & R_1 \\ -R_3 & (R_2 + R_4) \end{bmatrix} \begin{bmatrix} \dot{q}_1 \\ \dot{q}_2 \end{bmatrix} + \begin{bmatrix} S_1 & 0 \\ -S_1 & S_2 \end{bmatrix} \begin{bmatrix} q_1 \\ q_2 \end{bmatrix} = \begin{bmatrix} e \\ 0 \end{bmatrix}$$

from which we obtain the required set of first-order equations as

$$\begin{bmatrix} \dot{q}_1 \\ \dot{q}_2 \end{bmatrix} = - \begin{bmatrix} (R_1+R_3) & R_1 \\ -R_3 & (R_2+R_4) \end{bmatrix}^{-1} \begin{bmatrix} S_1 & 0 \\ -S_1 & S_2 \end{bmatrix} \begin{bmatrix} q_1 \\ q_2 \end{bmatrix}$$
$$+ \begin{bmatrix} (R_1+R_3) & R_1 \\ -R_3 & (R_2+R_4) \end{bmatrix}^{-1} \begin{bmatrix} e \\ 0 \end{bmatrix}$$

Example 5.10

Consider the mechanical network of Figure 5-37(a), having the associated oriented linear graph of Figure 5-37(b). Let v_i denote velocities and f_j denote forces with the subscripts denoting the component number as indicated on the linear graph, and let F_1 and F_2 be

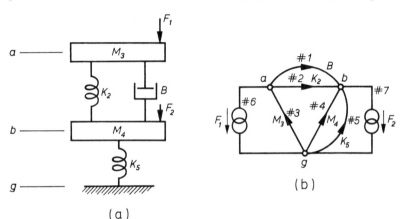

Figure 5-37

applied forces as shown in the schematic diagram. A set of state variables may be chosen as v_3, v_4, f_2 and f_5. The complete sets of system equations are as follows.

Component equations

$$M_3 \frac{dv_3}{dt} = f_3 \qquad M_4 \frac{dv_4}{dt} = f_4 \qquad f_1 = Bv_1$$

$$\frac{1}{K_2} \frac{df_2}{dt} = v_2 \qquad \frac{1}{K_5} \frac{df_5}{dt} = v_5$$

Circuit equations

$$v_3 + v_6 = 0 \qquad v_3 + v_2 - v_4 = 0 \qquad v_2 - v_1 = 0$$
$$v_4 - v_5 = 0 \qquad v_5 + v_7 = 0$$

Formulation of Canonical Equation Sets for Linear Networks 349

Vertex equations

$$f_3 - f_2 - f_1 = F_1 \qquad f_4 + f_5 + f_1 + f_2 = F_2$$

In order to obtain a set of first-order equations in the state variables, we first express f_3 and f_4 in terms of the chosen state variable set. This gives

$$f_3 = f_2 + f_1 + F_1 = f_2 + Bv_1 + F_1$$
$$= f_2 + Bv_2 + F_1$$
$$= f_2 + B(v_4 - v_3) + F_1$$
$$f_4 = -f_2 - f_1 - f_5 + F_2 = -f_2 - Bv_1 - f_5 + F_2$$
$$= -f_2 - B(v_4 - v_3) - f_5 + F_2$$

We thus have

$$M_3 \frac{dv_3}{dt} = f_2 + Bv_4 - Bv_3 + F_1$$

$$M_4 \frac{dv_4}{dt} = -f_2 - Bv_4 + Bv_3 - f_5 + F_2$$

From the spring component equations we get

$$\frac{1}{K_2} \frac{df_2}{dt} = v_2 = v_4 - v_3$$

$$\frac{1}{K_5} \frac{df_5}{dt} = v_5 = v_4$$

Collecting these and arranging in matrix form gives

$$\frac{d}{dt}\begin{bmatrix} v_3 \\ v_4 \\ f_2 \\ f_5 \end{bmatrix} = \begin{bmatrix} -\frac{B}{M_3} & \frac{B}{M_3} & \frac{1}{M_3} & 0 \\ \frac{B}{M_4} & -\frac{B}{M_4} & -\frac{1}{M_4} & -\frac{1}{M_4} \\ -K_2 & K_2 & 0 & 0 \\ 0 & K_5 & 0 & 0 \end{bmatrix} \begin{bmatrix} v_3 \\ v_4 \\ f_2 \\ f_5 \end{bmatrix} + \begin{bmatrix} \frac{F_1}{M_3} \\ \frac{F_2}{M_4} \\ 0 \\ 0 \end{bmatrix}$$

Formulation of State Equations: General Treatment [B1], [B11], [M3]

For a given network choose a state tree. Then group and number the branches in the following sequence: branch capacitors, branch resistors, and branch inductors; also group and number the chord elements in the following sequence: chord inductors, chord resistors, and chord capacitors. For this grouping and numbering of components we may partition equation (5–129) to get

$$\begin{bmatrix} \mathbf{i}_{st} \\ \mathbf{i}_{nt} \\ \mathbf{i}_{et} \\ \mathbf{v}_{sc} \\ \mathbf{v}_{nc} \\ \mathbf{v}_{ec} \end{bmatrix} = \begin{bmatrix} 0 & 0 & 0 & \mathbf{D}_{ss} & \mathbf{D}_{sn} & \mathbf{D}_{se} \\ 0 & 0 & 0 & \mathbf{D}_{ns} & \mathbf{D}_{nn} & \mathbf{D}_{a} \\ 0 & 0 & 0 & \mathbf{D}_{es} & \mathbf{D}_{b} & \mathbf{D}_{c} \\ -\mathbf{D}_{ss}^{t} & -\mathbf{D}_{ns}^{t} & -\mathbf{D}_{es}^{t} & 0 & 0 & 0 \\ -\mathbf{D}_{sn}^{t} & -\mathbf{D}_{nn}^{t} & -\mathbf{D}_{b}^{t} & 0 & 0 & 0 \\ -\mathbf{D}_{se}^{t} & -\mathbf{D}_{a}^{t} & -\mathbf{D}_{c}^{t} & 0 & 0 & 0 \end{bmatrix} \begin{bmatrix} \mathbf{v}_{st} \\ \mathbf{v}_{nt} \\ \mathbf{v}_{et} \\ \mathbf{i}_{sc} \\ \mathbf{i}_{nc} \\ \mathbf{i}_{ec} \end{bmatrix} + \begin{bmatrix} \mathbf{I}_{s} \\ \mathbf{I}_{n} \\ \mathbf{I}_{e} \\ \mathbf{E}_{s} \\ \mathbf{E}_{n} \\ \mathbf{E}_{e} \end{bmatrix} \quad (5\text{--}132)$$

where

(a) $\mathbf{i}_{st}, \mathbf{v}_{st}$ are vectors whose components are respectively the currents and voltages of the tree state passive elements, that is of the tree capacitors.

(b) $\mathbf{i}_{nt}, \mathbf{v}_{nt}$ are vectors whose components are respectively the currents and voltages of the tree non-state passive elements, that is of the tree resistors.

(c) $\mathbf{i}_{et}, \mathbf{v}_{et}$ are vectors whose components are respectively the currents and voltages of the tree excess elements, that is of the tree inductors.

(d) $\mathbf{v}_{sc}, \mathbf{i}_{sc}$ are vectors whose components are respectively the voltages and currents of the chord state elements, that is of the chord inductors.

(e) $\mathbf{v}_{nc}, \mathbf{i}_{nc}$ are vectors whose components are respectively the voltages and currents of the chord non-state elements, that is the chord resistors.

(f) $\mathbf{v}_{ec}, \mathbf{i}_{ec}$ are vectors whose components are respectively the voltages and currents of the chord excess elements, that is of the chord capacitors.

(g) $\mathbf{D}_{ss}, \mathbf{D}_{sn}, \mathbf{D}_{se}, \mathbf{D}_{ns}, \mathbf{D}_{nn}, \mathbf{D}_{es}, \mathbf{D}_{a}, \mathbf{D}_{b}$ and $\mathbf{D}_{c}$ are matrices obtained by the appropriate partitioning of the matrix $\mathbf{D}$ corresponding to a state tree.

(h) $\mathbf{I}_{s}, \mathbf{I}_{n}, \mathbf{I}_{e}$ and $\mathbf{E}_{s}, \mathbf{E}_{n}, \mathbf{E}_{e}$ are vectors obtained by the appropriate partitioning of the vectors $\mathbf{I}$ and $\mathbf{E}$ corresponding to the same state tree.

It is an immediate consequence of property (ST 3) of the state tree that the submatrices $\mathbf{D}_{a}$ and $\mathbf{D}_{c}$ must both be zero; and an immediate consequence of the property (ST 4) that the submatrix $\mathbf{D}_{b}$ must be zero. Equation (5–132) therefore gives, after multiplying out and rearranging

$$\mathbf{i}_{st} = \mathbf{D}_{ss}\mathbf{i}_{sc} + \mathbf{D}_{sn}\mathbf{i}_{nc} + \mathbf{D}_{se}\mathbf{i}_{ec} + \mathbf{I}_{s}$$

$$\mathbf{v}_{sc} = -\mathbf{D}_{ss}^{t}v_{st} - \mathbf{D}_{ns}^{t}v_{nt} - \mathbf{D}_{es}^{t}\mathbf{v}_{et} + \mathbf{E}_{s}$$

$$\mathbf{i}_{nt} = \mathbf{D}_{ns}\mathbf{i}_{sc} + \mathbf{D}_{nn}\mathbf{i}_{nc} + \mathbf{I}_{n}$$

$$\mathbf{v}_{nc} = -\mathbf{D}_{sn}^{t}v_{st} - \mathbf{D}_{nn}^{t}\mathbf{v}_{nt} + \mathbf{E}_{n}$$

$$\mathbf{i}_{et} = \mathbf{D}_{es}\mathbf{i}_{sc} + \mathbf{I}_{e}$$

$$\mathbf{v}_{ec} = -\mathbf{D}_{se}^{t}\mathbf{v}_{st} + \mathbf{E}_{e}$$

Formulation of Canonical Equation Sets for Linear Networks 351

Arranging these in matrix form we have

$$\begin{bmatrix} \mathbf{i}_{st} \\ \mathbf{v}_{sc} \end{bmatrix} = \begin{bmatrix} 0 & \mathbf{D}_{ss} \\ -\mathbf{D}_{ss}^t & 0 \end{bmatrix} \begin{bmatrix} \mathbf{v}_{st} \\ \mathbf{i}_{sc} \end{bmatrix} + \begin{bmatrix} 0 & \mathbf{D}_{sn} \\ -\mathbf{D}_{ns}^t & 0 \end{bmatrix} \begin{bmatrix} \mathbf{v}_{nt} \\ \mathbf{i}_{nc} \end{bmatrix}$$
$$+ \begin{bmatrix} 0 & \mathbf{D}_{se} \\ -\mathbf{D}_{es}^t & 0 \end{bmatrix} \begin{bmatrix} \mathbf{v}_{et} \\ \mathbf{i}_{ec} \end{bmatrix} + \begin{bmatrix} \mathbf{I}_s \\ \mathbf{E}_s \end{bmatrix} \quad (5\text{-}133)$$

$$\begin{bmatrix} \mathbf{i}_{nt} \\ \mathbf{v}_{nc} \end{bmatrix} = \begin{bmatrix} 0 & \mathbf{D}_{ns} \\ -\mathbf{D}_{sn}^t & 0 \end{bmatrix} \begin{bmatrix} \mathbf{v}_{st} \\ \mathbf{i}_{sc} \end{bmatrix} + \begin{bmatrix} 0 & \mathbf{D}_{nn} \\ -\mathbf{D}_{nn}^t & 0 \end{bmatrix} \begin{bmatrix} \mathbf{v}_{nt} \\ \mathbf{i}_{nc} \end{bmatrix} + \begin{bmatrix} \mathbf{I}_n \\ \mathbf{E}_n \end{bmatrix} \quad (5\text{-}134)$$

$$\begin{bmatrix} \mathbf{i}_{et} \\ \mathbf{v}_{ec} \end{bmatrix} = \begin{bmatrix} 0 & \mathbf{D}_{es} \\ -\mathbf{D}_{se}^t & 0 \end{bmatrix} \begin{bmatrix} \mathbf{v}_{st} \\ \mathbf{i}_{sc} \end{bmatrix} + \begin{bmatrix} \mathbf{I}_e \\ \mathbf{E}_e \end{bmatrix} \quad (5\text{-}135)$$

In order to obtain a set of first-order differential equations for the general linear electrical network from the equation set (5-133) to (5-135) we must now incorporate the appropriate element constitutive relationships in matrix form. We therefore introduce the following set of matrices.

(a) $\mathbf{C}_s$, a diagonal matrix having diagonal elements equal to the capacitances of the tree capacitors.
(b) $\mathbf{C}_e$, a diagonal matrix having diagonal elements equal to the capacitances of the chord capacitors.
(c) $\mathbf{G}$, a diagonal matrix having diagonal elements equal to the conductances of the tree resistors.
(d) $\mathbf{R}$, a diagonal matrix having diagonal elements equal to the resistances of the chord resistors.
(e) $\mathbf{L}$, a symmetrical matrix of inductance and mutual inductance values such that

$$\mathbf{v}_l = \mathbf{L} \frac{d\mathbf{i}_l}{dt} \quad (5\text{-}136)$$

where $\mathbf{v}_l, \mathbf{i}_l$ are respectively the voltages and currents of the network inductors. In equation (5-136) group, as specified above with respect to the state tree, the network inductors into branch inductors and chord inductors, and partition equation (5-136) to get

$$\begin{bmatrix} \mathbf{v}_{et} \\ \mathbf{v}_{sc} \end{bmatrix} = \begin{bmatrix} \mathbf{L}_e & \mathbf{L}_{es} \\ \mathbf{L}_{es}^t & \mathbf{L}_s \end{bmatrix} \begin{bmatrix} \dfrac{d\mathbf{i}_{et}}{dt} \\ \dfrac{d\mathbf{i}_{sc}}{dt} \end{bmatrix} \quad (5\text{-}137)$$

where the matrices $\mathbf{L}_e$, $\mathbf{L}_s$ and $\mathbf{L}_{es}$ are defined by the partitioning of the matrix $\mathbf{L}$.

Using these component matrices we proceed to obtain from equations (5-133) to (5-135) inclusive a set of first-order differential equations in $\mathbf{v}_{st}$ and $\mathbf{i}_{sc}$. This is done by expressing all the variables in equation (5-133) in terms of $\mathbf{v}_{st}$ and $\mathbf{i}_{sc}$ and the source outputs. Consider the various terms of equation (5-133) one at a time.

(a) Take the left-hand side of the equation first. We have

$$\mathbf{i}_{st} = \mathbf{C}_s \frac{d\mathbf{v}_{st}}{dt}$$

$$\mathbf{v}_{sc} = \mathbf{L}_{es}^t \frac{d\mathbf{i}_{et}}{dt} + \mathbf{L}_s \frac{d\mathbf{i}_{sc}}{dt}$$

so that

$$\begin{bmatrix} \mathbf{i}_{st} \\ \mathbf{v}_{sc} \end{bmatrix} = \begin{bmatrix} \mathbf{C}_s & 0 \\ 0 & \mathbf{L}_s \end{bmatrix} \frac{d}{dt} \begin{bmatrix} \mathbf{v}_{st} \\ \mathbf{i}_{sc} \end{bmatrix} + \begin{bmatrix} 0 \\ \mathbf{L}_{es}^t \frac{d\mathbf{i}_{et}}{dt} \end{bmatrix}$$

Differentiating the top row of equation (5-135) gives

$$\frac{d\mathbf{i}_{et}}{dt} = \mathbf{D}_{es} \frac{d\mathbf{i}_{sc}}{dt} + \frac{d\mathbf{I}_e}{dt}$$

$$\mathbf{L}_{es}^t \frac{d\mathbf{i}_{et}}{dt} = \mathbf{L}_{es}^t \mathbf{D}_{es} \frac{d\mathbf{i}_{sc}}{dt} + \mathbf{L}_{es}^t \frac{d\mathbf{I}_e}{dt}$$

Thus we have

$$\begin{bmatrix} \mathbf{i}_{st} \\ \mathbf{v}_{sc} \end{bmatrix} = \begin{bmatrix} \mathbf{C}_s & 0 \\ 0 & \mathbf{L}_s \end{bmatrix} \frac{d}{dt} \begin{bmatrix} \mathbf{v}_{st} \\ \mathbf{i}_{sc} \end{bmatrix} + \begin{bmatrix} 0 \\ \mathbf{L}_{es}^t \mathbf{D}_{es} \frac{d\mathbf{i}_{sc}}{dt} + \mathbf{L}_{es}^t \frac{d\mathbf{I}_e}{dt} \end{bmatrix}$$

$$= \begin{bmatrix} \mathbf{C}_s & 0 \\ 0 & \mathbf{L}_s + \mathbf{L}_{es}^t \mathbf{D}_{es} \end{bmatrix} \frac{d}{dt} \begin{bmatrix} \mathbf{v}_{st} \\ \mathbf{i}_{sc} \end{bmatrix} + \begin{bmatrix} 0 \\ \mathbf{L}_{es}^t \frac{d\mathbf{I}_e}{dt} \end{bmatrix}$$

(b) Now consider the second term on the right-hand side. We have

$$\begin{bmatrix} \mathbf{i}_{nt} \\ \mathbf{v}_{nc} \end{bmatrix} = \begin{bmatrix} \mathbf{G} & 0 \\ 0 & \mathbf{R} \end{bmatrix} \begin{bmatrix} \mathbf{v}_{nt} \\ \mathbf{i}_{nc} \end{bmatrix}$$

Thus equation (5-134) gives

$$\begin{bmatrix} \mathbf{G} & 0 \\ 0 & \mathbf{R} \end{bmatrix} \begin{bmatrix} \mathbf{v}_{nt} \\ \mathbf{i}_{nc} \end{bmatrix} = \begin{bmatrix} 0 & \mathbf{D}_{ns} \\ -\mathbf{D}_{sn}^t & 0 \end{bmatrix} \begin{bmatrix} \mathbf{v}_{st} \\ \mathbf{i}_{sc} \end{bmatrix} + \begin{bmatrix} 0 & \mathbf{D}_{nn} \\ -\mathbf{D}_{nn}^t & 0 \end{bmatrix} \begin{bmatrix} \mathbf{v}_{nt} \\ \mathbf{i}_{nc} \end{bmatrix} + \begin{bmatrix} \mathbf{I}_n \\ \mathbf{E}_n \end{bmatrix}$$

Formulation of Canonical Equation Sets for Linear Networks   353

so that we have

$$\begin{bmatrix} G & -D_{nn} \\ D_{nn}^t & R \end{bmatrix} \begin{bmatrix} v_{nt} \\ i_{nc} \end{bmatrix} = \begin{bmatrix} 0 & D_{ns} \\ -D_{sn}^t & 0 \end{bmatrix} \begin{bmatrix} v_{st} \\ i_{sc} \end{bmatrix} + \begin{bmatrix} I_n \\ E_n \end{bmatrix}$$

giving that

$$\begin{bmatrix} v_{nt} \\ i_{nc} \end{bmatrix} = \begin{bmatrix} G & -D_{nn} \\ D_{nn}^t & R \end{bmatrix}^{-1} \begin{bmatrix} 0 & D_{ns} \\ -D_{sn}^t & 0 \end{bmatrix} \begin{bmatrix} v_{st} \\ i_{sc} \end{bmatrix} + \begin{bmatrix} G & -D_{nn} \\ D_{nn}^t & R \end{bmatrix}^{-1} \begin{bmatrix} I_n \\ E_n \end{bmatrix}$$

provided that the matrix being inverted is non-singular. To investigate this let a and b be vectors of appropriate order and consider the quadratic form

$$[a^t b^t] \begin{bmatrix} G & -D_{nn} \\ D_{nn}^t & R \end{bmatrix} \begin{bmatrix} a \\ b \end{bmatrix} = a^t G a - a^t D_{nn} b + b^t D_{nn}^t a + b^t R b$$
$$= a^t G a + b^t R b$$

Thus the matrix concerned is positive definite and therefore non-singular since all the entries in G and R are positive for a passive network. Thus

$$\begin{bmatrix} 0 & D_{sn} \\ -D_{ns}^t & 0 \end{bmatrix} \begin{bmatrix} v_{nt} \\ i_{nc} \end{bmatrix}$$

$$= \begin{bmatrix} 0 & D_{sn} \\ -D_{ns}^t & 0 \end{bmatrix} \begin{bmatrix} G & -D_{nn} \\ D_{nn}^t & R \end{bmatrix}^{-1} \begin{bmatrix} 0 & D_{ns} \\ -D_{sn}^t & 0 \end{bmatrix} \begin{bmatrix} v_{st} \\ i_{sc} \end{bmatrix}$$

$$+ \begin{bmatrix} 0 & D_{sn} \\ -D_{ns}^t & 0 \end{bmatrix} \begin{bmatrix} G & -D_{nn} \\ D_{nn}^t & R \end{bmatrix}^{-1} \begin{bmatrix} I_n \\ E_n \end{bmatrix} \quad (5\text{-}138)$$

(c) Now consider the third term on the right-hand side. We have

$$v_{et} = L_e \frac{di_{et}}{dt} + L_{es} \frac{di_{sc}}{dt}$$

$$i_{ec} = C_e \frac{dv_{ec}}{dt}$$

so that

$$\begin{bmatrix} v_{et} \\ i_{ec} \end{bmatrix} = \begin{bmatrix} L_e & 0 \\ 0 & C_e \end{bmatrix} \frac{d}{dt} \begin{bmatrix} i_{et} \\ v_{ec} \end{bmatrix} + \begin{bmatrix} L_{es} \frac{di_{sc}}{dt} \\ 0 \end{bmatrix} \quad (5\text{-}139)$$

Differentiating equation (5-135) gives

$$\frac{d}{dt} \begin{bmatrix} i_{et} \\ v_{ec} \end{bmatrix} = \begin{bmatrix} 0 & D_{es} \\ -D_{se}^t & 0 \end{bmatrix} \frac{d}{dt} \begin{bmatrix} v_{st} \\ i_{sc} \end{bmatrix} + \frac{d}{dt} \begin{bmatrix} I_e \\ E_e \end{bmatrix}$$

and substituting this in equation (5–139) we get

$$\begin{bmatrix} \mathbf{v}_{et} \\ \mathbf{i}_{ec} \end{bmatrix} = \begin{bmatrix} \mathbf{L}_e & 0 \\ 0 & \mathbf{C}_e \end{bmatrix} \begin{bmatrix} 0 & \mathbf{D}_{es} \\ -\mathbf{D}_{se}^t & 0 \end{bmatrix} \frac{d}{dt} \begin{bmatrix} \mathbf{v}_{st} \\ \mathbf{i}_{sc} \end{bmatrix} + \begin{bmatrix} \mathbf{L}_e & 0 \\ 0 & \mathbf{C}_e \end{bmatrix} \frac{d}{dt} \begin{bmatrix} \mathbf{I}_e \\ \mathbf{E}_e \end{bmatrix}$$
$$+ \begin{bmatrix} \mathbf{L}_{es} \dfrac{d\mathbf{i}_{sc}}{dt} \\ 0 \end{bmatrix}$$

Thus

$$\begin{bmatrix} 0 & \mathbf{D}_{se} \\ -\mathbf{D}_{es}^t & 0 \end{bmatrix} \begin{bmatrix} \mathbf{v}_{et} \\ \mathbf{i}_{ec} \end{bmatrix} = \begin{bmatrix} 0 & \mathbf{D}_{se} \\ -\mathbf{D}_{es}^t & 0 \end{bmatrix} \begin{bmatrix} \mathbf{L}_e & 0 \\ 0 & \mathbf{C}_e \end{bmatrix} \begin{bmatrix} 0 & \mathbf{D}_{es} \\ -\mathbf{D}_{se}^t & 0 \end{bmatrix} \frac{d}{dt} \begin{bmatrix} \mathbf{v}_{st} \\ \mathbf{i}_{sc} \end{bmatrix}$$
$$+ \begin{bmatrix} 0 & \mathbf{D}_{se} \\ -\mathbf{D}_{es}^t & 0 \end{bmatrix} \begin{bmatrix} \mathbf{L}_e & 0 \\ 0 & \mathbf{C}_e \end{bmatrix} \frac{dt}{dt} \begin{bmatrix} \mathbf{I}_e \\ \mathbf{E}_e \end{bmatrix}$$
$$+ \begin{bmatrix} 0 & \mathbf{D}_{se} \\ -\mathbf{D}_{es}^t & 0 \end{bmatrix} \begin{bmatrix} \mathbf{L}_{es} \dfrac{d\mathbf{i}_{sc}}{dt} \\ 0 \end{bmatrix}$$

so that

$$\begin{bmatrix} 0 & \mathbf{D}_{se} \\ -\mathbf{D}_{es}^t & 0 \end{bmatrix} \begin{bmatrix} \mathbf{v}_{et} \\ \mathbf{i}_{ec} \end{bmatrix} = \begin{bmatrix} -\mathbf{D}_{se}\mathbf{C}_e\mathbf{D}_{se}^t & 0 \\ 0 & -\mathbf{D}_{es}^t\mathbf{L}_e\mathbf{D}_{es} \end{bmatrix} \frac{d}{dt} \begin{bmatrix} \mathbf{v}_{st} \\ \mathbf{i}_{sc} \end{bmatrix}$$
$$+ \begin{bmatrix} 0 & \mathbf{D}_{se}\mathbf{C}_e \\ -\mathbf{D}_{es}^t\mathbf{L}_e & 0 \end{bmatrix} \frac{d}{dt} \begin{bmatrix} \mathbf{I}_e \\ \mathbf{E}_e \end{bmatrix} + \begin{bmatrix} 0 \\ -\mathbf{D}_{es}^t\mathbf{L}_{es}\dfrac{d\mathbf{i}_{sc}}{dt} \end{bmatrix}$$

and we may put

$$\begin{bmatrix} 0 \\ -\mathbf{D}_{es}^t\mathbf{L}_{es}\dfrac{d\mathbf{i}_{sc}}{dt} \end{bmatrix} = \begin{bmatrix} 0 & 0 \\ 0 & -\mathbf{D}_{es}^t\mathbf{L}_{es} \end{bmatrix} \frac{d}{dt} \begin{bmatrix} \mathbf{v}_{st} \\ \mathbf{i}_{sc} \end{bmatrix}$$

so that

$$\begin{bmatrix} 0 & \mathbf{D}_{se} \\ -\mathbf{D}_{es}^t & 0 \end{bmatrix} \begin{bmatrix} \mathbf{v}_{et} \\ \mathbf{i}_{ec} \end{bmatrix} = \begin{bmatrix} -\mathbf{D}_{se}\mathbf{C}_e\mathbf{D}_{se}^t & 0 \\ 0 & -\mathbf{D}_{es}^t\mathbf{L}_e\mathbf{D}_{es} - \mathbf{D}_{es}^t\mathbf{L}_{es} \end{bmatrix} \frac{d}{dt} \begin{bmatrix} \mathbf{v}_{st} \\ \mathbf{i}_{sc} \end{bmatrix}$$
$$+ \begin{bmatrix} 0 & \mathbf{D}_{se}\mathbf{C}_e \\ -\mathbf{D}_{es}^t\mathbf{L}_e & 0 \end{bmatrix} \frac{d}{dt} \begin{bmatrix} \mathbf{I}_e \\ \mathbf{E}_e \end{bmatrix} \qquad (5\text{–}140)$$

Formulation of Canonical Equation Sets for Linear Networks

Combining these results, we may write equation (5-133) in the following form

$$\begin{bmatrix} C_s & 0 \\ 0 & L_s + L_{es}^t D_{es} \end{bmatrix} \frac{d}{dt} \begin{bmatrix} v_{st} \\ i_{sc} \end{bmatrix} + \begin{bmatrix} 0 & 0 \\ L_{es}^t & 0 \end{bmatrix} \frac{d}{dt} \begin{bmatrix} I_e \\ E_e \end{bmatrix}$$

$$= \begin{bmatrix} 0 & D_{ss} \\ -D_{ss}^t & 0 \end{bmatrix} \begin{bmatrix} v_{st} \\ i_{sc} \end{bmatrix} + \begin{bmatrix} 0 & D_{sn} \\ -D_{ns}^t & 0 \end{bmatrix} \begin{bmatrix} G & -D_{nn} \\ D_{nn}^t & R \end{bmatrix}^{-1}$$

$$\cdot \begin{bmatrix} 0 & D_{ns} \\ -D_{sn}^t & 0 \end{bmatrix} \begin{bmatrix} v_{st} \\ i_{sc} \end{bmatrix} + \begin{bmatrix} 0 & D_{sn} \\ -D_{ns}^t & 0 \end{bmatrix} \begin{bmatrix} G & -D_{nn} \\ D_{nn}^t & R \end{bmatrix}^{-1} \begin{bmatrix} I_n \\ E_n \end{bmatrix}$$

$$+ \begin{bmatrix} -D_{se}C_e D_{se}^t & 0 \\ 0 & -D_{es}^t L_e D_{es} - D_{es}^t L_{es} \end{bmatrix} \frac{d}{dt} \begin{bmatrix} v_{st} \\ i_{sc} \end{bmatrix}$$

$$+ \begin{bmatrix} 0 & D_{se}C_e \\ -D_{es}^t L_e & 0 \end{bmatrix} \frac{d}{dt} \begin{bmatrix} I_e \\ E_e \end{bmatrix} + \begin{bmatrix} I_s \\ E_s \end{bmatrix}$$

Thus, after collecting terms we have

$$\begin{bmatrix} C_s + D_{se}C_e D_{se}^t & 0 \\ 0 & L_s + L_{es}^t D_{es} + D_{es}^t L_{es} + D_{es}^t L_e D_{es} \end{bmatrix} \frac{d}{dt} \begin{bmatrix} v_{st} \\ i_{sc} \end{bmatrix}$$

$$= \left\{ \begin{bmatrix} 0 & D_{ss} \\ -D_{ss}^t & 0 \end{bmatrix} + \begin{bmatrix} 0 & D_{sn} \\ -D_{ns}^t & 0 \end{bmatrix} \begin{bmatrix} G & -D_{nn} \\ D_{nn}^t & R \end{bmatrix}^{-1} \right.$$

$$\left. \cdot \begin{bmatrix} 0 & D_{ns} \\ -D_{sn}^t & 0 \end{bmatrix} \right\} \begin{bmatrix} v_{st} \\ i_{sc} \end{bmatrix} + \begin{bmatrix} I_s \\ E_s \end{bmatrix} + \begin{bmatrix} 0 & D_{sn} \\ -D_{ns}^t & 0 \end{bmatrix} \begin{bmatrix} G & -D_{nn} \\ D_{nn}^t & R \end{bmatrix}^{-1}$$

$$\cdot \begin{bmatrix} I_n \\ E_n \end{bmatrix} + \begin{bmatrix} 0 & D_{se}C_e \\ -L_{es}^t - D_{es}^t L_e & 0 \end{bmatrix} \frac{d}{dt} \begin{bmatrix} I_e \\ E_e \end{bmatrix} \quad (5\text{-}141)$$

as a general form for the equations of a general linear electrical network expressed as a set of simultaneous first-order differential equations. Such a set of equations will be termed a set of *state space equations* for the network. The state space equations will normally be written in the form

$$\dot{x} = Ax + f \quad (5\text{-}142)$$

where **x** is the network state vector, **A** the network A-matrix and **f** the state-forcing vector. The form of equation (5–141) is more easily discussed if certain substitutions are made to reduce it to a more manageable form.

We have

$$\begin{bmatrix} G & -D_{nn} \\ D_{nn}^t & R \end{bmatrix}^{-1}$$

$$= \begin{bmatrix} (G+D_{nn}R^{-1}D_{nn}^t)^{-1} & G^{-1}D_{nn}(R+D_{nn}^tG^{-1}D_{nn})^{-1} \\ -R^{-1}D_{nn}^t(G+D_{nn}R^{-1}D_{nn}^t)^{-1} & (R+D_{nn}^tG^{-1}D_{nn})^{-1} \end{bmatrix} \quad (5\text{–}143)$$

where we have already shown that this matrix is non-singular. Now put

$$C^* = C_s + D_{se}C_eD_{se}^t \qquad S^* = (C^*)^{-1}$$

$$L^* = L_s + L_{es}^tD_{es} + D_{es}^tL_{es} + D_{es}^tL_eD_{es} \qquad \Gamma^* = (L^*)^{-1}$$

$$R^* = (G+D_{nn}R^{-1}D_{nn}^t)^{-1} \qquad G^* = (R+D_{nn}^tG^{-1}D_{nn})^{-1}$$

Then we see that the network A-matrix may be written in the form

$$A = \begin{bmatrix} S^* & 0 \\ 0 & \Gamma^* \end{bmatrix} \left\{ \begin{bmatrix} 0 & D_{ss} \\ -D_{ss}^t & 0 \end{bmatrix} \right.$$

$$+ \begin{bmatrix} 0 & D_{sn} \\ -D_{ns}^t & 0 \end{bmatrix} \begin{bmatrix} R^* & G^{-1}D_{nn}G^* \\ -R^{-1}D_{nn}^tR^* & G^* \end{bmatrix} \begin{bmatrix} 0 & D_{ns} \\ -D_{sn}^t & 0 \end{bmatrix} \right\}$$

$$= \begin{bmatrix} S^* & 0 \\ 0 & \Gamma^* \end{bmatrix} \begin{bmatrix} -D_{sn}G^*D_{sn}^t & D_{ss} - D_{sn}R^{-1}D_{nn}^tR^*D_{ns} \\ -D_{ss}^t + D_{ns}^tG^{-1}D_{nn}G^*D_{sn}^t & -D_{ns}^tR^*D_{ns} \end{bmatrix}$$

$$= \begin{bmatrix} -S^*D_{sn}G^*D_{sn}^t & S^*D_{ss} - S^*D_{sn}R^{-1}D_{nn}^tR^*D_{ns} \\ -\Gamma^*D_{ss}^t + \Gamma^*D_{ns}^tG^{-1}D_{nn}G^*D_{sn}^t & -\Gamma^*D_{ns}^tR^*D_{ns} \end{bmatrix} \quad (5\text{–}144)$$

A detailed comparison of the pair of matrices

$$D_{sn}R^{-1}D_{nn}^tR^*D_{ns} \quad \text{and} \quad D_{ns}^tG^{-1}D_{nn}G^*D_{sn}^t$$

shows that, for any given pair of diagonal matrices **R** and **G**, one is the transpose of the other. It therefore follows from equation (5–144) that the A-matrix for the general electrical network is of the form

$$A = \begin{bmatrix} S^* & 0 \\ 0 & \Gamma^* \end{bmatrix} \begin{bmatrix} -\tilde{G} & \tilde{D} \\ -\tilde{D}^t & -\tilde{R} \end{bmatrix} \quad (5\text{–}145)$$

where $\tilde{G}$ is a matrix whose elements have the dimensions of conduc-

Formulation of Canonical Equation Sets for Linear Networks

tance, $\tilde{\mathbf{R}}$ is a matrix whose elements have the dimensions of resistance, and $\tilde{\mathbf{D}}$ is a matrix whose elements are dimensionless.
The general form of the state-forcing function vector $\mathbf{f}$ will be

$$\mathbf{f} = \begin{bmatrix} \mathbf{S}^* & 0 \\ 0 & \mathbf{\Gamma}^* \end{bmatrix} \begin{bmatrix} \mathbf{I}_s \\ \mathbf{E}_s \end{bmatrix}$$

$$+ \begin{bmatrix} \mathbf{S}^* & 0 \\ 0 & \mathbf{\Gamma}^* \end{bmatrix} \begin{bmatrix} 0 & \mathbf{D}_{sn} \\ -\mathbf{D}_{ns}^t & 0 \end{bmatrix} \begin{bmatrix} \mathbf{R}^* & \mathbf{G}^{-1}\mathbf{D}_{nn}\mathbf{G}^* \\ -\mathbf{R}^{-1}\mathbf{D}_{nn}^t\mathbf{R}^* & \mathbf{G}^* \end{bmatrix} \begin{bmatrix} \mathbf{I}_n \\ \mathbf{E}_n \end{bmatrix}$$

$$+ \begin{bmatrix} \mathbf{S}^* & 0 \\ 0 & \mathbf{\Gamma}^* \end{bmatrix} \begin{bmatrix} 0 & \mathbf{D}_{se}\mathbf{C}_e \\ -\mathbf{L}_{es}^t - \mathbf{D}_{es}^t\mathbf{L}_e & 0 \end{bmatrix} \frac{d}{dt} \begin{bmatrix} \mathbf{I}_e \\ \mathbf{E}_e \end{bmatrix}$$

that is

$$\mathbf{f} = \begin{bmatrix} \mathbf{S}^* & 0 \\ 0 & \mathbf{\Gamma}^* \end{bmatrix} \left\{ \begin{bmatrix} \mathbf{I}_s \\ \mathbf{E}_s \end{bmatrix} + \begin{bmatrix} -\mathbf{D}_{sn}\mathbf{R}^{-1}\mathbf{D}_{nn}^t\mathbf{R}^* & \mathbf{D}_{sn}\mathbf{G}^* \\ -\mathbf{D}_{ns}^t\mathbf{R}^* & -\mathbf{D}_{ns}^t\mathbf{G}^{-1}\mathbf{D}_{nn}\mathbf{G}^* \end{bmatrix} \begin{bmatrix} \mathbf{I}_n \\ \mathbf{E}_n \end{bmatrix} \right.$$

$$\left. + \begin{bmatrix} 0 & \mathbf{D}_{se}\mathbf{C}_e \\ (-\mathbf{L}_{es}^t - \mathbf{D}_{es}^t\mathbf{L}_e) & 0 \end{bmatrix} \frac{d}{dt} \begin{bmatrix} \mathbf{I}_e \\ \mathbf{E}_e \end{bmatrix} \right\} \quad (5\text{-}146)$$

Simple Forms of State Space Equations

The state space equations of a network reduce to a particularly simple form for networks for which:

(a) there exist no circuits consisting entirely of capacitors, i.e. there are no excess capacitors after a choice of state tree and thus $\mathbf{D}_{se} = 0$;
(b) there exist no cut-sets consisting entirely of inductors, i.e. there are no excess inductors after a choice of state co-tree and thus $\mathbf{D}_{es} = 0$;
(c) the circuits formed by the insertion of chord resistors into the chosen state tree contain no tree branch resistors that is $\mathbf{D}_{nn} = 0$.

Any network which satisfies such a set of topological constraints will be termed a simple network.

For simple networks the above form of state space equations reduces to

$$\begin{bmatrix} \mathbf{C}_s & 0 \\ 0 & \mathbf{L}_s \end{bmatrix} \frac{d}{dt} \begin{bmatrix} \mathbf{v}_{st} \\ \mathbf{i}_{cs} \end{bmatrix} = \begin{bmatrix} -\mathbf{D}_{sn}\mathbf{R}^{-1}\mathbf{D}_{sn}^t & \mathbf{D}_{ss} \\ -\mathbf{D}_{ss}^t & -\mathbf{D}_{ns}^t\mathbf{G}^{-1}\mathbf{D}_{ns} \end{bmatrix} \begin{bmatrix} \mathbf{v}_{st} \\ \mathbf{i}_{sc} \end{bmatrix}$$

$$+ \begin{bmatrix} \mathbf{I}_s \\ \mathbf{E}_s \end{bmatrix} + \begin{bmatrix} 0 & \mathbf{D}_{sn}\mathbf{R}^{-1} \\ -\mathbf{D}_{ns}^t\mathbf{G}^{-1} & 0 \end{bmatrix} \begin{bmatrix} \mathbf{I}_n \\ \mathbf{E}_n \end{bmatrix}$$

Thus if we put

$$\mathbf{S} = \mathbf{C}_s^{-1} \quad \mathbf{\Gamma} = \mathbf{L}_s^{-1}$$

we may write the state space equations in the form

$$\frac{d}{dt}\begin{bmatrix}\mathbf{v}_{st}\\ \mathbf{i}_{sc}\end{bmatrix} = \begin{bmatrix}-\mathbf{SD}_{sn}\mathbf{R}^{-1}\mathbf{D}_{sn}^t & \mathbf{SD}_{ss}\\ -\mathbf{\Gamma D}_{ss}^t & -\mathbf{\Gamma D}_{ns}^t\mathbf{G}^{-1}\mathbf{D}_{ns}\end{bmatrix}\begin{bmatrix}\mathbf{v}_{st}\\ \mathbf{i}_{sc}\end{bmatrix}$$

$$+ \begin{bmatrix}\mathbf{SI}_s + \mathbf{SD}_{sn}\mathbf{R}^{-1}\mathbf{E}_n\\ \mathbf{\Gamma E}_s - \mathbf{\Gamma D}_{ns}^t\mathbf{G}^{-1}\mathbf{I}_n\end{bmatrix} \quad (5\text{–}147)$$

Alternatively we may write the state space equations in terms of the capacitor charges and inductor flux-linkages by putting

$$\mathbf{q}_{st} = \mathbf{C}_s\mathbf{v}_{st} \qquad \mathbf{\lambda}_{sc} = \mathbf{L}_s\mathbf{i}_{sc}$$

where $\mathbf{q}_{st}, \mathbf{\lambda}_{sc}$ are respectively vectors whose components are the charges of the tree capacitors and the flux-linkages of the chord inductors. This gives

$$\mathbf{v}_{st} = \mathbf{S}\mathbf{q}_{st} \qquad \mathbf{i}_{sc} = \mathbf{\Gamma}\mathbf{\lambda}_{sc}$$

so that, substituting these in equation (5–147) we get

$$\frac{d}{dt}\begin{bmatrix}\mathbf{S}\mathbf{q}_{st}\\ \mathbf{\Gamma}\mathbf{\lambda}_{sc}\end{bmatrix} = \begin{bmatrix}-\mathbf{SD}_{sn}\mathbf{R}^{-1}\mathbf{D}_{sn}^t & \mathbf{SD}_{ss}\\ -\mathbf{\Gamma D}_{ss}^t & -\mathbf{\Gamma D}_{ns}^t\mathbf{G}^{-1}\mathbf{D}_{ns}\end{bmatrix}\begin{bmatrix}\mathbf{S}\mathbf{q}_{st}\\ \mathbf{\Gamma}\mathbf{\lambda}_{sc}\end{bmatrix}$$

$$+ \begin{bmatrix}\mathbf{SI}_s + \mathbf{SD}_{sn}\mathbf{R}^{-1}\mathbf{E}_n\\ \mathbf{\Gamma E}_s - \mathbf{\Gamma D}_{ns}^t\mathbf{G}^{-1}\mathbf{I}_n\end{bmatrix}$$

whence

$$\frac{d}{dt}\begin{bmatrix}\mathbf{q}_{st}\\ \mathbf{\lambda}_{sc}\end{bmatrix} = \begin{bmatrix}-\mathbf{D}_{sn}\mathbf{R}^{-1}\mathbf{D}_{sn}^t\mathbf{S} & \mathbf{D}_{ss}\mathbf{\Gamma}\\ -\mathbf{D}_{ss}^t\mathbf{S} & -\mathbf{D}_{ns}^t\mathbf{G}^{-1}\mathbf{D}_{ns}\mathbf{\Gamma}\end{bmatrix}\begin{bmatrix}\mathbf{q}_{st}\\ \mathbf{\lambda}_{sc}\end{bmatrix}$$

$$+ \begin{bmatrix}\mathbf{I}_s + \mathbf{D}_{sn}\mathbf{R}^{-1}\mathbf{E}_n\\ \mathbf{E}_s - \mathbf{D}_{ns}^t\mathbf{G}^{-1}\mathbf{I}_n\end{bmatrix} \quad (5\text{–}148)$$

Examples of Formation of Canonical Equation Sets for Linear Networks
We will now illustrate the above relationships by a few examples.

Example 5.11
First consider the network of Figure 5–38(a) with the associated oriented linear graph of Figure 5–38(b). In drawing the linear graph, the voltage source e has been associated with the inductor L. The network dynamical transformation matrix $\mathbf{D}$ is formed by drawing up a table with entries:

	Chord 1	Chord 2
Tree-branch 1	+1	−1
Tree-branch 2	+1	0

Formulation of Canonical Equation Sets for Linear Networks 359

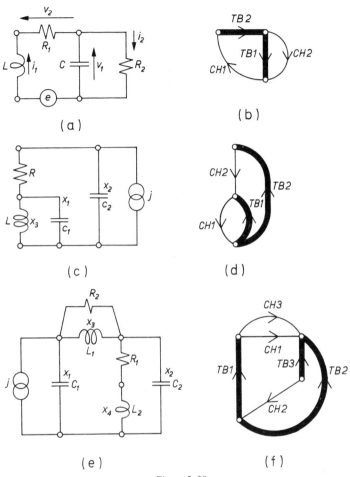

Figure 5-38

giving
$$\mathbf{D} = \begin{bmatrix} +1 & -1 \\ +1 & 0 \end{bmatrix}$$

The voltage vector **E** is given by

$$\mathbf{E} = \mathbf{e}_c + \mathbf{D}^t \mathbf{e}_t \quad \text{and} \quad \mathbf{e}_t = 0 \quad \text{and} \quad \mathbf{e}_c = \begin{bmatrix} e \\ 0 \end{bmatrix}$$

Thus
$$\mathbf{E} = \begin{bmatrix} e \\ 0 \end{bmatrix}$$

We therefore obtain from equation (5–129), for the variable numbering and orientations shown on the diagram:

$$\begin{bmatrix} C\dfrac{dv_1}{dt} \\ G_1 v_2 \\ L\dfrac{di_1}{dt} \\ R_2 i_2 \end{bmatrix} = \begin{bmatrix} 0 & 0 & 1 & -1 \\ 0 & 0 & 1 & 0 \\ -1 & -1 & 0 & 0 \\ 1 & 0 & 0 & 0 \end{bmatrix} \begin{bmatrix} v_1 \\ v_2 \\ i_1 \\ i_2 \end{bmatrix} + \begin{bmatrix} 0 \\ 0 \\ e \\ 0 \end{bmatrix}$$

where $$G_1 = \dfrac{1}{R_1}$$

Eliminating the non-state variables v_2 and i_2 then gives

$$\begin{bmatrix} \dfrac{dv_1}{dt} \\ \dfrac{di_1}{dt} \end{bmatrix} = \begin{bmatrix} -SG_2 & +S \\ -\Gamma & -\Gamma R_1 \end{bmatrix} \begin{bmatrix} v_1 \\ i_1 \end{bmatrix} + \begin{bmatrix} 0 \\ \Gamma e \end{bmatrix}$$

where $$S = \dfrac{1}{C}, \quad \Gamma = \dfrac{1}{L} \quad \text{and} \quad G_2 = \dfrac{1}{R_2}$$

Example 5.12

For the network of Figure 5–38(c) we have the oriented linear graph of Figure 5–38(d) for which the current source j is associated with capacitor C_2. This numbering and orientation of chords and tree-branches gives

$$\mathbf{D} = \begin{bmatrix} +1 & \vdots & -1 \\ 0 & \vdots & +1 \end{bmatrix} \equiv [\mathbf{D}_{ss} \ \vdots \ \mathbf{D}_{sn}]$$

The current source vector $\mathbf{I}$ is given by

$$\mathbf{I} = \mathbf{j}_t - \mathbf{D}\mathbf{j}_c \quad \text{where} \quad \mathbf{j}_t = \begin{bmatrix} 0 \\ j \end{bmatrix} \quad \text{and} \quad \mathbf{j}_c = \mathbf{0}$$

so that $$\mathbf{I} = \begin{bmatrix} 0 \\ j \end{bmatrix} \equiv \mathbf{I}_s$$

Formulation of Canonical Equation Sets for Linear Networks

Take as the network state variables

$$x_1 = \text{voltage across } C_1$$
$$x_2 = \text{voltage across } C_2$$
$$x_3 = \text{current through } L$$

Then, using equation (5–147) to form the canonical equation set, we have

$$-\mathbf{SD}_{sn}\mathbf{R}^{-1}\mathbf{D}_{sn}^t = -\begin{bmatrix} S_1 & 0 \\ 0 & S_2 \end{bmatrix}\begin{bmatrix} -1 \\ 1 \end{bmatrix}[G][-1 \quad 1] = \begin{bmatrix} -S_1 G & S_1 G \\ S_2 G & -S_2 G \end{bmatrix}$$

$$\mathbf{SD}_{ss} = \begin{bmatrix} S_1 & 0 \\ 0 & S_2 \end{bmatrix}\begin{bmatrix} 1 \\ 0 \end{bmatrix} = \begin{bmatrix} S_1 \\ 0 \end{bmatrix}$$

$$-\mathbf{\Gamma D}_{ss}^t = [-\Gamma][1 \quad 0] = [-\Gamma \quad 0]$$

$$\mathbf{SI}_s = \begin{bmatrix} S_1 & 0 \\ 0 & S_2 \end{bmatrix}\begin{bmatrix} 0 \\ j \end{bmatrix} = \begin{bmatrix} 0 \\ S_2 j \end{bmatrix}$$

where

$$S_1 = \frac{1}{C_1}, \quad S_2 = \frac{1}{C_2}, \quad G = \frac{1}{R} \quad \text{and} \quad \Gamma = \frac{1}{L}$$

Inserting these in the formula of equation (5–147) we have

$$\frac{d}{dt}\begin{bmatrix} x_1 \\ x_2 \\ x_3 \end{bmatrix} = \begin{bmatrix} -S_1 G & S_1 G & S_1 \\ S_2 G & -S_2 G & 0 \\ -\Gamma & 0 & 0 \end{bmatrix}\begin{bmatrix} x_1 \\ x_2 \\ x_3 \end{bmatrix} + \begin{bmatrix} 0 \\ S_2 j \\ 0 \end{bmatrix}$$

Example 5.13

The network of Figure 5–38(e) has the oriented linear graph of Figure 5–38(f) for which the current source j is associated with the capacitor C_1. The numbering and orientation of chords and tree-branches shown gives

$$\mathbf{D} = \begin{bmatrix} 1 & 0 & 1 \\ -1 & 1 & -1 \\ 0 & -1 & 0 \end{bmatrix} \equiv \begin{bmatrix} \mathbf{D}_{ss} & \mathbf{D}_{sn} \\ \mathbf{D}_{ns} & 0 \end{bmatrix}$$

The current source vector $\mathbf{I}$ is given by

$$\mathbf{I} = \mathbf{j}_t - \mathbf{D}\mathbf{j}_c \quad \text{with} \quad \mathbf{j}_t = \begin{bmatrix} j \\ 0 \\ 0 \end{bmatrix} \quad \text{and} \quad \mathbf{j}_c = 0$$

so that
$$\mathbf{I} = \begin{bmatrix} j \\ 0 \\ 0 \end{bmatrix} \quad \mathbf{I}_s = \begin{bmatrix} j \\ 0 \end{bmatrix}$$

Take as the network state variables

$x_1 = $ voltage across C_1

$x_2 = $ voltage across C_2

$x_3 = $ current through L_1

$x_4 = $ current through L_2

Again using equation (5–147) to form the network canonical equation set we have

$$-\mathbf{SD}_{sn}\mathbf{R}^{-1}\mathbf{D}_{sn}^t = -\begin{bmatrix} S_1 & 0 \\ 0 & S_2 \end{bmatrix}\begin{bmatrix} 1 \\ -1 \end{bmatrix}[G_2][1 \quad -1]$$

$$= \begin{bmatrix} -S_1 G_2 & S_1 G_2 \\ S_2 G_2 & -S_2 G_2 \end{bmatrix}$$

$$\mathbf{SD}_{ss} = \begin{bmatrix} S_1 & 0 \\ 0 & S_2 \end{bmatrix}\begin{bmatrix} 1 & 0 \\ -1 & 1 \end{bmatrix} = \begin{bmatrix} S_1 & 0 \\ -S_2 & S_2 \end{bmatrix}$$

$$-\mathbf{\Gamma D}_{ss}^t = -\begin{bmatrix} \Gamma_1 & 0 \\ 0 & \Gamma_2 \end{bmatrix}\begin{bmatrix} 1 & -1 \\ 0 & 1 \end{bmatrix} = \begin{bmatrix} -\Gamma_1 & \Gamma_1 \\ 0 & -\Gamma_2 \end{bmatrix}$$

$$-\mathbf{\Gamma D}_{ns}^t\mathbf{G}^{-1}\mathbf{D}_{ns} = -\begin{bmatrix} \Gamma_1 & 0 \\ 0 & \Gamma_2 \end{bmatrix}\begin{bmatrix} 0 \\ -1 \end{bmatrix}[R_1][0 \quad -1] = \begin{bmatrix} 0 & 0 \\ 0 & -\Gamma_2 R_1 \end{bmatrix}$$

$$\mathbf{SI}_s = \begin{bmatrix} S_1 & 0 \\ 0 & S_2 \end{bmatrix}\begin{bmatrix} j \\ 0 \end{bmatrix} = \begin{bmatrix} S_1 j \\ 0 \end{bmatrix}$$

Using these in equation (5–147) we obtain the network canonical equation set as

$$\frac{d}{dt}\begin{bmatrix} x_1 \\ x_2 \\ x_3 \\ x_4 \end{bmatrix} = \begin{bmatrix} -S_1 G_2 & S_1 G_2 & S_1 & 0 \\ S_2 G_2 & -S_2 G_2 & -S_2 & S_2 \\ -\Gamma_1 & \Gamma_1 & 0 & 0 \\ 0 & -\Gamma_2 & 0 & -\Gamma_2 R_1 \end{bmatrix}\begin{bmatrix} x_1 \\ x_2 \\ x_3 \\ x_4 \end{bmatrix} + \begin{bmatrix} S_1 j \\ 0 \\ 0 \\ 0 \end{bmatrix}$$

where

$$S_1 = \frac{1}{C_1}, \quad S_2 = \frac{1}{C_2}, \quad \Gamma_1 = \frac{1}{L_1}, \quad \Gamma_2 = \frac{1}{L_2}, \quad G_2 = \frac{1}{R_2}$$

5.13 Formulation of State Space Equations for Nonlinear Networks

In what follows, resistors and sources are regarded as essentially the same kind of component since both are specified by voltage-current constitutive characteristics; for both the term *converter* will be used [B6].

For a given network choose a state tree. Then group and number the tree components in the following sequence: tree capacitors, tree converters and tree inductors. Also group and number the chord elements in the following sequence: chord inductors, chord converters and chord capacitors. For this grouping and numbering of components, equations (5–22) and (5–23) may be combined and partitioned to give

$$\begin{bmatrix} \mathbf{i}_{st} \\ \mathbf{i}_{nt} \\ \mathbf{i}_{et} \\ \mathbf{v}_{sc} \\ \mathbf{v}_{nc} \\ \mathbf{v}_{ec} \end{bmatrix} = \begin{bmatrix} 0 & 0 & 0 & \mathbf{D}_{ss} & \mathbf{D}_{sn} & \mathbf{D}_{se} \\ 0 & 0 & 0 & \mathbf{D}_{ns} & \mathbf{D}_{nn} & \mathbf{D}_{a} \\ 0 & 0 & 0 & \mathbf{D}_{es} & \mathbf{D}_{b} & \mathbf{D}_{c} \\ -\mathbf{D}_{ss}^{t} & -\mathbf{D}_{ns}^{t} & -\mathbf{D}_{es}^{t} & 0 & 0 & 0 \\ -\mathbf{D}_{sn}^{t} & -\mathbf{D}_{nn}^{t} & -\mathbf{D}_{b}^{t} & 0 & 0 & 0 \\ -\mathbf{D}_{se}^{t} & -\mathbf{D}_{a}^{t} & -\mathbf{D}_{c}^{t} & 0 & 0 & 0 \end{bmatrix} \begin{bmatrix} \mathbf{v}_{st} \\ \mathbf{v}_{nt} \\ \mathbf{v}_{et} \\ \mathbf{i}_{sc} \\ \mathbf{i}_{nc} \\ \mathbf{i}_{ec} \end{bmatrix} \quad (5\text{–}149)$$

where

(a) $\mathbf{i}_{st}, \mathbf{v}_{st}$ are vectors whose components are respectively the currents and voltages of the tree state passive elements, that is the tree capacitors.

(b) $\mathbf{i}_{nt}, \mathbf{v}_{nt}$ are vectors whose components are respectively the currents and voltages of the tree non-state elements, that is the tree converters.

(c) $\mathbf{i}_{et}, \mathbf{v}_{et}$ are vectors whose components are respectively the currents and voltages of the tree excess elements, that is the tree inductors.

(d) $\mathbf{v}_{sc}, \mathbf{i}_{sc}$ are vectors whose components are respectively the voltages and currents of the chord state elements, that is the chord inductors.

(e) $\mathbf{v}_{nc}, \mathbf{i}_{nc}$ are vectors whose components are respectively the voltages and currents of the chord non-state elements, that is the chord converters.

(f) $\mathbf{v}_{ec}, \mathbf{i}_{ec}$ are vectors whose components are respectively the voltages and currents of the chord excess elements, that is the chord capacitors.

(g) $\mathbf{D}_{ss}, \mathbf{D}_{sn}, \mathbf{D}_{se}, \mathbf{D}_{ns}, \mathbf{D}_{nn}, \mathbf{D}_{es}, \mathbf{D}_{a}, \mathbf{D}_{b}$ and $\mathbf{D}_{c}$ are matrices obtained by the appropriate partitioning of the matrix $\mathbf{D}$ corresponding to a state tree.

It is an immediate consequence of property (ST 3) of the state tree (given in Section 5.12) that the sub-matrices $\mathbf{D}_{a}$ and $\mathbf{D}_{c}$ must both be zero; and an immediate consequence of property (ST 4) (given in Section 5.12) that the sub-matrix $\mathbf{D}_{b}$ must also be zero. Equation (5–149)

Network Models

therefore gives, after multiplying out and rearranging

$$\mathbf{i}_{st} - \mathbf{D}_{se}\mathbf{i}_{ec} = \mathbf{D}_{ss}\mathbf{i}_{sc} + \mathbf{D}_{sn}\mathbf{i}_{nc} \qquad (5\text{-}150)$$

$$\mathbf{i}_{nt} = \mathbf{D}_{ns}\mathbf{i}_{sc} + \mathbf{D}_{nn}\mathbf{i}_{nc} \qquad (5\text{-}151)$$

$$\mathbf{i}_{et} = \mathbf{D}_{es}\mathbf{i}_{sc} \qquad (5\text{-}152)$$

$$\mathbf{v}_{sc} + \mathbf{D}_{es}^t \mathbf{v}_{et} = -\mathbf{D}_{ss}^t \mathbf{v}_{st} - \mathbf{D}_{ns}^t \mathbf{v}_{nt} \qquad (5\text{-}153)$$

$$\mathbf{v}_{nc} = -\mathbf{D}_{sn}^t \mathbf{v}_{st} - \mathbf{D}_{nn}^t \mathbf{v}_{nt} \qquad (5\text{-}154)$$

$$\mathbf{v}_{ec} = -\mathbf{D}_{se}^t \mathbf{v}_{st} \qquad (5\text{-}155)$$

Now let:

(a) $\mathbf{q}_{st}$ be a vector whose components are the state tree capacitor charges.
(b) $\mathbf{q}_{ec}$ be a vector whose components are the excess capacitor charges.
(c) $\boldsymbol{\lambda}_{sc}$ be a vector whose components are the chord inductor flux-linkages.
(d) $\boldsymbol{\lambda}_{et}$ be a vector whose components are the tree inductor flux-linkages.

Then we may choose as a set of network state variables the components of a pair of vectors $\mathbf{q}$ and $\boldsymbol{\lambda}$ where

$$\mathbf{q} \triangleq \mathbf{q}_{st} - \mathbf{D}_{se}\mathbf{q}_{ec} \qquad (5\text{-}156)$$

$$\boldsymbol{\lambda} \triangleq \boldsymbol{\lambda}_{sc} + \mathbf{D}_{es}^t \boldsymbol{\lambda}_{et} \qquad (5\text{-}157)$$

To show that these are indeed a set of network state variables requires that we show that all the network variables are determined by a set of differential equations in the components of vectors $\mathbf{q}$ and $\boldsymbol{\lambda}$. Using equations (5-156) and (5-157) in equations (5-150) and (5-153) gives

$$\frac{d\mathbf{q}}{dt} = \mathbf{D}_{ss}\mathbf{i}_{sc} + \mathbf{D}_{sn}\mathbf{i}_{nc} \qquad (5\text{-}158)$$

$$\frac{d\boldsymbol{\lambda}}{dt} = -\mathbf{D}_{ss}^t \mathbf{v}_{st} - \mathbf{D}_{ns}^t \mathbf{v}_{nt} \qquad (5\text{-}159)$$

Equations (5-158), (5-159), (5-151), (5-152), (5-154) and (5-155) may be combined with the appropriate component relationships to give the signal-flow graph scheme of causal dependence shown in Figure 5-39. This clearly shows that the components of $\mathbf{q}$ and $\boldsymbol{\lambda}$ form a set of state variables for a nonlinear electrical network provided that a suitable set of restrictions are placed on the component relationships to ensure a unique solution of the network differential equation set.

Formulation of State Space Equations for Nonlinear Networks 365

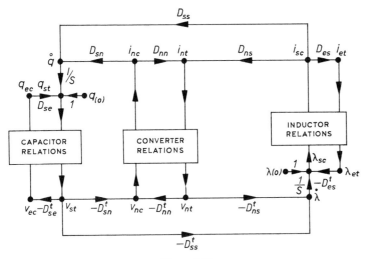

Figure 5-39

An inspection of the network relationships summarized by Figure 5-39 shows that the network equations will have a unique solution if the component characteristics satisfy a set of Lipschitz conditions (discussed in Chapter 6) and in addition:

(a) The voltages of the state tree capacitors are single-valued functions of their charges.
(b) The charges of the chord capacitors for the state tree are single-valued functions of their voltages.
(c) The currents of any non-mutually-coupled state tree inductors are single-valued functions of their flux-linkages.
(d) The flux-linkages of any non-mutually-coupled state tree inductors are singled-valued functions of their currents.
(e) The voltages of the tree converters are single-valued functions of their currents.
(f) The currents of the chord converters are single-valued functions of their voltages.
(g) Any sets of mutually-coupled inductors are such that the currents of those incorporated in the chords of the state tree are single-valued functions of the chord inductor flux-linkages and the state tree inductor currents, and such that the flux-linkages of those incorporated in the state tree are single-valued functions of the currents of the tree inductors and the flux-linkages of the chord inductors.

This discussion shows how we may select a computing structure for the solution of the state space equations governing the behaviour of a given network. In this configuration the source outputs may be time-dependent. In what follows we derive a canonical form of equations which can be used for analytical investigations but which only holds for time-invariant source outputs.

Component Relationships

The nonlinear network components are taken to be defined by constitutive characteristics of the types shown in Figures 5–40 to 5–44 inclusive. Capacitors, inductors and resistors will be assumed passive,

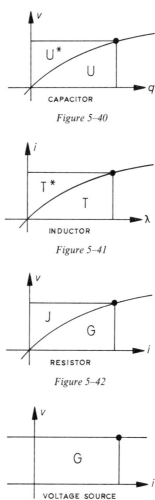

Figure 5–40

Figure 5–41

Figure 5–42

Figure 5–43

Formulation of State Space Equations for Nonlinear Networks 367

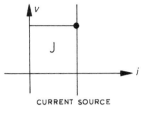

CURRENT SOURCE

Figure 5-44

their constitutive characteristics passing through the origin. Any non-passive component, whose constitutive characteristic does not pass through the origin, may be replaced by a suitable combination of passive components and ideal source.

Content and Co-content for General Components

Let **v** and **i** be vectors whose entries are the voltages and currents of the components of a general nonlinear electrical network. The total network power is then given by the scalar product

$$P = \langle \mathbf{i}, \mathbf{v} \rangle \tag{5-160}$$

Thus for voltage and current variations given by **dv** and **di** respectively, the corresponding differential in power is

$$\begin{aligned} dP &= \langle \mathbf{i}, \mathbf{dv} \rangle + \langle \mathbf{v}, \mathbf{di} \rangle \\ &= dG + dJ \end{aligned} \tag{5-161}$$

where we may call dG and dJ the differentials in total network content and co-content respectively. For any network component, the changes in content and co-content associated with a change of voltage of v_1 to v_2, and a corresponding change of current of i_1 to i_2, are given by the integrals

$$G(i) = G_2 - G_1 = \int_{i_1}^{i_2} v\, di \tag{5-162}$$

$$J(v) = J_2 - J_1 = \int_{v_1}^{v_2} i\, dv \tag{5-163}$$

For converters, these quantities may be interpreted in terms of areas on constitutive characteristics. For inductors and capacitors no such direct interpretation arises; this is of no consequence however since for such components we are only really concerned with changes in content and co-content, and not with absolute values.

Total Content and Total Co-content as Line Integrals in State Space
Consider the integrals

$$G_t = \int_\Gamma \langle \mathbf{v}, \mathbf{di} \rangle = \int_\Gamma dG \qquad (5\text{-}164)$$

and

$$J_t = \int_\Gamma \langle \mathbf{i}, \mathbf{dv} \rangle = \int_\Gamma dJ \qquad (5\text{-}165)$$

where Γ is a specified curvilinear arc in the network state space, and $\mathbf{v}$ and $\mathbf{i}$ are vectors whose entries are the voltages and currents of all the network components. All the network component voltages may be expressed in terms of the voltages and currents of the state components or are known explicitly (if a source is involved). The line integrals (5-164) and (5-165) are thus well-defined along any specified curvilinear arc Γ for any network of time-invariant components driven by constant-voltage and constant-current sources.

The network may be considered as a collection of components joined to a box inside which all the connections are made. These connections are a set of *workless constraints* on the component variables. The components of any *admissible differential* vectors $\mathbf{di}$ and $\mathbf{dv}$ must be such that, for any given choice of tree

$$\mathbf{di}_t = \mathbf{D}\mathbf{di}_c \qquad (5\text{-}166)$$

$$\mathbf{dv}_c = -\mathbf{D}^t \mathbf{dv}_t \qquad (5\text{-}167)$$

where

(a) $\mathbf{di}_t$ is a vector whose entries are the differentials in the currents of the tree components.
(b) $\mathbf{di}_c$ is a vector whose entries are the differentials in the currents of the chord components.
(c) $\mathbf{dv}_c$ is a vector whose entries are the differentials in the voltages of the chord components.
(d) $\mathbf{dv}_t$ is a vector whose entries are the differentials in the voltages of the tree components.

In computing the line integrals (5-164) and (5-165) for the interconnected network, the differential constraints specified by equations (5-166) and (5-167) must be satisfied. We therefore have

$$G_t = \int_\Gamma \langle \mathbf{v}, \mathbf{di} \rangle$$

$$= \int_\Gamma \langle \mathbf{v}_c, \mathbf{di}_c \rangle + \int_\Gamma \langle \mathbf{v}_t, \mathbf{di}_t \rangle \qquad (5\text{-}168)$$

Formulation of State Space Equations for Nonlinear Networks

Using equations (5-23) and (5-166) in equation (5-168) gives

$$G_t = \int_\Gamma \langle -\mathbf{D}^t\mathbf{v}_t, \mathbf{di}_c \rangle + \int_\Gamma \langle \mathbf{v}_t, \mathbf{Ddi}_c \rangle$$

$$= -\int_\Gamma \langle \mathbf{v}_t, \mathbf{Ddi}_c \rangle + \int_\Gamma \langle \mathbf{v}_t, \mathbf{Ddi}_c \rangle$$

$$= 0 \tag{5-169}$$

Again
$$J_t = \int_\Gamma \langle \mathbf{i}, \mathbf{dv} \rangle$$

$$= \int_\Gamma \langle \mathbf{i}_t, \mathbf{dv}_t \rangle + \int_\Gamma \langle \mathbf{i}_c, \mathbf{dv}_c \rangle \tag{5-170}$$

Using equations (5-22) and (5-167) in equation (5-170) gives

$$J_t = \int_\Gamma \langle \mathbf{Di}_c, \mathbf{dv}_t \rangle + \int_\Gamma \langle \mathbf{i}_c, -\mathbf{D}^t\mathbf{dv}_t \rangle$$

$$= \int_\Gamma \langle \mathbf{Di}_c, \mathbf{dv}_t \rangle - \int_\Gamma \langle \mathbf{Di}_c, \mathbf{dv}_t \rangle$$

$$= 0 \tag{5-171}$$

It follows from the above argument that the total network content and co-content, defined by integrals (5-164) and (5-165) respectively, are zero when evaluated along any curvilinear arc in the state space.

5.13.1 Integral Invariants [W2]

The dynamical behaviour of a nonlinear electrical network may be described by a set of nonlinear *state space equations*

$$\frac{dx_i}{dt} = \phi_i(x_1, x_2, \ldots, x_n; t) + f_i$$

$$i = 1, 2, \ldots, n \tag{5-172}$$

where the variables $x_1, x_2, \ldots, x_n$ are the network state variables, and the variables $f_1, \ldots, f_n$ represent the effect of inputs. Any set of first-order equations having a particularly useful or simple structure is called a *canonical* set of equations. The network state variables may be taken as the coordinates of a vector referred to the standard basis in a Euclidean space of n dimensions, called the network state space, and the network solution discussed in vector terms. Each solution of the equation set (5-172) determines a trajectory for a point in the state space; provided that the functions ϕ_i and f_i satisfy certain conditions

(Lipschitz conditions), the equation set will possess a unique solution through every point of the state space. The totality of solutions then specifies a *flow* in the state space analogous to the flow of a compressible fluid in a physical space.

Consider a set of points occupying a p-dimensional region Γ' of the state space at some time t'. Through each point passes a trajectory of the flow so that, at some subsequent time t, they will have moved with the flow to some other p-dimensional region Γ. A p-tuple integral over Γ is said to be an integral invariant of the system if it has the same value for all time t; p is called the order of the integral invariant. In what follows we are concerned with integral invariants of order one, that is with line integrals invariant under the state space flow. The relationships involved are illustrated in Figure 5-45 for a second-order system.

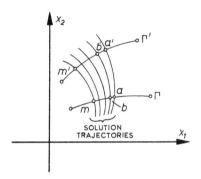

Figure 5-45

Γ' and Γ are curvilinear arcs in the state space through which pass a set of solution trajectories such that a set of points $a', b', \ldots, m'$ on Γ' at time t' is carried under the flow to the set of points $a, b, \ldots, m$ on Γ at time t. Let some differential form

$$M_1 \, dx_1 + M_2 \, dx_2 + \cdots + M_n \, dx_n$$

be defined on the state state space. This enables us to evaluate the two line integrals

$$I_1 = \int_{\Gamma'} \sum_k M_k \, dx_k$$

$$I_2 = \int_{\Gamma} \sum_k M_k \, dx_k$$

If the flow carries Γ' into Γ and these two line integrals are equal, the corresponding line integral is an integral invariant of the motion. It follows immediately from the arguments above leading to equations (5–169) and (5–171), that both the total network content and the total network co-content are integral invariants of the network state space equation set.

5.13.2 Derivation of Canonical Equation Set

We first consider the case when no excess components are present.

Case When no Excess Components Present

Group the network components into converters, inductors and capacitors and let the vectors associated with the various types of component be identified by the following subscripts

$$\rho = \text{converter}$$
$$\lambda = \text{inductor}$$
$$\gamma = \text{capacitor}$$

For any curvilinear arc Γ in the state space we have the following integral invariants of the flow

$$\int_\Gamma dG = 0 \qquad (5\text{–}173)$$

$$\int_\Gamma dJ = 0 \qquad (5\text{–}174)$$

which holds for any network of time-invariant components driven by constant-voltage and constant-current sources.

First consider equation (5–173). Splitting up the components into their respective types, we have

$$\int_\Gamma \langle \mathbf{v}_\rho, \mathbf{di}_\rho \rangle + \int_\Gamma \langle \mathbf{v}_\lambda, \mathbf{di}_\lambda \rangle + \int_\Gamma \langle \mathbf{v}_\gamma, \mathbf{di}_\gamma \rangle = 0 \qquad (5\text{–}175)$$

By the normal rules for product differentiation

$$d\langle \mathbf{v}_\gamma, \mathbf{i}_\gamma \rangle = \langle \mathbf{v}_\gamma, \mathbf{di}_\gamma \rangle + \langle \mathbf{i}_\gamma, \mathbf{dv}_\gamma \rangle$$

so that

$$\int_\Gamma \langle \mathbf{v}_\gamma, \mathbf{di}_\gamma \rangle = \langle \mathbf{v}_\gamma, \mathbf{i}_\gamma \rangle|_\Gamma - \int_\Gamma \langle \mathbf{i}_\gamma, \mathbf{dv}_\gamma \rangle \qquad (5\text{–}176)$$

Substituting equation (5–176) into equation (5–175) gives

$$\int_\Gamma \langle \mathbf{i}_\gamma, \mathbf{dv}_\gamma \rangle - \int_\Gamma \langle \mathbf{v}_\lambda, \mathbf{di}_\lambda \rangle = \langle \mathbf{v}_\gamma, \mathbf{i}_\gamma \rangle|_\Gamma + \int_\Gamma \langle \mathbf{v}_\rho, \mathbf{di}_\rho \rangle$$

$$= \langle \mathbf{v}_\gamma, \mathbf{i}_\gamma \rangle|_\Gamma + G_{t\rho} \qquad (5\text{–}177)$$

where

(a) $\langle \mathbf{v}_\gamma, \mathbf{i}_\gamma \rangle|_\Gamma$ denotes the difference in the values of $\langle \mathbf{v}_\gamma, \mathbf{i}_\gamma \rangle$ evaluated at the end points of Γ.
(b) G_{tp} denotes the difference in the total converter content evaluated at the end points of Γ. G_{tp} depends only on the end-points since *either* the voltage *or* the current of a converter is fixed by the state variable set and the other variable then fixed by the constitutive characteristic of the converter. Thus, whatever variation of state variables is involved in Γ, the converter constitutive characteristics are always traced out.

Since all the converter variables are determined by the state variables together with the converter constitutive characteristic, it follows that the expression on the right-hand side of equation (5–177) is a scalar function of $\mathbf{v}_\gamma$ and $\mathbf{i}_\lambda$ which we may denote by $Q(\mathbf{v}_\gamma, \mathbf{i}_\lambda)$. This gives

$$\int_\Gamma \langle \mathbf{i}_\gamma, d\mathbf{v}_\gamma \rangle - \int_\Gamma \langle \mathbf{v}_\lambda, d\mathbf{i}_\lambda \rangle = Q(\mathbf{v}_\gamma, \mathbf{i}_\lambda) \qquad (5\text{–}178)$$

from which we conclude that the left-hand side line integrals depend only on the initial and final points of Γ. Differentiating equation (5–178) we then obtain

$$\mathbf{i}_\gamma = \left(\frac{\partial Q}{\partial \mathbf{v}_\gamma}\right)^t \qquad (5\text{–}179)$$

$$\mathbf{v}_\lambda = -\left(\frac{\partial Q}{\partial \mathbf{i}_\lambda}\right)^t \qquad (5\text{–}180)$$

If we let E^* denote the total co-energy stored in the inductors and capacitors of the network we have

$$\mathbf{q}_\gamma = \left(\frac{\partial E^*}{\partial \mathbf{v}_\gamma}\right)^t \qquad (5\text{–}181)$$

$$\boldsymbol{\lambda}_\lambda = \left(\frac{\partial E^*}{\partial \mathbf{i}_\lambda}\right)^t \qquad (5\text{–}182)$$

where

(a) $\mathbf{q}_\gamma$ is a vector whose entries are the capacitor charges.
(b) $\boldsymbol{\lambda}_\lambda$ is a vector whose entries are the inductor flux-linkages.

Now

$$\mathbf{i}_\gamma = \frac{d}{dt}\mathbf{q}_\gamma \qquad (5\text{–}183)$$

$$\mathbf{v}_\lambda = \frac{d}{dt}\boldsymbol{\lambda}_\lambda \qquad (5\text{–}184)$$

Formulation of State Space Equations for Nonlinear Networks 373

Using equations (5–183) and (5–184) in equations (5–179) and (5–180) we obtain the canonical form of network equations

$$\frac{d}{dt}\left(\frac{\partial E^*}{\partial \mathbf{v}_\gamma}\right)^t = \left(\frac{\partial Q}{\partial \mathbf{v}_\gamma}\right)^t \quad (5\text{-}185)$$

$$\frac{d}{dt}\left(\frac{\partial E^*}{\partial \mathbf{i}_\lambda}\right)^t = -\left(\frac{\partial Q}{\partial \mathbf{i}_\lambda}\right)^t \quad (5\text{-}186)$$

For any instantaneous value of state we may put

$$\mathbf{i}_\gamma = \mathbf{C}(\mathbf{v}_\gamma)\frac{d\mathbf{v}_\gamma}{dt} \quad (5\text{-}187)$$

$$\mathbf{v}_\lambda = \mathbf{L}(\mathbf{i}_\lambda)\frac{d\mathbf{i}_\lambda}{dt} \quad (5\text{-}188)$$

where $\mathbf{C}(\mathbf{v}_\gamma)$ and $\mathbf{L}(\mathbf{i}_\lambda)$ are matrices whose elements are respectively the appropriately evaluated incremental capacitance and incremental inductance values. Substituting from equations (5–187) and (5–188) into equations (5–179) and (5–180) gives

$$\mathbf{C}(\mathbf{v}_\gamma)\frac{d\mathbf{v}_\gamma}{dt} = \left(\frac{\partial Q}{\partial \mathbf{v}_\gamma}\right)^t \quad (5\text{-}189)$$

$$\mathbf{L}(\mathbf{i}_\lambda)\frac{d\mathbf{i}_\lambda}{dt} = -\left(\frac{\partial Q}{\partial \mathbf{i}_\lambda}\right)^t \quad (5\text{-}190)$$

which is the Brayton–Moser form of equations [B6], [B7].

Now consider equation (5–174). Again grouping the components into their respective types we have

$$\int_\Gamma \langle \mathbf{i}_\rho, d\mathbf{v}_\rho \rangle + \int_\Gamma \langle \mathbf{i}_\lambda, d\mathbf{v}_\lambda \rangle + \int_\Gamma \langle \mathbf{i}_\gamma, d\mathbf{v}_\gamma \rangle = 0 \quad (5\text{-}191)$$

By the usual rule for product differentiation we have

$$d\langle \mathbf{v}_\lambda, \mathbf{i}_\lambda \rangle = \langle \mathbf{v}_\lambda, d\mathbf{i}_\lambda \rangle + \langle \mathbf{i}_\lambda, d\mathbf{v}_\lambda \rangle$$

so that

$$\int_\Gamma \langle \mathbf{i}_\lambda, d\mathbf{v}_\lambda \rangle = \langle \mathbf{i}_\lambda, \mathbf{v}_\lambda \rangle|_\Gamma - \int_\Gamma \langle \mathbf{v}_\lambda, d\mathbf{i}_\lambda \rangle \quad (5\text{-}192)$$

Substituting from equation (5–192) into equation (5–191) gives

$$\int_\Gamma \langle \mathbf{v}_\lambda, d\mathbf{i}_\lambda \rangle - \int_\Gamma \langle \mathbf{i}_\gamma, d\mathbf{v}_\gamma \rangle = \langle \mathbf{i}_\lambda, \mathbf{v}_\lambda \rangle|_\Gamma + \int_\Gamma \langle \mathbf{i}_\rho, d\mathbf{v}_\rho \rangle$$

$$= \langle \mathbf{i}_\lambda, \mathbf{v}_\lambda \rangle|_\Gamma + J_{t\rho} \quad (5\text{-}193)$$

where:

(a) $\langle \mathbf{i}_\lambda, \mathbf{v}_\lambda \rangle|_\Gamma$ denotes the difference in the values of $\langle \mathbf{i}_\lambda, \mathbf{v}_\lambda \rangle$ evaluated at the end-point of Γ.

(b) J_{tp} denotes the difference in the total converter content evaluated at the end-points of Γ. (For reasons similar to those governing G_{tp}, J_{tp} will depend only on the end-points of Γ.)

The right-hand side of equation (5–193) will be, using a similar argument to that adopted for equation (5–177) above, a scalar function of $\mathbf{v}_\gamma$ and $\mathbf{i}_\lambda$, which a comparison of the left-hand sides of equations (5–178) and (5–193) shows must be $-Q(\mathbf{v}_\gamma, \mathbf{i}_\lambda)$. To show this directly we may use the fact that the total network power vanishes identically. Thus

$$G_{tp} + J_{tp} + \langle \mathbf{v}_\gamma, \mathbf{i}_\gamma \rangle|_\Gamma + \langle \mathbf{v}_\lambda, \mathbf{i}_\lambda \rangle|_\Gamma = 0$$

so that
$$\langle \mathbf{v}_\lambda, \mathbf{i}_\lambda \rangle|_\Gamma + J_{tp} = -\langle \mathbf{v}_\gamma, \mathbf{i}_\gamma \rangle|_\Gamma - G_{tp} \quad (5\text{–}194)$$

This shows that the same form of network canonical equations is obtained by using either equation (5–173) or equation (5–174). In particular, the dual forms of integral invariant principle do not lead to dual forms of canonical network equation.

Case when Excess Components are Present

If excess inductors and capacitors are present we may group the state and excess components together and partition $\mathbf{i}_\gamma, \mathbf{v}_\gamma, \mathbf{i}_\lambda$ and $\mathbf{v}_\lambda$ as follows

$$\mathbf{i}_\lambda = \begin{bmatrix} \mathbf{i}_{et} \\ \mathbf{i}_{sc} \end{bmatrix} \qquad \mathbf{v}_\gamma = \begin{bmatrix} \mathbf{v}_{ec} \\ \mathbf{v}_{st} \end{bmatrix}$$

$$\mathbf{i}_\gamma = \begin{bmatrix} \mathbf{i}_{ec} \\ \mathbf{i}_{st} \end{bmatrix} \qquad \mathbf{v}_\lambda = \begin{bmatrix} \mathbf{v}_{et} \\ \mathbf{v}_{sc} \end{bmatrix}$$

where $\mathbf{i}_{et}, \mathbf{i}_{sc}, \mathbf{v}_{ec}, \mathbf{v}_{st}, \mathbf{i}_{ec}, \mathbf{i}_{st}, \mathbf{v}_{et}$ and $\mathbf{v}_{sc}$ are as defined previously.

The subscripts e and s denote excess and state component quantities respectively. Proceeding as before, we obtain equation (5–178) and then express it in terms of the above partitioned vectors as

$$\int_\Gamma \langle \mathbf{i}_{ec}, \mathbf{dv}_{ec} \rangle + \int_\Gamma \langle \mathbf{i}_{st}, \mathbf{dv}_{st} \rangle - \int_\Gamma \langle \mathbf{v}_{et}, \mathbf{di}_{et} \rangle - \int_\Gamma \langle \mathbf{v}_{sc}, \mathbf{di}_{sc} \rangle$$
$$= Q(\mathbf{v}_{st}, \mathbf{i}_{sc}) \quad (5\text{–}195)$$

Formulation of State Space Equations for Nonlinear Networks 375

Using equations (5–152) and (5–155) we have

$$\int_\Gamma \langle \mathbf{i}_{ec}, -\mathbf{D}_{se}^t \, d\mathbf{v}_{st} \rangle + \int_\Gamma \langle \mathbf{i}_{st}, d\mathbf{v}_{st} \rangle - \int_\Gamma \langle \mathbf{v}_{et}, \mathbf{D}_{es} \, d\mathbf{i}_{sc} \rangle - \int_\Gamma \langle \mathbf{v}_{sc}, d\mathbf{i}_{sc} \rangle$$

$$= \int_\Gamma \langle -\mathbf{D}_{se}\mathbf{i}_{ec}, d\mathbf{v}_{st} \rangle + \int_\Gamma \langle \mathbf{i}_{st}, d\mathbf{v}_{st} \rangle$$

$$- \int_\Gamma \langle \mathbf{D}_{es}^t \mathbf{v}_{et}, d\mathbf{i}_{sc} \rangle - \int_\Gamma \langle \mathbf{v}_{sc}, d\mathbf{i}_{sc} \rangle$$

$$= Q(\mathbf{v}_{st}, \mathbf{i}_{sc}) \qquad (5\text{–}196)$$

Differentiating equation (5–196) then gives

$$\mathbf{i}_{st} - \mathbf{D}_{se}\mathbf{i}_{ec} = \left(\frac{\partial Q}{\partial \mathbf{v}_{st}}\right)^t \qquad (5\text{–}197)$$

$$\mathbf{v}_{sc} + \mathbf{D}_{es}^t \mathbf{v}_{et} = -\left(\frac{\partial Q}{\partial \mathbf{i}_{sc}}\right)^t \qquad (5\text{–}198)$$

If we denote the total co-energy of the state components by E_s^* and the total co-energy of the excess components by E_e^* then we have

$$\mathbf{v}_{et} = \frac{d}{dt}\left(\frac{\partial E_e^*}{\partial \mathbf{i}_{et}}\right)^t \qquad (5\text{–}199)$$

$$\mathbf{v}_{sc} = \frac{d}{dt}\left(\frac{\partial E_s^*}{\partial \mathbf{i}_{sc}}\right)^t \qquad (5\text{–}200)$$

$$\mathbf{i}_{ec} = \frac{d}{dt}\left(\frac{\partial E_e^*}{\partial \mathbf{v}_{ec}}\right)^t \qquad (5\text{–}201)$$

$$\mathbf{i}_{st} = \frac{d}{dt}\left(\frac{\partial E_s^*}{\partial \mathbf{v}_{st}}\right)^t \qquad (5\text{–}202)$$

Substituting from equations (5–199) to (5–202) inclusive into equations (5–197) and (5–198) gives

$$\frac{d}{dt}\left(\frac{\partial E_s^*}{\partial \mathbf{v}_{st}}\right)^t - \mathbf{D}_{se}\frac{d}{dt}\left(\frac{\partial E_e^*}{\partial \mathbf{v}_{ec}}\right)^t = \left(\frac{\partial Q}{\partial \mathbf{v}_{st}}\right)^t \qquad (5\text{–}203)$$

$$\frac{d}{dt}\left(\frac{\partial E_s^*}{\partial \mathbf{i}_{sc}}\right)^t + \mathbf{D}_{es}^t \frac{d}{dt}\left(\frac{\partial E_e^*}{\partial \mathbf{i}_{et}}\right)^t = -\left(\frac{\partial Q}{\partial \mathbf{i}_{sc}}\right)^t \qquad (5\text{–}204)$$

376   Network Models

Substituting from equations (5–152) and (5–155) into equation (5–203) then gives

$$\frac{d}{dt}\left(\frac{\partial E_s^*}{\partial \mathbf{v}_{st}}\right)^t - \frac{d}{dt}\left[\mathbf{D}_{se}\left(\frac{\partial E_e^*}{\partial [-\mathbf{D}_{se}^t \mathbf{v}_{st}]}\right)^t\right] = \left(\frac{\partial Q}{\partial \mathbf{v}_{st}}\right)^t$$

therefore

$$\frac{d}{dt}\left(\frac{\partial E_s^*}{\partial \mathbf{v}_{st}}\right)^t + \frac{d}{dt}\left(\frac{\partial E_e^*}{\partial \mathbf{v}_{st}}\right)^t = \left(\frac{\partial Q}{\partial \mathbf{v}_{st}}\right)^t$$

therefore

$$\frac{d}{dt}\left(\frac{\partial E^*}{\partial \mathbf{v}_{st}}\right)^t = \left(\frac{\partial Q}{\partial \mathbf{v}_{st}}\right)^t \quad (5\text{–}205)$$

where $E^* = E_s^* + E_e^*$

is the total network co-energy.

Substituting from equations (5–152) and (5–155) into equation (5–204) gives

$$\frac{d}{dt}\left(\frac{\partial E_s^*}{\partial \mathbf{i}_{sc}}\right)^t + \frac{d}{dt}\left[\mathbf{D}_{es}^t\left(\frac{\partial E_e^*}{\partial \mathbf{D}_{es}\mathbf{i}_{sc}}\right)^t\right] = -\left(\frac{\partial Q}{\partial \mathbf{i}_{sc}}\right)^t$$

therefore

$$\frac{d}{dt}\left(\frac{\partial E_s^*}{\partial \mathbf{i}_{sc}}\right)^t + \frac{d}{dt}\left(\frac{\partial E_e^*}{\partial \mathbf{i}_{sc}}\right)^t = -\left(\frac{\partial Q}{\partial \mathbf{i}_{sc}}\right)^t$$

therefore

$$\frac{d}{dt}\left(\frac{\partial E^*}{\partial \mathbf{i}_{sc}}\right)^t = -\left(\frac{\partial Q}{\partial \mathbf{i}_{sc}}\right)^t \quad (5\text{–}206)$$

Equations (5–205) and (5–206) show that the canonical form of equations holds for the general case.

As in the previous section, incremental inductance and capacitance matrices may be introduced. This gives

$$\mathbf{L}_e \frac{d\mathbf{i}_{et}}{dt} = \mathbf{v}_{et} \quad (5\text{–}207)$$

$$\mathbf{L}_s \frac{d\mathbf{i}_{sc}}{dt} = \mathbf{v}_{sc} \quad (5\text{–}208)$$

$$\mathbf{C}_e \frac{d\mathbf{v}_{ec}}{dt} = \mathbf{i}_{ec} \quad (5\text{–}209)$$

$$\mathbf{C}_s \frac{d\mathbf{v}_{st}}{dt} = \mathbf{i}_{st} \quad (5\text{–}210)$$

Formulation of State Space Equations for Nonlinear Networks 377

Substituting from equations (5–207) to (5–210) inclusive into equations (5–197) and (5–198) gives

$$\mathbf{C}_s \frac{d\mathbf{v}_{st}}{dt} - \mathbf{D}_{se}\mathbf{C}_e \frac{d\mathbf{v}_{ec}}{dt} = \left(\frac{\partial Q}{\partial \mathbf{v}_{st}}\right)^t \quad (5\text{–}211)$$

$$\mathbf{L}_s \frac{d\mathbf{i}_{sc}}{dt} - \mathbf{D}_{es}^t \mathbf{L}_e \frac{d\mathbf{i}_{et}}{dt} = -\left(\frac{\partial Q}{\partial \mathbf{i}_{sc}}\right)^t \quad (5\text{–}212)$$

and, finally, using equations (5–152) and (5–155) we get

$$[\mathbf{C}_s + \mathbf{D}_{se}\mathbf{C}_e\mathbf{D}_{se}^t]\frac{d\mathbf{v}_{st}}{dt} = \left(\frac{\partial Q}{\partial \mathbf{v}_{st}}\right)^t \quad (5\text{–}213)$$

$$[\mathbf{L}_s + \mathbf{D}_{es}^t\mathbf{L}_e\mathbf{D}_{es}]\frac{d\mathbf{i}_{sc}}{dt} = -\left(\frac{\partial Q}{\partial \mathbf{i}_{sc}}\right)^t \quad (5\text{–}214)$$

which is the appropriately extended form of the Brayton–Moser equations.

5.13.3 Construction of Scalar Functions

In order to use this canonical form of equations we must derive the scalar function Q for a given network. The starting point of the curvilinear arc Γ may be taken as the origin equilibrium point of the state space and we then have the following alternative forms of Q

$$Q = G_{t\rho} + \langle \mathbf{v}_\gamma, \mathbf{i}_\gamma \rangle \quad (5\text{–}215)$$

$$Q = -J_{t\rho} - \langle \mathbf{v}_\lambda, \mathbf{i}_\lambda \rangle \quad (5\text{–}216)$$

where $G_{t\rho}$ and $J_{t\rho}$ are respectively the total content and the total co-content of the network converters. To use the canonical form of network equations therefore the key step is to express the total network converter content and the scalar product of the capacitor voltage and current vectors in terms of a complete independent set of capacitor voltages and inductor currents. Alternatively we may express the total network co-content and the scalar product of the inductor current and voltage vectors in terms of such a set of state variables.

In the general case this will have to be done by inspection and manipulation of the network variable relationships. There is a useful formula however which may be derived giving an explicit expression for Q for networks in which no excess elements are present. Consider the class of networks such that all network capacitors may be included in a chosen tree and all network inductors in the chord for the chosen tree. Put

$$G_{t\rho} = G_{\text{tree}} + G_{\text{chords}} \quad (5\text{–}217)$$

378   Network Models

and
$$J_{tp} = J_{tree} + J_{chords} \quad (5\text{–}218)$$

where G_{tree} and G_{chords} are respectively the total content of the tree converters and chord converters; and J_{tree} and J_{chords} are the total co-content of the tree converters and chord converters. We may also put

$$G_{chords} = \langle \mathbf{v}_{nc}, \mathbf{i}_{nc} \rangle - J_{chords} \quad (5\text{–}219)$$

This gives, using these in equation (5–215)

$$Q = G_{tree} - J_{chords} + \langle \mathbf{v}_{nc}, \mathbf{i}_{nc} \rangle + \langle \mathbf{v}_{st}, \mathbf{i}_{st} \rangle \quad (5\text{–}220)$$

Using equations (5–151), (5–152), (5–154) and (5–155) gives

$$\langle \mathbf{v}_{nc}, \mathbf{i}_{nc} \rangle + \langle \mathbf{v}_{st}, \mathbf{i}_{st} \rangle = \langle -\mathbf{D}_{sn}^t \mathbf{v}_{st}, \mathbf{i}_{nc} \rangle + \langle \mathbf{v}_{st}, (\mathbf{D}_{ss}\mathbf{i}_{sc} + \mathbf{D}_{sn}\mathbf{i}_{nc}) \rangle$$
$$= -\mathbf{v}_{st}^t \mathbf{D}_{sn}\mathbf{i}_{nc} + \mathbf{v}_{st}^t \mathbf{D}_{ss}\mathbf{i}_{sc} + \mathbf{v}_{st}^t \mathbf{D}_{sn}\mathbf{i}_{nc}$$
$$= \mathbf{v}_{st}^t \mathbf{D}_{ss}\mathbf{i}_{sc} \quad (5\text{–}221)$$

Using equation (5–221) in equation (5–220) then gives the following explicit expression for Q

$$Q = \mathbf{v}_{st}^t \mathbf{D}_{ss}\mathbf{i}_{sc} + G_{tree} - J_{chords} \quad (5\text{–}222)$$

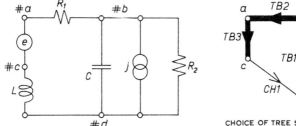

CHOICE OF TREE SHOWN BY HEAVY LINES

Figure 5–46

Example 5.14

Consider the simple nonlinear network of Figure 5–46 in which the passive elements have the following constitutive characteristics

$$v_{R1} = i_{R1}^3$$
$$i_{R2} = 2v_{R2}^3$$
$$\lambda = i_L$$
$$q = 4v_{cap}$$
(5–223)

Formulation of State Space Equations for Nonlinear Networks 379

For the oriented linear graph and state tree shown in Figure 5–41 we have

$$\mathbf{D} = \begin{bmatrix} \mathbf{D}_{ss} & \mathbf{D}_{sn} \\ \mathbf{D}_{ns} & \mathbf{D}_{nn} \end{bmatrix} = \begin{bmatrix} 1 & \vdots & 1 & 1 \\ \hdashline 1 & \vdots & 0 & 0 \\ 1 & \vdots & 0 & 0 \end{bmatrix} \quad (5\text{–}224)$$

Computing G_{tree} and J_{chords} we get

$$G_{\text{tree}} = \int_0^{i_{R1}} i_{R1}^3 \, di_{R1} + ei_L = \tfrac{1}{4} i_{R1}^4 + ei_L = \tfrac{1}{4} i_L^4 + ei_L$$

$$J_{\text{chords}} = \int_0^{v_{R2}} 2v_{R2}^3 \, dv_{R2} + jv_{\text{cap}} = \tfrac{1}{2} v_{R2}^4 + jv_{\text{cap}} = -\tfrac{1}{2} v_{\text{cap}}^4 + jv_{\text{cap}}$$

Since $\mathbf{D}_{ss} = +1$ we get

$$Q = v_{\text{cap}} i_L + \tfrac{1}{4} i_L^4 + ei_L + jv_{\text{cap}} - \tfrac{1}{2} v_{\text{cap}}^4$$

and thus

$$\frac{\partial Q}{\partial v_{\text{cap}}} = i_L - 2v_{\text{cap}}^3 + j$$

$$-\frac{\partial Q}{\partial i_L} = -v_{\text{cap}} - i_L^3 - e$$

Computing the total network co-energy we have

$$E^* = \int_0^{v_{\text{cap}}} 4v_{\text{cap}} \, dv + \int_0^{i_L} i_L \, di$$

$$= 2v_{\text{cap}}^2 + \tfrac{1}{2} i_L^2$$

so that

$$\frac{\partial E^*}{\partial v_{\text{cap}}} = 4v_{\text{cap}} \qquad \frac{\partial E^*}{\partial i_L} = i_L$$

giving the network state space equation set as

$$\frac{d}{dt}(4v_{\text{cap}}) = i_L - 2v_{\text{cap}}^3 + j$$

$$\frac{dv_{\text{cap}}}{dt} = \tfrac{1}{4} i_L - \tfrac{1}{2} v_{\text{cap}}^3 + \tfrac{1}{4} j \quad (5\text{–}225)$$

$$\frac{di_L}{dt} = -v_{\text{cap}} - i_L^3 - e$$

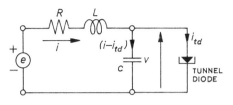

Figure 5-47

Example 5.15

Consider next the tunnel diode circuit of Figure 5-47 in which the resistor, inductor and capacitor are all linear and the tunnel diode has a constitutive characteristic

$$i_{td} = f(v_{td})$$

In this example we will construct the mixed potential function Q using equation (5-215). This gives

$$Q = -\int_0^i e\,di + \int_0^i (Ri)\,di + \int_0^v v\,d[f(v)] + v[i - f(v)]$$

$$= -ei + \tfrac{1}{2}i^2 R - \int_0^v f(v)\,dv + vi$$

The Brayton–Moser equations then give

$$L\frac{di}{dt} = -\frac{\partial Q}{\partial i} = e - iR - v$$

$$C\frac{dv}{dt} = \frac{\partial Q}{\partial v} = -f(v) + i$$

that is

$$L\frac{di}{dt} + iR + v = e$$

$$C\frac{dv}{dt} + f(v) = i$$

which can be checked by a simple inspection of the relationships of Figure 5-47.

Example 5.16

Finally consider the twin tunnel diode circuit of Figure 5-48, in which all the components are linear except the two tunnel diodes which

Formulation of State Space Equations for Nonlinear Networks

have constitutive characteristics

$$i_{td1} = f(v_{c1})$$
$$i_{td2} = f(v_{c2})$$

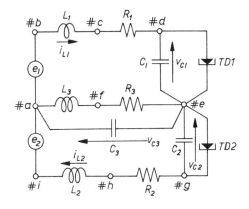

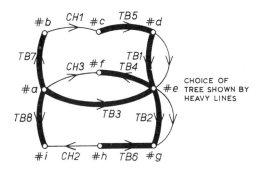

Figure 5–48

The mixed potential function may be obtained from equation (5–222). From the linear graph shown we may construct $\mathbf{D}_{ss}$ as follows:

	Inductor 1	Inductor 2	Inductor 3
Capacitor 1	+1	0	0
Capacitor 2	0	+1	0
Capacitor 3	−1	+1	−1

Also we have
$$G_{\text{tree}} = \tfrac{1}{2}R_2 i_{L2}^2 - e_2 i_{L2} + \tfrac{1}{2}R_3 i_{L3}^2 - e_1 i_{L1} + \tfrac{1}{2}R_1 i_{L1}^2$$
$$J_{\text{chords}} = \int_0^{v_{c1}} f(v_{c1})\, dv_{c1} + \int_0^{v_{c2}} f(v_{c2})\, dv_{c2}$$

Since
$$\mathbf{v}_{st}^t \mathbf{D}_{ss} \mathbf{i}_{sc} = [v_{c1}\ \ v_{c2}\ \ v_{c3}] \begin{bmatrix} 1 & 0 & 0 \\ 0 & 1 & 0 \\ -1 & 1 & -1 \end{bmatrix} \begin{bmatrix} i_{L1} \\ i_{L2} \\ i_{L3} \end{bmatrix}$$
$$= v_{c1} i_{L1} + v_{c2} i_{L2} - v_{c3} i_{L1} + v_{c3} i_{L2} - v_{c3} i_{L3}$$

Thus
$$Q = \mathbf{v}_{st}^t \mathbf{D}_{ss} \mathbf{i}_{sc} + G_{\text{tree}} - J_{\text{chords}}$$

giving
$$Q = v_{c1} i_{L1} + v_{c2} i_{L2} - v_{c3} i_{L1} + v_{c3} i_{L2} - v_{c3} i_{L3} + \tfrac{1}{2}R_2 i_{L2}^2$$
$$- e_2 i_{L2} + \tfrac{1}{2}R_3 i_{L3}^2 - e_1 i_{L1} + \tfrac{1}{2}R_1 i_{L1}^2$$
$$- \int_0^{v_{c1}} f(v_{c1})\, dv_{c1} - \int_0^{v_{c2}} f(v_{c2})\, dv_{c2}$$

The Brayton–Moser equations are thus
$$L_1 \frac{di_{L1}}{dt} = -\frac{\partial Q}{\partial i_{L1}} = -v_{c1} + v_{c3} + e_1 - R_1 i_{L1}$$
$$L_2 \frac{di_{L2}}{dt} = -\frac{\partial Q}{\partial i_{L2}} = -v_{c2} - v_{c3} - R_2 i_{L2} + e_2$$
$$L_3 \frac{di_{L3}}{dt} = -\frac{\partial Q}{\partial i_{L3}} = v_{c3} - R_3 i_{L3}$$
$$C_1 \frac{dv_{c1}}{dt} = \frac{\partial Q}{\partial v_{c1}} = i_{L1} - f(v_{c1})$$
$$C_2 \frac{dv_{c2}}{dt} = \frac{\partial Q}{\partial v_{c2}} = i_{L2} - f(v_{c2})$$
$$C_3 \frac{dv_{c3}}{dt} = \frac{\partial Q}{\partial v_{c3}} = -i_{L1} + i_{L2} - i_{L3}$$

CHAPTER SIX

State Models

The essential intuitive idea behind the formulation of dynamical problems in terms of a linear vector space is that of the state of the system. *State.* A state of a dynamical system is any complete independent set of system model variables whose knowledge enables one to predict completely the future behaviour of the system. For a large class of dynamical models, the state variables could specify the energy stored in a set of independent energy storage elements; knowledge of this set of variables would enable one to determine all the energy stored in the system model. Since the behaviour of each individual storage element in the model is governed by a first-order differential equation, it follows that the behaviour of the entire interconnected model may be determined from a set of first order differential equations

$$\frac{dx_i}{dt} = f_i(x_1, x_2, \ldots, x_n; u_1, u_2, \ldots, u_r; t)$$

$$i = 1, 2, \ldots, n$$

(6–1)

where the variables $\{x_1, x_2, \ldots, x_n\}$ are a set of state variables and the variables $\{u_1, u_2, \ldots, u_r\}$ represent the effects of source forcing functions and external disturbances to the system. We have shown in Chapter 5 how canonical forms of first-order differential equation sets may be formed for network models. Since in addition many differential equations may be replaced by an equivalent set of first-order differential equations by using the simple substitution of variables technique due to Moigno, the set of equations (6–1) may be taken as a standard form for the discussion of certain general features of dynamical model behaviour. The importance of the state concept in the formal structure of dynamical theory was first realized by Poincaré. Following the researches of Cauchy and Moigno [M16] into differential equation theory, he based the whole of his approach to dynamics in his famous *Méthodes Nouvelles de la Mécanique Céleste* (Gauthier Villars, Paris, 1892–99) on the description of dynamical system behaviour by sets of first-order differential equations. This may be done in a straightforward way, and without loss of generality, by adopting Moigno's device of introducing auxiliary variables. For a simple example of this procedure consider the fourth-order differential equation

$$\frac{d^4y}{dt^4}+6\frac{d^3y}{dt^3}+3\frac{d^2y}{dt^2}+9\frac{dy}{dt}+12y = \cos \omega t$$

Define the four auxiliary variables x_1, x_2, x_3 and x_4 as

$$x_1 = y \quad x_2 = \frac{dy}{dt} \quad x_3 = \frac{d^2y}{dt^2} \quad x_4 = \frac{d^3y}{dt^3}$$

Since the specification of the values of these variables at a given reference time will determine a unique solution to the given equation, they are taken as the system state variables. A simple substitution of variables then gives a set of four first-order differential equations which are equivalent to the given fourth-order equation. These are

$$\frac{dx_1}{dt} = x_2$$

$$\frac{dx_2}{dt} = x_3$$

$$\frac{dx_3}{dt} = x_4$$

$$\frac{dx_4}{dt} = -12x_1 - 9x_2 - 3x_3 - 6x_4 + \cos \omega t$$

The state variables of a dynamical system model may be taken as the coordinates of a vector referred to the standard basis in a Euclidean space of n-dimensions, and the entire dynamical model behaviour considered using the apparatus of matrix algebra and linear vector space theory. The linear vector space used for this purpose is called the *state space* for the dynamical system model. The solutions of the equation set (6–1) determine a *trajectory* for the state point in the n-dimensional Euclidean space [P5]. If certain conditions discussed below are satisfied, this equation set will define a *unique* solution through every point in the state space. The *totality* of solutions then specifies a flow in the space analogous to the flow of an ideal compressible fluid in a physical space. The essential feature of the uniqueness of the equation solutions is that the *entire* future behaviour of the system is completely determined by a specifications of the system state at a given time, together with the time variation of the function set $\{u_1(t), u_2(t), \ldots, u_r(t)\}$, which represent the effects of sources and disturbances. In particular, if the set of functions $\{u_1(t), \ldots, u_r(t)\}$ is identically zero, the specification of an initial value of state will completely determine the resultant future

Analytical Aspects of State Space Equation Sets 385

system behaviour. State-determinate ideas of this sort are of great value in dynamical work despite the facts that no physical system may be completely isolated from random outside influences which will disturb the behaviour, and that the instantaneous state can never be determined with sufficient accuracy to predict exact behaviour over an arbitrarily long period of time.

A particularly valuable feature of the use of n-dimensional Euclidean vector spaces in the analysis of dynamical system behaviour is that it enables all the analytical and algebraic manipulations of a theoretical investigation to be interpreted and, in some cases, guided by the use of geometrical intuition. For such geometrical purposes, the definition of the distance between two points in the space is essential, i.e. a metric must be introduced. Certain obvious simple generalizations of familiar two- and three-dimensional point sets such as the line, plane, sphere and so on are much employed in dynamical investigations in an appropriate metric space. A further advantage of the systematic use of first-order sets of differential equations for the study of dynamical system behaviour is that the majority of automatic computing facilities require differential equation sets to be arranged in this form; the results of such computations may thus be interpreted from a vector space point of view.

6.1 Analytical Aspects of State Space Equation Sets

In this chapter we are chiefly concerned with a set of fundamental properties of the state space model. Before embarking on such detailed considerations however we must study certain analytical aspects of the state space equations. We start with the crucial question of the uniqueness of the solutions.

6.1.1 Existence and Uniqueness of Solutions

Consider the vector differential equation

$$\frac{d\mathbf{x}}{dt} = \mathbf{f}(\mathbf{x}, t) \tag{6-2}$$

and let $\mathbf{f}(\mathbf{x}, t)$ be a known vector function of $\mathbf{x}$ and t which is differentiable and has all its first derivatives bounded for bounded values of $\mathbf{x}$ and t. The method of successive approximations introduced by Picard then gives the following theorem [P8].

Existence theorem. A unique solution vector $\mathbf{x}(t)$ exists which satisfies the differential equation (6-2) together with an initial condition

$$\mathbf{x}(t_0) = \mathbf{x}_0$$

if the following conditions hold

$$\|f(x, t)\|_E < M$$

$$\left|\frac{\partial f_i}{\partial x_j}\right| < N$$

for
$$\|x - x_0\|_E < A$$

and
$$|t - t_0| < a$$

where M, N, A and a are real constants. For this set of constants the unique solution exists for

$$t > t_0 \qquad |t - t_0| < k$$

where k is the smaller of a and A/M.

If the above conditions are satisfied f will be said to satisfy *a Lipschitz condition*.

Any solution of equation (6–2) must satisfy the integral equation

$$x(t) = x_0 + \int_{t_0}^{t} f[x(t), t]\, dt \tag{6-3}$$

and any solution of (6–3) will satisfy both the differential equation (6–2) and the initial condition. Consider the following sequence of functions introduced by Picard:

$$x_1(t) = x_0$$

$$x_2(t) = x_0 + \int_{t_0}^{t} f[x_1(t), t]\, dt$$

$$x_3(t) = x_0 + \int_{t_0}^{t} f[x_2(t), t]\, dt$$

$$\vdots$$

$$x_{m+1}(t) = x_0 + \int_{t_0}^{t} f[x_m(t), t]\, dt$$

For this sequence we have:

$$\|x_2(t) - x_1(t)\|_E \leq \int_{t_0}^{t} \|f(x_1)\|_E\, dt$$

$$\|x_3(t) - x_2(t)\|_E \leq \int_{t_0}^{t} \|f(x_2) - f(x_1)\|_E\, dt$$

$$\vdots$$

$$\|x_{m+1}(t) - x_m(t)\|_E \leq \int_{t_0}^{t} \|f(x_m) - f(x_{m-1})\|_E\, dt \tag{6-4}$$

Now using the mean-value theorem for differentiation we have that

$$\|f(x_m) - f(x_{m-1})\|_E^2 = \sum_{i=1}^{n} \|f_i(x_m) - f_i(x_{m-1})\|_E^2$$

$$= \sum_{i=1}^{n} \left| \sum_{j=1}^{n} \frac{\partial f_i}{\partial x_j}(\xi_{ij})(x_{m_j} - x_{m-1_j}) \right|^2$$

where ξ_{ij} is a suitable set of values of the components of x at which the derivatives $\partial f_i / \partial x_j$ are evaluated. This gives that, for some suitable constant K

$$\|f(x_m) - f(x_{m-1})\|_E^2 \leq K^2 \|x_m - x_{m-1}\|_E^2 \tag{6-5}$$

From equations (6-4) and (6-5) we have

$$\|x_{m+1} - x_m\|_E \leq \int_{t_0}^{t} \|f(x_m) - f(x_{m-1})\|_E \, dt \leq K \int_{t_0}^{t} \|x_m - x_{m-1}\|_E \, dt \tag{6-6}$$

and for $m = 1$ we have that

$$\|x_2 - x_1\|_E \leq \int_{t_0}^{t} \|f(x_1, t)\|_E \, dt < \int_{t_0}^{t} M \, dt$$

giving
$$\|x_2 - x_1\|_E \leq M(t - t_0) \tag{6-7}$$

An inspection of equation (6-7) shows that the relationship

$$\|x_{m+1} - x_m\|_E \leq \frac{M K^{m-1}(t - t_0)^m}{m!} \tag{6-8}$$

holds for $m = 1$. Suppose it held for $m = r - 1$, i.e.

$$\|x_r - x_{r-1}\|_E \leq \frac{M K^{r-2}(t - t_0)^{r-1}}{(r-1)!} \tag{6-9}$$

Then we would have, using equation (6-6)

$$\|x_{r+1} - x_r\|_E \leq K \int_{t_0}^{t} \|x_r - x_{r-1}\|_E \, dt$$

so that, using equation (6-9) gives

$$\|x_{r+1} - x_r\|_E \leq K \int_{t_0}^{t} \frac{M K^{r-2}(t - t_0)^{r-1}}{(r-1)!} \, dt$$

that is
$$\|x_{r+1} - x_r\|_E \leq \frac{M K^{r-1}(t - t_0)^r}{r!}$$

Thus if equation (6–8) holds for $m = r-1$ it also holds for $m = r$. Since it holds for $m = 1$, by induction it holds for all values of m. Thus

$$\|\mathbf{x}_{r+1}-\mathbf{x}_r\|_E \to 0 \quad \text{as} \quad r \to \infty$$

and the Picard sequence of functions will converge uniformly in the interval of t over which the Lipschitz conditions hold. Since the convergence is uniform, the limit of the sequence of integrals of the Picard sequence will be the integral of the limit of the sequence of functions integrated. Thus the limit vector of the Picard sequence exists, satisfies the integral equation and therefore satisfies the differential equation (6–2). It only remains to consider the uniqueness of the solution.

Suppose that two solutions of the differential equation existed with the same initial condition $\mathbf{x}_0$ and denote them by $\mathbf{y}(t)$ and $\mathbf{z}(t)$. Then

$$\|\mathbf{y}(t)-\mathbf{z}(t)\|_E = \left\|\int_{t_0}^t \{\mathbf{f}[\mathbf{y}(t), t]-\mathbf{f}[\mathbf{z}(t), t]\} \, dt\right\|_E$$

since both would satisfy the integral equation (6–3). Thus we would have, using equation (6–5)

$$\|\mathbf{y}(t)-\mathbf{z}(t)\|_E \leqslant K \int_{t_0}^t \|\mathbf{y}(t)-\mathbf{z}(t)\|_E |dt| \qquad (6\text{–}10)$$

Now let m be the maximum value attained by $\mathbf{y}(t)-\mathbf{z}(t)$ over an interval

$$|t-t_0| \leqslant \frac{1}{2K}$$

and let this maximum value be attained at $t = t_1$. Then, from equation (6–10),

$$0 \leqslant \|\mathbf{y}-\mathbf{z}\|_E \leqslant m \leqslant K \int_{t_0}^{t_1} m|dt|$$

so that

$$m \leqslant mK|t_1-t_0| \qquad (6\text{–}11)$$

but

$$|t_1-t_0| \leqslant \frac{1}{2K} \qquad (6\text{–}12)$$

so that equations (6–11) and (6–12) taken together imply that

$$m \leqslant \frac{m}{2}$$

which can only be true if $m = 0$. This shows that $\mathbf{y}(t) = \mathbf{z}(t)$ over the interval considered. The uniqueness of the solution guaranteed by the Lipschitz conditions can be extended over the whole interval in which the solution exists by repetition of the above argument.

In all the discussions of sets of differential equations which follow it is assumed that the conditions necessary for a unique solution to exist are satisfied. In order to avoid pointless repetition, no explicit statement of this will be made each time an equation set is introduced.

6.1.2 Singular Points and the Liapunov First-Approximation Matrix

Consider the differential equation set

$$\frac{dx_i}{dt} = f_i(x_1, x_2, \ldots, x_n) \qquad i = 1, 2, \ldots, n \qquad (6-13)$$

corresponding to a system free from external disturbance or to a feedback system in which all the source outputs may be expressed in terms of state variables.

Singular point. A point in the state space at which

$$f_i(x_1, \ldots, x_n) = 0 \qquad i = 1, 2, \ldots, n$$

is called a singular point of the equation set (6–13). Since all derivatives of the state variables vanish at a singular point, the dynamical system is in a state of local equilibrium.

Let $\{c_1, c_2, \ldots, c_n\}$ be the coordinates of a fixed state space vector **c** and put

$$\mathbf{x} = \mathbf{c} + \mathbf{y}$$

where **y** is a vector representing the deviation of the state vector **x** from the fixed vector **c**. Using the multivariable form of the Taylor expansion theorem gives

$$f_i(x_1, x_2, \ldots, x_n) = f_i(c_1, c_2, \ldots, c_n) + \sum_{r=1}^{n} \left[\frac{\partial f_i}{\partial x_r}\right]_{\mathbf{x}=\mathbf{c}} (x_r - c_r)$$

$$+ \frac{1}{2!} \sum_{r=1}^{n} \sum_{s=1}^{n} \left[\frac{\partial^2 f_i}{\partial x_r \partial x_s}\right]_{\mathbf{x}=\mathbf{c}} (x_r - c_r)(x_s - c_s) + \cdots$$

where the derivatives are evaluated for

$$x_i = c_i \qquad i = 1, 2, \ldots, n$$

Thus

$$\frac{d}{dt}(y_i + c_i) = \frac{dy_i}{dt}$$

$$= f_i(c_1, \ldots, c_n) + \sum_{r=1}^{n} \frac{\partial f_i}{\partial x_r} y_r + \frac{1}{2} \sum_{r=1}^{n} \sum_{s=1}^{n} \frac{\partial^2 f_i}{\partial x_r \partial x_s} y_r y_s + \cdots$$

Now let **c** be a singular point of the equation set (6–13) so that

$$f_i(c_1, \ldots, c_n) = 0 \qquad i = 1, 2, \ldots, n$$

and thus in a sufficiently small neighbourhood of the singular point

$$\frac{dy_i}{dt} \approx \sum_{r=1}^{n} \left[\frac{\partial f_i}{\partial x_r}\right]_{x=c} y_r$$

This locally-linearized first approximation to the behaviour in the vicinity of a critical point may be written in matrix form as

$$\frac{d\mathbf{y}}{dt} \approx \mathbf{Jy} \qquad (6\text{–}14)$$

where **J** is a matrix having elements

$$J_{rs} = \frac{\partial f_r}{\partial x_s}$$

evaluated at the singular point. The matrix **J** will be called the Liapunov first-approximation matrix for motion in the vicinity of the singular point. The determinant of **J** is called the Jacobian of the equation set. For any region of the state space in which the equation set Jacobian does not vanish, the equation set (6–13) defines a unique reciprocal relationship between a velocity and a position vector in the state space.

Classification of Singular Points for Second-order Systems

The various types of solution trajectory behaviour in the vicinity of the singular points of a second-order system may be classified in terms of the nature of the eigenvalues of the Liapunov first-approximation matrix **J** evaluated at the singular point. (See 6.1.4.)

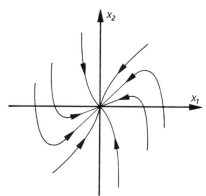

Figure 6–1 Stable node.

(a) *Stable node.* If both eigenvalues of **J** are real and negative, the singular point is called a stable node. The corresponding flow pattern in the state plane is shown in Figure 6–1.

(b) *Unstable node.* If both eigenvalues of the matrix **J** are real and positive, the singular point is called an unstable node. The state plane flow pattern is then of the form shown in Figure 6–2.

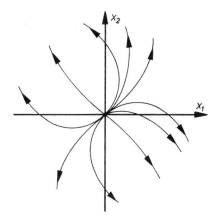

Figure 6–2 Unstable node.

(c) *Saddle point.* If both the eigenvalues of the matrix **J** are real and of opposite sign, the corresponding singular point is called a saddle point and is associated with the state plane flow pattern of Figure 6–3.

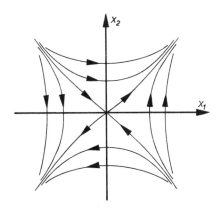

Figure 6–3 Saddle point.

(d) *Stable focus.* When the eigenvalues of **J** are complex conjugate with negative real part, the state plane flow pattern is of the form shown in Figure 6–4, and the singular point is called a stable focus.

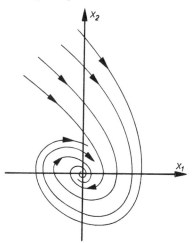

Figure 6–4 Stable focus.

(e) *Unstable focus.* When the eigenvalues of **J** are complex conjugate with positive real part, the singular point is called an unstable focus and is associated with the flow pattern of Figure 6–5.

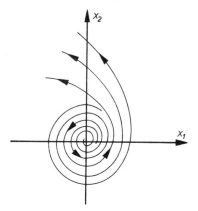

Figure 6–5 Unstable focus.

(f) *Vortex.* If both the eigenvalues of **J** are pure imaginary, the singular point is called a vortex; the corresponding flow pattern is shown in Figure 6–6.

Analytical Aspects of State Space Equation Sets 393

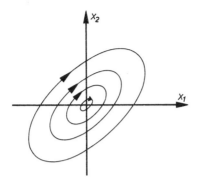

Figure 6-6 Vortex.

6.1.3 Simple Trajectory Properties in the State Plane

For a second-order system it is often useful to sketch out a selection of solution trajectories and thus establish the main features of the system behaviour. The most direct procedure is to simply calculate the velocity vector $\dot{x}$ at a large number of points and thus establish the directions of the tangents to the trajectories at these points. Certain procedures now considered however both reduce the amount of computation required and introduce certain ideas of considerable theoretical interest.

Contact curves. Let $V(x_1, x_2)$ be a scalar function defined at all points of the state plane. The total time derivative of V for an autonomous system of equations, that is one for which the functions f_i are not explicit functions of t in equations (6-13), may be evaluated along a trajectory as

$$\frac{dV}{dt} = \frac{\partial V}{\partial x_1}\dot{x}_1 + \frac{\partial V}{\partial x_2}\dot{x}_2$$

For those values of $\{x_1, x_2\}$ at which dV/dt vanishes, the trajectories are tangent to a curve along which the function V is constant. Thus the equation

$$\frac{\partial V}{\partial x_1}f_1(x_1, x_2) + \frac{\partial V}{\partial x_2}f_2(x_1, x_2) = 0$$

defines a set of curves called *contact curves* for a given function V along which the solution trajectories contact the constant-value-of-V contours. At every point of a contact curve, the corresponding trajectory is tangent to the $V = $ constant contour passing through this point, and thus a carefully chosen set of V-functions with constant-value contours plotted on the state plane may be used in conjunction with the corresponding contact curves to accurately sketch the solution trajectories.

Isoclines. The simplest choice of *V*-function is the linear function
$$V = ax_1 + x_2$$
where a is a real constant. For this case the constant-value contours are straight lines of equal slope and the corresponding contact curves thus define point sets through which pass trajectories of constant slope. These particular contact curves are called *isoclines*.

For example consider the system of equations
$$\begin{bmatrix} \dot{x}_1 \\ \dot{x}_2 \end{bmatrix} = \begin{bmatrix} 0 & 1 \\ -\omega_n^2 & -2\zeta\omega_n \end{bmatrix} \begin{bmatrix} x_1 \\ x_2 \end{bmatrix}$$
and put $V = ax_1 + x_2$. Then
$$\frac{\partial V}{\partial x_1} = a \qquad \frac{\partial V}{\partial x_2} = 1$$
so that the isocline equation is
$$ax_2 + [-\omega_n^2 x_1 - 2\zeta\omega_n x_2] = 0$$
$$\omega_n^2 x_1 + (2\zeta\omega_n - a)x_2 = 0$$

For various values of a this defines a family of straight lines through the origin.

Limit cycles. A limit cycle is a closed trajectory in the state plane to which solution trajectories tend with increasing time.

Bendixon's first theorem. The first of Bendixon's theorems is a criterion for the nonexistence of limit cycles in the state plane. Consider, as shown in Figure 6-7, a closed curve C in the state plane, and let M be a representative point on this curve. Let $\delta\mathbf{c}$ be a vector tangent to the

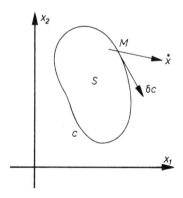

Figure 6-7

curve C at the point M, and consider the system

$$\dot{x}_1 = f_1(x_1, x_2)$$
$$\dot{x}_2 = f_2(x_1, x_2)$$

If the solution trajectory through the point M lies along the curve C, then the component of the vector $\dot{x}$ perpendicular to the vector δc will be zero. Thus a necessary condition for a closed curve C to be a solution trajectory is that the line integral of the vector product $\dot{x}$ and δc taken round the curve C is zero, that is

$$\int_C [f_1 \, dx_2 - f_2 \, dx_1] = 0$$

Using Stokes' Theorem, this line integral may also be evaluated as a double integral over the area S enclosed by the curve C, that is as the integral [J1]

$$\int\int_S \left[\frac{\partial f_1}{\partial x_1} + \frac{\partial f_2}{\partial x_2} \right] dx_1 \, dx_2$$

It therefore allows that if the quantity

$$\frac{\partial f_1}{\partial x_1} + \frac{\partial f_2}{\partial x_2}$$

has a constant sign over the region enclosed by a curve C, neither C nor any closed curve interior to C can be a solution trajectory. This is Bendixon's first theorem which states that no limit cycle can exist in any region of the state plane in which

$$\frac{\partial f_1}{\partial x_1} + \frac{\partial f_2}{\partial x_2}$$

has a constant sign.

Bendixon's second theorem. This is an existence theorem for limit cycles which follows directly from the uniqueness property of solution trajectories. It states that a trajectory which remains in a bounded region of the state plane for $t_0 < t < \infty$ is either a limit cycle or approaches a limit cycle asymptotically.

Index of a closed curve. Again, as in Figure 6–7, let C be a closed curve in the generalized state plane and let M be a representative point on it. The number of revolutions of the local velocity vector $\dot{x}_M$ at point M, as the point M is taken once round the curve C in an anti-clockwise direction, is called the index of the curve C.

Poincaré's index theorems establish certain facts about the nature of system singular points in the state plane in terms of a knowledge of the velocity vector $\dot{x}$ at all points on a closed curve in the state plane.

396   State Models

Poincaré's first index theorem. Let m be the index of a closed curve C in the state plane. Then if the system has one singular point in the state plane:

(a) If $m = 0$, curve C encloses no singular points.
(b) If $m = 1$, curve C encloses a node, focus or vortex.
(c) If $m = -1$, curve C encloses a saddle point.

In general $m = N - N^1$ where N is the total number of nodes, foci and vortices enclosed by the curve C, and N^1 is the number of saddle points enclosed by C.

Poincaré's second index theorem. A necessary condition for some closed curve C in the state plane to be a limit cycle is that its index be $+1$. Consequently a limit cycle must always enclose a singular point of the equation set.

6.1.4 Analytical Solutions of Linear Equation Sets

The homogeneous linear n-vector equation

$$\frac{d\mathbf{x}(t)}{dt} = \mathbf{A}(t)\mathbf{x}(t) \qquad (6\text{-}15)$$

has the crucially important property that its solutions form a linear vector space of dimension n. This property that any solution vector may be expressed as a linear combination of a set of n linearly independent solution vectors is conventionally called the *Principle of Superposition*.

To establish this fundamental result we first note that if $\mathbf{x}_1(t)$ and $\mathbf{x}_2(t)$ are solutions of equation (6–15) then

$$\frac{d}{dt}[c_1\mathbf{x}_1(t) + c_2\mathbf{x}_2(t)] = c_1\frac{d\mathbf{x}_1}{dt} + c_2\frac{d\mathbf{x}_2}{dt}$$

$$= c_1\mathbf{A}(t)\mathbf{x}_1(t) + c_2\mathbf{A}(t)\mathbf{x}_2(t)$$

$$= \mathbf{A}(t)[c_1\mathbf{x}_1(t) + c_2\mathbf{x}_2(t)]$$

so that $[c_1\mathbf{x}_1(t) + c_2\mathbf{x}_2(t)]$ is also a solution where c_1 and c_2 are real constants. This shows that the solutions form a linear vector space, and it only remains to show that there exist n linearly independent solution vectors. Take, in the n-dimensional state space, any set of n linearly independent constant vectors $\{\mathbf{x}_k : k = 1, \ldots, n\}$ and let them be initial condition vectors for a set of n corresponding solutions of equation set (6–15) over some time interval $t_0 \leqslant t \leqslant T$. Denote this set of n solutions by $\{x_k(t) : k = 1, 2, \ldots, n\}$ and define an $n \times n$ matrix

Analytical Aspects of State Space Equation Sets

$\mathbf{X}(t)$ such that $\mathbf{x}_k(t)$ is the kth column of $\mathbf{X}(t)$. Thus

$$\mathbf{X}(t) = \begin{bmatrix} x_{11}(t) & \cdots & x_{1n}(t) \\ x_{21}(t) & \cdots & x_{2n}(t) \\ \vdots & & \vdots \\ x_{n1}(t) & \cdots & x_{nn}(t) \end{bmatrix}$$

where $x_{ij}(t)$ is the ith component of $\mathbf{x}_j(t)$. Differentiate the determinant of $\mathbf{X}(t)$ with respect to t one row at a time and substitute from equation (6–15) for the differentiated elements. This gives

$$\frac{d}{dt}[\det \mathbf{X}(t)] = \begin{vmatrix} \frac{dx_{11}}{dt} & \cdots & \frac{dx_{1n}}{dt} \\ x_{21} & \cdots & x_{2n} \\ \vdots & & \vdots \\ x_{n1} & \cdots & x_{nn} \end{vmatrix} + \begin{vmatrix} x_{11} & \cdots & x_{1n} \\ \frac{dx_{21}}{dt} & \cdots & \frac{dx_{2n}}{dt} \\ \vdots & & \vdots \\ x_{n1} & \cdots & x_{nn} \end{vmatrix}$$

$$+ \cdots + \begin{vmatrix} x_{11} & \cdots & x_{1n} \\ x_{21} & \cdots & x_{2n} \\ \vdots & & \vdots \\ \frac{dx_{n1}}{dt} & \cdots & \frac{dx_{nn}}{dt} \end{vmatrix}$$

$$= \begin{vmatrix} \sum_{r=1}^{n} a_{1r}x_{r1} & \cdots & \sum_{r=1}^{n} a_{1r}x_{rn} \\ x_{21} & \cdots & x_{2n} \\ \vdots & & \\ x_{n1} & \cdots & x_{nn} \end{vmatrix}$$

$$+ \begin{vmatrix} x_{11} & \cdots & x_{1n} \\ \sum_{r=1}^{n} a_{2r}x_{r1} & \cdots & \sum_{r=1}^{n} a_{2r}x_{rn} \\ \vdots & & \vdots \\ x_{n1} & \cdots & x_{nn} \end{vmatrix}$$

$$+ \cdots + \begin{vmatrix} x_{11} & \cdots & x_{1n} \\ x_{21} & \cdots & x_{2n} \\ \vdots & & \vdots \\ \sum_{r=1}^{n} a_{nr}x_{r1} & \cdots & \sum_{r=1}^{n} a_{nr}x_{rn} \end{vmatrix}$$

Now consider a typical term in this sum. We have for the second term:

$$\begin{vmatrix} x_{11} & x_{12} & \cdots \\ (a_{21}x_{11}+a_{22}x_{21}+\cdots+a_{2n}x_{n1}) & (a_{21}x_{12}+a_{22}x_{22}+\cdots+a_{2n}x_{n2}) & \cdots \\ x_{31} & x_{32} & \cdots \\ \vdots & \vdots & \\ x_{n1} & x_{n2} & \cdots \end{vmatrix}$$

Now the value of this determinant will be unaltered by subtracting any multiple of any row from any other row. Therefore subtracting a_{21} times the first row, a_{23} times the third row, $\ldots$, a_{2n} times the nth row from the second row gives that it has the same value as the determinant:

$$\begin{vmatrix} x_{11} & x_{12} & \cdots & x_{1n} \\ a_{22}x_{21} & a_{22}x_{22} & \cdots & a_{22}x_{2n} \\ x_{31} & x_{32} & \cdots & x_{3n} \\ \vdots & \vdots & & \vdots \\ x_{n1} & x_{n2} & \cdots & x_{nn} \end{vmatrix}$$

$$= a_{22} \det \mathbf{X}$$

It follows from this that

$$\frac{d}{dt}[\det(\mathbf{X})] = \sum_{r=1}^{n} a_{rr} \det(\mathbf{X}) = \text{trace}(\mathbf{A}) \det(\mathbf{X})$$

and so

$$\det[\mathbf{X}(t)] = \det[\mathbf{X}(t_0)] \exp\left[\int_{t_0}^{t} \text{trace}(\mathbf{A})\, dt\right] \qquad (6\text{--}16)$$

Now if the set of initial vectors $\{\mathbf{x}_k(t_0): k = 1, 2, \ldots, n\}$ are linearly independent we will have

$$\det[\mathbf{X}(t_0)] \neq 0$$

and it immediately follows from equation (6–16) that

$$\det[\mathbf{X}(t)] \neq 0$$

for all t. Therefore the n solution vectors $\mathbf{x}_k(t)$ are linearly independent for all t and the dimension of the solution space must be greater than or equal to n.

Now let $\mathbf{x}(t)$ be any solution of equation (6–15), for some initial time t_0. Since $\det \mathbf{X}(t)$ does not vanish we can express $\mathbf{x}(t_0)$ as the linear

Analytical Aspects of State Space Equation Sets

combination

$$x(t_0) = c_1 x_1(t_0) + c_2 x_2(t_0) + \cdots + c_n x_n(t_0)$$

and since the solution $x(t)$ is unique we must therefore have

$$x(t) = c_1 x_1(t) + c_2 x_2(t) + \cdots + c_n x_n(t)$$

for all t. Thus the set $\{x_1(t), \ldots, x_n(t)\}$ is a basis for the solution space which is therefore of dimension n.

Fundamental matrix. The matrix $X(t)$ is called a fundamental matrix. A fundamental matrix has a basis set of n linearly independent solution vectors as its columns, is non-singular and thus possesses an inverse $X^{-1}(t)$ defined at all times t.

Note that an immediate consequence of the above discussion of the fundamental matrix properties is that if $Y(t)$ is any matrix whose columns are solutions of equation (6–15) then there exists a matrix of constants C such that

$$Y(t) = X(t)C$$

Now consider the inhomogeneous system

$$\frac{dx}{dt} = Ax + f$$

$$x(t_0) = x_0 \tag{6–17}$$

Let X be a known fundamental matrix for the homogeneous system. Then

$$\frac{d}{dt}(X^{-1}X) = \frac{dX^{-1}}{dt}X + X^{-1}\frac{dX}{dt} = 0$$

Thus

$$\frac{dX^{-1}}{dt} = -X^{-1}\frac{dX}{dt}X^{-1}$$

So that

$$\frac{d}{dt}(X^{-1}x) = \frac{dX^{-1}}{dt}x + X^{-1}\frac{dx}{dt}$$

$$= -X^{-1}\frac{dX}{dt}X^{-1}x + X^{-1}(Ax + f)$$

$$= -X^{-1}AXX^{-1}x + X^{-1}Ax + X^{-1}f$$

$$= -X^{-1}Ax + X^{-1}Ax + X^{-1}f$$

$$= X^{-1}f \tag{6–18}$$

Integration of equation (6–18) thus gives

$$\mathbf{X}^{-1}(t)\mathbf{x}(t) = \mathbf{X}^{-1}(t_0)\mathbf{x}(t_0) + \int_{t_0}^{t} \mathbf{X}^{-1}(\tau)\mathbf{f}(\tau)\,d\tau$$

so that
$$\mathbf{x}(t) = \mathbf{X}(t)\mathbf{X}^{-1}(t_0)\mathbf{x}(t_0) + \int_{t_0}^{t} \mathbf{X}(t)\mathbf{X}^{-1}(\tau)\mathbf{f}(\tau)\,d\tau \quad (6\text{–}19)$$

Transition matrix. The nonhomogeneous system solution of equation (6–19) may be written as

$$\mathbf{x}(t) = \mathbf{\Phi}(t, t_0)\mathbf{x}(t_0) + \int_{t_0}^{t} \mathbf{\Phi}(t, \tau)\mathbf{f}(\tau)\,d\tau \quad (6\text{–}20)$$

where
$$\mathbf{\Phi}(t, \tau) \triangleq \mathbf{X}(t)\mathbf{X}^{-1}(\tau) \quad (6\text{–}21)$$

is called the *transition matrix* for the set of equations (6–17). It follows directly from the above definition that

(a) $\quad \mathbf{\Phi}(t, t) = \mathbf{X}(t)\mathbf{X}^{-1}(t) = \mathbf{I} \quad$ for all t.

(b) $\quad \mathbf{\Phi}(t, \sigma)\mathbf{\Phi}(\sigma, \tau) = \mathbf{X}(t)\mathbf{X}^{-1}(\sigma)\mathbf{X}(\sigma)\mathbf{X}^{-1}(\tau)$

$$= \mathbf{X}(t)\mathbf{X}^{-1}(\tau) = \mathbf{\Phi}(t, \tau)$$

(c) $\quad \mathbf{\Phi}^{-1}(t, \tau) = [\mathbf{X}(t)\mathbf{X}^{-1}(\tau)]^{-1} = \mathbf{X}(\tau)\mathbf{X}^{-1}(t) = \mathbf{\Phi}(\tau, t)$

and so the transition matrix is never singular.

(d) For constant coefficient systems

$$\mathbf{\Phi}(t, \tau) = \mathbf{\Phi}(t-\tau, 0) \triangleq \mathbf{\Phi}(t-\tau) \quad \text{or simply} \quad \mathbf{\Phi}(t) \quad \text{if} \quad \tau = 0$$

and we may unambiguously denote the transition matrix for a constant coefficient system in this way.

The transition matrix is sometimes called the resolvent of the set of equations (6–17). In the original developments of the subject it was called the matrizant [G1].

Constant Coefficient Linear Systems
Since

$$\frac{d}{dt}\exp(\mathbf{A}t) = \mathbf{A}\exp(\mathbf{A}t)$$

and
$$\exp(\mathbf{A}0) = \mathbf{I}$$

it follows that a fundamental matrix of the homogeneous system of equations

$$\frac{d\mathbf{x}}{dt} = \mathbf{A}\mathbf{x} \quad (6\text{–}22)$$

Analytical Aspects of State Space Equation Sets

where $\mathbf{A}$ is a matrix of constant coefficients is

$$\mathbf{X}(t) = \exp(\mathbf{A}t)$$

and that this is the particular fundamental matrix which becomes equal to the identity matrix when $t = 0$. The general solution of the corresponding inhomogeneous system

$$\frac{d\mathbf{x}}{dt} = \mathbf{A}\mathbf{x} + \mathbf{f}(t) \qquad (6\text{-}23)$$

is then obtained from equation (6-19) as

$$\mathbf{x}(t) = \exp(\mathbf{A}t)\mathbf{I}\mathbf{x}(0) + \int_0^t \exp(\mathbf{A}t)\exp(-\mathbf{A}\tau)\mathbf{f}(\tau)\,d\tau$$

giving

$$\mathbf{x}(t) = [\exp(\mathbf{A}t)]\mathbf{x}(0) + \int_0^t \exp[\mathbf{A}(t-\tau)]\mathbf{f}(\tau)\,d\tau \qquad (6\text{-}24)$$

so that the transition matrix for this system is

$$\boldsymbol{\Phi}(t, \tau) = \exp \mathbf{A}(t-\tau) = \sum_{i=0}^{\infty} \frac{\mathbf{A}^i(t-\tau)^i}{i!} \qquad (6\text{-}25)$$

It follows by direct differentiation that $\boldsymbol{\Phi}$ satisfies the equation

$$\frac{d\boldsymbol{\Phi}}{dt} = \mathbf{A}\boldsymbol{\Phi} \qquad (6\text{-}25\text{a})$$

Spectral Form for Constant Coefficient System
Consider the free motion of the linear constant coefficient system

$$\frac{d\mathbf{x}}{dt} = \mathbf{A}\mathbf{x}$$

Let the eigenvalues of the matrix $\mathbf{A}$ be distinct and denoted by $\{\lambda_1, \lambda_2, \ldots, \lambda_n\}$ and let $\{\mathbf{u}_1, \mathbf{u}_2, \ldots, \mathbf{u}_n\}$ be a corresponding eigenvector set. The state vector $\mathbf{x}$ may be expressed as a linear combination of the eigenvector set of $\mathbf{A}$ as

$$\mathbf{x} = \sum_{i=1}^{n} c_i \mathbf{u}_i$$

where $\{c_1, c_2, \ldots, c_n\}$ are the components of vector $\mathbf{x}$ referred to the eigenvector set as a basis. Note that the set $\{c_i : i = 1, 2, \ldots, n\}$ will be functions of time whereas the eigenvector set is constant. We therefore have

$$\frac{d\mathbf{x}}{dt} = \sum_{i=1}^{n} \frac{dc_i}{dt} \mathbf{u}_i$$

so that
$$\mathbf{A}\mathbf{x} = \mathbf{A}\sum_{i=1}^{n} c_i \mathbf{u}_i = \sum_{i=1}^{n} \frac{dc_i}{dt} \mathbf{u}_i \quad (6\text{-}26)$$

Now
$$\mathbf{A}\sum_{i=1}^{n} c_i \mathbf{u}_i = \sum_{i=1}^{n} c_i \mathbf{A}\mathbf{u}_i = \sum_{i=1}^{n} c_i \lambda_i \mathbf{u}_i \quad (6\text{-}27)$$

so that from equations (6–26) and (6–27) we have

$$\frac{dc_i}{dt} = \lambda_i c_i \quad i = 1, 2, \ldots, n$$

giving
$$c_i(t) = \exp(\lambda_i t) c_i(0) \quad i = 1, 2, \ldots, n$$

The time variation of the state vector for free motion of the system may thus be expressed in terms of the eigenvector set of $\mathbf{A}$ as

$$\mathbf{x}(t) = \sum_{i=1}^{n} c_i(0)[\exp(\lambda_i t)]\mathbf{u}_i$$

If $\{\mathbf{v}_1, \mathbf{v}_2, \ldots, \mathbf{v}_n\}$ is a dual basis set to the eigenvector set $\{\mathbf{u}_1, \mathbf{u}_2, \ldots, \mathbf{u}_n\}$ then, using the Projection Theorem of Chapter 2, the eigenvector-basis components $\{c_i(0)\}$ are obtained from the initial condition vector $\mathbf{x}(0)$ by

$$c_i(0) = \langle \mathbf{v}_i, \mathbf{x}(0) \rangle \quad i = 1, 2, \ldots, n$$

and thus the free motion may be expressed in the form

$$\mathbf{x}(t) = \sum_{i=1}^{n} \langle \mathbf{v}_i, \mathbf{x}(0) \rangle [\exp(\lambda_i t)]\mathbf{u}_i \quad (6\text{-}28)$$

This is called the spectral form of the solution for the free motion of the linear constant-coefficient system.

Complex Conjugate Eigenvalues

Since the matrix $\mathbf{A}$ is a matrix of real constants, the eigenvalues of $\mathbf{A}$ will be real or will occur in complex conjugate pairs. For a real eigenvalue there will be a real eigenvector, and for a pair of complex conjugate eigenvalues there will be a pair of complex conjugate eigenvectors. When complex conjugate eigenvectors are present, it is usually more convenient to discuss the resultant motion in terms of wholly real vectors by grouping the characteristic solutions together in complex conjugate pairs.

Let λ_1 and λ_2 be a pair of complex conjugate eigenvalues and let the corresponding pair of complex conjugate eigenvectors be $\mathbf{u}_1$ and $\mathbf{u}_2$ with a corresponding dual eigenvector set $\mathbf{v}_1$ and $\mathbf{v}_2$. Let $\mathbf{s}(t)$ be the sum of the two terms in the spectral form of solution of equations (6–28) which correspond to the eigenvalues λ_1 and λ_2, that is

$$\mathbf{s}(t) = \langle \mathbf{v}_1, \mathbf{x}(0) \rangle [\exp(\lambda_1 t)]\mathbf{u}_1 + \langle \mathbf{v}_2, \mathbf{x}(0) \rangle [\exp(\lambda_2 t)]\mathbf{u}_2$$

Now put
$$\mathbf{u}_1 = \mathbf{u}' + j\mathbf{u}''$$
$$\mathbf{u}_2 = \mathbf{u}' - j\mathbf{u}''$$
$$\mathbf{v}_1 = \mathbf{v}' + j\mathbf{v}''$$
$$\mathbf{v}_2 = \mathbf{v}' - j\mathbf{v}''$$

where $\mathbf{u}'$, $\mathbf{u}''$, $\mathbf{v}'$ and $\mathbf{v}''$ are vectors whose components are wholly real. Also put
$$\lambda_1 = \sigma_1 + j\omega_1$$
$$\lambda_2 = \sigma_1 - j\omega_1$$

We then have:
$$\begin{aligned}\mathbf{s}(t) &= \langle \mathbf{v}_1, \mathbf{x}(0)\rangle [\exp(\lambda_1 t)](\mathbf{u}' + j\mathbf{u}'') + \langle \mathbf{v}_2, \mathbf{x}(0)\rangle [\exp(\lambda_2 t)](\mathbf{u}' - j\mathbf{u}'') \\ &= \langle \mathbf{v}_1, \mathbf{x}(0)\rangle [\exp(\sigma_1 t)\exp(j\omega_1 t)](\mathbf{u}' + j\mathbf{u}'') \\ &\quad + \langle \mathbf{v}_2, \mathbf{x}(0)\rangle [\exp(\sigma_1 t)\exp(-j\omega_1 t)](\mathbf{u}' - j\mathbf{u}'') \\ &= \exp(\sigma_1 t)[(\alpha+\beta)\mathbf{u}' + j(\alpha-\beta)\mathbf{u}'']\end{aligned} \quad (6\text{-}29)$$

where $\alpha \pm \beta = \langle \mathbf{v}_1, \mathbf{x}(0)\rangle \exp(j\omega_1 t) \pm \langle \mathbf{v}_2, \mathbf{x}(0)\rangle \exp(-j\omega_1 t)$

In the expressions for α and β inner products occur in which the vectors concerned have complex entries. The scalar product of two complex vectors $\mathbf{x}$ and $\mathbf{y}$ is defined to be

$$\langle \mathbf{x}, \mathbf{y}\rangle = \sum_{i=1}^{n} x_i^* y_i \quad (6\text{-}30)$$

where the asterisk denotes complex conjugation. This convention is adopted so that the self-scalar product of vectors with complex entries does not vanish, and thus all complex vectors with non-zero entries have non-zero norms. It follows from this definition that if $\mathbf{p}$ and $\mathbf{q}$ are real vectors

$$\langle(\mathbf{p}+j\mathbf{q}), \mathbf{x}(0)\rangle = \langle \mathbf{p}, \mathbf{x}(0)\rangle - j\langle \mathbf{q}, \mathbf{x}(0)\rangle$$
and
$$\langle(\mathbf{p}-j\mathbf{q}), \mathbf{x}(0)\rangle = \langle \mathbf{p}, \mathbf{x}(0)\rangle + j\langle \mathbf{q}, \mathbf{x}(0)\rangle$$

Multiplying out the expressions for $\alpha \pm \beta$ thus gives

$$\alpha \pm \beta = (a-jb)(\cos\omega_1 t + j\sin\omega_1 t) \pm (a+jb)(\cos\omega_1 t - j\sin\omega_1 t)$$

where $a = \langle \mathbf{v}', \mathbf{x}(0)\rangle$ and $b = \langle \mathbf{v}'', \mathbf{x}(0)\rangle$. Multiplying out and collecting terms then gives

$$\alpha + \beta = 2a\cos\omega_1 t + 2b\sin\omega_1 t$$
$$\alpha - \beta = j2a\sin\omega_1 t - j2b\cos\omega_1 t$$

Substituting these expressions in equation (6–29) then gives

$$\mathbf{s}(t) = \exp(\sigma_1 t)\{[2a \cos\omega_1 t + 2b \sin\omega_1 t]\mathbf{u}'$$
$$+ j[j2a \sin\omega_1 t - j2b \cos\omega_1 t]\mathbf{u}''\}$$
$$= 2\exp(\sigma_1 t)\{[\langle \mathbf{v}', \mathbf{x}(0)\rangle \cos\omega_1 t + \langle \mathbf{v}'', \mathbf{x}(0)\rangle \sin\omega_1 t]\mathbf{u}'$$
$$+ [\langle \mathbf{v}'', \mathbf{x}(0)\rangle \cos\omega_1 t - \langle \mathbf{v}', \mathbf{x}(0)\rangle \sin\omega_1 t]\mathbf{u}''\} \quad (6\text{–}31)$$

which may be written in polar form

$$\mathbf{s}(t) = 2\exp(\sigma_1 t)\{\gamma \cos(\omega_1 t - \phi)\mathbf{u}' + \gamma \sin(\omega_1 t - \phi)\mathbf{u}''\}$$

where $\gamma \cos\phi = \langle \mathbf{v}', \mathbf{x}(0)\rangle$ and $\gamma \sin\phi = \langle \mathbf{v}'', \mathbf{x}(0)\rangle$.

This shows that the net contribution to the system free motion from a pair of complex conjugate eigenvectors may be expressed as a sum of oscillatory motions directed along two real vectors, $\mathbf{u}'$ and $\mathbf{u}''$, whose components are respectively the real and imaginary parts of the vector $\mathbf{u}_1$. It is important to note that *both* the amplitude and the phase of these oscillations depend on the initial state vector $\mathbf{x}(0)$. For release from any initial state in the plane spanned by the real vectors $\mathbf{u}'$ and $\mathbf{u}''$, the subsequent motion remains in this plane and consists of a spiral trajectory about the origin, having the form of a stable focus if σ_1 is negative, and an unstable focus if σ_1 is positive.

Suppose a matrix $\mathbf{A}$ has n independent eigenvectors corresponding to n distinct eigenvalues, some of which are real and the remainder are complex conjugate pairs. Let the real eigenvalues be $\lambda_1, \lambda_2, \ldots, \lambda_q$ corresponding to the real eigenvectors $\mathbf{u}_1, \mathbf{u}_2, \ldots, \mathbf{u}_q$ and let the complex conjugate eigenvalues be

$$\mu_1 = \sigma_1 + j\omega_1 \qquad \mu_1^* = \sigma_1 - j\omega_1$$
$$\mu_2 = \sigma_2 + j\omega_2 \qquad \mu_2^* = \sigma_2 - j\omega_2$$
$$\vdots$$
$$\mu_s = \sigma_s + j\omega_s \qquad \mu_s^* = \sigma_s - j\omega_s$$

where $2s + q = n$.

Let the corresponding complex conjugate eigenvectors be

$$\mathbf{w}_1 = \mathbf{r}_1 + j\mathbf{s}_1 \qquad \mathbf{w}_1^* = \mathbf{r}_1 - j\mathbf{s}_1$$
$$\mathbf{w}_2 = \mathbf{r}_2 + j\mathbf{s}_2 \qquad \mathbf{w}_2^* = \mathbf{r}_2 - j\mathbf{s}_2$$
$$\vdots$$
$$\mathbf{w}_s = \mathbf{r}_s + j\mathbf{s}_s \qquad \mathbf{w}_s^* = \mathbf{r}_s - j\mathbf{s}_s$$

Analytical Aspects of State Space Equation Sets

We will then have a set of relations of the form

$$\mathbf{A}\mathbf{u}_i = \lambda_i \mathbf{u}_i \qquad i = 1, 2, \ldots, q \qquad (6\text{-}32)$$

$$\mathbf{A}\mathbf{w}_i = \mu_i \mathbf{w}_i \qquad i = 1, 2, \ldots, s \qquad (6\text{-}33)$$

$$\mathbf{A}\mathbf{w}_i^* = \mu_i^* \mathbf{w}_i^* \qquad i = 1, 2, \ldots, s \qquad (6\text{-}34)$$

Equation (6–33) gives

$$\mathbf{A}(\mathbf{r}_i + j\mathbf{s}_i) = (\sigma_i + j\omega_i)(\mathbf{r}_i + j\mathbf{s}_i)$$

from which, by equating real and imaginary parts, we obtain the pair of relationships

$$\begin{aligned} \mathbf{A}\mathbf{r}_i &= \sigma_i \mathbf{r}_i - \omega_i \mathbf{s}_i \\ \mathbf{A}\mathbf{s}_i &= \omega_i \mathbf{r}_i + \sigma_i \mathbf{s}_i \end{aligned} \qquad (6\text{-}35)$$

This pair of relationships may also be obtained from equation (6–34). The set of equations (6–32) and (6–35) may be written in the form

$$\mathbf{A}\mathbf{U} = \mathbf{U}\boldsymbol{\Lambda} \qquad (6\text{-}36)$$

where $\mathbf{U}$ is composed of column vectors as follows

$$\mathbf{U} = [\mathbf{u}_1 \mathbf{u}_2 \cdots \mathbf{u}_q \mathbf{r}_1 \mathbf{s}_1 \mathbf{r}_2 \mathbf{s}_2 \ldots \mathbf{r}_s \mathbf{s}_s] \qquad (6\text{-}37)$$

and the block-diagonal matrix $\boldsymbol{\Lambda}$ has the form

$$\boldsymbol{\Lambda} = \begin{bmatrix} \lambda_1 & & & & & & & & & \\ & \lambda_2 & & & & & & & & \\ & & \ddots & & & & & 0 & & \\ & & & \lambda_q & & & & & & \\ & & & & \sigma_1 & \omega_1 & & & & \\ & & & & -\omega_1 & \sigma_1 & & & & \\ & & & & & & \sigma_2 & \omega_2 & & \\ & & & & & & -\omega_2 & \sigma_2 & & \\ & & 0 & & & & & & \ddots & \\ & & & & & & & & \sigma_s & \omega_s \\ & & & & & & & & -\omega_s & \sigma_s \end{bmatrix} \qquad (6\text{-}38)$$

Postmultiplying both sides of equation (6–36) by $\mathbf{U}^{-1}$ gives

$$\mathbf{A} = \mathbf{U}\boldsymbol{\Lambda}\mathbf{U}^{-1} \qquad (6\text{-}39)$$

and we may take Λ as defined in equation (6–38) as a natural canonical form for this case.

Now consider the spectral form of solution of the free motion system

$$\dot{\mathbf{x}} = \mathbf{A}\mathbf{x}$$

for this case. Express the state vector $\mathbf{x}$ as a linear combination of the vectors $\mathbf{u}_i$, $\mathbf{r}_i$ and $\mathbf{s}_i$, that is put

$$\mathbf{x} = \sum_{i=1}^{q} c_i \mathbf{u}_i + \sum_{i=1}^{s} p_i \mathbf{r}_i + \sum_{i=1}^{s} l_i \mathbf{s}_i \qquad (6\text{–}40)$$

so that

$$\dot{\mathbf{x}} = \sum_{i=1}^{q} \dot{c}_i \mathbf{u}_i + \sum_{i=1}^{s} \dot{p}_i \mathbf{r}_i + \sum_{i=1}^{s} \dot{l}_i \mathbf{s}_i \qquad (6\text{–}41)$$

and

$$\mathbf{A}\mathbf{x} = \sum_{i=1}^{q} c_i \mathbf{A}\mathbf{u}_i + \sum_{i=1}^{s} p_i \mathbf{A}\mathbf{r}_i + \sum_{i=1}^{s} l_i \mathbf{A}\mathbf{s}_i \qquad (6\text{–}42)$$

Using the relationships (6–32) and (6–35) in equation (6–42) then gives

$$\mathbf{A}\mathbf{x} = \sum_{i=1}^{q} c_i \lambda_i \mathbf{u}_i + \sum_{i=1}^{s} p_i(\sigma_i \mathbf{r}_i - \omega_i \mathbf{s}_i) + \sum_{i=1}^{s} l_i(\omega_i \mathbf{r}_i + \sigma_i \mathbf{s}_i)$$

$$\dot{\mathbf{x}} = \sum_{i=1}^{q} c_i \lambda_i \mathbf{u}_i + \sum_{i=1}^{s} (\sigma_i p_i + \omega_i l_i)\mathbf{r}_i + \sum_{i=1}^{s} (-\omega_i p_i + \sigma_i l_i)\mathbf{s}_i \qquad (6\text{–}43)$$

Equating the expressions for $\dot{\mathbf{x}}$ given by equations (6–43) and (6–41) then gives

$$\dot{c}_i = \lambda_i c_i$$
$$\dot{p}_i = \sigma_i p_i + \omega_i l_i$$
$$\dot{l}_i = -\omega_i p_i + \sigma_i l_i$$

which may be written in matrix form as

$$\frac{d}{dt}\begin{bmatrix} c_i \\ p_i \\ l_i \end{bmatrix} = \begin{bmatrix} \lambda_i & 0 & 0 \\ 0 & \sigma_i & \omega_i \\ 0 & -\omega_i & \sigma_i \end{bmatrix} \begin{bmatrix} c_i \\ p_i \\ l_i \end{bmatrix}$$

from which we obtain

$$c_i(t) = \exp(\lambda_i t) c_i(0) \qquad i = 1, 2, \ldots, q$$

$$\begin{bmatrix} p_i \\ l_i \end{bmatrix} = \exp(\mathbf{\Omega}_i t) \begin{bmatrix} p_i(0) \\ l_i(0) \end{bmatrix} \qquad i = 1, 2, \ldots, s$$

where

$$\mathbf{\Omega}_i = \begin{bmatrix} \sigma_i & \omega_i \\ -\omega_i & \sigma_i \end{bmatrix}$$

Analytical Aspects of State Space Equation Sets 407

The spectral form of solution may be written as

$$\mathbf{x}(t) = \sum_{i=1}^{q} \langle \mathbf{v}_i, \mathbf{x}(0) \rangle [\exp(\lambda_i t)] \mathbf{u}_i + \sum_{i=1}^{s} \langle \mathbf{a}_i, \mathbf{x}(0) \rangle [\exp(\mu_i t)] \mathbf{w}_i$$

$$+ \sum_{i=1}^{s} \langle \mathbf{a}_i^*, \mathbf{x}(0) \rangle [\exp(\mu_i^* t)] \mathbf{w}_i^* \quad (6\text{-}43a)$$

where $\mathbf{a}_i$ and $\mathbf{a}_i^*$ are the reciprocal basis vectors corresponding to the eigenvectors $\mathbf{w}_i$ and $\mathbf{w}_i^*$.

If $(\mathbf{r}_i + j\mathbf{s}_i)$ is a complex eigenvector corresponding to the complex eigenvalue $(\sigma_i + j\omega_i)$ then we have, as above,

$$\begin{aligned} \mathbf{A}\mathbf{r}_i &= \sigma_i \mathbf{r}_i - \omega_i \mathbf{s}_i \\ \mathbf{A}\mathbf{s}_i &= \omega_i \mathbf{r}_i + \sigma_i \mathbf{s}_i \end{aligned} \quad i = 1, 2, \ldots, s \quad (6\text{-}44)$$

Suppose some vector $\mathbf{x}$ is a linear combination of the vectors $\mathbf{r}_i$ and $\mathbf{s}_i$, that is lies in the plane spanned by $\mathbf{r}_i$ and $\mathbf{s}_i$, so that

$$\mathbf{x} = \alpha \mathbf{r}_i + \beta \mathbf{s}_i$$

where α and β are known constants. Then

$$\begin{aligned} \mathbf{A}\mathbf{x} &= \mathbf{A}(\alpha \mathbf{r}_i + \beta \mathbf{s}_i) \\ &= \alpha \mathbf{A}\mathbf{r}_i + \beta \mathbf{A}\mathbf{s}_i \\ &= \alpha(\sigma_i \mathbf{r}_i - \omega_i \mathbf{s}_i) + \beta(\omega_i \mathbf{r}_i + \sigma_i \mathbf{s}_i) \\ &= \gamma \mathbf{r}_i + \delta \mathbf{s}_i \end{aligned}$$

where γ and δ are another pair of constants. This shows that the plane spanned by $\mathbf{r}_i$ and $\mathbf{s}_i$ is invariant under the action of the operator $\mathbf{A}$. Thus any initial condition lying in this plane gives rise, as stated previously, to a free motion trajectory which remains in this plane.

Equations (6-44) may be written as

$$\begin{bmatrix} \mathbf{A} & \mathbf{0} \\ \mathbf{0} & \mathbf{A} \end{bmatrix} \begin{bmatrix} \mathbf{r}_i \\ \mathbf{s}_i \end{bmatrix} = \begin{bmatrix} \sigma_i \mathbf{I} & -\omega_i \mathbf{I} \\ \omega_i \mathbf{I} & \sigma_i \mathbf{I} \end{bmatrix} \begin{bmatrix} \mathbf{r}_i \\ \mathbf{s}_i \end{bmatrix} \quad (6\text{-}45)$$

where $\mathbf{I}$ and $\mathbf{0}$ respectively denote unit and null matrices of appropriate order. This is useful in determining the complex eigenvectors as shown in the example below.

Example 6.2
Consider the matrix

$$A = \begin{bmatrix} 0 & 1 \\ -8 & -4 \end{bmatrix}$$

408   State Models

We first determine its complex eigenvalues by solving the characteristic equation

$$\begin{vmatrix} -\lambda & 1 \\ -8 & -4-\lambda \end{vmatrix} = \lambda(4+\lambda)+8 = \lambda^2+4\lambda+8 = 0$$

This gives the pair of complex conjugate eigenvalues

$$\lambda_1 = -2+j2 \qquad \lambda_2 = -2-j2$$

Denote the corresponding complex eigenvector pair by $\mathbf{r} \pm j\mathbf{s}$. Using equation (6–45) gives

$$\begin{bmatrix} 0 & 1 & 0 & 0 \\ -8 & -4 & 0 & 0 \\ 0 & 0 & 0 & 1 \\ 0 & 0 & -8 & -4 \end{bmatrix} \begin{bmatrix} r_1 \\ r_2 \\ s_1 \\ s_2 \end{bmatrix} = \begin{bmatrix} -2 & 0 & -2 & 0 \\ 0 & -2 & 0 & -2 \\ 2 & 0 & -2 & 0 \\ 0 & 2 & 0 & -2 \end{bmatrix} \begin{bmatrix} r_1 \\ r_2 \\ s_1 \\ s_2 \end{bmatrix}$$

where r_1, r_2 and s_1, s_2 are the components of vectors $\mathbf{r}$ and $\mathbf{s}$ respectively. We thus have

$$\begin{bmatrix} 2 & 1 & 2 & 0 \\ -8 & -2 & 0 & 2 \\ -2 & 0 & 2 & 1 \\ 0 & -2 & -8 & -2 \end{bmatrix} \begin{bmatrix} r_1 \\ r_2 \\ s_1 \\ s_2 \end{bmatrix} = 0 \qquad (6\text{–}46)$$

The matrix on the left-hand side of this equation is of rank two. In other words, the number of unknowns exceeds the number of independent linear equations available by two and thus two of the four quantities r_1, r_2, s_1 and s_2 may be given arbitrary values. The reason for this arbitrariness in the determination of vectors $\mathbf{r}$ and $\mathbf{s}$ arises, as shown in the previous discussion, from the fact that we need only find any pair of vectors lying in the appropriate plane which is invariant under the action of $\mathbf{A}$. If we choose to allocate values to r_1 and s_1 then we may delete the first and third rows from equation (6–46). This gives

$$\begin{bmatrix} -8 & -2 & 0 & 2 \\ 0 & -2 & -8 & -2 \end{bmatrix} \begin{bmatrix} r_1 \\ r_2 \\ s_1 \\ s_2 \end{bmatrix} = 0$$

which may be rearranged to give

$$\begin{bmatrix} -2 & 2 \\ -2 & -2 \end{bmatrix} \begin{bmatrix} r_2 \\ s_2 \end{bmatrix} = \begin{bmatrix} 8 & 0 \\ 0 & 8 \end{bmatrix} \begin{bmatrix} r_1 \\ s_1 \end{bmatrix}$$

$$\begin{bmatrix} r_2 \\ s_2 \end{bmatrix} = \begin{bmatrix} -2 & 2 \\ -2 & -2 \end{bmatrix}^{-1} \begin{bmatrix} 8 & 0 \\ 0 & 8 \end{bmatrix} \begin{bmatrix} r_1 \\ s_1 \end{bmatrix} = 2 \begin{bmatrix} -1 & -1 \\ 1 & -1 \end{bmatrix} \begin{bmatrix} r_1 \\ s_1 \end{bmatrix}$$

In order to illustrate the effects of allocating different values to r_1 and s_1, we will consider two different choices and denote them by Case 1 and Case 2 respectively.

Case 1
Let $r_1 = s_1 = 1$. We then get

$$\begin{bmatrix} r_2 \\ s_2 \end{bmatrix} = 2 \begin{bmatrix} -1 & -1 \\ 1 & -1 \end{bmatrix} \begin{bmatrix} 1 \\ 1 \end{bmatrix} = \begin{bmatrix} -4 \\ 0 \end{bmatrix}$$

and so a pair of complex conjugate eigenvectors is given by

$$\begin{bmatrix} r_1 \\ r_2 \end{bmatrix} \pm j \begin{bmatrix} s_1 \\ s_2 \end{bmatrix} = \begin{bmatrix} 1 \pm j1 \\ -4 \pm j0 \end{bmatrix}$$

To verify that these are indeed eigenvectors for the eigenvalues $(-2 \pm j2)$ we may substitute them back into

$$[\mathbf{A} - \lambda \mathbf{I}]\mathbf{w} = \mathbf{0} \qquad (6\text{-}47)$$

This gives for the first eigenvector

$$\left\{ \begin{bmatrix} 0 & 1 \\ -8 & -4 \end{bmatrix} - \begin{bmatrix} -2+j2 & 0 \\ 0 & -2-j2 \end{bmatrix} \right\} \begin{bmatrix} 1+j1 \\ -4+j0 \end{bmatrix}$$

$$= \begin{bmatrix} +2-j2 & 1 \\ -2 & -2-j2 \end{bmatrix} \begin{bmatrix} 1+j1 \\ -4+j0 \end{bmatrix} = \begin{bmatrix} 0 \\ 0 \end{bmatrix}$$

and the validity of the second eigenvector may be checked in the same way.

Case 2
Let $r_1 = 0$ and $s_1 = 1$.
Then we have

$$\begin{bmatrix} r_2 \\ s_2 \end{bmatrix} = 2 \begin{bmatrix} -1 & -1 \\ 1 & -1 \end{bmatrix} \begin{bmatrix} 0 \\ 1 \end{bmatrix} = \begin{bmatrix} -2 \\ -2 \end{bmatrix}$$

giving us as eigenvectors the pair

$$\mathbf{w}, \mathbf{w}^* = \begin{bmatrix} 0 \pm j1 \\ -2 \mp j2 \end{bmatrix}$$

To verify that these are also eigenvectors we may again substitute in equation (6–47). This gives for the first eigenvector

$$\left\{ \begin{bmatrix} 0 & 1 \\ -8 & -4 \end{bmatrix} - \begin{bmatrix} -2+j2 & 0 \\ 0 & -2+j2 \end{bmatrix} \right\} \begin{bmatrix} 0+j1 \\ -2-j2 \end{bmatrix}$$

$$= \begin{bmatrix} j2+2-2-j2 \\ -j8+j8 \end{bmatrix} = \begin{bmatrix} 0 \\ 0 \end{bmatrix}$$

and the validity of the second eigenvector may be checked in the same way.

The eigenvectors obtained in Cases 1 and 2 are related to each other by a scalar complex multiplier. For example the first eigenvector in Case 1 is

$$\begin{bmatrix} 1+j1 \\ -4+j0 \end{bmatrix} = \begin{bmatrix} j1 \\ -2-j2 \end{bmatrix} - j \begin{bmatrix} j1 \\ -2-j2 \end{bmatrix} = (1-j1) \begin{bmatrix} 0+j1 \\ -2-j2 \end{bmatrix}$$

that is $(1-j1)$ times the first eigenvector in Case 2.

Finally we will use this example to illustrate the spectral solution form in the complex conjugate eigenvector case. We use equation (6–43a) and again consider the same two illustrative cases to show that the same solution is obtained irrespective of the arbitrariness in the choice leading to the complex eigenvectors.

Case 1

$$\mu = -2+j2 \qquad \mu^* = -2-j2$$

$$\mathbf{w} = \begin{bmatrix} 1+j1 \\ -4+j0 \end{bmatrix} \qquad \mathbf{w}^* = \begin{bmatrix} 1-j1 \\ -4+j0 \end{bmatrix}$$

To find the complex reciprocal eigenvector set, we form a matrix $\mathbf{W}$ from the columns $\mathbf{w}$, $\mathbf{w}^*$ and invert it. This gives

$$\mathbf{W} = \begin{bmatrix} 1+j1 & 1-j1 \\ -4+j0 & -4+j0 \end{bmatrix} \qquad \mathbf{W}^{-1} = \begin{bmatrix} 0-j\tfrac{1}{2} & -\tfrac{1}{8}-j\tfrac{1}{8} \\ 0+j\tfrac{1}{2} & -\tfrac{1}{8}+j\tfrac{1}{8} \end{bmatrix}$$

from which we get the reciprocal eigenvector set below. Because of the complex inner product convention explained on page 403, the reciprocal eigenvectors are taken as conjugates of the row entries of $\mathbf{W}^{-1}$.

$$\mathbf{a} = \begin{bmatrix} 0+j\tfrac{1}{2} \\ -\tfrac{1}{8}+j\tfrac{1}{8} \end{bmatrix}, \qquad \mathbf{a}^* = \begin{bmatrix} 0-j\tfrac{1}{2} \\ -\tfrac{1}{8}-j\tfrac{1}{8} \end{bmatrix}$$

Take an initial condition vector $\mathbf{x}(0) = \begin{bmatrix} 1 \\ 0 \end{bmatrix}$. We then have

$$\langle \mathbf{a}, \mathbf{x}(0) \rangle = -j\tfrac{1}{2}$$
$$\langle \mathbf{a^*}, \mathbf{x}(0) \rangle = +j\tfrac{1}{2}$$

Using equation (6-43), the spectral form of solution is thus

$$\mathbf{x}(t) = -\frac{j}{2}\exp(-2+j2)t \begin{bmatrix} 1+j1 \\ -4+j0 \end{bmatrix} + \frac{j}{2}\exp(-2-j2)t \begin{bmatrix} 1-j1 \\ -4+j0 \end{bmatrix}$$

which will be found to give

$$\mathbf{x}(t) = \exp(-2t) \begin{bmatrix} \cos 2t + \sin 2t \\ 4\sin 2t \end{bmatrix}$$

Case 2

$$\mathbf{w} = \begin{bmatrix} 0+j1 \\ -2-j2 \end{bmatrix} \qquad \mathbf{w^*} = \begin{bmatrix} 0-j1 \\ -2+j2 \end{bmatrix}$$

In a similar way to Case 1 we get

$$\mathbf{a} = \begin{bmatrix} -\tfrac{1}{2}+j\tfrac{1}{2} \\ -\tfrac{1}{4}+j0 \end{bmatrix} \qquad \mathbf{a^*} = \begin{bmatrix} -\tfrac{1}{2}-j\tfrac{1}{2} \\ -\tfrac{1}{4}+j0 \end{bmatrix}$$

The same initial condition vector gives

$$\langle \mathbf{a}, \mathbf{x}(0) \rangle = -\tfrac{1}{2}-j\tfrac{1}{2} \qquad \langle \mathbf{a^*}, \mathbf{x}(0) \rangle = -\tfrac{1}{2}+j\tfrac{1}{2}$$

so that we get

$$\mathbf{x}(t) = (-\tfrac{1}{2}-j\tfrac{1}{2})\exp(-2+j2)t \begin{bmatrix} 0+j1 \\ -2-j2 \end{bmatrix}$$
$$+(-\tfrac{1}{2}+j\tfrac{1}{2})\exp(-2-j2)t \begin{bmatrix} 0-j1 \\ -2+j2 \end{bmatrix}$$

which will be found to give the same answer as obtained above for Case 1. This shows that the arbitrariness in eigenvector determination is compensated by a suitable and automatic adjustment in the values of the corresponding reciprocal eigenvector set.

Repeated Eigenvalues

If any of the eigenvalues of the matrix **A** are repeated, the standard form of the matrix under a similarity transformation is the Jordan canonical form:

$$\mathbf{J} = \begin{bmatrix} \mathbf{J}_1 & 0 & 0 & \cdots & 0 \\ 0 & \mathbf{J}_2 & 0 & \cdots & 0 \\ \vdots & & & & \\ 0 & 0 & 0 & \cdots & \mathbf{J}_k \end{bmatrix}$$

where the submatrices $\mathbf{J}_j$ are $n_j \times n_j$ matrices of the form:

$$\mathbf{J}_j = \begin{bmatrix} \lambda_j & 1 & 0 & 0 & \cdots & 0 \\ 0 & \lambda_j & 1 & 0 & \cdots & 0 \\ 0 & 0 & \lambda_j & 1 & \cdots & 0 \\ \vdots & & & & & \vdots \\ 0 & 0 & 0 & 0 & \cdots & 1 \\ 0 & 0 & 0 & 0 & \cdots & \lambda_j \end{bmatrix}$$

having all diagonal entries equal to a single eigenvalue λ_j; all superdiagonal entries equal to $+1$; and all other entries zero. We may write

$$\mathbf{J}_j = \lambda_j \mathbf{I}_j + \mathbf{H}_j$$

where $\mathbf{I}_j$ is a unit matrix of appropriate order n_j and

$$\mathbf{H}_j = \begin{bmatrix} 0 & 1 & 0 & 0 & 0 & \cdots & 0 \\ 0 & 0 & 1 & 0 & 0 & \cdots & 0 \\ 0 & 0 & 0 & 1 & 0 & \cdots & 0 \\ \vdots & & & & & & \vdots \\ 0 & 0 & 0 & 0 & 0 & \cdots & 1 \\ 0 & 0 & 0 & 0 & 0 & \cdots & 0 \end{bmatrix}$$

so that $\mathbf{H}_j^{n_j} = 0$ since each multiplication by $\mathbf{H}_j$ on itself moves the diagonal row of units up one row, till they vanish at the top right-hand corner.

An extension of the diagonalization procedure discussed in Chapter 2 gives that

$$\exp(\mathbf{A}t) = \mathbf{U} \exp(\mathbf{J}t) \mathbf{U}^{-1}$$

where, as before, $\mathbf{U}$ is a matrix whose columns are a set of linearly independent eigenvectors. It is then found that

$$\exp(\mathbf{J}t) = \begin{bmatrix} \exp(\mathbf{J}_1 t) & & & \mathbf{0} \\ & \exp(\mathbf{J}_2 t) & & \\ & & \cdot & \\ & & & \cdot \\ \mathbf{0} & & & \exp(\mathbf{J}_k t) \end{bmatrix}$$

where each of the diagonal blocks $\exp(\mathbf{J}_j t)$ is of rank n_j and may be

expanded to give
$$\exp(\mathbf{J}_j t) = \exp(\lambda_j \mathbf{I}_j + \mathbf{H}_j)t$$
$$= \exp(\lambda_j t)\exp(\mathbf{H}_j t)$$

Now $\mathbf{H}_j^{n_j} = \mathbf{0}$

and so the exponential series for $\exp(\mathbf{H}_j t)$ will terminate after n_j terms. Therefore

$$\exp(\mathbf{H}_j t) = \mathbf{I} + \mathbf{H}_j t + \frac{1}{2!}\mathbf{H}_j^2 t^2 + \cdots + \frac{1}{(n_j-1)!}\mathbf{H}_j^{n_j-1} t^{n_j-1} + \mathbf{0} + \mathbf{0} + \cdots.$$

This shows that $\exp(\mathbf{J}_j t)$ will consist of products of the scalar exponential function $\exp(\lambda_j t)$ and powers of t of order less than n_j.

Laplace Transform Solution of Linear Constant Coefficient State Space Equations

Consider the linear constant coefficient system of equations

$$\dot{\mathbf{x}} = \mathbf{Ax} + \mathbf{Bu}$$
$$\mathbf{y} = \mathbf{Cx} \tag{6-48}$$

where $\mathbf{x}$ is the system state vector, $\mathbf{u}$ is a vector representing the effects of forcing inputs, and $\mathbf{y}$ is a vector whose components are output observations, assumed to be linear combinations of the state variables. Taking Laplace transforms in equation (6-48) gives

$$s\mathbf{x}(s) - \mathbf{x}(0) = \mathbf{A}\mathbf{x}(s) + \mathbf{B}\mathbf{u}(s)$$
$$\mathbf{y}(s) = \mathbf{C}\mathbf{x}(s)$$

Thus $[s\mathbf{I} - \mathbf{A}]\mathbf{x}(s) = \mathbf{x}(0) + \mathbf{B}\mathbf{u}(s)$

$$\mathbf{x}(s) = [s\mathbf{I} - \mathbf{A}]^{-1}\mathbf{x}(0) + [s\mathbf{I} - \mathbf{A}]^{-1}\mathbf{B}\mathbf{u}(s)$$

and $\mathbf{y}(s) = \mathbf{C}[s\mathbf{I} - \mathbf{A}]^{-1}\mathbf{x}(0) + \mathbf{C}[s\mathbf{I} - \mathbf{A}]^{-1}\mathbf{B}\mathbf{u}(s)$

For zero initial conditions, the input and output transform vectors are related by the transfer function matrix $\mathbf{G}(s)$ where

$$\mathbf{G}(s) = \mathbf{C}[s\mathbf{I} - \mathbf{A}]^{-1}\mathbf{B} \tag{6-49}$$

so that $\mathbf{y}(s) = \mathbf{G}(s)\mathbf{u}(s)$

In order to compute the transfer function matrix $\mathbf{G}(s)$ we must determine the inverse matrix $[s\mathbf{I} - \mathbf{A}]^{-1}$. This is most conveniently done using the following algorithm. We have

$$[s\mathbf{I} - \mathbf{A}]^{-1} = \frac{\mathrm{adj}\,[s\mathbf{I} - \mathbf{A}]}{\det\,[s\mathbf{I} - \mathbf{A}]} \tag{6-50}$$

Let

$$\det[s\mathbf{I} - \mathbf{A}] = \Delta(s) = s^n + d_1 s^{n-1} + d_2 s^{n-2} + \cdots + d_{n-1} s + d_n$$

$$\operatorname{adj}[s\mathbf{I} - \mathbf{A}] = \mathbf{F}_0 s^{n-1} + \mathbf{F}_1 s^{n-2} + \mathbf{F}_2 s^{n-3} + \cdots + \mathbf{F}_{n-2} s + \mathbf{F}_{n-1}$$

then the algorithm obtains the scalar coefficients $d_1, d_2, \ldots, d_n$ and the matrices $\mathbf{F}_0, \mathbf{F}_1, \ldots, \mathbf{F}_{n-1}$ by the following recursion:

$$\mathbf{F}_0 = \mathbf{I}$$

$$\begin{aligned}
d_1 &= -\operatorname{trace}(\mathbf{A}\mathbf{F}_0) & \mathbf{F}_1 &= \mathbf{A}\mathbf{F}_0 + d_1 \mathbf{I} \\
d_2 &= -\tfrac{1}{2}\operatorname{trace}(\mathbf{A}\mathbf{F}_1) & \mathbf{F}_2 &= \mathbf{A}\mathbf{F}_1 + d_2 \mathbf{I} \\
d_3 &= -\tfrac{1}{3}\operatorname{trace}(\mathbf{A}\mathbf{F}_2) & \mathbf{F}_3 &= \mathbf{A}\mathbf{F}_2 + d_3 \mathbf{I} \\
&\vdots & &\vdots \\
d_n &= -\tfrac{1}{n}\operatorname{trace}(\mathbf{A}\mathbf{F}_{n-1}) & \mathbf{F}_{n-1} &= \mathbf{A}\mathbf{F}_{n-2} + d_{n-1} \mathbf{I} \\
& & \mathbf{F}_n &= \mathbf{A}\mathbf{F}_{n-1} + d_n \mathbf{I} = \mathbf{0}
\end{aligned} \qquad (6\text{--}51)$$

The proof of this algorithm depends on the fact that the trace of the adjoint matrix of $(s\mathbf{I} - \mathbf{A})$ is equal to the derivative with respect to s of the characteristic equation of $\mathbf{A}$.
That is

$$\operatorname{trace}[\operatorname{adj}(s\mathbf{I} - \mathbf{A})] = \frac{d}{ds}\Delta(s) \qquad (6\text{--}52)$$

Consider the right-hand side of equation (6–52). We have

$$\frac{d}{ds} \begin{vmatrix} (s-a_{11}) & -a_{12} & \cdots & -a_{1n} \\ -a_{21} & (s-a_{22}) & \cdots & -a_{2n} \\ \vdots & & & \\ -a_{n1} & -a_{n2} & \cdots & (s-a_{nn}) \end{vmatrix}$$

$$= \begin{vmatrix} 1 & -a_{12} & \cdots & -a_{1n} \\ 0 & (s-a_{22}) & \cdots & -a_{2n} \\ \vdots & & & \\ 0 & -a_{n2} & \cdots & (s-a_{nn}) \end{vmatrix} + \begin{vmatrix} (s-a_{11}) & 0 & \cdots & -a_{1n} \\ -a_{21} & 1 & \cdots & -a_{2n} \\ \vdots & \vdots & & \\ -a_{n1} & 0 & \cdots & (s-a_{nn}) \end{vmatrix}$$

$$+ \cdots + \begin{vmatrix} (s-a_{11}) & -a_{12} & \cdots & 0 \\ -a_{21} & (s-a_{22}) & \cdots & 0 \\ \vdots & \vdots & & \vdots \\ -a_{n1} & -a_{n2} & \cdots & 1 \end{vmatrix}$$

Analytical Aspects of State Space Equation Sets

Expanding each of the determinants in this sum by the first, second, ..., nth column respectively gives

$$\frac{d}{ds}\Delta(s) = \Delta_{11}(s) + \Delta_{22}(s) + \cdots + \Delta_{nn}(s) = \sum_{i=1}^{n} \Delta_{ii}(s)$$

where $\Delta_{ii}(s)$ is the cofactor of the element at the intersection of the ith row and ith column of $\Delta(s)$. Now the elements $\Delta_{ii}(s)$ are also the diagonal elements of the matrix adj $(s\mathbf{I} - \mathbf{A})$ and so we have

$$\text{trace}\,[\text{adj}\,(s\mathbf{I} - \mathbf{A})] = \sum_{i=1}^{n} \Delta_{ii}(s) = \frac{d}{ds}\Delta(s)$$

Now consider the identiy

$$(s\mathbf{I} - \mathbf{A})\,\text{adj}\,(s\mathbf{I} - \mathbf{A}) = \Delta(s)\mathbf{I} \qquad (6\text{--}53)$$

We first note that this immediately gives $\mathbf{F}_0 = \mathbf{I}$.

Taking the trace of both sides of equation (6–53) gives

$$\text{trace}\,\{s[\text{adj}\,(s\mathbf{I} - \mathbf{A})]\} - \text{trace}\,\{\mathbf{A}[\text{adj}\,(s\mathbf{I} - \mathbf{A})]\} = n\Delta(s) \qquad (6\text{--}54)$$

and using equation (6–52) we have

$$s\frac{d\Delta(s)}{ds} - \text{trace}\,\{\mathbf{A}[\text{adj}\,(s\mathbf{I} - \mathbf{A})]\} = n\Delta(s)$$

so that

$$s\frac{d\Delta(s)}{ds} - \text{trace}\,\{\mathbf{A}[\mathbf{F}_0 s^{n-1} + \mathbf{F}_1 s^{n-2} + \cdots + \mathbf{F}_j s^{n-j-1} + \cdots + \mathbf{F}_{n-1}]\}$$

$$= ns^n + nd_1 s^{n-1} + nd_2 s^{n-2} + \cdots + nd_j s^{n-j} + \cdots + nd_n \qquad (6\text{--}55)$$

where

$$\frac{d\Delta(s)}{ds} = ns^{n-1} + (n-1)d_1 s^{n-2}$$
$$+ (n-2)d_2 s^{n-3} + \cdots + (n-j)d_j s^{n-j-1} + \cdots + d_{n-1}$$

and thus

$$s\frac{d\Delta(s)}{ds} = ns^n + (n-1)d_1 s^{n-1} + (n-2)d_2 s^{n-2}$$
$$+ \cdots + (n-j)d_j s^{n-j} + \cdots + d_{n-1}s \qquad (6\text{--}56)$$

Combining equations (6–55) and (6–56) we have

$$\text{trace}\,(\mathbf{AF}_0)s^{n-1} + \text{trace}\,(\mathbf{AF}_1)s^{n-2}$$
$$+ \cdots + \text{trace}\,(\mathbf{AF}_{j-1})s^{n-j} + \cdots + \text{trace}\,(\mathbf{AF}_{n-1})$$
$$= -d_1 s^{n-1} - 2d_2 s^{n-2} - \cdots - jd_j s^{n-j} \cdots - nd_n \qquad (6\text{--}57)$$

Equating the coefficients of like powers of s on both sides of this identity gives

$$d_1 = -\text{trace}(\mathbf{AF}_0)$$
$$d_2 = -\tfrac{1}{2}\text{trace}(\mathbf{AF}_1)$$
$$\vdots$$
$$d_j = -\frac{1}{j}\text{trace}(\mathbf{AF}_{j-1})$$
$$\vdots$$
$$d_n = -\frac{1}{n}\text{trace}(\mathbf{AF}_{n-1})$$

From the identity of equation (6–53) we have

$$(s\mathbf{I} - \mathbf{A})[\mathbf{F}_0 s^{n-1} + \mathbf{F}_1 s^{n-2} + \cdots + \mathbf{F}_{n-2} s + \mathbf{F}_{n-1}]$$
$$= \mathbf{F}_0 s^n + \mathbf{F}_1 s^{n-1} + \cdots + \mathbf{F}_{n-2} s^2 + \mathbf{F}_{n-1} s - \mathbf{AF}_0 s^{n-1} - \mathbf{AF}_1 s^{n-2}$$
$$\cdots - \mathbf{AF}_{n-2} s - \mathbf{AF}_{n-1}$$
$$= s^n + (\mathbf{F}_1 - \mathbf{AF}_0) s^{n-1} + (\mathbf{F}_2 - \mathbf{AF}_1) s^{n-2}$$
$$+ \cdots + (\mathbf{F}_{n-1} - \mathbf{AF}_{n-2}) s - \mathbf{AF}_{n-1}$$
$$= s^n \mathbf{I} + d_1 s^{n-1} \mathbf{I} + d_2 s^{n-2} \mathbf{I} + \cdots + d_{n-1} s \mathbf{I} + d_n \mathbf{I} \qquad (6\text{–}58)$$

Equating the coefficients of like powers on both sides of the identity (6–58) we finally obtain

$$\mathbf{F}_1 = \mathbf{AF}_0 + d_1 \mathbf{I}$$
$$\mathbf{F}_2 = \mathbf{AF}_1 + d_2 \mathbf{I}$$
$$\vdots$$
$$\mathbf{F}_{n-2} = \mathbf{AF}_{n-3} + d_{n-2} \mathbf{I}$$
$$\mathbf{F}_{n-1} = \mathbf{AF}_{n-2} + d_{n-1} \mathbf{I} \qquad \mathbf{AF}_{n-1} + d_n \mathbf{I} = \mathbf{0} \qquad (6\text{–}59)$$

From the final term of this sequence, equation (6–59), we have the useful relationship

$$\mathbf{A}^{-1} = -\frac{1}{d_n} \mathbf{F}_{n-1}$$

In evaluating recursively the sequence of matrices $\mathbf{F}_1, \mathbf{F}_2, \ldots$, etc., the fact that $\mathbf{F}_n = \mathbf{0}$ provides a useful check on the arithmetic involved.

Cayley–Hamilton Theorem

If we successively multiply the equations in the set (6–59) by $A^{n-1}, A^{n-2}, \ldots, A^0$ and rearrange them we have

$$d_1 A^{n-1} = A^{n-1} F_1 - A^n F_0$$
$$d_2 A^{n-2} = A^{n-2} F_2 - A^{n-1} F_1$$
$$\vdots \qquad\qquad (6\text{–}59a)$$
$$d_{n-2} A^2 = A^2 F_{n-2} - A^3 F_{n-3}$$
$$d_{n-1} A = A F_{n-1} - A^2 F_{n-2}$$
$$d_n I = -A F_{n-1}$$

Adding the set of equations (6–59a) together gives

$$d_1 A^{n-1} + d_2 A^{n-2} + \cdots + d_{n-1} A + d_n I = -A^n F_0 = -A^n$$

and so we have

$$A^n + d_1 A^{n-1} + d_2 A^{n-2} + \cdots + d_{n-1} A + d_n I = 0 \qquad (6\text{–}60)$$

This establishes the important Cayley–Hamilton theorem: *that every matrix satisfies its own characteristic equation.*

Sylvester Expansion

By the Cayley–Hamilton theorem above, any matrix A must satisfy its own characteristic equation and we can thus always express A^n in terms of lower-order powers of A. Suppose we have some polynomial $f(A)$ in which a power A^s occurs where

$$s > n \quad \text{and} \quad n+l = s$$

This can be replaced as follows

$$A^s = A^l A^n = A^l[\phi(A)] = \psi(A)$$

where $\phi(A)$ is computed using the Cayley–Hamilton theorem. Any terms in $\psi(A)$ of higher than nth order can be treated in a similar way. Arguing in this fashion we see that any given polynomial $f(A)$ can be replaced by an equivalent polynomial of not more than nth order. Therefore given any polynomial $f(A)$, where A is a matrix with distinct eigenvalues, we may write

$$f(A) = \sum_{i=1}^{n} \alpha_i \prod_{\substack{j=1 \\ j \neq i}}^{n} (A - \lambda_j I) \qquad (6\text{–}61)$$

where $\{\alpha_1, \alpha_2, \ldots, \alpha_n\}$ is a set of constants which may be determined in the following way. Multipy both sides of equation (6–61) on the right by $\mathbf{u}_k$, where $\mathbf{u}_k$ is the eigenvector corresponding to the eigenvalue λ_k. This gives

$$\mathbf{f(A)}\mathbf{u}_k = \mathbf{f}(\lambda_k)\mathbf{u}_k = \sum_{i=1}^{n} \alpha_i \prod_{\substack{j=1 \\ j \neq i}}^{n} (\mathbf{A} - \lambda_j \mathbf{I})\mathbf{u}_k$$

$$= \alpha_k \prod_{\substack{j=1 \\ j \neq k}}^{n} (\lambda_k - \lambda_j)\mathbf{u}_k$$

from which the constants are found as

$$\alpha_k = \frac{f(\lambda_k)}{\prod_{\substack{j=1 \\ j \neq k}}^{n} (\lambda_k - \lambda_j)} \qquad k = 1, 2, \ldots, n \tag{6-62}$$

Substituting from equation (6–62) into equation (6–61) then gives

$$\mathbf{f(A)} = \sum_{i=1}^{n} f(\lambda_i)\mathbf{Z}(\lambda_i) \tag{6-63}$$

where

$$\mathbf{Z}(\lambda_i) = \prod_{\substack{j=1 \\ i \neq j}}^{n} \frac{(\lambda_j \mathbf{I} - \mathbf{A})}{(\lambda_j - \lambda_i)}$$

This form of $\mathbf{f(A)}$ is usually called Sylvester's interpolation formula.

Since we can expand $\exp(\mathbf{A}t)$ as a convergent power series it follows that we can express this matrix exponential in the form

$$\exp(At) = \sum_{i=1}^{n} \exp(\lambda_i t) \prod_{\substack{j=1 \\ j \neq i}}^{n} \frac{(\lambda_j \mathbf{I} - \mathbf{A})}{(\lambda_j - \lambda_i)} \tag{6-64}$$

We may also express the resolvent matrix $(s\mathbf{I} - \mathbf{A})^{-1}$ in the form

$$(s\mathbf{I} - \mathbf{A})^{-1} = \sum_{i=1}^{n} (s - \lambda_i)^{-1} \prod_{\substack{j=1 \\ j \neq i}}^{n} \frac{(\lambda_j \mathbf{I} - \mathbf{A})}{(\lambda_j - \lambda_i)}$$

$$= \sum_{i=1}^{n} \frac{\mathbf{Z}_i}{(s - \lambda_i)} \tag{6-65}$$

where

$$\mathbf{Z}_i = \prod_{\substack{j=1 \\ i \neq j}}^{n} \frac{(\lambda_j \mathbf{I} - \mathbf{A})}{(\lambda_j - \lambda_i)} \tag{6-66}$$

and it has been assumed that the eigenvalues of $\mathbf{A}$ are distinct. This form of resolvent matrix expansion leads to the matrix equivalent of

Analytical Aspects of State Space Equation Sets

Heaviside's expansion theorem. We have

$$(s-\lambda_k)(s\mathbf{I}-\mathbf{A})^{-1} = (s-\lambda_k)\sum_{i=1}^{k-1}\frac{\mathbf{Z}_i}{(s-\lambda_i)}+\mathbf{Z}_k+(s-\lambda_k)\sum_{i=k+1}^{n}\frac{\mathbf{Z}_i}{(s-\lambda_i)}$$

from which we obtain

$$\mathbf{Z}_k = \lim_{s \to \lambda_k}[(s-\lambda_k)(s\mathbf{I}-\mathbf{A})^{-1}] \qquad (6\text{-}67)$$

Scaling of linear equation sets. In the neighbourhood of any specified point in the state space, the behaviour of a dynamical system model may be approximated by the set of equations

$$\frac{d\mathbf{x}}{dt} = \mathbf{J}\mathbf{x}+\mathbf{g}$$

where $\mathbf{J}$ is the Liapunov first-approximation matrix evaluated at the state about which the linearized behaviour is being investigated and $\mathbf{g}$ is a vector representing the effect of forcing functions. In investigations of the behaviour of models representing practical systems, a very wide range of values of the elements of the matrix $\mathbf{J}$ is encountered, and it is often essential, particularly in analog computing studies, to deal with equation sets such that as many as possible of the element values of the $\mathbf{J}$ matrix lie in the magnitude range 0·1 to 10 over the region of state space in which the behaviour is investigated. Any modification of the system equations to ensure this will be termed a scaling operation.

(a) *Time scaling.* The time scale of model system behaviour is changed by putting

$$t = k_0\tau$$

where k_0 is a real constant called the time-scaling constant. This gives

$$\frac{d\mathbf{x}}{d\tau} = k_0\mathbf{J}\mathbf{x}+k_0\mathbf{g}$$

A time-scaling operation multiplies all the eigenvalues of the matrix $\mathbf{J}$ by the same constant k_0 and thus is used to bring the speed of model response into a desired range, usually the operational speed range of an analogue computer.

(b) *Amplitude scaling.* If all of the element values of $\mathbf{J}$ are of the same order of magnitude, a time-scaling operation may suffice. The element values, however, are normally of widely differing magnitude and an amplitude scaling operation is also required. This is carried out by the state variable transformation

$$\bar{\mathbf{x}} = \mathbf{K}\mathbf{x} \qquad \mathbf{x} = \mathbf{K}^{-1}\bar{\mathbf{x}}$$

420 State Models

where **K** is a non-singular square matrix of scaling coefficients and $\bar{\mathbf{x}}$ is the amplitude-scaled state vector. We thus have

$$\dot{\bar{\mathbf{x}}} = \mathbf{K}\dot{\mathbf{x}}$$

so that
$$\dot{\bar{\mathbf{x}}} = \mathbf{KJx} + \mathbf{Kg}$$
$$= \mathbf{KJK}^{-1}\bar{\mathbf{x}} + \mathbf{Kg}$$

The elements of the amplitude-scaling matrix **K** are chosen to bring as many elements of the matrix $\mathbf{KJK}^{-1}$ into the desired range of magnitude (which is usually the range 0·1 to 10 for analog computer work). The eigenvalues of the matrix **J** are unaltered under the similarity transformation **K**, and so amplitude scaling does not affect the time scale of system behaviour.

Combined time- and amplitude-scaling operations give

$$\frac{d\bar{\mathbf{x}}}{d\tau} = k_0 \mathbf{KJK}^{-1}\bar{\mathbf{x}} + k_0 \mathbf{Kg} \qquad (6\text{--}68)$$

In practice **K** is usually confined to a diagonal matrix, but such a restriction is not essential. The matrix **K** is usually chosen to bring all of the system matrix elements to the same order of magnitude and then the time-scaling constant k_0 chosen to shift this to the desired range of magnitudes.

6.2 Stability

The equation

$$\Delta(s) = s^n + d_1 s^{n-1} + d_2 s^{n-2} + \cdots + d_{n-1} s + d_n = 0 \qquad (6\text{--}69)$$

is called the characteristic equation of the system. When it has been obtained, a direct test may be made for the stability of the system by solving for its roots. The system will be stable if all the roots of the characteristic equation have negative real parts. If a digital computer and suitable standard programs are available, then this may be the most expedient thing to do; alternatively one can obtain the roots as eigenvalues of the system **A**-matrix using an eigenanalysis program. If access to a computer is limited or not available, certain algebraic criteria are useful which provide information about the presence of roots with positive real parts in terms of the equation coefficients. The most useful of these is the following simple necessary *but not sufficient* condition, which should always be checked whether a computer is available or not.

Stodola Criterion

If the characteristic equation is to have all its roots with negative real part, it is necessary that all its coefficients are nonzero and of the same sign. In particular, if the coefficient of s^n is $+1$ (which may always be arranged without loss of generality by dividing equation (6–69) through by the coefficient of s^n), then it is necessary that

$$d_1 > 0 \quad d_2 > 0 \quad \cdots \quad d_n > 0$$

Thus if the coefficients of the characteristic equation exhibit a sign change, or if any coefficient is missing then the system will be unstable or have some root or roots which are pure imaginary. If this direct inspection test is satisfied, then further tests may be made which provide necessary and sufficient conditions for all the roots to have negative real part. The Stodola criterion is a simple corollary of the Routh criterion below.

Maxwell–Vyschnegradskii Criterion

In 1868 Maxwell published an investigation into the stability of systems controlled by the Watt flyball governor [M17]. He was concerned with the special case $n = 3$ and noted that the corresponding system would be stable if and only if

$$d_1 > 0 \quad d_3 > 0 \quad d_1 d_2 > d_3$$

A similar criterion resulting from an independent study of the Watt governor was put forward by the Russian engineer Vyschnegradskii in 1876. In some Russian literature, a normalized form of third-order equation is used with $d_3 = 1$. A plot on the corresponding d_1, d_2 plane is called the Vyschnegradskii diagram with a stable area corresponding to

$$d_1 > 0 \quad d_2 > 0 \quad d_1 d_2 > 1$$

Routh's Stability Criterion [R5]

After reading his paper on governors, Maxwell posed the general problem for an nth-order characteristic equation since Hermite's solution, given below, was not widely known at the time. A solution was put forward by Routh in his Adams Prize Essay of 1877 which is well suited to hand computation. This set of necessary and sufficient conditions for all the roots of equation (6–69) to have negative real part is that all the elements of the first column of an array of numbers formed in a particular way specified below are positive. The array consists of several rows, and the first row has elements c_{ij} defined in terms of the even and odd coefficients of equation (6–69) as

$$c_{11} = 1$$
$$c_{ij} = d_{2(j-1)} \quad \text{for} \quad 2 \leqslant j \leqslant m+1$$
$$c_{ij} = 0 \quad \text{for} \quad j > m+1$$

The second row has elements

$$c_{2j} = d_{2j-1} \quad \text{for} \quad 1 \leq j \leq m$$
$$c_{2j} = 0 \quad \text{for} \quad j > m$$
$$\text{if } n = 2m$$

$$c_{2j} = d_{2j-1} \quad \text{for} \quad 1 \leq j \leq m+1$$
$$c_{2j} = 0 \quad \text{for} \quad j > m+1$$
$$\text{if } n = 2m+1$$

The elements of subsequent rows are defined by the recurrence relationship

$$c_{ij} = c_{(i-2),(j+1)} - \frac{c_{(i-2),1} c_{(i-1),(j+1)}}{c_{(i-1),1}}$$

$$\text{for} \quad i = 3, 4, 5, \ldots, (n+1)$$

$$j = 1, 2, 3, \ldots$$

Routh's Tabular Algorithm

Routh's criterion is easily applied in the following tabular form

	1 d_1	d_2 d_3	d_4 d_5	d_6 d_7	etc
$\alpha_1 = \dfrac{1}{d_1}$	$\beta_1 = d_2 - \alpha_1 d_3$	$\beta_2 = d_4 - \alpha_1 d_5$	$\beta_3 = d_6 - \alpha_1 d_7$	$\ldots$	$\ldots$
$\alpha_2 = \dfrac{d_1}{\beta_1}$	$\gamma_1 = d_3 - \alpha_2 \beta_2$	$\gamma_2 = d_5 - \alpha_2 \beta_3$	$\ldots$		
$\alpha_3 = \dfrac{\beta_1}{\gamma_1}$	$\delta_1 = \beta_2 - d_3 \gamma_2$	$\ldots$			
$\alpha_4 = \dfrac{\gamma_1}{\delta_1}$	$\ldots$				

All of the roots of equation (6–69) will have negative real parts if and only if $d_1, \beta_1, \gamma_1, \delta_1, \ldots$ in the above table are all positive.

Hurwitz's Criterion [H6]

Hurwitz's criterion, published in 1895, states that a necessary and sufficient condition for all the roots of equation (6–69) to have negative real parts is that the series of determinants $\Delta_1, \Delta_2, \Delta_3, \ldots, \Delta_n$ are all positive where

$$\Delta_1 = d_1$$

$$\Delta_2 = \begin{vmatrix} d_1 & 1 \\ d_3 & d_2 \end{vmatrix}$$

$$\Delta_3 = \begin{vmatrix} d_1 & 1 & 0 \\ d_3 & d_2 & d_1 \\ d_5 & d_4 & d_3 \end{vmatrix}$$

$$\Delta_4 = \begin{vmatrix} d_1 & 1 & 0 & 0 \\ d_3 & d_2 & d_1 & 1 \\ d_5 & d_4 & d_3 & d_2 \\ d_7 & d_6 & d_5 & d_4 \end{vmatrix}$$

with the general form given by

$$\Delta_r = \begin{vmatrix} d_1 & 1 & 0 & 0 & 0 & \cdots & 0 \\ d_3 & d_2 & d_1 & 1 & 0 & \cdots & 0 \\ \vdots & & & & & & \\ d_{2r-1} & d_{2r-2} & d_{2r-3} & d_{2r-4} & d_{2r-5} & \cdots & d_r \end{vmatrix}$$

where in the general form Δ_r the element d_s is zero if $s > n$.

Lienard–Chipart Criteria [L5]

The requirement that all the Hurwitz determinants are positive involves some redundant computation. Lienard and Chipart discovered four *alternative* sets of conditions which involve only computing about half the number of Hurwitz determinants. These four alternative sets of conditions that all the roots of equation (6–69) have negative real parts are

(a)
$d_n > 0 \quad d_{n-2} > 0 \quad d_{n-4} > 0 \quad \cdots$
$\Delta_1 > 0 \quad \Delta_3 > 0 \quad \Delta_5 > 0 \quad \cdots$

(b)
$d_n > 0 \quad d_{n-2} > 0 \quad d_{n-4} > 0 \quad \cdots$
$\Delta_2 > 0 \quad \Delta_4 > 0 \quad \Delta_6 > 0 \quad \cdots$

(c)
$d_n > 0 \quad d_{n-1} > 0 \quad d_{n-3} > 0 \quad \cdots$
$\Delta_1 > 0 \quad \Delta_3 > 0 \quad \Delta_5 > 0 \quad \cdots$

(d)
$d_n > 0 \quad d_{n-1} > 0 \quad d_{n-3} > 0 \quad \cdots$
$\Delta_2 > 0 \quad \Delta_4 > 0 \quad \Delta_6 > 0 \quad \cdots$

Hermite's Criterion [H7]

The first known set of necessary and sufficient conditions are due to Hermite and were published in 1856. Hermite's criterion states that a necessary and sufficient condition that all the roots of equation

(6–69) have negative real part is that the following set of symmetrical determinants are all positive

$$d_1 > 0$$

$$\begin{vmatrix} (-d_3 + d_1 d_2) & 0 \\ 0 & d_1 \end{vmatrix} > 0$$

$$\begin{vmatrix} (d_5 - d_1 d_4 + d_2 d_3) & 0 & d_3 \\ 0 & (-d_3 + d_1 d_2) & 0 \\ d_3 & 0 & d_1 \end{vmatrix} > 0$$

$$\begin{vmatrix} (-d_7 + d_1 d_6 - d_2 d_5 + d_3 d_4) & 0 & (-d_5 + d_1 d_4) & 0 \\ 0 & (d_5 - d_1 d_4 + d_2 d_3) & 0 & d_3 \\ (-d_5 + d_1 d_4) & 0 & (-d_3 + d_1 d_2) & 0 \\ 0 & d_3 & 0 & d_1 \end{vmatrix} > 0$$

and so on up to the $n \times n$ determinant with elements p_{ij} (where $1 \leq i \leq n$ and $1 \leq j \leq n$) such that

$$p_{ij} = \sum_{k=0}^{n} (-1)^{k+n-i} d_k d_{2n-i-j-k+1} \quad \text{where } i \geq j \text{ and } (i+j) \text{ is even}$$

$$p_{ij} = 0 \quad \text{where } (i+j) \text{ is odd}$$

$$p_{ij} = p_{ji} \quad \text{where } i < j$$

and where $d_s = 0$ if $s > n$.

These determinants are all symmetrical and may be expressed as the product of two lower-order symmetrical determinants. Thus the highest-order determinant to be evaluated is of order

$$\frac{n}{2} \times \frac{n}{2} \quad \text{or} \quad \left(\frac{n+1}{2}\right) \times \left(\frac{n+1}{2}\right)$$

according as n is even or odd respectively.

6.2.1 Liapunov Stability Theory [L6]

Liapunov originated two distinct approaches to the problem of stability determination; these two approaches are normally termed his *indirect* method and his *direct* method respectively. The indirect method establishes the stability or instability of a singular point of a nonlinear differential equation set in terms of the stability or instability of the singular point for the corresponding set of locally-approximating linear equations. It is simple to apply and should always be used as a criterion of instability; the only stability information which may normally be obtained by use of the indirect method is the asymptotic stability of a singular point within an arbitrarily small region of asymptotic stability surrounding it. The proof of the indirect method relies on the more general concepts of the direct method.

The direct method investigates the stability of a system without recourse to the solution of the governing differential equation set. It uses a scalar function (Liapunov function) in the system state space whose level surfaces (constant-value contours) form closed shells about the singular point under investigation (or closed tubes about a trajectory under investigation). Stability or instability is then determined from the sign of the time derivative of this scalar function, computed with respect to the system state, since the sign of this time derivative determines whether state trajectories are crossing the shells towards or away from the singular point (trajectory).

The behaviour of the dynamical systems being investigated is assumed to be governed by the set of state space equations

$$\frac{dx_i}{dt} = f_i(x_1, x_2, \ldots, x_n; t) \qquad i = 1, 2, \ldots, n \qquad (6\text{-}70)$$

where $x_1, x_2, \ldots, x_n$ are a set of state variables. These are zero-source equations; we are either concerned with small perturbations about an equilibrium point, or with an equivalent free-motion system for small perturbations about a driven state. It is convenient to write this set of equations as

$$\frac{d\mathbf{x}}{dt} = \mathbf{f}(\mathbf{x}, t) \qquad (6\text{-}71)$$

where $\mathbf{x}$ and $\mathbf{f}$ are n-vectors. If the components f_i all satisfy a Lipschitz condition for all $\mathbf{x}$ and $t > t_0$, then equation (6-71) has a *unique* solution $\mathbf{x}(t, \mathbf{x}_0, t_0)$ such that $\mathbf{x}(t_0) = \mathbf{x}_0$.

A point $\mathbf{x}'$ in the state space at which the functions

$$f_i(x'_1, x'_2, \ldots, x'_n, t) = 0 \qquad i = 1, 2, \ldots, n$$

is termed a *singular point* of the set of equations (6-70) and (6-71). Since all derivatives of the state variables vanish at a singular point, it corresponds to an equilibrium condition of the dynamical system. In what follows we will always take a singular point under investigation to be the origin of coordinates in the state space, since this can always be accomplished by a change of coordinates.

Stability of a singular point. A singular point at the origin of equation (6-54) is stable in the sense of Liapunov if, given a real constant $\epsilon > 0$,

there exists a real constant $\delta(\epsilon, t_0) > 0$ such that the inequality
$$\|\mathbf{x}_0\|_E < \delta$$
implies the inequality
$$\|\mathbf{x}(t, \mathbf{x}_0, t_0)\|_E < \epsilon$$
for all $\quad t > t_0$

The stability is said to be uniform if δ is independent of t_0.

Asymptotic stability of a singular point. A singular point at the origin of equation (6–71) is asymptotically stable, if it is stable and if in addition the inequality
$$\|\mathbf{x}_0\|_E < \delta_0 \quad \text{for some} \quad \delta_0 > 0$$
implies that for each $\eta > 0$ there exists a real number $\tau(\eta, \mathbf{x}_0, t_0)$ such that
$$\|\mathbf{x}(t, \mathbf{x}_0, t_0)\|_E < \eta \quad \text{when} \quad t > t_0 + \tau$$

When τ can be chosen independently of t_0 and $\mathbf{x}_0$, and the stability is uniform, the asymptotic stability is said to be uniform. The above inequalities ensure that
$$\mathbf{x}(t, t_0, \mathbf{x}_0) \to 0 \quad \text{as} \quad t \to \infty$$

Instability. A system is said to be unstable if it is not stable; just as a system can be stable in a number of different senses, so we get a variety of senses in which a system may be said to be unstable.

Region of asymptotic stability. The set of all release points $\mathbf{x}_0$ for which the subsequent motion is such that
$$\mathbf{x}(t, \mathbf{x}_0, t_0) \quad \text{is bounded and} \to 0 \text{ as} \quad t \to \infty$$
is called the region of asymptotic stability for the origin singular point.

If the region of asymptotic stability is the whole state space, the origin singular point is said to be asymptotically stable in the large, or completely stable.

It is important to realize that the concept of asymptotic stability is rather 'theoretical' in that very large excursions of state variables, which may be unacceptable in practice, may occur before the system 'settles down'. Nor does asymptotic stability imply insensitivity to small variations in release coordinates. Both these points are illustrated by the following example. The one-dimensional system
$$\frac{dx}{dt} = g(t)x$$
with
$$g(t) = \log_e 10 \quad \text{for} \quad 0 \leqslant t \leqslant 10$$
$$g(t) = -1 \quad \text{for} \quad t > 10$$

has the solution

$$x(t) = 10^t x_0 \quad \text{for} \quad 0 \leqslant t \leqslant 10$$
$$x(t) = 10^{10} \exp(10-t) x_0 \quad \text{for} \quad t > 10$$

It is thus asymptotically stable. Despite this $x(t)$ becomes extremely large for $t = 10$ and a change of only 10^{-5} in the value of x_0 gives a change in $x(20)$ of $10^{10} e^{-10} 10^{-5} \approx 4 \cdot 5$.

Positive definite function. A function $V(\mathbf{x})$ is said to be positive definite if

(a) $V(\mathbf{0}) = 0$
(b) $V(\mathbf{x}) > 0$ for $\mathbf{x} \neq \mathbf{0}$ in R

where R is a spherical neighbourhood containing the origin of the state space.

(c) $V(\mathbf{x})$ and $\partial V/\partial x_1, \partial V/\partial x_2, \ldots, \partial V/\partial x_n$ are continuous in R.

Such a function will have an isolated minimum at the origin and in some region sufficiently close to the origin the level surfaces $V =$ constant will form closed shells round the origin.

The function $V(\mathbf{x}, t)$ is said to be positive definite if it is continuous together with its first partial derivatives in some half-cylindrical region Φ about the t-axis; if $V(\mathbf{0}, t) = 0$; and $V(\mathbf{x}, t) > W(\mathbf{x})$ where $W(\mathbf{x})$ is positive definite in Φ.

Positive semi-definite function. A function $V(\mathbf{x})$ is said to be positive semi-definite if condition (b) above is relaxed to

(b') $V(\mathbf{x}) \geqslant 0$ for $\mathbf{x} \neq \mathbf{0}$ in R

Negative definite and negative semi-definite functions. If the negative of a function is positive definite or positive semi-definite, the function is said to be negative definite or negative semi-definite respectively.

Liapunov Stability Theorems

Liapunov function. A function $V(\mathbf{x})$ is said to be a Liapunov function in a region R provided

(a) it is positive definite in R
(b) its time derivative with respect to equation (6–71)

$$\frac{dV}{dt} = \sum_{i=1}^{n} \frac{\partial V}{\partial x_i} \frac{dx_i}{dt} \qquad (6\text{–}72)$$

is continuous and negative semi-definite in R. A function $V(\mathbf{x}, t)$ is said to be a Liapunov function in a region in event space Φ provided

(a) it is positive definite in Φ
(b) its time derivative with respect to equation (6–54)

$$\frac{dV}{dt} = \frac{\partial V}{\partial t} + \sum_{i=1}^{n} \frac{\partial V}{\partial x_i} \frac{dx_i}{dt} \qquad (6\text{–}73)$$

is continuous and negative semi-definite in Φ.

Stability of singular point. The existence of a Liapunov function $V(\mathbf{x}, t)$ in Φ is a sufficient condition for the origin point of equation (6–71) to be stable in the sense of Liapunov.

Suppose a Liapunov function V exists for the equation set in the region Φ. Let C_1 be a real constant such that the surface

$$V(\mathbf{x}, t) = C_1$$

is wholly contained in Φ. Let Ω be the region in $X \times T$ such that

$$V(\mathbf{x}, t) < C_1$$

Since $V(\mathbf{x}, t)$ is a Liapunov function in Φ

$$\frac{\partial V}{\partial t} + \left\langle \left[\frac{\partial V(\mathbf{x}, t)}{\partial \mathbf{x}}\right]^t, \dot{\mathbf{x}} \right\rangle \leqslant 0 \quad \text{in} \quad \Phi$$

or

$$\left\langle \left(\frac{\partial V}{\partial \mathbf{y}}\right)^t, \dot{\mathbf{y}} \right\rangle \leqslant 0 \quad \text{in} \quad \Phi$$

where

$$\mathbf{y} = \begin{bmatrix} t \\ \mathbf{x} \end{bmatrix} \quad \dot{\mathbf{y}} = \begin{bmatrix} 1 \\ \dot{\mathbf{x}} \end{bmatrix}$$

Now $(\partial V/\partial \mathbf{y})^t$ is a vector in the $(n+1)$-space containing Φ, having the direction of an outward normal to the surface $V(\mathbf{y}) = C$, and $\dot{\mathbf{y}}$ is another vector having the direction of the trajectory through the point $\mathbf{y}$. The inequality above thus implies that through all points in the surface $V = C_1$ pass trajectories which are either directed from Φ into Ω or are tangent to the surface. Thus *no* trajectory exists such that the motion of a state point is from Ω into Φ.

Let N_1 and N_2 be half-cylindrical regions of radius ϵ and δ respectively centred on the origin, such that N_1 lies wholly in Ω and N_2 lies wholly in Φ. It is then an immediate consequence of the above that for all release points for which

$$\|\mathbf{x}_0\|_E < \delta \quad \text{at} \quad t = t_0$$

the subsequent motion is such that

$$\|\mathbf{x}(t, \mathbf{x}_0, t_0)\|_E < \epsilon \quad \text{for} \quad 0 < t < \infty$$

and the origin point in the state space is thus stable in the sense of Liapunov.

Asymptotic stability of singular point. The origin point of equation (6–71) is asymptotically stable if there exists a Liapunov function $V(\mathbf{x}, t)$ in Φ such that

$$R(\mathbf{x}) \geqslant V(\mathbf{x}, t)$$

where $R(\mathbf{x})$ is positive definite, and if dV/dt is negative definite.

Stability

Liapunov's indirect method theorem. If the equation (6–70) is expanded in a Taylor series about the origin point, as in Section 6.1.2, we get

$$\frac{d}{dt}(x_i+0) = f_i(0,0,\ldots,0;t) + \sum_{r=1}^{n} \frac{\partial f_i}{\partial x_r} x_r + \frac{1}{2!} \sum_{r=1}^{n} \sum_{s=1}^{n} \frac{\partial^2 f_i}{\partial x_r \partial x_s} x_r x_s + \cdots$$

and thus for an origin singular point we may write equation (6–71) as

$$\frac{d\mathbf{x}}{dt} = \mathbf{Jx} + \mathbf{N(x)} \qquad (6\text{–}74)$$

where $\mathbf{J}$ is the system Liapunov first-approximation matrix for motion in a small neighbourhood of the origin singular point, and $\mathbf{N(x)}$ is a vector whose components represent higher order terms in the series expansion.

Consider the scalar function

$$V(\mathbf{x}) = \sum_{i=1}^{n} \tfrac{1}{2} x_i^2$$

This function is globally positive definite and has

$$\left(\frac{\partial V}{\partial \mathbf{x}}\right)^t = \mathbf{x}$$

Thus its derivative with respect to equation set (6–74) is

$$\frac{dV}{dt} = \left\langle \left(\frac{\partial V}{\partial \mathbf{x}}\right)^t, \dot{\mathbf{x}} \right\rangle = \mathbf{x}^t \mathbf{J} \mathbf{x} + \mathbf{x}^t \mathbf{N(x)}$$

Since none of the components of the vector $\mathbf{N(x)}$ is of less than second-order in the state variable components, there will exist a real constant ϵ, such that in a spherical neighbourhood of radius ϵ about the origin

$$|\mathbf{x}^t \mathbf{J} \mathbf{x}| > |\mathbf{x}^t \mathbf{N(x)}|$$

The origin singular point of the equation set will thus be asymptotically stable if the quadratic form $\mathbf{x}^t \mathbf{J} \mathbf{x}$ is negative definite in this spherical region, since the scalar function $V(\mathbf{x})$ will then be a Liapunov function for the system in this region. A sufficient condition for asymptotic stability of the linearized first-approximation set of zero-source equations

$$\dot{\mathbf{x}} = \mathbf{J} \mathbf{x}$$

is that all the eigenvalues of $\mathbf{J}$ have negative real part. These are also the necessary and sufficient conditions for the quadratic form $\mathbf{x}^t \mathbf{J} \mathbf{x}$ to be globally negative definite. It therefore follows that

(a) if the origin singular point of the first-approximation set of equations is asymptotically stable then so is the origin singular point of the corresponding nonlinear equation set;

(b) if the origin singular point of the approximation set of equations is unstable then so is the origin singular point of the nonlinear set.

It may also be shown that:

(c) if matrix **J** has any eigenvalues which are pure imaginary, then the stability or instability of the origin singular point is determined by the nature of the components of the neglected higher-order terms in **N(x)**.

Instability of singular point (Autonomous Case). If equation (6–70) is autonomous and there exists a function $V(\mathbf{x})$ such that

(a) $V(\mathbf{0}) = 0$
(b) V and $\partial V/\partial x_i$ $(i = 1, \ldots, n)$ are continuous in a region Ω
(c) dV/dt is positive definite in Ω
(d) V may assume positive values arbitrarily close to the origin in state space, then the origin is unstable.

Existence of Liapunov function. A Liapunov $V(\mathbf{x}, t)$ function exists in Φ if

(a) the functions f_i $i = 1, 2, \ldots, n$ are continuous in Φ,
(b) $\partial f_i/\partial x_j$ $i, j = 1, 2, \ldots, n$ are continuous in Φ, and
(c) The origin singular point is stable in the sense of Liapunov.

Example 6.3

Consider the free motion of the linear system

$$\frac{d\mathbf{x}}{dt} = \mathbf{A}\mathbf{x} \qquad (6-75)$$

with quadratic scalar function

$$V = \mathbf{x}^t \mathbf{B} \mathbf{x}$$

where **B** is a symmetric matrix. We then have

$$\frac{dV}{dt} = \mathbf{x}^t[\mathbf{A}^t\mathbf{B} + \mathbf{B}\mathbf{A}]\mathbf{x}$$

$$= -\mathbf{x}^t\mathbf{C}\mathbf{x}$$

where $-\mathbf{C} = \mathbf{A}^t\mathbf{B} + \mathbf{B}\mathbf{A}$.

A detailed examination of the solution of this equation shows that, for given matrices **A** and **C**, the matrix **B** is unique and positive definite provided **C** is positive definite and

(a) $\lambda_i \neq 0$ $\quad i = 1, 2, \ldots, n$
(b) $\lambda_i + \lambda_j \neq 0$ $\quad i, j = 1, 2, \ldots, n$
(c) $\text{Re}(\lambda_i) < 0$ $\quad i = 1, 2, \ldots, n$

where λ_i are the eigenvalues of **A**. Thus if these conditions are satisfied

V is a Liapunov function for the system, and the system origin singular point is asymptotically stable.

Now consider the controlled linear system

$$\frac{d\mathbf{x}}{dt} = \mathbf{A}\mathbf{x} + \mathbf{B}\mathbf{u}$$

with
$$\mathbf{x}(t_0) = \mathbf{x}_0$$

Suppose the system response is evaluated by the performance index

$$P(\mathbf{x}_0, \mathbf{u}) = \int_0^\infty [\mathbf{x}^t \mathbf{C} \mathbf{x} + \mathbf{u}^t \mathbf{R} \mathbf{u}]\, dt$$

and let
$$V(\mathbf{x}) = \mathbf{x}^t \mathbf{Q} \mathbf{x}$$

where
$$\mathbf{A}^t \mathbf{Q} + \mathbf{Q}\mathbf{A} = -\mathbf{C}$$

Then, in the absence of control

$$\frac{dV}{dt} = -\mathbf{x}^t \mathbf{C} \mathbf{x}$$

Integrating this expression directly then gives

$$P(\mathbf{x}_0, \mathbf{0}) = V(\mathbf{x}_0)$$

Now consider the case when control is present; our object is to choose the components of u to increase dV/dt, i.e. make the system 'more stable'. Under these conditions it can be verified directly that

$$\frac{dV}{dt} = -\mathbf{x}^t \mathbf{C} \mathbf{x} + 2\mathbf{u}^t \mathbf{B}^t \mathbf{Q} \mathbf{x}$$

Thus for the specific choice of linear control $\mathbf{u}$ such that

$$\mathbf{u} = -2\mathbf{P}^{-1} \mathbf{B}^t \mathbf{Q} \mathbf{x}$$

where $\mathbf{P}$ is a positive definite matrix, we have

$$\frac{dV}{dt} = -\mathbf{x}^t \mathbf{C} \mathbf{x} - \mathbf{u}^t \mathbf{P} \mathbf{u}$$

and the system is 'more stable'.

If this last expression is integrated between 0 and ∞ we get, after some manipulation

$$P(\mathbf{x}_0, \mathbf{0}) - P(\mathbf{x}_0, \mathbf{u}) = \int_0^\infty \mathbf{u}^t [\mathbf{P} - \mathbf{R}] \mathbf{u}\, dt$$

so that, if the matrix $[\mathbf{P} - \mathbf{R}]$ is positive definite, the value of the performance index with control will be less than the value of the performance index without control.

Example 6.4

Consider the stability of the equilibrium condition of the output angular position of a control system for which the controlled output angle satisfies the zero-source equation

$$\frac{d^2\theta}{dt^2} + \left[2\zeta\omega_n + \left(\frac{d\theta}{dt}\right)^2\right]\frac{d\theta}{dt} + \omega_n^2\theta = 0$$

choosing as state variables

$$x_1 = \theta \qquad x_2 = \frac{d\theta}{dt}$$

gives the state-space equation set

$$\begin{bmatrix} \dfrac{dx_1}{dt} \\ \dfrac{dx_2}{dt} \end{bmatrix} = \begin{bmatrix} 0 & 1 \\ -\omega_n^2 & -2\zeta\omega_n \end{bmatrix} \begin{bmatrix} x_1 \\ x_2 \end{bmatrix} + \begin{bmatrix} 0 \\ -x_2^3 \end{bmatrix}$$

Inspection of this set of equations shows that there is one singular point located at the origin of the state space. The stability of this singular point may be considered by both the indirect and direct methods.

(a) *Indirect Method*

The Liapunov first-approximation set of equations is

$$\begin{bmatrix} \dfrac{dx_1}{dt} \\ \dfrac{dx_2}{dt} \end{bmatrix} = \begin{bmatrix} 0 & 1 \\ -\omega_n^2 & -2\zeta\omega_n \end{bmatrix} \begin{bmatrix} x_1 \\ x_2 \end{bmatrix}$$

The corresponding characteristic equation is

$$s^2 + 2\zeta\omega_n s + \omega_n^2 = 0$$

both roots of which will have negative real part if

$$\zeta > 0 \quad \text{and} \quad \omega_n > 0$$

in which case the first-approximation set of equations will have an asymptotically stable origin singular point. We therefore conclude that the origin singular point of the original nonlinear set of equations will be asymptotically stable, at least in a small neighbourhood of the origin, if

$$\zeta > 0 \quad \text{and} \quad \omega_n > 0$$

(b) *Direct Method*
The function

$$V(\mathbf{x}) = \tfrac{1}{2}(\omega_n^2 x_1^2 + x_2^2)$$

is globally positive definite and satisfies the continuity requirements for a candidate Liapunov function. It has

giving
$$\left(\frac{\partial V}{\partial \mathbf{x}}\right)^t = \begin{bmatrix} \omega_n^2 x_1 \\ x^2 \end{bmatrix}$$

$$\frac{dV}{dt} = -2\zeta\omega_n x_2^2 - x_2^4$$

which will be globally negative definite if $\zeta > 0$ and $\omega_n > 0$. $V(\mathbf{x})$ is thus a Liapunov function for the system which establishes sufficient conditions for asymptotic stability in the large.

Use of Direct Method to Establish Existence and Location of Limit Cycles for Second-order Systems

For certain types of second-order equation, the direct method may be used to establish the existence and approximate location of limit cycles; this is best demonstrated by an example. Suppose the zero-source behaviour of an angular variable in a dynamical system is governed by the differential equation

$$\frac{d^2\theta}{dt^2} + \omega_n\left[\alpha\left(\frac{d\theta}{dt}\right)^2 + \beta\theta^2 - 1\right]\frac{d\theta}{dt} + \omega_n^2\theta = 0$$

where α, β and ω_n are positive real constants. Taking

$$x_1 = \theta \qquad x_2 = \frac{d\theta}{dt}$$

as state variables gives the system state space equations as

$$\frac{dx_1}{dt} = x_2$$

$$\frac{dx_2}{dt} = -\omega_n^2 x_1 - \alpha\omega_n x_2^3 - \beta\omega_n x_1^2 x_2 + \omega_n x_2$$

An inspection of this set of equations shows that there is one singular point located at the origin of coordinates; the stability of this singular

point may be investigated by the direct method. Take
$$V(\mathbf{x}) = \tfrac{1}{2}(\omega_n^2 x_1^2 + x_2^2)$$
as a trial Liapunov function. It is globally positive-definite, satisfies the required continuity conditions and has

$$\left(\frac{\partial V}{\partial \mathbf{x}}\right)^t = \begin{bmatrix} \omega_n^2 x_1 \\ x_2 \end{bmatrix}$$

thus giving
$$\frac{dV}{dt} = \omega_n^2 x_2^2 (1 - \alpha x_2^2 - \beta x_1^2)$$

Let Γ be the closed, bounded curve in the state plane having equation
$$\beta x_1^2 + \alpha x_2^2 = 1$$
so that (remembering that the system is autonomous) $dV/dt < 0$ outside the curve Γ and $dV/dt > 0$ inside the curve Γ.

For the set of closed, bounded curves
$$V(x_1, x_2) = \tfrac{1}{2}(\omega_n^2 x_1^2 + x_2^2) = C_k \quad k = 1, 2, \ldots,$$
let C_1 be the smallest value of real constant C_k such that curve
$$V = C_k$$
is wholly exterior to Γ and let C_2 be the largest value of C_k such that
$$V = C_k$$
is wholly interior to Γ. Let Ω be the region enclosed between $V = C_1$ and $V = C_2$. Then

(a) Since $dV/dt < 0$ at all points on the curve $V = C_1$, it follows that all trajectories through points on this curve pass from outside the curve into the region Ω. Thus no trajectory through a point in the region Ω can penetrate outside of the curve $V = C_1$, and it follows from this that the origin singular point is stable in the sense of Liapunov.

(b) Since $dV/dt > 0$ at all points on the curve $V = C_2$, it follows that all trajectories through points on this curve pass from inside the curve to the region Ω. It therefore follows from both the above conclusions that all trajectories through points in the region Ω remain in the region Ω as $t \to \infty$. It therefore follows (from Bendixon's Second Theorem) that a limit cycle must exist in the region Ω.

6.3 Modality

For a linear, time-invariant system with distinct eigenvalues, it has been shown in Section 6.1.4 above that the free motion may be

expressed in the form
$$x(t) = \sum_{i=1}^{n} \langle v_i, x(0) \rangle [\exp(\lambda_i t)] u_i$$
where $\{\lambda_i : i = 1, 2, \ldots, n\}$ are the eigenvalues of the matrix A governing the system motion
$$\dot{x} = Ax$$
and $\{u_i : i = 1, 2, \ldots, n\}$, $\{v_i : i = 1, 2, \ldots, n\}$ are the corresponding eigenvector set and dual eigenvector set respectively. This shows that the free motion of the linear, time-invariant dynamical system may be regarded as a sum of characteristic motions directed along straight lines through the origin of the state space. These rectilinear characteristic motions in the state space define the *modes* of the system. If all the eigenvalues of the matrix A are real, the modes are associated with rectilinear motions along eigenvectors corresponding to real eigenvalues. The discussion of the complex eigenvalue case in Section 6.1.4 shows that the corresponding mode will be confined to a plane in the state space through the origin spanned by a pair of vectors whose components are respectively the real and imaginary parts of the corresponding complex eigenvector. The modes have the following properties:

(a) The excitation of each mode of the free motion of a system depends only on the initial state.
(b) Each mode may be excited independently of any other mode.
(c) Any free motion of the system may be expressed in a unique way as a superposition of system modes. Put in an alternative way, the vector set defining the system modes spans the state space.

A system mode is said to be excited if the free motion is of the form
$$x(t) = c_i \exp(\lambda_i t) u_i$$
where c_i is a constant and λ_i is a real eigenvalue, or if $x(t)$ is of the form given in equation (6–31) in the case of a pair of complex conjugate eigenvalues.

In order to discuss the excitation of system modes by a forcing input, it is useful to introduce the dyadic representation of a matrix. If we represent any vector x as a linear combination of eigenvectors u_i, the Projection Theorem of Chapter 2 gives that:
$$x = \sum_{j=1}^{n} \langle x, v_j \rangle u_j$$
$$= \sum_{j=1}^{n} \sum_{i=1}^{n} x_i v_{ji} u_j$$
$$= \sum_{j=1}^{n} u_j \sum_{i=1}^{n} v_{ji} x_i$$

436   State Models

so that
$$x = \sum_{j=1}^{n} u_j v_j^t x \qquad (6\text{-}76)$$

where v_{ji} is the ith component of vector v_j.
Multiplying both sides of equation (6-76) by A gives

$$Ax = \sum_{j=1}^{n} A u_j v_j^t x = \sum_{j=1}^{n} \lambda_j u_j v_j^t x \qquad (6\text{-}77)$$

which shows that the operator A may be expressed in the form

$$A = \sum_{j=1}^{n} \lambda_j u_j v_j^t$$

A linear transformation of the form $u_j v_j^t$ is called a *dyad* and has a one-dimensional range spanned by the vector u_j since for any vector a

$$u_j v_j^t a = u_j \langle v_j, a \rangle = \alpha u_j$$

where α is a constant. It is conventional to denote the outer vector product involved in the dyad by

$$u_j v_j^t \equiv u_j \rangle \langle v_j$$

The corresponding representation of A

$$A = \sum_{j=1}^{n} \lambda_j u_j \rangle \langle v_j \qquad (6\text{-}78)$$

is known as the *spectral representation* of A. It gives an immediate geometric interpretation of the effect of the operator A on a vector x in terms of the projection of x into the basis set formed by the operator eigenvectors. It is an immediate consequence of this that we may express the matrix exponential function in the form

$$\exp(At) = \sum_{j=1}^{n} \exp(\lambda_j t) u_j \rangle \langle v_j \qquad (6\text{-}79)$$

Now consider the single input linear, time-invariant system governed by the state space equation

$$\dot{x} = Ax + bu(t) \qquad (6\text{-}80)$$

where x is the system state vector, A an $n \times n$ matrix of constant coefficients, b an $n \times 1$ constant vector and $u(t)$ the system input. Using equation (6-24), we see that the system response is

$$x(t) = [\exp(At)]x(0) + \int_0^t \exp[A(t-\tau)]bu(\tau)\,d\tau \qquad (6\text{-}81)$$

If the matrix A has distinct eigenvalues we may use the spectral representation of equation (6–79) to express equation (6–81) in the form

$$x(t) = \exp(At)\left\{x(0) + \int_0^t \exp(-A\tau)bu(\tau)\,d\tau\right\}$$

$$= \sum_{j=1}^n \exp(\lambda_j t)u_j\rangle\langle v_j\left\{x(0) + \int_0^t \sum_{j=1}^n \exp(-\lambda_j \tau)u_j\rangle\langle v_j, b\rangle u(\tau)\,d\tau\right\}$$

$$= \sum_{j=1}^n \exp(\lambda_j t)u_j\left\{\langle v_j, x(0)\rangle + \int_0^t \sum_{j=1}^n \exp(-\lambda_j \tau)\langle v_j, b\rangle u(\tau)\,d\tau\right\}$$

(6–82)

This shows that the space response to a forcing function input may be expressed as a linear combination of the system modes. The excitation of the jth mode by the input $u(t)$ gives rise to the term

$$\exp(\lambda_j t)\langle v_j, b\rangle \int_0^t \exp(-\lambda_j \tau)u(\tau)\,d\tau\, u_j$$

which is proportional, at any time t, to $\langle v_j, b\rangle$, i.e. to the component of b along u_j. We thus see that for any forcing input the excitation of any mode will depend solely on the amount of the component of b along the eigenvector defining the mode. In particular, if $\langle v_j, b\rangle$ is zero, then the corresponding mode will not be excited and is said to be uncoupled from the input.

6.4 Discrete Model Approximation of Linear Constant Coefficient Systems

It follows from equation (6–24) that the solution of the linear constant coefficient system

$$\dot{x} = Ax + Bu$$

$$y = Cx + Du$$

has the solution for the state component x

$$x(t) = [\exp(At)]x(0) + \int_0^t \exp[A(t-\tau)]Bu(\tau)\,d\tau \qquad (6\text{–}83)$$

Suppose that, given the system described by equation (6–83) we approximate the input forcing function $u(t)$ by a piecewise-constant input having the constant value $u(kT)$ over the interval of time kT to $(k+1)T$. We may then derive a discrete system which will approximate the behaviour of the continuous system at uniformly spaced intervals of time separated by a time increment T. We have, using equation (6–83)

$$x[(k+1)T] = \exp(AT)x(kT) + \int_{kT}^{(k+1)T} \exp A(kT+T-\tau)Bu(kT)\,d\tau$$

Thus the approximate discrete system is given by

$$\mathbf{x}_{k+1} = \mathbf{F}\mathbf{x}_k + \mathbf{G}\mathbf{u}_k$$
$$\mathbf{y}_k = \mathbf{C}\mathbf{x}_k + \mathbf{D}\mathbf{u}_k \tag{6-84}$$

where
$$\mathbf{F} = \exp(\mathbf{A}T) \tag{6-85}$$

$$\mathbf{G} = \int_{kT}^{(k+1)T} \exp \mathbf{A}(kT + T - \tau)\mathbf{B}\, d\tau \tag{6-86}$$

The equation set (6–84) may be solved by taking the z-transform of both sides and using equation (3–107). This gives

$$z\mathbf{X}(z) - z\mathbf{x}(0) = \mathbf{F}\mathbf{X}(z) + \mathbf{G}\mathbf{U}(z)$$
$$\mathbf{Y}(z) = \mathbf{C}\mathbf{X}(z) + \mathbf{D}\mathbf{U}(z)$$

where $\mathbf{X}(z)$, $\mathbf{U}(z)$, $\mathbf{Y}(z)$ are respectively the z-transforms of the sequences $\mathbf{x}(k)$, $\mathbf{u}(k)$ and $\mathbf{y}(k)$.

Rearranging this we get

$$[z\mathbf{I} - \mathbf{F}]\mathbf{X}(z) = z\mathbf{x}(0) + \mathbf{G}\mathbf{U}(z)$$

and so
$$\mathbf{X}(z) = [z\mathbf{I} - \mathbf{F}]^{-1} z\mathbf{x}(0) + [z\mathbf{I} - \mathbf{F}]^{-1} \mathbf{G}\mathbf{U}(z) \tag{6-87}$$

where $\mathbf{I}$ is a unit matrix of appropriate order. The z-transform of the observed output variable sequence is thus

$$\mathbf{Y}(z) = \mathbf{C}[z\mathbf{I} - \mathbf{F}]^{-1} z\mathbf{x}(0) + \{\mathbf{C}[z\mathbf{I} - \mathbf{F}]^{-1}\mathbf{G} + \mathbf{D}\}\mathbf{U}(z) \tag{6-88}$$

In the absence of initial conditions, the input and output sequence vectors are thus related by the z-transfer function matrix

$$\psi(z) = \mathbf{C}[z\mathbf{I} - \mathbf{F}]^{-1}\mathbf{G} + \mathbf{D} \tag{6-89}$$

If we write the first of the equations in the set (6–84) as

$$\mathbf{x}_{k+1} = \mathbf{F}\mathbf{x}_k + \mathbf{f}_k \tag{6-90}$$

then the solution may also be obtained in the following simple iterative manner. We have

$$\mathbf{x}_1 = \mathbf{F}\mathbf{x}_0 + \mathbf{f}_0$$
$$\mathbf{x}_2 = \mathbf{F}\mathbf{x}_1 + \mathbf{f}_1 = \mathbf{F}(\mathbf{F}\mathbf{x}_0 + \mathbf{f}_0) + \mathbf{f}_1 = \mathbf{F}^2\mathbf{x}_0 + \mathbf{F}\mathbf{f}_0 + \mathbf{f}_1$$
$$\mathbf{x}_3 = \mathbf{F}\mathbf{x}_2 + \mathbf{f}_2 = \mathbf{F}(\mathbf{F}^2\mathbf{x}_0 + \mathbf{F}\mathbf{f}_0 + \mathbf{f}_1) + \mathbf{f}_2$$
$$= \mathbf{F}^3\mathbf{x}_0 + \mathbf{F}^2\mathbf{f}_0 + \mathbf{F}\mathbf{f}_1 + \mathbf{f}_2$$
$$= \mathbf{F}^3\mathbf{x}_0 + \sum_{h=0}^{3-1} \mathbf{F}^{3-h-1}\mathbf{f}_h$$

Functional Matrices 439

where $\mathbf{F}^0 = \mathbf{I}$, a unit matrix of appropriate order. Proceeding in this way the solution is obtained as

$$\mathbf{x}_k = \mathbf{F}^k \mathbf{x}_0 + \sum_{h=0}^{k-1} \mathbf{F}^{k-h-1} \mathbf{f}_h \qquad (6\text{-}91)$$

As shown above we may write a matrix $\mathbf{A}$ and its associated matrix exponential function in dyadic form as, assuming distinct eigenvalues,

$$\mathbf{A} = \sum_{j=1}^{n} \lambda_j \mathbf{u}_j \rangle \langle \mathbf{v}_j$$

$$\exp(\mathbf{A}T) = \sum_{j=1}^{n} \exp(\lambda_j T) \mathbf{u}_j \rangle \langle \mathbf{v}_j$$

Now put $\qquad \lambda_j = \sigma_j + j\omega_j$

so that $\qquad \exp(\lambda_j T) = \exp(\sigma_j T)\{\cos \omega_j T + j \sin \omega_j T\}$

Consideration of the above relationships shows that

(a) If $\lambda_1, \lambda_2, \ldots, \lambda_n$ are the eigenvalues of $\mathbf{A}$, then the eigenvalues of $\exp(\mathbf{A}T)$ are $\exp(\lambda_1 T), \exp(\lambda_2 T), \ldots, \exp(\lambda_n T)$.

(b) If all the eigenvalues of $\mathbf{A}$ are in the left-half complex plane, $\sigma_1, \sigma_2, \ldots, \sigma_n$ will be negative so that

$$|\exp(\sigma_i T)| < 1 \quad \text{for} \quad i = 1, 2, \ldots, n$$

and so all the eigenvalues of $\exp(\mathbf{A}T)$ will be inside the unit circle centred on the origin in the complex plane.

If we consider the free motion solution obtained from equation (6-91) by setting $\mathbf{f}_k \equiv \mathbf{0}$, namely

$$\mathbf{x}_k = \mathbf{F}^k \mathbf{x}_0$$

and choose a transformation which diagonalizes $\mathbf{F}$, we get

$$\bar{\mathbf{x}}_k = \mathbf{\Lambda}^k \bar{\mathbf{x}}_0$$

with $\qquad \bar{\mathbf{x}}_k = \mathbf{T}\mathbf{x}_k \quad \text{and} \quad \mathbf{\Lambda} = \mathbf{T}^{-1}\mathbf{F}\mathbf{T}$

where $\mathbf{\Lambda}$ is a diagonal matrix whose entries are the eigenvalues of $\mathbf{F}$. It immediately follows that the free motion system will be asymptotically stable in the large, if all the eigenvalues of $\mathbf{F}$ lie in the open disc such that $|z| < 1$ in the complex plane.

6.5 Functional Matrices [M3]

Consider the system

$$\frac{d\mathbf{x}}{dt} = \mathbf{A}\mathbf{x} + \mathbf{f}$$

If the outputs of all system sources are identically zero, the free motion of the system will be determined by the equation

$$\frac{d\mathbf{x}}{dt} = \mathbf{A}\mathbf{x}$$

Integrating with respect to time then gives

$$\int_{\mathbf{x}(0)}^{\mathbf{x}(\infty)} d\mathbf{x} = \int_{0}^{\infty} \mathbf{A}\mathbf{x}\, dt = \mathbf{A}\int_{0}^{\infty} \mathbf{x}\, dt$$

thus giving

$$\int_{0}^{\infty} \mathbf{x}\, dt = \mathbf{A}^{-1}[\mathbf{x}(\infty) - \mathbf{x}(0)]$$

If the system is asymptotically stable, then

$$\mathbf{x}(\infty) \to 0$$

and we have

$$\int_{0}^{\infty} \mathbf{x}\, dt = -\mathbf{A}^{-1}\mathbf{x}(0)$$

so that

$$\int_{0}^{\infty} \mathbf{K}\mathbf{x}\, dt = -\mathbf{K}\mathbf{A}^{-1}\mathbf{x}(0)$$

where $\mathbf{K}$ is an arbitrary matrix of real constants. It follows that any linear functional of the free motion of the state variables of the system may be calculated in terms of the release coordinates by inverting the $\mathbf{A}$-matrix. Since any system variable can be expressed in terms of the state variables, any linear functional of system variables may be found in this way. For this reason we may call $\mathbf{A}$ the system *linear functional matrix*.

Let $P(\mathbf{x}, \mathbf{x}, \ldots, \mathbf{x})$ and $Q(\mathbf{x}, \mathbf{x}, \ldots, \mathbf{x})$ be time-independent homogeneous algebraic forms of order m in the coefficients of a network state vector $\mathbf{x}$ such that

$$\frac{d}{dt}P(\mathbf{x}, \mathbf{x}, \ldots, \mathbf{x}) = Q(\mathbf{x}, \mathbf{x}, \ldots, \mathbf{x})$$

where

$$P(\mathbf{x}, \mathbf{x}, \ldots, \mathbf{x}) = \sum_{\alpha=1}^{n} \sum_{\beta=1}^{n} \sum_{\gamma=1}^{n} \cdots \sum_{\mu=1}^{n} p_{\alpha\beta\gamma\ldots\mu} x_\alpha x_\beta x_\gamma \ldots x_\mu$$

and

$$Q(\mathbf{x}, \mathbf{x}, \ldots, \mathbf{x}) = \sum_{\alpha=1}^{n} \sum_{\beta=1}^{n} \sum_{\gamma=1}^{n} \cdots \sum_{\mu=1}^{n} q_{\alpha\beta\gamma\ldots\mu} x_\alpha x_\beta x_\gamma \ldots x_\mu$$

In an expansion of such algebraic forms no distinction is made between coefficients having differing arrangements of the same subscripts; thus the general term in an expansion is premultiplied by an appropriate

Functional Matrices

multinomial coefficient. Since

$$\frac{d}{dt}P(\mathbf{x},\mathbf{x},\ldots,\mathbf{x}) = \left\langle \operatorname{grad}_\mathbf{x} P(\mathbf{x},\mathbf{x},\ldots,\mathbf{x}), \frac{d\mathbf{x}}{dt} \right\rangle$$

$$= \langle \operatorname{grad}_\mathbf{x} P(\mathbf{x},\mathbf{x},\ldots,\mathbf{x}), \mathbf{Ax} \rangle$$

where $\quad \operatorname{grad}_\mathbf{x} P(\mathbf{x},\mathbf{x},\ldots,\mathbf{x}) \triangleq \left(\dfrac{\partial P}{\partial \mathbf{x}}\right)^t$

we have $\quad Q(\mathbf{x},\mathbf{x},\ldots,\mathbf{x}) = \left\langle \left(\dfrac{\partial P}{\partial \mathbf{x}}\right)^t, \mathbf{Ax} \right\rangle \quad$ (6–92)

from which the coefficients of the form $P(\mathbf{x},\mathbf{x},\ldots,\mathbf{x})$ may be determined for a given form $Q(\mathbf{x},\mathbf{x},\ldots,\mathbf{x})$ and a given matrix $\mathbf{A}$.

In order to determine general functionals of the free motion of network state variables in which arbitrary time-weighting functions may be introduced, a general functional matrix is sought which extends the properties of the linear functional or $\mathbf{A}$ matrix to the general case. To do this, a generalized state vector $\mathbf{x}_{(m)}$ is introduced; this generalized state vector is related to the algebraic forms of degree m by putting

$$P(\mathbf{x},\mathbf{x},\ldots,\mathbf{x}) = \langle \mathbf{p}, \mathbf{x}_{(m)} \rangle \quad (6\text{–}93)$$

and $\quad Q(\mathbf{x},\mathbf{x},\ldots,\mathbf{x}) = \langle \mathbf{q}, \mathbf{x}_{(m)} \rangle \quad$ (6–94)

where the vectors $\mathbf{p}$ and $\mathbf{q}$ are formed by arranging the coefficients of the algebraic forms $P(\mathbf{x},\mathbf{x},\ldots,\mathbf{x})$ and $Q(\mathbf{x},\mathbf{x},\ldots,\mathbf{x})$ in a defined order. The multinomial coefficients of the algebraic form expansions will be included in the components of the vector $\mathbf{x}_{(m)}$, and excluded from the vectors $\mathbf{p}$ and $\mathbf{q}$ whose general-component terms are defined to be

$$p_{\alpha\beta\gamma\ldots\mu} \quad \text{and} \quad q_{\alpha\beta\gamma\ldots\mu}$$

where $\quad \alpha \leqslant \beta \leqslant \gamma \cdots \leqslant \mu$

and $\quad \alpha + \beta + \gamma + \cdots + \mu = m, (m+1), (m+2), \ldots, mn$

The vector $\mathbf{x}_{(m)}$ thus has components of the general form

$$\frac{m!}{a!b!c!\ldots h!} x_1^a x_2^b x_3^c \ldots x_n^h$$

For example, with $m = 3$ and $n = 2$

$$\mathbf{x}_{(3)} = \begin{bmatrix} x_1^3 \\ 3x_1^2 x_2 \\ 3x_1 x_2^2 \\ x_2^3 \end{bmatrix} \quad \mathbf{p} = \begin{bmatrix} p_{111} \\ p_{112} \\ p_{122} \\ p_{222} \end{bmatrix} \quad \mathbf{q} = \begin{bmatrix} q_{111} \\ q_{112} \\ q_{122} \\ q_{222} \end{bmatrix}$$

442   State Models

Suppose that, associated with the generalized state vector $\mathbf{x}_{(m)}$, there existed a corresponding generalized functional matrix $\mathbf{M}$ such that, for free motion of the system,

$$\frac{d}{dt}\mathbf{x}_{(m)} = \mathbf{M}\mathbf{x}_{(m)} \qquad (6\text{-}95)$$

This would give

$$\frac{d}{dt}\langle \mathbf{p}, \mathbf{x}_{(m)}\rangle = \left\langle \mathbf{p}, \frac{d}{dt}\mathbf{x}_{(m)}\right\rangle$$
$$= \langle \mathbf{p}, \mathbf{M}\mathbf{x}_{(m)}\rangle = \langle \mathbf{M}^t\mathbf{p}, \mathbf{x}_{(m)}\rangle = \langle \mathbf{q}, \mathbf{x}_{(m)}\rangle \qquad (6\text{-}96)$$

and the matrix $\mathbf{M}$ would therefore relate the vectors $\mathbf{p}$ and $\mathbf{q}$ in such a way that

$$\mathbf{q} = \mathbf{M}^t\mathbf{p} \qquad (6\text{-}97)$$

Since equation (6–92) establishes a unique relationship between the coefficients of the vectors $\mathbf{p}$ and $\mathbf{q}$, the matrix $\mathbf{M}$ exists and may be uniquely determined from equations (6–92) and (6–97) in terms of the coefficients of the linear functional matrix $\mathbf{A}$. This straightforward, but tedious, operation has been carried out, for a range of orders of dynamical system and algebraic form, and the resulting $\mathbf{M}$ matrices are displayed in Appendix A (page 486).

Having established the existence and uniqueness of the matrix $\mathbf{M}$ we may now consider its use for the determination of functionals of system response. Since

$$\frac{d}{dt}\langle \mathbf{p}, \mathbf{x}_{(m)}\rangle = \langle \mathbf{p}, \mathbf{M}\mathbf{x}_{(m)}\rangle$$

we have $$[\langle \mathbf{p}, \mathbf{x}_{(m)}\rangle]_0^\infty = \int_0^\infty \langle \mathbf{p}, \mathbf{M}\mathbf{x}_{(m)}\rangle\, dt$$

whence $$\mathbf{M}\int_0^\infty \mathbf{x}_{(m)}\, dt = \mathbf{x}_{(m)}(\infty) - \mathbf{x}_{(m)}(0)$$

and thus, for the free motion of an asymptotically stable system

$$\int_0^\infty \mathbf{x}_{(m)}\, dt = -\mathbf{M}^{-1}\mathbf{x}_{(m)}(0) \qquad (6\text{-}98)$$

Again, since

$$\frac{d}{dt}\langle \mathbf{p}, t\mathbf{x}_{(m)}\rangle = \langle \mathbf{p}, \mathbf{x}_{(m)}\rangle + \langle \mathbf{q}, t\mathbf{x}_{(m)}\rangle$$

we have $$[\langle \mathbf{p}, t\mathbf{x}_{(m)}\rangle]_0^\infty = \int_0^\infty \langle \mathbf{p}, \mathbf{x}_{(m)}\rangle\, dt + \int_0^\infty \langle \mathbf{q}, t\mathbf{x}_{(m)}\rangle\, dt$$

Functional Matrices 443

For an asymptotically stable system the left-hand side of this equation will vanish, giving

$$\left\langle \mathbf{q}, \int_0^\infty t\mathbf{x}_{(m)}\,dt \right\rangle = -\left\langle \mathbf{p}, \int_0^\infty \mathbf{x}_{(m)}\,dt \right\rangle$$

Thus
$$\left\langle \mathbf{M}^t\mathbf{p}, \int_0^\infty t\mathbf{x}_{(m)}\,dt \right\rangle = -\left\langle \mathbf{p}, \int_0^\infty \mathbf{x}_{(m)}\,dt \right\rangle$$

whence
$$\int_0^\infty t\mathbf{x}_{(m)}\,dt = -\mathbf{M}^{-1}\int_0^\infty \mathbf{x}_{(m)}\,dt = \mathbf{M}^{-2}\mathbf{x}_{(m)}(0)$$

A continuation of this form of argument gives

$$\int_0^\infty \mathbf{K}t^r\mathbf{x}_{(m)}\,dt = (-1)^{(r+1)} r! \mathbf{K}\mathbf{M}^{-(r+1)}\mathbf{x}_{(m)}(0) \tag{6-99}$$

where $\mathbf{K}$ is a matrix of arbitrary constants.
Since, for free motion of the network,

$$\mathbf{x}_{(m)}(t) = \exp(\mathbf{M}t)\mathbf{x}_{(m)}(0) \tag{6-100}$$

we have, for a general weighting function $f(t)$,

$$\int_0^\infty f(t)\mathbf{x}_{(m)}(t)\,dt = \int_0^\infty f(t)\exp(\mathbf{M}t)\mathbf{x}_{(m)}(0)\,dt$$

$$= \int_0^\infty f(t)\exp(\mathbf{M}t)\,dt\,\mathbf{x}_{(m)}(0) \tag{6-101}$$

$$= \Phi(\mathbf{M})\mathbf{x}_{(m)}(0)$$

where the matrix $\Phi(\mathbf{M})$ is defined by

$$\Phi(\mathbf{M}) = \int_0^\infty f(t)\exp(\mathbf{M}t)\,dt \tag{6-102}$$

The form of the matrix function $\Phi(\mathbf{M})$ may be determined by use of equation (6-99). For the function $f(t) = t^r$, this gives

$$\Phi(\mathbf{M}) = (-1)^{(r+1)} r! \mathbf{M}^{-(r+1)} \tag{6-103}$$

For more general functions of time the corresponding form of the matrix $\Phi(\mathbf{M})$ may be determined by expanding them as a series in t. For example, for

$$f(t) = \exp(-at) = 1 - at + \frac{a^2 t^2}{2!} - \frac{a^3 t^3}{3!} + \cdots$$

the use of equation (6–99) gives that

$$\Phi(\mathbf{M}) = -\mathbf{M}^{-1} - a\mathbf{M}^{-2} - a^2\mathbf{M}^{-3} - a^3\mathbf{M}^{-4} - \cdots$$
$$= -[\mathbf{M} - a\mathbf{I}]^{-1}$$
$$= [(-\mathbf{M}) + a\mathbf{I}]^{-1} \qquad (6\text{–}104)$$

If a matrix $\mathbf{F}(\mathbf{M})$ is defined by

$$\mathbf{F}(\mathbf{M}) = \Phi(-\mathbf{M}) = \int_0^\infty f(t) \exp(-\mathbf{M}t)\, dt \qquad (6\text{–}105)$$

then we have

$$\int_0^\infty f(t)\mathbf{K}\mathbf{x}_{(m)}(t)\, dt = \mathbf{K}\mathbf{F}(-\mathbf{M})\mathbf{x}_{(m)}(0) \qquad (6\text{–}106)$$

where for

$$\begin{aligned}
f(t) &= 1 & \mathbf{F}(\mathbf{M}) &= \mathbf{M}^{-1} \\
f(t) &= t & \mathbf{F}(\mathbf{M}) &= \mathbf{M}^{-2} \\
f(t) &= t^r & \mathbf{F}(\mathbf{M}) &= r!\,\mathbf{M}^{-(r+1)} \\
f(t) &= \exp(at) & \mathbf{F}(\mathbf{M}) &= [\mathbf{M} + a\mathbf{I}]^{-1}
\end{aligned}$$

and $\mathbf{K}$ is a matrix of arbitrary constants.

A detailed study of the formation of the matrix $\mathbf{F}(\mathbf{M})$ using equation (6–99) in this way shows that $\mathbf{F}(\mathbf{M})$ may be formally considered as a matrix Laplace transform of the time-weighting function $f(t)$; it may thus be readily found for any Laplace transformable function of time by direct use of a table of transform pairs. Equation (6–106) may then be used to compute a very large class of time-weighted functionals of the free motion of a linear constant-coefficient dynamical system, in terms of a set of release coordinates, without an intermediate determination of an explicit solution for the system behaviour.

6.5.1 Equivalent Free-Motion Systems for Network Impulse, Step and Ramp Response

Impulse Response

If the source outputs, and therefore the components of the vector $\mathbf{f}$, are Dirac impulses, the components of the Laplace transform of vector $\mathbf{f}$ will be constants equal to the value of the impulses. Consideration of the Laplace transforms of the constituents of the $\mathbf{A}$-matrix equation then shows that the response of a system initially at rest, subjected to an input whose components are Dirac impulses, may be exactly reproduced by the free motion of the system with no input released from an

Functional Matrices 445

initial condition vector $\mathbf{x}(0)$, whose components are equal to the value of the component impulses applied to the rest system. Thus any method of computing functionals of the free motion of the system variables for a system released from known initial conditions may be used to compute functionals of system impulse response.

Step and Ramp Response

Any method developed for the determination of functionals of the free motion of a system may be used to determine functionals of the departure of network variables from their asymptotic values in response to vector step and ramp inputs.

Denote the asymptotic value of the system state variable response by $\mathbf{x}_a$, that is let

$$\mathbf{x}_a = \mathbf{x}(t) \quad \text{for} \quad t \to \infty$$

Let, in the system $\mathbf{A}$-matrix equation,

$$\mathbf{f} = \mathbf{0} \quad \text{for} \quad t \leq 0$$

and

$$\mathbf{f} = \mathbf{g} \quad \text{for} \quad t > 0$$

where $\mathbf{g}$ is a vector whose components are real constants. If all the eigenvalues of $\mathbf{A}$ have negative real part, then $\mathbf{x}(t) \to \mathbf{0}$ as $t \to \infty$ and

$$\mathbf{x}_a = -\mathbf{A}^{-1}\mathbf{g} \quad (6\text{--}107)$$

Put

$$\mathbf{x}(t) = \mathbf{r}(t) + \mathbf{x}_a$$

$$= \mathbf{r}(t) - \mathbf{A}^{-1}\mathbf{g} \quad (6\text{--}108)$$

that is let $\mathbf{r}(t)$ be a vector whose components represent the departure of $\mathbf{x}(t)$ from its asymptotic condition $\mathbf{x}_a$.

If $\mathbf{x}(0) = \mathbf{0}$ then $\mathbf{r}(0) = -\mathbf{x}_a$, and substituting from equation (6–108) into the equation

$$\frac{d\mathbf{x}}{dt} = \mathbf{A}\mathbf{x} + \mathbf{g} \quad (6\text{--}109)$$

gives

$$\frac{d\mathbf{r}}{dt} = \mathbf{A}\mathbf{r}$$

with

$$\mathbf{r}(0) = \mathbf{A}^{-1}\mathbf{g} \quad (6\text{--}110)$$

This is the equivalent free-motion system governing the departure of the step response from its asymptotic condition.

A similar procedure may be applied to derive an equivalent free motion for the computation of functionals of system variable departure from the asymptotic response to an input $\mathbf{f}$ vector whose components are ramp functions. In this case if

$$\mathbf{f} = \mathbf{0} \quad \text{for} \quad t \leqslant 0$$

and $\quad \mathbf{f} = t\mathbf{g} \quad \text{for} \quad t > 0$

then $\quad \mathbf{x}_a = -\mathbf{A}^{-1}t\mathbf{g} - \mathbf{A}^{-2}\mathbf{g}$

and the equivalent free-motion system is

$$\frac{d}{dt}\mathbf{r} = \mathbf{A}\mathbf{r}$$

with $\quad \mathbf{r}(0) = \mathbf{A}^{-2}\mathbf{g}$

6.6 Generalized Mohr Circles and their Use in Feedback Design
[M6], [M7]

The state space analysis of a multivariable system depends upon the concept of an operator on a linear vector space. This operator is usually characterized by its eigenvalues and eigenvectors, the eigenvectors defining modes of system behaviour, and the eigenvalues providing a link with the frequency domain concepts of dynamical theory. In other engineering disciplines in which such matrix operators are used, such as stress analysis, a more complete representation is available by making use of constructions such as Mohr's circle, first introduced in 1882 [M18]. Mohr's original construction was devised to show the effects of a rotation of coordinate axes on the components of stress and strain tensors; its wide utility stems from the fact that it not only gives the principal axis values (i.e. the eigenvector values) but gives a complete representation of planar stress components. The object of this section is to show that similar constructions can be used in the design of multivariable feedback control systems. In order to establish as close a link as possible with classical control concepts, the circle construction will be defined in terms of the familiar complex frequency plane.

The construction is illustrated in terms of a linear proportional regulator design. Let a multivariable linear feedback controller be applied to a linear constant coefficient dynamical system whose state space equations are

$$\dot{\mathbf{x}} = \mathbf{A}\mathbf{x} + \mathbf{B}\mathbf{u}$$
$$\mathbf{y} = \mathbf{C}\mathbf{x} \qquad (6\text{--}111)$$

where $\mathbf{x}$ is the system state vector (of order n), $\mathbf{u}$ is a vector of manipulated variables (of order l), and $\mathbf{y}$ is a vector of measured outputs (of order m). Proportional feedback is defined by

$$\mathbf{u} = \mathbf{K}(\mathbf{r} - \mathbf{y}) \qquad (6\text{--}112)$$

Generalized Mohr Circles and their Use in Feedback Design 447

where **r** is a vector of reference inputs (of order m). In these equations **A**, **B**, **C** and **K** are constant coefficient matrices of orders $n \times n$, $n \times l$, $m \times n$ and $l \times m$ respectively. The corresponding closed-loop state space equations are

$$\dot{\mathbf{x}} = (\mathbf{A} - \mathbf{BKC})\mathbf{x} + \mathbf{BKr}$$
$$\mathbf{y} = \mathbf{Cx}$$
(6–113)

If **r** is a vector of constant inputs and the system is asymptotically stable, then the steady state response is given by

$$\mathbf{y}(\infty) = -\mathbf{C}(\mathbf{A} - \mathbf{BKC})^{-1}\mathbf{BKr}(\infty)$$
$$= \mathbf{Er}(\infty)$$
(6–114)

where $\mathbf{y}(\infty)$ and $\mathbf{r}(\infty)$ denote the steady state values of $\mathbf{y}(t)$ and $\mathbf{r}(t)$, and **E** is a matrix of constants which will specify the accuracy of steady state regulation. The basic linear multivariable regulation problem may then be stated as: given **A**, **B**, **C** and an acceptable **E**, choose **K** so that all the eigenvalues of $(\mathbf{A} - \mathbf{BKC})$ lie in a specified region of the complex plane. If we put

$$\mathbf{F} = -\mathbf{BKC}$$
(6–115)

then it is convenient to call **F** the system feedback matrix. The stability and general dynamical behaviour of the closed-loop system are determined by the properties of the matrix $(\mathbf{A} + \mathbf{F})$, and the steady state accuracy depends on setting acceptably high values for the elements of **K**. The circle construction gives a means of graphically exhibiting, in an additive manner, the effects of the elements of **K** on the dynamical properties of $(\mathbf{A} + \mathbf{F})$.

6.6.1 Generalized Mohr Circles

Let $\mathbf{E}_1$ and $\mathbf{E}_2$ be a pair of orthonormal vectors emanating from the origin of the state space. For any state vector **x** lying in the plane spanned by $\mathbf{E}_1$ and $\mathbf{E}_2$ such that

$$\mathbf{x} = a\mathbf{E}_1 + b\mathbf{E}_2$$
(6–116)

define a related vector **y**, orthogonal to **x** and in this plane, by

$$\mathbf{y} = -b\mathbf{E}_1 + a\mathbf{E}_2$$
(6–117)

For any linear state space flow defined by

$$\dot{\mathbf{x}} = \mathbf{Ax}$$
(6–118)

let a complex number ρ be defined by

$$\rho = \frac{\langle \mathbf{x}, \mathbf{Ax} \rangle}{\langle \mathbf{x}, \mathbf{x} \rangle} + j\frac{\langle \mathbf{y}, \mathbf{Ax} \rangle}{\langle \mathbf{y}, \mathbf{y} \rangle}$$
(6–119)

The real and imaginary parts of ρ correspond to the components of instantaneous velocity along **x** (which we may call the recession component) and along **y** (which we may call the spin component in this plane). This is directly analogous to the resolution of stress into direct stress and shear stress in the conventional Mohr construction.

Equation (6–119) defines a mapping from the plane spanned by $\mathbf{E}_1$ and $\mathbf{E}_2$ in the state space on to the complex plane. If we denote, for any given plane in the state space, the complex numbers determined by $(\mathbf{A}+\mathbf{F})$, $\mathbf{A}$ and $\mathbf{F}$ by $\rho(\mathbf{A}+\mathbf{F})$, $\rho(\mathbf{A})$ and $\rho(\mathbf{F})$ respectively, we have

$$\rho(\mathbf{A}+\mathbf{F}) = \frac{\langle \mathbf{x}, (\mathbf{A}+\mathbf{F})\mathbf{x}\rangle}{\langle \mathbf{x}, \mathbf{x}\rangle} + j\frac{\langle \mathbf{y}, (\mathbf{A}+\mathbf{F})\mathbf{x}\rangle}{\langle \mathbf{y}, \mathbf{y}\rangle}$$

$$= \frac{\langle \mathbf{x}, \mathbf{A}\mathbf{x}\rangle}{\langle \mathbf{x}, \mathbf{x}\rangle} + j\frac{\langle \mathbf{y}, \mathbf{A}\mathbf{x}\rangle}{\langle \mathbf{y}, \mathbf{y}\rangle} + \frac{\langle \mathbf{x}, \mathbf{F}\mathbf{x}\rangle}{\langle \mathbf{x}, \mathbf{x}\rangle} + j\frac{\langle \mathbf{y}, \mathbf{F}\mathbf{x}\rangle}{\langle \mathbf{y}, \mathbf{y}\rangle}$$

$$= \rho(\mathbf{A}) + \rho(\mathbf{F}) \qquad (6\text{–}120)$$

It is this additive property of the mapping which is crucial to feedback controller synthesis, since it enables one to display graphically the effect of **F** on the closed-loop system matrix $(\mathbf{A}+\mathbf{F})$.

If any vector **x** in the selected plane is multiplied by a real positive constant k, then **y** by definition is also multiplied by the same constant, and we have the corresponding new value of ρ for a matrix **A** as

$$\rho = \frac{\langle k\mathbf{x}, \mathbf{A}k\mathbf{x}\rangle}{\langle k\mathbf{x}, k\mathbf{x}\rangle} + j\frac{\langle k\mathbf{y}, \mathbf{A}k\mathbf{x}\rangle}{\langle k\mathbf{y}, k\mathbf{y}\rangle}$$

$$= \frac{k^2\langle \mathbf{x}, \mathbf{A}\mathbf{x}\rangle}{k^2\langle \mathbf{x}, \mathbf{x}\rangle} + j\frac{k^2\langle \mathbf{y}, \mathbf{A}\mathbf{x}\rangle}{k^2\langle \mathbf{y}, \mathbf{y}\rangle}$$

This shows that the value of ρ is constant along any line emanating from the origin of the state space. Since the whole plane can be swept out by rotating such a ray about the origin, and since each ray maps into a point in the complex plane, it follows that the whole plane in the state space maps into a closed locus in the complex plane.

To find the form of this locus, let the state plane be swept out by taking

$$\mathbf{x} = \cos\theta E_1 + \sin\theta E_2$$

$$\mathbf{y} = -\sin\theta E_1 + \cos\theta E_2$$

where θ is varied continuously. Since in this case $\langle \mathbf{x}, \mathbf{x}\rangle$ and $\langle \mathbf{y}, \mathbf{y}\rangle$ are both equal to unity we have

$$\rho = \langle \mathbf{x}, \mathbf{A}\mathbf{x}\rangle + j\langle \mathbf{y}, \mathbf{A}\mathbf{x}\rangle$$

$$= \rho_R + j\rho_I \qquad (6\text{–}121)$$

Generalized Mohr Circles and their Use in Feedback Design

Simple calculations then give

$$\rho_R - \alpha_1 = \beta_1 \cos 2\theta + \beta_2 \sin 2\theta$$
$$\rho_I - \alpha_2 = -\beta_1 \sin 2\theta + \beta_2 \cos 2\theta \qquad (6\text{-}122)$$

where
$$\alpha_1 = \tfrac{1}{2}(\mathbf{E}_1^t \mathbf{AE}_1 + \mathbf{E}_2^t \mathbf{AE}_2)$$
$$\alpha_2 = \tfrac{1}{2}(\mathbf{E}_2^t \mathbf{AE}_1 - \mathbf{E}_1^t \mathbf{AE}_2)$$
$$\beta_1 = \tfrac{1}{2}(\mathbf{E}_1^t \mathbf{AE}_1 - \mathbf{E}_2^t \mathbf{AE}_2)$$
$$\beta_2 = \tfrac{1}{2}(\mathbf{E}_2^t \mathbf{AE}_1 + \mathbf{E}_1^t \mathbf{AE}_2)$$

Squaring and adding equations (6-122) then gives

$$(\rho_R - \alpha_1)^2 + (\rho_I - \alpha_2)^2 = \beta_1^2 + \beta_2^2 \qquad (6\text{-}123)$$

This shows that the ρ-locus in the complex plane is a circle with the following characteristics:
  (i) centre at (α_1, α_2)
  (ii) radius of $\sqrt{(\beta_1^2 + \beta_2^2)}$
  (iii) lines through the origin of the state space at an angle of θ to each other give points subtending an angle 2θ at the centre of the circle. Thus orthogonal lines in the state space determine diametrically opposite points on the circle.

6.6.2 Properties of Generalized Mohr Circles

Denote the symmetric and skew-symmetric parts of the matrix $\mathbf{A}$ by $\mathbf{A}_+$ and $\mathbf{A}_-$ respectively, where

$$\mathbf{A}_+ = \tfrac{1}{2}(\mathbf{A} + \mathbf{A}^t) \qquad \mathbf{A}_- = \tfrac{1}{2}(\mathbf{A} - \mathbf{A}^t)$$

From the additive property of the mapping we immediately have that

$$\rho(\mathbf{A}) = \rho(\mathbf{A}_+) + \rho(\mathbf{A}_-)$$

For the symmetric part we have

$$\alpha_2 = \tfrac{1}{2}(\mathbf{E}_2^t \mathbf{A}_+ \mathbf{E}_1 - \mathbf{E}_1^t \mathbf{A}_+ \mathbf{E}_2) = 0$$

since
$$\mathbf{E}_2^t \mathbf{A}_+ \mathbf{E}_1 = (\mathbf{E}_2^t \mathbf{A}_+ \mathbf{E}_1)^t = \mathbf{E}_1^t \mathbf{A}_+ \mathbf{E}_2$$

It follows from this that the symmetric matrices map into circles centred on the real axis of the complex plane. Such circles are precisely analogous to the Mohr circles obtained for plane stress from Cartesian tensors.

For the skew-symmetric part we have

$$\alpha_1 = 0 \quad \text{since} \quad E_1^t A_- E_1 = 0 \quad \text{and} \quad E_2^t A_- E_2 = 0$$

We also have that

$$E_2^t A_- E_1 = (E_2^t A_- E_1)^t = E_1^t A_-^t E_2 = -E_1^t A_- E_2$$

so that $\quad \beta_1 = 0 \quad \text{and} \quad \beta_2 = 0$

It follows from this that skew-symmetric matrices map into points on the imaginary axis in the complex plane.

Real Eigenvalues

Let u_1 be a real eigenvector of A associated with the real eigenvalue λ_i. We then have

$$\rho_R = \frac{\langle u_i, \lambda_i u_i \rangle}{\langle u_i, u_i \rangle} = \lambda_i \qquad \rho_I = 0$$

It follows from this that:
  (i) if the plane being mapped contains such an eigenvector, the eigenvector is mapped into the associated real eigenvalue.
  (ii) a plane spanned by two real eigenvectors u_i and u_j, associated with two real eigenvalues λ_i and λ_j, is mapped into a circle which cuts the real axis in the complex plane at the real eigenvalues λ_i and λ_j. This is directly analogous to the conventional Mohr circle cutting the direct stress axis at the principal values of stress.

Complex Eigenvalues

Let $r_i \pm js_i$ be a pair of complex conjugate eigenvectors associated with the pair of complex conjugate eigenvalues $\sigma_i \pm j\omega_i$. We then have

$$A(r_i + js_i) = (\sigma_i + j\omega_i)(r_i + js_i) \tag{6-124}$$

from which we obtain the pair of relations

$$\begin{aligned} Ar_i &= \sigma_i r_i - \omega_i s_i \\ As_i &= \omega_i r_i + \sigma_i s_i \end{aligned} \tag{6-125}$$

It is an immediate consequence of equations (6–125) that the plane spanned by vectors r_i and s_i is invariant under the action of the operator A. Consider the mapping of this invariant plane into the complex plane. Let the real vectors r_i and s_i be of unit length and let y_u and y_v be a pair of related orthonormal vectors as shown in Figure 6–8.

Generalized Mohr Circles and their Use in Feedback Design 451

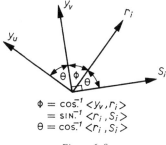

$$\phi = \cos^{-1} \langle y_v, r_i \rangle$$
$$= \sin^{-1} \langle r_i, S_i \rangle$$
$$\theta = \cos^{-1} \langle r_i, S_i \rangle$$

Figure 6–8

When $\mathbf{r}_i$ is mapped into the complex plane we have

$$\rho_R = \langle \mathbf{r}_i, (\sigma_i \mathbf{r}_i - \omega_i \mathbf{s}_i) \rangle = \sigma_i - \omega_i \langle \mathbf{r}_i, \mathbf{s}_i \rangle$$
$$\rho_I = \langle \mathbf{y}_u, (\sigma_i \mathbf{r}_i - \omega_i \mathbf{s}_i) \rangle = -\omega_i \langle \mathbf{y}_u, \mathbf{s}_i \rangle$$

When $\mathbf{s}_i$ is mapped into the complex plane we have

$$\rho_R = \langle \mathbf{s}_i, (\omega_i \mathbf{r}_i + \sigma_i \mathbf{s}_i) \rangle = \sigma_i + \omega_i \langle \mathbf{r}_i, \mathbf{s}_i \rangle$$
$$\rho_I = \langle \mathbf{y}_v, (\omega_i \mathbf{r}_i + \sigma_i \mathbf{s}_i) \rangle = \omega_i \langle \mathbf{y}_v, \mathbf{r}_i \rangle$$

An inspection of the relationships in Figure 6–8 shows that

$$\langle \mathbf{y}_v, \mathbf{r}_i \rangle = \cos \phi = \sin \theta$$
$$\langle \mathbf{y}_u, \mathbf{s}_i \rangle = \cos(2\theta + \phi) = \cos(90° + \theta) = -\sin \theta$$

This shows that the map of both $\mathbf{r}_i$ and $\mathbf{s}_i$ have the same imaginary part $\omega_i \sin \theta$ and have real parts $\sigma_i \pm \omega_i \cos \theta$, so that the invariant plane will map into a circle as shown in Figure 6–9.

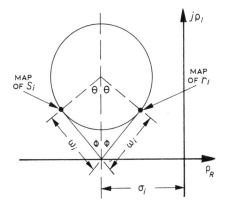

Eigenvalue Bounds

Let λ_i, μ_i ($i = 1, 2, \ldots, n$) be the eigenvalues of $\mathbf{A}, \mathbf{A}_+$ respectively, and suppose that all the respective eigenvalues have been ordered such that subscripts n and 1 denote the largest and smallest eigenvalues respectively. The inequalities of Section 2.11 in Chapter 2 then give that

$$\mu_1 \leqslant \mathrm{Re}\,(\lambda_i) \leqslant \mu_n \qquad i = 1, 2, \ldots, n \tag{6-126}$$

where $\mathrm{Re}\,(\lambda_i)$ denotes the real part of λ_i. The extremum relationships for symmetric matrices of Section 2.16 in Chapter 2 give

$$\begin{aligned} \mu_n &= \max_x \langle \mathbf{x}, \mathbf{A}_+ \mathbf{x} \rangle = \max_x \rho_R(\mathbf{A}_+) \\ \mu_1 &= \min_x \langle \mathbf{x}, \mathbf{A}_+ \mathbf{x} \rangle = \min_x \rho_R(\mathbf{A}_+) \end{aligned} \tag{6-127}$$

Combining these results we obtain the inequalities

$$\min_x \rho_R(\mathbf{A}_+) \leqslant \mathrm{Re}\,(\lambda_i) \leqslant \max_i \rho_R(\mathbf{A}_+) \qquad i = 1, 2, \ldots, n \tag{6-128}$$

Example 6.5 *Use of Circle Mappings in Design*

As an example of the circle mapping, consider first the current-regulating system shown in Figure 6–10. This maintains the inductor

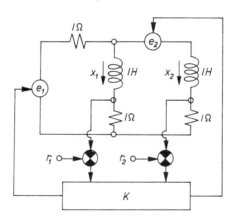

Figure 6–10

currents constant at a pair of specified values. Suppose we take as a provisional specification:

(i) no interaction between output variables in the steady state.
(ii) transient settling time of both outputs less than 0·5 seconds.
(iii) critical damping, for step change in references, on both outputs.

Generalized Mohr Circles and their Use in Feedback Design

(iv) Steady state error on both outputs less than 5%. The steady state accuracy matrix **E** is then given by

$$\mathbf{E} = \begin{bmatrix} 0.95 & 0 \\ 0 & 0.95 \end{bmatrix}$$

and the state space equations, for the component values shown, are

$$\begin{bmatrix} \dot{x}_1 \\ \dot{x}_2 \end{bmatrix} = \begin{bmatrix} -2 & -1 \\ -1 & -2 \end{bmatrix} \begin{bmatrix} x_1 \\ x_2 \end{bmatrix} + \begin{bmatrix} 1 & 0 \\ 1 & 1 \end{bmatrix} \begin{bmatrix} u_1 \\ u_2 \end{bmatrix}$$

$$\begin{bmatrix} y_1 \\ y_2 \end{bmatrix} = \begin{bmatrix} 1 & 0 \\ 0 & 1 \end{bmatrix} \begin{bmatrix} x_1 \\ x_2 \end{bmatrix}$$

As the system **B** and **C** matrices are both non-singular in this case, the matrix **K** is readily obtained from

$$\mathbf{K} = \mathbf{B}^{-1} \mathbf{A} \mathbf{C}^{-1} (\mathbf{I} - \mathbf{E}^{-1})^{-1}$$

giving the feedback matrix as

$$\mathbf{F} = -\mathbf{A} \mathbf{C}^{-1} \mathbf{E}^{*} \mathbf{C}$$

where

$$\mathbf{E}^{*} = (\mathbf{I} - \mathbf{E}^{-1})^{-1} \quad (6\text{–}129)$$

Inserting the appropriate numerical values gives

$$\mathbf{E}^{*} = \begin{bmatrix} -19 & 0 \\ 0 & -19 \end{bmatrix} \quad \mathbf{F} = \begin{bmatrix} -38 & -19 \\ -19 & -38 \end{bmatrix} \quad \mathbf{A} + \mathbf{F} = \begin{bmatrix} -40 & -20 \\ -20 & -40 \end{bmatrix}$$

The open-loop system matrix **A** maps into the circle shown in Figure 6–11(a). The closed-loop matrix (**A** + **F**) maps into a circle with centre at (−40, 0) and radius $\sqrt{(0^2 + 20^2)}$ as shown in Figure 6–11(b). The closed-loop system eigenvalues are given by the intersection of this

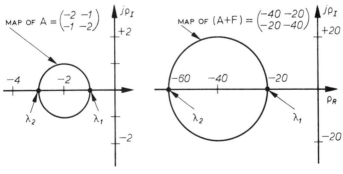

Figure 6–11(a)　　　　　Figure 6–11(b)

latter circle with the real axis and are therefore -60 and -20. Thus the steady state accuracy and settling time requirements have been achieved, but the system response will not be critically damped. To determine whether this latter requirement can be satisfied, we can use the circle mapping to determine the required form of matrix **F**. For a specific choice of **E**, say

$$\mathbf{E} = \begin{bmatrix} \alpha & 0 \\ 0 & \beta \end{bmatrix}$$

we get $\quad \mathbf{E}^* = \begin{bmatrix} v & 0 \\ 0 & \mu \end{bmatrix} \quad \text{and} \quad \mathbf{F} = \begin{bmatrix} 2v & \mu \\ v & 2\mu \end{bmatrix}$

where $\quad v = \left(\dfrac{\alpha}{\alpha-1}\right) \quad \text{and} \quad \mu = \left(\dfrac{\beta}{\beta-1}\right)$

The matrix **A** is symmetric and the matrix **F** decomposes into

$$\mathbf{F}_+ = \frac{1}{2}\begin{bmatrix} 4v & \mu+v \\ \mu+v & 4\mu \end{bmatrix} \qquad \mathbf{F}_- = \frac{1}{2}\begin{bmatrix} 0 & \mu-v \\ v-\mu & 0 \end{bmatrix}$$

The matrix $(\mathbf{A}+\mathbf{F}_+)$ maps into a circle with

$$\text{centre} = (-2+v+\mu, 0)$$

and $\quad\quad\quad \text{radius} = \text{modulus}\left(v-\mu, \dfrac{\mu+v-2}{2}\right)$

The matrix $\mathbf{F}_-$ maps into a point on the imaginary axis at $[0, \frac{1}{2}(v-\mu)]$. Consideration of the additive combination of $(\mathbf{A}+\mathbf{F}_++\mathbf{F}_-)$ then shows that, for critical damping, the circle must be tangent to the real axis in the complex plane, and so we must have that

$$\dfrac{v-\mu}{2} = \text{modulus}\left(v-\mu, \dfrac{\mu+v-2}{2}\right)$$

If we take both steady state accuracies to be the same, then we have that

$$v = \mu = 1$$

Now these values of v and μ will not give an acceptable steady state accuracy. The circle mapping has therefore shown that the original specification is unattainable since good regulation, for this particular system, is incompatible with a critically damped response. If the critical damping requirement is removed then a non-interacting (in the steady state) design is readily obtained for the required accuracy with a gain

matrix
$$K = \begin{bmatrix} 38 & 19 \\ -19 & 19 \end{bmatrix}$$

6.7 Controllability and Observability

The concepts of controllability and observability are closely related to the effects of a linear transformation of coordinates in the state space. Consider the system having state space equation set

$$\dot{x} = Ax + Bu$$
$$y = Cx \qquad (6-130)$$

and let the following coordinate transformation be carried out

$$\bar{x} = Tx \qquad x = T^{-1}\bar{x} \qquad (6-131)$$

where T is a non-singular transformation matrix. The state space equations in the new state vector $\bar{x}$ are found as follows

$$\dot{\bar{x}} = T\dot{x} = TAx + TBu$$
$$= TAT^{-1}\bar{x} + TBu \qquad (6-132)$$

This gives the required new state space equation set as

$$\dot{\bar{x}} = TAT^{-1}\bar{x} + TBu$$
$$y = CT^{-1}\bar{x} \qquad (6-133)$$

As shown in Section 6.1 above, the transfer function matrix relating input and output transforms for system (6–130) is

$$G(s) = C(sI - A)^{-1}B \qquad (6-134)$$

Let us now determine the transfer function matrix for system (6–133). Taking Laplace transforms of both sides of (6–133) with zero initial conditions gives

$$s\bar{x}(s) = TAT^{-1}\bar{x}(s) + TBu(s)$$
$$y(s) = CT^{-1}\bar{x}(s)$$

so that $(sI - TAT^{-1})\bar{x}(s) = TBu(s)$

giving $\bar{x}(s) = (sI - TAT^{-1})^{-1}TBu(s)$

thus $y(s) = CT^{-1}(sI - TAT^{-1})^{-1}TBu(s)$

and so the required transfer function matrix is

$$\overline{G}(s) = CT^{-1}(sI - TAT^{-1})^{-1}TB \qquad (6-135)$$

Now we have that

$$[T^{-1}(sI-TAT^{-1})^{-1}T]^{-1} = T^{-1}(sI-TAT^{-1})T = (sI-A)$$

so that $T^{-1}(sI-TAT^{-1})^{-1}T = (sI-A)^{-1}$

and we therefore have

$$\overline{G}(s) = C(sI-A)^{-1}B = G(s)$$

The two systems (6–130) and (6–133) related by the linear coordinate transformation T are said to be *algebraically equivalent*. The above argument shows that algebraically equivalent systems have the same transfer function matrix. All algebraically equivalent systems for a given transfer function matrix would therefore constitute *equally valid* representations of a transfer function matrix obtained from input–output measurements of a given physical system.

6.7.1 Controllability [G3], [K6]

A linear constant coefficient dynamical system is said to be *completely controllable* if it is not algebraically equivalent to any system of the form

$$\frac{d}{dt}\begin{bmatrix} x^1 \\ x^2 \end{bmatrix} = \begin{bmatrix} A^{11} & A^{12} \\ 0 & A^{22} \end{bmatrix} \begin{bmatrix} x^1 \\ x^2 \end{bmatrix} + \begin{bmatrix} B^1 \\ 0 \end{bmatrix} u$$

$$y = \begin{bmatrix} C^1 & C^2 \end{bmatrix} \begin{bmatrix} x^1 \\ x^2 \end{bmatrix} \quad (6\text{–}136)$$

where:

(a) x^1 and x^2 are vectors of order n_1 and $n_2 = (n-n_1)$ respectively.
(b) $A^{11}, A^{12}, A^{22}; B^1; C^1$ and C^2 are matrices obtained by partitioning A, B, and C respectively.

A system is thus said to be completely controllable if it is not possible to find a coordinate system in the state space in which the state variables x_i are separated into two groups $x^1 = (x_1, x_2, \ldots, x_{n_1})$ and $x^2 = (x_{n_1+1}, \ldots, x_n)$ such that the second group of state variables is not affected either by the first group or by the inputs to the system. If such a coordinate system existed we would have the relationship between input vector and state components shown in Figure 6–12. Controllability is a system property which depends only on the matrices A and B and is completely independent of the way in which the system outputs are formed.

Consider the single-input system

$$\dot{x} = Ax + bu \quad (6\text{–}137)$$

where b is a column vector. Suppose the eigenvalues of A are distinct.

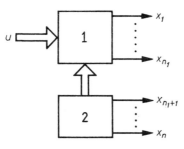

Figure 6-12

Then there exists a non-singular transformation matrix **V**, formed (as discussed in Chapter 2) of rows of dual eigenvectors, which transforms **A** into the form of a diagonal matrix Λ such that

$$\Lambda = \text{diag}(\lambda_1, \lambda_2, \ldots, \lambda_n)$$

Thus we have

$$\bar{\mathbf{x}} = \mathbf{V}\mathbf{x} \qquad \mathbf{x} = \mathbf{V}^{-1}\bar{\mathbf{x}} = \mathbf{U}\bar{\mathbf{x}}$$

where **U** is a matrix whose columns are the eigenvector set whose dual set form the rows of **V**.

This gives

$$\dot{\bar{\mathbf{x}}} = \mathbf{V}\mathbf{A}\mathbf{V}^{-1}\bar{\mathbf{x}} + \mathbf{V}\mathbf{b}u$$

$$= \Lambda\bar{\mathbf{x}} + \boldsymbol{\beta}u \qquad (6\text{-}138)$$

where $\boldsymbol{\beta} = \mathbf{V}\mathbf{b}$ and $\Lambda = \mathbf{V}\mathbf{A}\mathbf{V}$

When equation (6-138) is written out in full we have:

$$\dot{\bar{x}}_i = \lambda_i \bar{x}_i + \beta_i u \qquad i = 1, 2, \ldots, n \qquad (6\text{-}139)$$

We therefore conclude that every mode will contribute to $\bar{\mathbf{x}}$ and thus to **x** (since **V** is non-singular) if and only if

$$\beta_i \neq 0 \qquad i = 1, 2, \ldots, n \qquad (6\text{-}140)$$

Let a matrix **M** be defined by

$$\mathbf{M} = [\boldsymbol{\beta} \quad \Lambda\boldsymbol{\beta} \quad \Lambda^2\boldsymbol{\beta} \quad \ldots \quad \Lambda^{n-1}\boldsymbol{\beta}]$$

$$= \begin{bmatrix} \beta_1 & \lambda_1\beta_1 & \lambda_1^2\beta_1 & \ldots & \lambda_1^{n-1}\beta_1 \\ \beta_2 & \lambda_2\beta_2 & \lambda_2^2\beta_2 & \ldots & \lambda_2^{n-1}\beta_2 \\ \vdots & \vdots & \vdots & & \vdots \\ \beta_n & \lambda_n\beta_n & \lambda_n^2\beta_n & \ldots & \lambda_n^{n-1}\beta_n \end{bmatrix}$$

Since the eigenvalues of **A** are distinct, the matrix **M** will be non-singular if and only if condition (6–140) is satisfied. In terms of the matrix **M**, the nullity of **M** will be the number of unexcited system modes. Now we have

$$\boldsymbol{\beta} = \mathbf{Vb}$$
$$\Lambda\boldsymbol{\beta} = \mathbf{VAV}^{-1}\mathbf{Vb} = \mathbf{VAb}$$
$$\Lambda^2\boldsymbol{\beta} = \mathbf{VAV}^{-1}\mathbf{VAV}^{-1}\mathbf{Vb} = \mathbf{VA}^2\mathbf{b}$$
$$\vdots$$
$$\Lambda^{n-1}\boldsymbol{\beta} = \mathbf{VA}^{n-1}\mathbf{b}$$

so that we may write the matrix **M** in the form

$$\mathbf{M} = \mathbf{V}[\mathbf{b} \quad \mathbf{Ab} \quad \mathbf{A}^2\mathbf{b} \quad \ldots \quad \mathbf{A}^{n-1}\mathbf{b}]$$

It follows from this that, since **V** is non-singular, the rank of **M** is equal to the rank of

$$\mathbf{P} = [\mathbf{b} \quad \mathbf{Ab} \quad \mathbf{A}^2\mathbf{b} \quad \ldots \quad \mathbf{A}^{n-1}\mathbf{b}] \quad (6\text{–}141)$$

This leads to the following result:

The necessary and sufficient condition for a single input $u(t)$ to excite all the modes of the system of equation (6–137) is that the matrix **P** defined in equation (6–141) be non-singular.

If the system has r inputs $u_1(t), \ldots, u_r(t)$ then the corresponding state-space equation set will be

$$\dot{\mathbf{x}} = \mathbf{Ax} + \mathbf{Bu}$$
$$= \mathbf{Ax} + \sum_{i=1}^{r} \mathbf{b}_i u_i(t) \quad (6\text{–}142)$$

where $\mathbf{b}_i$ is the ith column of the input matrix **B**.
We thus have that:

(a) The necessary and sufficient condition for *all* the modes of the linear constant coefficient system associated with equation (6–142) to be excited by *all* the components of the input vector **u** is that *every* matrix

$$\mathbf{P}_i = [\mathbf{b}_i \quad \mathbf{Ab}_i \quad \mathbf{A}^2\mathbf{b}_i \quad \ldots \quad \mathbf{A}^{n-1}\mathbf{b}_i] \quad i = 1, 2, \ldots, r$$

be non-singular.

(b) The necessary and sufficient condition for the system to be completely controllable, that is such that *all* the modes be excited by *some* component or components of **u** is that the matrix

$$\bar{\mathbf{P}} = [\mathbf{B} \quad \mathbf{AB} \quad \mathbf{A}^2\mathbf{B} \quad \ldots \quad \mathbf{A}^{n-1}\mathbf{B}] \quad (6\text{–}143)$$

have rank n.

6.7.2 Observability

The dual concept to controllability is observability; this depends only on the system outputs and is independent of inputs.

A linear constant coefficient dynamical system is said to be *completely observable* if it is not algebraically equivalent to any system of the type

$$\frac{d}{dt}\begin{bmatrix} \mathbf{x}^1 \\ \mathbf{x}^2 \end{bmatrix} = \begin{bmatrix} \mathbf{A}^{11} & 0 \\ \mathbf{A}^{21} & \mathbf{A}^{22} \end{bmatrix} \begin{bmatrix} \mathbf{x}^1 \\ \mathbf{x}^2 \end{bmatrix} + \begin{bmatrix} \mathbf{B}^1 \\ \mathbf{B}^2 \end{bmatrix} \mathbf{u}$$

$$\mathbf{y} = [\mathbf{C}^1 \quad 0]\begin{bmatrix} \mathbf{x}^1 \\ \mathbf{x}^2 \end{bmatrix}$$

(6-144)

where:

(a) $\mathbf{x}^1$ and $\mathbf{x}^2$ are vectors of order n_1 and $n_2 = (n - n_1)$ respectively.
(b) $\mathbf{A}^{11}, \mathbf{A}^{21}, \mathbf{A}^{22}, \mathbf{B}^1, \mathbf{B}^2$, and $\mathbf{C}^1$ are matrices obtained by partitioning $\mathbf{A}, \mathbf{B}$, and $\mathbf{C}$ respectively.

A system is thus said to be completely observable if it is not possible to find a coordinate system in the state space in which the state variables x_i are separated into two groups such that the second group does not affect either the first group or the outputs of the system. If such a coordinate system existed we would have the relationship between output vector and state components shown in Figure (6-13).

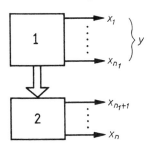

Figure 6-13

We now consider under what conditions all the modes of a linear, constant-coefficient dynamical system may be observed in the output. Let the system state space equations be

$$\dot{\mathbf{x}} = \mathbf{A}\mathbf{x} + \mathbf{B}\mathbf{u}$$
$$y = \mathbf{c}^t \mathbf{x}$$

(6-145)

where y is a single observed output which is a linear combination of the states, and $\mathbf{c}^t$ is a single-row matrix. Suppose the eigenvalues of $\mathbf{A}$

are distinct, then we can again write
$$\bar{x} = Vx \qquad x = V^{-1}\bar{x}$$
and express the output vector as
$$y = \gamma^t \bar{x}$$
$$= \sum_{i=1}^{n} \gamma_i \bar{x}_i$$
where $\qquad \gamma^t = c^t V^{-1} \quad$ and $\quad \gamma = (V^t)^{-1} c$

It is evident that *all* the modes will appear in the observed outputs if and only if
$$\gamma_i \neq 0 \quad \text{for} \quad i = 1, 2, \ldots, n$$
Now let a matrix N be defined by:
$$N^t = \begin{bmatrix} \gamma^t \\ \gamma^t \Lambda \\ \gamma^t \Lambda^2 \\ \vdots \\ \gamma^t \Lambda^{n-1} \end{bmatrix}$$
where Λ is a diagonal matrix of eigenvalues as previously defined. This gives
$$N = \begin{bmatrix} \gamma_1 & \lambda_1 \gamma_1 & \cdots & \lambda_1^{n-1} \gamma_1 \\ \gamma_2 & \lambda_2 \gamma_2 & \cdots & \lambda_2^{n-1} \gamma_2 \\ \vdots & \vdots & & \vdots \\ \gamma_n & \lambda_n \gamma_n & \cdots & \lambda_n^{n-1} \gamma_n \end{bmatrix}$$

As the eigenvalues are all distinct, this matrix will be non-singular if and only if none of the $\{\gamma_i : i = 1, 2, \ldots, n\}$ are zero; that is, if and only if all the system modes contribute to the output. Now we have that
$$\gamma = (V^t)^{-1} c$$
$$\Lambda \gamma = \Lambda^t \gamma = (V^t)^{-1} A^t V^t (V^t)^{-1} c = (V^t)^{-1} A^t c$$
$$\Lambda^2 \gamma = \Lambda^t \Lambda^t \gamma = (V^t)^{-1} A^t V^t (V^t)^{-1} A^t c = (V^t)^{-1} (A^t)^2 c$$
$$\vdots$$
$$\Lambda^{n-1} \gamma = (V^t)^{-1} (A^t)^{n-1} c$$
It follows from this that
$$N = (V^t)^{-1} [c \quad A^t c \quad (A^t)^2 c \quad \cdots \quad (A^t)^{n-1} c]$$

Since **V** is non-singular, the rank of **N** is equal to the rank of

$$Q = [c \quad A^t c \quad (A^t)^2 c \quad \ldots \quad (A^t)^{n-1} c] \quad (6\text{-}146)$$

This leads to the result that:
The necessary and sufficient condition for all the modes of the system (6-145) to contribute to the single output y is that the rank of the matrix **Q** is n.

In the general case we will have m outputs say and the system equations will be

$$\dot{x} = Ax + Bu$$
$$y = Cx \quad (6\text{-}147)$$

where **y** is an m-vector of observed outputs and **C** is an $m \times n$ matrix. We then have that:
The necessary and sufficient condition for *all* the modes of system (6-82) to be observed in *every* output is that every matrix

$$Q_i = [c_i \quad A^t c_i \quad (A^t)^2 c_i \quad \ldots \quad (A^t)^{n-1} c_i] \quad i = 1, 2, \ldots, m$$

be of rank n, where c_i is the transpose of the ith row of the matrix **C**.

The necessary and sufficient condition for the system to be completely observable, that is such that *all* the modes may be observed in *some* output, is that the matrix

$$\overline{Q} = [c^t \quad A^t c^t \quad (A^t)^2 c^t \quad \ldots \quad (A^t)^{n-1} c^t] \quad (6\text{-}148)$$

be of rank n.

6.7.3 Decomposition of State Space Systems [Z1], [K8]

In general a linear, constant coefficient dynamical system will be neither completely controllable nor completely observable. It will therefore in general be possible to arrange the state vector components, referred to a suitable coordinate system into the following four mutually exclusive parts.

Part (α): Completely controllable but unobservable.
Part (β): Completely controllable and completely observable.
Part (γ): Uncontrollable and unobservable.
Part (δ): Uncontrollable but observable.

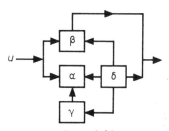

Figure 6-14

This canonical decomposition is due to Kalman and is shown in Figure 6–14. For such a choice of coordinates the state space equations assume the Kalman canonical form [K6]

$$\frac{d}{dt}\begin{bmatrix} \mathbf{x}^\alpha \\ \mathbf{x}^\beta \\ \mathbf{x}^\gamma \\ \mathbf{x}^\delta \end{bmatrix} = \begin{bmatrix} \mathbf{A}^{\alpha\alpha} & \mathbf{A}^{\alpha\beta} & \mathbf{A}^{\alpha\gamma} & \mathbf{A}^{\alpha\delta} \\ \mathbf{0} & \mathbf{A}^{\beta\beta} & \mathbf{0} & \mathbf{A}^{\beta\delta} \\ \mathbf{0} & \mathbf{0} & \mathbf{A}^{\gamma\gamma} & \mathbf{A}^{\gamma\delta} \\ \mathbf{0} & \mathbf{0} & \mathbf{0} & \mathbf{A}^{\delta\delta} \end{bmatrix} \begin{bmatrix} \mathbf{x}^\alpha \\ \mathbf{x}^\beta \\ \mathbf{x}^\gamma \\ \mathbf{x}^\delta \end{bmatrix} + \begin{bmatrix} \mathbf{B}^\alpha \\ \mathbf{B}^\beta \\ \mathbf{0} \\ \mathbf{0} \end{bmatrix} \mathbf{u}$$

$$\mathbf{y} = \begin{bmatrix} \mathbf{0} & \mathbf{C}^\beta & \mathbf{0} & \mathbf{C}^\delta \end{bmatrix} \begin{bmatrix} \mathbf{x}^\alpha \\ \mathbf{x}^\beta \\ \mathbf{x}^\gamma \\ \mathbf{x}^\delta \end{bmatrix}$$

(6–149)

6.7.4 Duality and Adjoint Systems

For zero initial conditions, the response of the system (6–130) is obtained from equation (6–20) as

$$\mathbf{y}(t) = \mathbf{C} \int_{t_0}^{t} \mathbf{\Phi}(t, \tau)\mathbf{B}\mathbf{u}(\tau)\, d\tau \qquad (6\text{–}150)$$

If we interchange the roles of input and output by making the substitutions

$$\mathbf{y} \to \mathbf{u} \qquad \mathbf{u} \to \mathbf{y}$$
$$\mathbf{B} \to \mathbf{C}^t \qquad \mathbf{C} \to \mathbf{B}^t$$
$$\tau \to t \qquad t \to \tau$$
$$\mathbf{\Phi}(t, \tau) \to \mathbf{\Phi}^t(\tau, t)$$

we get
$$\mathbf{u}(\tau) = \mathbf{B}^t \int_{\tau_0}^{\tau} \mathbf{\Phi}^t(\tau, t)\mathbf{C}^t\mathbf{y}(t)\, dt \qquad (6\text{–}151)$$

The convolution integral (6–150) corresponds to the differential equation set

$$\frac{d\mathbf{x}}{dt} = \mathbf{A}\mathbf{x} + \mathbf{B}\mathbf{u}$$
$$\mathbf{y} = \mathbf{C}\mathbf{x} \qquad (6\text{–}152)$$

To find the differential equation set corresponding to the convolution integral (6–151) we consider the transition matrix $\mathbf{\Phi}^t(\tau, t)$. We have

$$\Phi^t(\tau, t) = \exp \mathbf{A}^t(\tau - t)$$
$$= \exp \mathbf{A}^t(-t + \tau) \qquad (6\text{-}153)$$

Now the transition matrix corresponding to the equation set (6–152) is

$$\Phi(t, \tau) = \exp \mathbf{A}(t - \tau) \qquad (6\text{-}154)$$

It follows from this that the differential equation set corresponding to the convolution integral (6–151) is obtained by making the following substitutions in the differential equation set (6–152):

$$t - t_0 \to t_0 - t$$
$$\mathbf{A} \to \mathbf{A}^t$$
$$\mathbf{B} \to \mathbf{C}^t$$
$$\mathbf{C} \to \mathbf{B}^t$$

giving

$$\frac{d\bar{\mathbf{x}}}{dt} = -\mathbf{A}^t\bar{\mathbf{x}} - \mathbf{C}^t\mathbf{u}$$
$$\mathbf{y} = \mathbf{B}^t\bar{\mathbf{x}} \qquad (6\text{-}155)$$

The systems (6–152) and (6–155) are said to be dual or adjoint systems. It is most important to note the *time reversal* involved which arises from the interchange of t and τ in the transition matrices.

It follows immediately from the interchange of the roles of input and output in such dual (or adjoint) systems that all theorems on controllability may be 'dualized' by means of these transformations to give analogous results on observability and vice versa.

Example 6.5
Consider the system

$$\begin{bmatrix} \dot{x}_1 \\ \dot{x}_2 \end{bmatrix} = \begin{bmatrix} 1 & 1 \\ 0 & -1 \end{bmatrix} \begin{bmatrix} x_1 \\ x_2 \end{bmatrix} + \begin{bmatrix} 1 \\ -2 \end{bmatrix} u$$

Forming the matrix **P** and testing its rank we have

$$\mathbf{P} = [\mathbf{b} \quad \mathbf{Ab}] = \begin{bmatrix} 1 & -1 \\ -2 & 2 \end{bmatrix}$$

which is singular and of rank 1. The system is therefore not completely controllable.

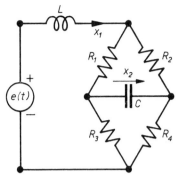

Figure 6–15

Example 6.6

Consider the electrical bridge network of Figure 6–15. Taking the state variables x_1 and x_2 as the inductor current and capacitor voltage respectively the system state space equations have **A** and **B** matrices given by

$$\mathbf{A} = \begin{bmatrix} \dfrac{-1}{L}\left(\dfrac{R_1 R_2}{R_1 + R_2} + \dfrac{R_3 R_4}{R_3 + R_4}\right) & \dfrac{1}{L}\left(\dfrac{R_1}{R_1 + R_2} - \dfrac{R_3}{R_3 + R_4}\right) \\ \dfrac{-1}{C}\left(\dfrac{R_1}{R_1 + R_2} - \dfrac{R_3}{R_3 + R_4}\right) & \dfrac{-1}{C}\left(\dfrac{1}{R_3 + R_4} + \dfrac{1}{R_1 + R_2}\right) \end{bmatrix}$$

$$\mathbf{B} = \begin{bmatrix} \dfrac{1}{L} \\ 0 \end{bmatrix}$$

Forming the matrix **P** we have

$$\mathbf{P} = [\mathbf{B} \quad \mathbf{AB}] = \begin{bmatrix} \dfrac{1}{L} & \dfrac{-1}{L^2}\left(\dfrac{R_1 R_2}{R_1 + R_2} + \dfrac{R_3 R_4}{R_3 + R_4}\right) \\ 0 & \dfrac{-1}{LC}\left(\dfrac{R_1}{R_1 + R_2} - \dfrac{R_3}{R_3 + R_4}\right) \end{bmatrix}$$

The matrix **P** will be singular if and only if

$$\frac{R_1}{R_1 + R_2} = \frac{R_3}{R_3 + R_4}$$

that is if $R_1 R_4 = R_2 R_3$. For these particular conditions the bridge is said to be balanced; this gives a good physical example of an uncontrollable state variable.

Controllability and Observability

Example 6.7
Suppose we have a system such that
$$\mathbf{A} = \begin{bmatrix} 0 & 1 \\ 2 & -1 \end{bmatrix} \quad \text{and} \quad \mathbf{C} = [c_1 \quad c_2]$$
Forming the matrix $\mathbf{Q}$ we will have
$$\mathbf{Q} = [\mathbf{C}^t \quad \mathbf{A}^t\mathbf{C}^t] = \begin{bmatrix} c_1 & 2c_2 \\ c_2 & c_1 - c_2 \end{bmatrix}$$
The determinant of $\mathbf{Q}$ is given by
$$\det(\mathbf{Q}) = c_1(c_1 - c_2) - 2c_2^2 = (c_1 + c_2)(c_1 - 2c_2)$$
The system will therefore not be completely observable if $c_1 = -c_2$ or if $c_1 = 2c_2$. The transition matrix corresponding to this system is
$$\Phi(t) = \mathscr{L}^{-1}(s\mathbf{I} - \mathbf{A})^{-1} = \frac{1}{3}\begin{bmatrix} (2e^t + e^{-2t}) & (e^t - e^{-2t}) \\ (2e^t - 2e^{-2t}) & (e^t + 2e^{-2t}) \end{bmatrix}$$
The system has one stable and one unstable mode. For some arbitrary set of initial conditions, the output will be given by
$$y(t) = c_1 x_1(t) + c_2 x_2(t)$$
$$= \tfrac{1}{3}[(2c_1 + 2c_2)x_1(0) + (c_1 + c_2)x_2(0)]e^t$$
$$+ \tfrac{1}{3}[(c_1 - 2c_2)x_1(0) + (-c_1 + 2c_2)x_2(0)]e^{-2t}$$
An inspection of this expression shows that, when the unobservability arises from the condition $c_1 = -c_2$, only the stable mode appears in the output and that, when the unobservability arises from the condition $c_1 = 2c_2$, only the unstable mode appears in the output.

Irreducible and Minimal Representations
In the Kalman canonical structure representation above, let
$$n_\alpha = \text{dimension of vector } \mathbf{x}^\alpha$$
$$n_\beta = \text{dimension of vector } \mathbf{x}^\beta$$
and so on. We then have that
$$n_\alpha + n_\beta = \text{rank}\,[\mathbf{B}\,\mathbf{AB}\,\ldots\,\mathbf{A}^{n-1}\mathbf{B}]$$
$$n_\gamma + n_\delta = \text{rank}\,[\mathbf{C}^t\,\mathbf{A}^t\mathbf{C}^t\,\ldots\,(\mathbf{A}^t)^{n-1}\mathbf{C}^t]$$
If $n_\beta = n$ and $n_\alpha = n_\gamma = n_\delta = 0$ the system is said to be *irreducible*. The state space equations of an irreducible system cannot be replaced, for purposes of an external description relating inputs and outputs, by

466   State Models

a set of equations of lower order. If a system is not completely controllable and completely observable then it may be replaced, for the purpose of relating input and output variables, by a system of lower order. The lowest-order systems giving the same input–output relations (that is the same transfer function matrix) are said to be *minimal* representations of the given input–output description.

6.7.5 Determination of the Controllable Part of a given Representation [M8]

If a given vector $\boldsymbol{\xi}$ lies in the controllable subspace for a given system then there will exist a time $T > 0$ and a control $\mathbf{u}^*(t)$ such that the control $\mathbf{u}^*$ transfers the system state from $\boldsymbol{\xi}$ to the origin in time T. Thus we have

$$0 = [\exp(\mathbf{A}T)\boldsymbol{\xi}] + \int_0^T \exp[\mathbf{A}(T-\tau)]\mathbf{B}\mathbf{u}^*(\tau)\,d\tau$$

from which we obtain

$$-\boldsymbol{\xi} = \int_0^T \exp(-\mathbf{A}\tau)\mathbf{B}\mathbf{u}^*(\tau)\,d\tau \qquad (6\text{–}156)$$

where

$$\exp(-\mathbf{A}\tau) = \sum_{n=0}^{\infty} (-1)^n \mathbf{A}^n \frac{\tau^n}{n!} \qquad (6\text{–}157)$$

Now, by the Cayley–Hamilton theorem, every matrix $\mathbf{A}$ satisfies its own characteristic equation. There thus exists a set of constants $c_1, c_2, \ldots, c_n$ such that

$$(-1)^n \mathbf{A}^n + c_1 \mathbf{A}^{n-1} + c_2 \mathbf{A}^{n-2} + \cdots + c_n \mathbf{I} = \mathbf{0}$$

In other words, of the infinite sequence of matrices $\mathbf{A}^0, \mathbf{A}^1, \mathbf{A}^2, \ldots,$ $\mathbf{A}^k, \ldots$, only the set $\mathbf{A}^0, \mathbf{A}^1, \ldots, \mathbf{A}^{n-1}$ are independent in the sense that no set of scalar multipliers exist which give a vanishing sum. Since the right-hand side of equation (6–157) can be expressed in terms of a finite sum in this way, so must the left-hand side. It follows therefore that there must exist a set of scalar-valued functions of τ such that

$$\exp(-\mathbf{A}\tau) = \sum_{\alpha=0}^{n-1} \psi_\alpha(\tau) \mathbf{A}^\alpha \qquad (6\text{–}158)$$

Inserting this into equation (6–156) we get

$$-\boldsymbol{\xi} = \sum_{\alpha=0}^{n-1} \int_0^T \psi_\alpha(\tau) \mathbf{A}^\alpha \mathbf{B}\mathbf{u}^*(\tau)\,d\tau$$

Now let $\mathbf{u}^*(\tau)$ be expressed as a linear combination of the standard basis vectors $\{\mathbf{e}_\beta : \beta = 1, 2, \ldots, n\}$ in the state space

$$\mathbf{u}^*(\tau) = \sum_{\beta=1}^{r} \mu_\beta(\tau) \mathbf{e}_\beta$$

This then gives

$$-\boldsymbol{\xi} = \sum_{\alpha=0}^{n-1} \int_0^T \psi_\alpha(\tau) \mathbf{A}^\alpha \mathbf{B} \sum_{\beta=1}^{r} \mu_\beta(\tau) \mathbf{e}_\beta \, d\tau$$

$$= \sum_{\alpha=0}^{n-1} \int_0^T \psi_\alpha(\tau) \mathbf{A}^\alpha \sum_{\beta=1}^{r} \mu_\beta(\tau) \mathbf{B} \mathbf{e}_\beta \, d\tau$$

$$= \sum_{\alpha=0}^{n-1} \int_0^T \psi_\alpha(\tau) \mathbf{A}^\alpha \sum_{\beta=1}^{r} \mu_\beta(\tau) \mathbf{b}_\beta \, d\tau$$

where $\{\mathbf{b}_\beta : \beta = 1, 2, \ldots, r\}$ are the columns of matrix $\mathbf{B}$. We thus have

$$-\boldsymbol{\xi} = \sum_{\alpha=0}^{n-1} \sum_{\beta=1}^{r} \left[\int_0^T \psi_\alpha(\tau) \mu_\beta(\tau) \, d\tau \right] \mathbf{A}^\alpha \mathbf{b}_\beta$$

from which we conclude that the set of vectors

$$\mathbf{A}^\alpha \mathbf{b}_\beta \quad \begin{aligned} \alpha &= 0, 1, 2, \ldots, n-1 \\ \beta &= 1, 2, \ldots, n \end{aligned} \quad (6\text{-}159)$$

must span the controllable subspace of the state space.

Let $\mathbf{S}$ be a matrix whose columns $\mathbf{s}_1, \mathbf{s}_2, \ldots, \mathbf{s}_{n_k}$ are the maximum number of linearly independent vectors of the set $\mathbf{A}^\alpha \mathbf{b}_\beta$; the column vectors of $\mathbf{S}$ then span the controllable subspace of the system

$$\begin{aligned} \dot{\mathbf{x}} &= \mathbf{A}\mathbf{x} + \mathbf{B}\mathbf{u} \\ \mathbf{y} &= \mathbf{C}\mathbf{x} \end{aligned} \quad (6\text{-}160)$$

Suppose we carry out the coordinate transformation

$$\bar{\mathbf{x}} = \mathbf{T}\mathbf{x} \quad (6\text{-}161)$$

where

$$\mathbf{T}^{-1} = [\mathbf{S} \quad \mathbf{X}] \quad (6\text{-}162)$$

and $\mathbf{X}$ is any $n \times (n - n_k)$ matrix such that the matrix $\mathbf{T}$ is non-singular. Construct a matrix $\mathbf{Y}$ which is a left-inverse of matrix $\mathbf{S}$ so that

$$\mathbf{Y}\mathbf{S} = \mathbf{I}_{n_k} \quad (6\text{-}163)$$

and let the appropriately partitioned form of the matrix $\mathbf{T}$ be

$$\mathbf{T} = \begin{bmatrix} \mathbf{Y} \\ \mathbf{W} \end{bmatrix} \quad (6\text{-}164)$$

so that

$$TT^{-1} = \begin{bmatrix} Y \\ W \end{bmatrix} [S \quad X] = \begin{bmatrix} YS & YX \\ WS & WX \end{bmatrix} = \begin{bmatrix} I_{n_k} & 0 \\ 0 & I_{(n-n_k)} \end{bmatrix}$$

and so

$$YS = I_{n_k} \qquad WX = I_{(n-n_k)} \qquad YX = 0_{n_k,(n-n_k)} \qquad WS = 0_{(n-n_k),n_k}$$

Therefore, since $WS = 0$, the rows of W will be orthogonal to the column vectors of S.

The result of applying the coordinate transformation represented by T is to generate a system

$$\frac{d}{dt}\begin{bmatrix} x^1 \\ x^2 \end{bmatrix} = \begin{bmatrix} Y \\ W \end{bmatrix} A[S \quad X] + \begin{bmatrix} Y \\ W \end{bmatrix} Bu$$

$$y = C[S \quad X]\begin{bmatrix} x^1 \\ x^2 \end{bmatrix}$$

that is such that

$$\frac{d}{dt}\begin{bmatrix} x^1 \\ x^2 \end{bmatrix} = \begin{bmatrix} YAS & YAX \\ WAS & WAX \end{bmatrix}\begin{bmatrix} x^1 \\ x^2 \end{bmatrix} + \begin{bmatrix} YB \\ WB \end{bmatrix} u$$

$$y = [CS \quad CX]\begin{bmatrix} x^1 \\ x^2 \end{bmatrix}$$

(6–165)

Consideration of the convolution integral of equation (6–20) shows that the controllable subspace of a system must be invariant under Φ, since an initially controllable state cannot be transferred by the free motion of the system into an uncontrollable state. It follows from equation (6–25a) that the controllable subspace must also be invariant under A. Thus for any vector s_i in the controllable subspace the vector As_i must also lie in the controllable subspace. It follows from this that all the column vectors of the matrix AS must lie in the controllable subspace. However, since

$$WS = 0 \qquad (6\text{–}166)$$

we have that all the rows of W are orthogonal to all the column vectors of S and thus to any vector lying in the controllable subspace of the system, so that we must have

$$WAS = 0 \qquad (6\text{–}167)$$

Also the column vectors of B belong to the controllable subspace so that similar reasoning gives that

$$WB = 0 \qquad (6\text{–}168)$$

Inserting these in equation (6–165) gives

$$\frac{d}{dt}\begin{bmatrix} \mathbf{x}^1 \\ \mathbf{x}^2 \end{bmatrix} = \begin{bmatrix} \mathbf{YAS} & \mathbf{YAX} \\ \mathbf{0} & \mathbf{WAX} \end{bmatrix}\begin{bmatrix} \mathbf{x}^1 \\ \mathbf{x}^2 \end{bmatrix} + \begin{bmatrix} \mathbf{YB} \\ \mathbf{0} \end{bmatrix}\mathbf{u}$$

$$\mathbf{y} = [\mathbf{CS} \quad \mathbf{CX}]\begin{bmatrix} \mathbf{x}^1 \\ \mathbf{x}^2 \end{bmatrix}$$

(6–169)

and it immediately follows that the controllable part of the given system is given by

$$\frac{d\mathbf{x}^1}{dt} = \mathbf{YASx}^1 + \mathbf{YBu}$$

$$\mathbf{y} = \mathbf{CSx}^1$$

(6–170)

6.7.6 Determination of the Observable Part of a Given Representation [M8]

Using the duality relationships and the arguments for the controllable-part case we have that:

(a) The vectors $\mathbf{c}_1, \mathbf{c}_2, \ldots, \mathbf{c}_r$, $\mathbf{A}^t\mathbf{c}_1, \ldots, \mathbf{A}^t\mathbf{c}_r$, $(\mathbf{A}^t)^2\mathbf{c}_1, \ldots, (\mathbf{A}^t)^{n-1}\mathbf{c}_r$ span the observable subspace where $\mathbf{c}_1, \ldots, \mathbf{c}_r$ are the columns of matrix $\mathbf{C}$.

(b) The observable subspace is invariant under $\mathbf{A}^t$.

Construct a matrix $\mathbf{N}^t$ whose columns are the maximum number of linearly independent vectors of the set $\{\mathbf{C}_\beta(\mathbf{A}^t)^\alpha : \alpha = 1, 2, \ldots, (n-1); \beta = 1, 2, \ldots, r\}$; the column vectors of $\mathbf{N}^t$ then span the observable subspace of system (6–160).

Again carry out a coordinate transformation

$$\bar{\mathbf{x}} = \mathbf{Tx}$$

where

$$\mathbf{T} = \begin{bmatrix} \mathbf{N} \\ \mathbf{W} \end{bmatrix}$$

(6–171)

and $\mathbf{N}$ has been constructed as stated above so that its columns span the observable subspace. $\mathbf{W}$ is any $(n - n_k) \times n$ matrix such that the transformation matrix $\mathbf{T}$ is non-singular. Let

$$\mathbf{T}^{-1} = [\mathbf{S} \quad \mathbf{M}]$$

(6–172)

where $\mathbf{S}$ and $\mathbf{M}$ are the corresponding appropriate partitionings of $\mathbf{T}^{-1}$. We thus have

$$\mathbf{TT}^{-1} = \begin{bmatrix} \mathbf{N} \\ \mathbf{W} \end{bmatrix}[\mathbf{S} \quad \mathbf{M}] = \begin{bmatrix} \mathbf{NS} & \mathbf{NM} \\ \mathbf{WS} & \mathbf{WM} \end{bmatrix} = \begin{bmatrix} \mathbf{I} & \mathbf{0} \\ \mathbf{0} & \mathbf{I} \end{bmatrix}$$

470   State Models

so that $\quad$ $NS = I_{n_k}$ $\qquad$ $WM = I_{(n-n_k)}$

$\qquad$ $NM = 0_{n_k,(n-n_k)}$ $\qquad$ $WS = 0_{(n-n_k),n_k}$

Since $NM = 0$, the column vectors of M are orthogonal to the row vectors of N and thus to the column vectors of N^t. But the column vectors of N^t span the observable subspace, so it follows that the column vectors of M are orthogonal to the observable subspace. Also we have that the observable subspace is invariant under A^t so that the column vectors of $A^t N^t$ and thus the row vectors of NA lie in the observable subspace. We must therefore have that

$$NAM = 0 \qquad (6\text{-}173)$$

Finally, since the rows of C lie in the observable subspace we must also have that

$$CM = 0 \qquad (6\text{-}174)$$

The result of applying the coordinate transformation represented by T is to generate a system

$$\frac{d}{dt}\begin{bmatrix} x^1 \\ x^2 \end{bmatrix} = \begin{bmatrix} NAS & NAM \\ WAS & WAM \end{bmatrix} \begin{bmatrix} x^1 \\ x^2 \end{bmatrix} + \begin{bmatrix} NB \\ WB \end{bmatrix} u$$

$$y = [CS \quad CM] \begin{bmatrix} x^1 \\ x^2 \end{bmatrix}$$

$(6\text{-}175)$

Substituting from equations (6-173) and (6-174) then gives the form

$$\frac{d}{dt}\begin{bmatrix} x^1 \\ x^2 \end{bmatrix} = \begin{bmatrix} NAS & 0 \\ WAS & WAM \end{bmatrix} \begin{bmatrix} x^1 \\ x^2 \end{bmatrix} + \begin{bmatrix} NB \\ WB \end{bmatrix} u$$

$$y = [CS \quad 0] \begin{bmatrix} x^1 \\ x^2 \end{bmatrix}$$

$(6\text{-}176)$

and it immediately follows that the observable part of the given system is given by

$$\frac{dx^1}{dt} = NASx^1 + NBu$$

$$y = CSx^1$$

$(6\text{-}177)$

Controllability and Observability

6.7.7 Determination of Transfer Function Matrix Representations

Consider the system having the **A**, **B**, **C** matrices

$$\mathbf{A} = \begin{bmatrix} 0 & 0 & 0 & \cdots & 0 & 0 \\ 0 & 0 & 1 & \cdots & 0 & 0 \\ \vdots & \vdots & \vdots & & \vdots \\ 0 & 0 & 0 & & 0 & 1 \\ -b_1 & -b_2 & -b_3 & \cdots & -b_{n-1} & -b_n \end{bmatrix}$$

(6-178)

$$\mathbf{B} = \begin{bmatrix} 0 \\ 0 \\ \vdots \\ 0 \\ 1 \end{bmatrix} \qquad \mathbf{C} = [a_1 \quad a_2 \quad \cdots \quad a_{n-1} \quad a_n]$$

The corresponding transfer function relating the single input and single output is

$$g(s) = \frac{a_n s^{n-1} + \cdots + a_1}{s^n + b_n s^{n-1} + \cdots + b_1}$$

(6-179)

The same transfer function relating a single input and single output is also realized by the system having **A**, **B**, **C** matrices

$$\mathbf{A} = \begin{bmatrix} 0 & 0 & \cdots & 0 & -b_1 \\ 1 & 0 & \cdots & 0 & -b_2 \\ 0 & 1 & \cdots & 0 & -b_3 \\ \vdots & \vdots & & \vdots & \vdots \\ 0 & 0 & \cdots & 0 & -b_n \end{bmatrix}$$

(6-180)

$$\mathbf{B} = \begin{bmatrix} a_1 \\ a_2 \\ \vdots \\ a_n \end{bmatrix} \qquad \mathbf{C} = [0 \quad 0 \quad \cdots \quad 0 \quad 1]$$

An inspection of the form of the matrix

$$\mathbf{P} = [\mathbf{B} \quad \mathbf{AB} \quad \mathbf{A}^2 \mathbf{B} \quad \cdots \quad \mathbf{A}^{n-1} \mathbf{B}]$$

shows that the realization (6–178) is completely controllable. It follows from duality arguments (or from forming and testing the appropriate **Q** matrix) that the system (6–180) is completely observable. Now consider the overall transfer function matrix

$$\mathbf{G}(s) = \begin{bmatrix} g_{11}(s) & g_{12}(s) & \cdots & g_{1m}(s) \\ g_{21}(s) & g_{22}(s) & \cdots & g_{2m}(s) \\ \vdots & \vdots & & \vdots \\ g_{r1}(s) & g_{r2}(s) & & g_{rm}(s) \end{bmatrix} \quad (6\text{–}181)$$

having a typical element $g_{ij}(s)$. Let $b^i(s)$ be the least common multiple of the denominators of the elements of the ith row of $\mathbf{G}(s)$. This ith row can be written as

$$\begin{bmatrix} \dfrac{a^{i1}(s)}{b^i(s)} & \dfrac{a^{i2}(s)}{b^i(s)} & \cdots & \dfrac{a^{im}(s)}{b^i(s)} \end{bmatrix}$$

with a typical element

$$g_{ij}(s) = \frac{a^{ij}(s)}{b^i(s)} = \frac{a_{n_i}^{ij} s^{n_i-1} + \cdots + a_2^{ij} s + a_1^{ij}}{s^{n_i} + b_{n_i}^i s^{n_i-1} + \cdots + b_2^i s + b_1^i}$$

A set of completely observable state space realizations of the elements of the ith row of the transfer function matrix can thus be formed by insertion of the appropriate values into **A**, **B**, **C** matrices of the form given by equation (6–180). Since all the row elements have been arranged to have the same denominator terms, the corresponding realizations for the row elements will have the same **A** and **C** matrices and only differ in their **B** matrices, so that we get the corresponding realizations as given by

$$\mathbf{A}_i = \begin{bmatrix} 0 & \cdots & 0 & -b_1^i \\ 1 & \cdots & 0 & -b_2^i \\ \vdots & & & \vdots \\ 0 & \cdots & 1 & -b_{n_i}^i \end{bmatrix}$$

$$\mathbf{C}_i = \begin{bmatrix} 0 & 0 & \cdots & 1 \end{bmatrix} \quad (6\text{–}182)$$

$$\mathbf{B}_{ij} = \begin{bmatrix} a_{n_i}^{ij} \\ a_{n_i-1}^{ij} \\ \vdots \\ a_2^{ij} \\ a_1^{ij} \end{bmatrix}$$

Similar considerations apply to the other rows. In terms of these row element realizations we may write the overall transfer function matrix as

$$G(s)$$

$$= \begin{bmatrix} C_1(sI-A_1)^{-1}B_{11} & C_1(sI-A_1)^{-1}B_{12} & \cdots & C_1(sI-A_1)^{-1}B_{1m}^1 \\ C_2(sI-A_2)^{-1}B_{21} & C_2(sI-A_2)^{-1}B_{22} & \cdots & C_2(sI-A_2)^{-1}B_{2m} \\ \vdots & \vdots & & \vdots \\ C_r(sI-A_r)^{-1}B_{r1} & C_r(sI-A_r)^{-1}B_{r2} & \cdots & C_r(sI-A_r)^{-1}B_{rm} \end{bmatrix}$$

$$= [C_1 C_2 \ldots C_r] \begin{bmatrix} (sI-A_1)^{-1} & 0 & \cdots & 0 \\ 0 & (sI-A_2)^{-1} & \cdots & 0 \\ \vdots & \vdots & & \vdots \\ 0 & 0 & \cdots & (sI-A_r)^{-1} \end{bmatrix}$$

$$\cdot \begin{bmatrix} B_{11} & B_{12} & \cdots & B_{1m} \\ B_{21} & B_{22} & \cdots & B_{2m} \\ \vdots & \vdots & & \vdots \\ B_{r1} & B_{r2} & \cdots & B_{rm} \end{bmatrix}$$

It immediately follows from this that a state space realization of the overall transfer function matrix is given by

$$A = \begin{bmatrix} A_1 & 0 & 0 & \cdots & 0 \\ 0 & A_2 & 0 & \cdots & 0 \\ \vdots & & & & \\ 0 & 0 & 0 & \cdots & A_r \end{bmatrix} \qquad C = [C_1 \ C_2 \ \cdots \ C_r]$$

$$B = \begin{bmatrix} B_{11} & B_{12} & \cdots & B_{1m} \\ B_{21} & B_{22} & \cdots & B_{2m} \\ \vdots & \vdots & & \vdots \\ B_{r1} & B_{r2} & \cdots & B_{rm} \end{bmatrix}$$

(6-183)

where the blocks of these partitioned matrices are determined from the transfer function matrix entries by using equation (6-182). Since each of the individual transfer function realizations is completely observable, it follows that the overall representation is completely observable.

A completely controllable representation may be constructed in a similar way by using the columns of $G(s)$. If $b^j(s)$ is the least common multiple of the denominators of the elements of the jth column of $\mathbf{G}(s)$, we may write this column as

$$\left[\frac{a^{1j}(s)}{b^j(s)} \quad \frac{a^{2j}(s)}{b^j(s)} \quad \cdots \quad \frac{a^{rj}(s)}{b^j(s)}\right]^t$$

with a typical element

$$g_{ij}(s) = \frac{\bar{a}^{ij}(s)}{b^j(s)} = \frac{\bar{a}^{ij}_{n_j}s^{n_j-1} + \cdots + \bar{a}^{ij}_1}{s^{n_j} + b^j_{n_j}s^{n_j-1} + \cdots + b^j_1}$$

The state space realization of such an element is given by

$$\mathbf{A}_j = \begin{bmatrix} 0 & 1 & 0 & \cdots & 0 \\ 0 & 0 & 1 & \cdots & 0 \\ \vdots & & & & \vdots \\ 0 & 0 & 0 & \cdots & 1 \\ -b^j_1 & -b^j_2 & -b^j_3 & \cdots & -b^j_{n_j} \end{bmatrix}$$

$$\mathbf{B}_j = \begin{bmatrix} 0 \\ 0 \\ \vdots \\ 0 \\ 1 \end{bmatrix} \quad \mathbf{C}_{ij} = \begin{bmatrix} C_{11} & C_{12} & \cdots & C_{1m} \\ C_{21} & C_{22} & \cdots & C_{2m} \\ \vdots & & & \\ C_{r1} & C_{r2} & \cdots & C_{rm} \end{bmatrix}$$

(6–184)

The state space realizations of the elements of the other columns of the transfer function matrix may be treated in a similar way. We may thus write the overall transfer function matrix as

$$\mathbf{G}(s) = \begin{bmatrix} C_{11} & C_{12} & \cdots & C_{1m} \\ C_{21} & C_{22} & \cdots & C_{2m} \\ \vdots & & & \\ C_{r1} & C_{r2} & \cdots & C_{rm} \end{bmatrix}$$

$$\cdot \begin{bmatrix} (s\mathbf{I}-\mathbf{A}_1)^{-1} & 0 & \cdots & 0 \\ 0 & (s\mathbf{I}-\mathbf{A}_2)^{-1} & \cdots & 0 \\ \vdots & \vdots & & \vdots \\ 0 & 0 & \cdots & (s\mathbf{I}-\mathbf{A}_m)^{-1} \end{bmatrix} \begin{bmatrix} \mathbf{B}_1 \\ \mathbf{B}_2 \\ \vdots \\ \mathbf{B}_m \end{bmatrix}$$

(6–185)

The submatrices $\mathbf{A}_1, \mathbf{A}_2 \ldots$ realized in this way are not of course in general the same as those obtained by the previous procedure. It

Controllability and Observability

immediately follows from this that a state space realization of the overall transfer function matrix is given by

$$A = \begin{bmatrix} A_1 & 0 & 0 & \cdots & 0 \\ 0 & A_2 & 0 & \cdots & 0 \\ \vdots & \vdots & \vdots & & \\ 0 & 0 & 0 & \cdots & A_m \end{bmatrix} \quad C = \begin{bmatrix} C_{11} & C_{12} & \cdots & C_{1m} \\ C_{21} & C_{22} & \cdots & C_{2m} \\ \vdots & & & \\ C_{r1} & C_{r2} & \cdots & C_{rm} \end{bmatrix}$$

$$B = \begin{bmatrix} B_1 \\ \vdots \\ B_m \end{bmatrix} \quad (6\text{-}186)$$

where the blocks of these partitioned matrices are determined from the transfer function matrix entries by using equation (6-184). Since each of the individual transfer function realizations is completely controllable, it follows that the overall representation is completely controllable.

To determine a minimal realization of a given transfer function matrix two straightforward methods of procedure follows from the arguments presented above:

(a) determine a completely observable representation using equations (6-183) and then determine its controllable part; or
(b) determine a completely controllable representation using equations (6-186) and then determine its observable part.

Example 6.8
Consider the transfer function matrix

$$G(s) = \begin{bmatrix} \dfrac{1}{(s+1)} & \dfrac{1}{(s+1)} \\ \dfrac{1}{(s+1)} & \dfrac{1}{(s+1)} \end{bmatrix}$$

Using equations (6-183) we find that a completely observable realization is given by

$$A = \begin{bmatrix} -1 & 0 \\ 0 & -1 \end{bmatrix} \quad B = \begin{bmatrix} 1 & 1 \\ 1 & 1 \end{bmatrix} \quad C = \begin{bmatrix} 1 & 0 \\ 0 & 1 \end{bmatrix}$$

To determine the controllable part of this realization we must first determine a maximum number of linearly independent vectors from the set $\{A^\alpha b_\beta : \alpha = 1, 2; \beta = 1, 2\}$ where the b_β are the columns of B. We have that

$$\mathbf{b}_1 = \begin{bmatrix} 1 \\ 1 \end{bmatrix} \quad \mathbf{b}_2 = \begin{bmatrix} 1 \\ 1 \end{bmatrix} \quad \mathbf{A}\mathbf{b}_1 = \begin{bmatrix} -1 \\ -1 \end{bmatrix} \quad \mathbf{A}\mathbf{b}_2 = \begin{bmatrix} -1 \\ -1 \end{bmatrix}$$

$$\mathbf{A}^2\mathbf{b}_1 = \begin{bmatrix} 1 \\ 1 \end{bmatrix} \quad \mathbf{A}^2\mathbf{b}_2 = \begin{bmatrix} 1 \\ 1 \end{bmatrix}$$

The maximum number of linearly independent vectors is one and we may take the matrix **S** as

$$\mathbf{S} = \begin{bmatrix} 1 \\ 1 \end{bmatrix}$$

with a suitable **Y**-matrix as $\mathbf{Y} = [1 \ 0]$. The controllable part of the realization, and thus a minimal realization of the transfer function matrix is thus given by

$$\mathbf{A}_k = \mathbf{YAS} = [1 \ 0] \begin{bmatrix} -1 & 0 \\ 0 & -1 \end{bmatrix} \begin{bmatrix} 1 \\ 1 \end{bmatrix} = -1$$

$$\mathbf{B}_k = \mathbf{YB} = [1 \ 0] \begin{bmatrix} 1 & 1 \\ 1 & 1 \end{bmatrix} = [1 \ 1]$$

$$\mathbf{C}_k = \mathbf{CS} = \begin{bmatrix} 1 & 0 \\ 0 & 1 \end{bmatrix} \begin{bmatrix} 1 \\ 1 \end{bmatrix} = \begin{bmatrix} 1 \\ 1 \end{bmatrix}$$

6.8 Reduction [M11]

Given a set of linear differential equations representing the behaviour of a linear, constant coefficient, dynamical system, it is often desired to trade the accuracy of representation available against the order of the system model. In the reduction techniques considered here a particular matrix plays a central role; its properties are thus first considered. Consider the system having state space equations

$$\begin{aligned} \dot{\mathbf{x}} &= \mathbf{Ax} + \mathbf{b}u \\ \mathbf{y} &= \mathbf{Cx} \end{aligned} \quad (6\text{--}187)$$

where **b** is a column vector and the single input variable u is a unit Dirac impulse $\delta(t)$. Denote the state response to this input by $\mathbf{r}(t)$ and define a matrix $\mathbf{W}(t)$ by

$$\mathbf{W}(t) \triangleq \int_0^t \mathbf{r}(\tau)\mathbf{r}^t(\tau) \, d\tau \quad (6\text{--}188)$$

Reduction

Let the eigenvalues of **A** be distinct and put

$$\mathbf{r}(\tau) = \mathbf{U}\mathbf{s}(\tau)$$

where $\mathbf{s}(\tau)$ is the impulse response referred to the system eigenvector set as a basis. Then

$$\mathbf{W}(t) = \int_0^t \mathbf{r}(\tau)\mathbf{r}^t(\tau)\, d\tau = \int_0^t \mathbf{U}\mathbf{s}(\tau)\mathbf{s}^t(\tau)\mathbf{U}^t\, d\tau$$

$$= \mathbf{U} \int_0^t \mathbf{s}(\tau)\mathbf{s}^t(\tau)\, d\tau\, \mathbf{U}^t$$

$$= \mathbf{U}\mathbf{G}(t)\mathbf{U}^t \qquad (6\text{--}189)$$

where $\mathbf{G}(t)$ is a matrix having elements

$$[G]_{ij} = \int_0^t s_i(\tau) s_j(\tau)\, d\tau$$

where $s_i(\tau)$ and $s_j(\tau)$ are respectively the ith and jth elements of the vector $\mathbf{s}(\tau)$. Thus if β_i and β_j are the ith and jth elements respectively of the vector $\boldsymbol{\beta}$ introduced in equation (6–138) we have

$$[G]_{ij} = \int_0^t \beta_i \beta_j \exp(\lambda_i + \lambda_j)\tau\, d\tau \qquad (6\text{--}190)$$

and thus

$$[G]_{ij} = \left[\frac{\beta_i \beta_j}{(\lambda_i + \lambda_j)} \exp(\lambda_i + \lambda_j)\tau \right]_{\tau=0}^{\tau=t}$$

giving

$$[G]_{ij} = \frac{-\beta_i \beta_j}{(\lambda_i + \lambda_j)}[1 - \exp(\lambda_i + \lambda_j)t] \qquad (6\text{--}191)$$

Thus we have:

$$\mathbf{W}(t) = [\mathbf{u}_1 \;\vdots\; \mathbf{u}_2 \;\vdots\; \ldots \;\vdots\; \mathbf{u}_n][\mathbf{G}(t)] \begin{bmatrix} \mathbf{u}_1^t \\ \hline \mathbf{u}_2^t \\ \hline \vdots \\ \hline \mathbf{u}_n^t \end{bmatrix}$$

$$= \sum_{i,j}^n \frac{-\beta_i \beta_j}{(\lambda_i + \lambda_j)}[1 - \exp(\lambda_i + \lambda_j)t]\mathbf{u}_i \mathbf{u}_j^t \qquad (6\text{--}192)$$

where $\{\mathbf{u}_1, \mathbf{u}_2, \ldots, \mathbf{u}_n\}$ are the eigenvectors of **A** corresponding to the eigenvalue set $\{\lambda_1, \lambda_2, \ldots, \lambda_n\}$. As discussed in Section 6.3 above, the range of the dyad $\mathbf{u}_i \mathbf{u}_j^t$ is the subspace spanned by $\mathbf{u}_i$; further the range of a sum of matrices is the direct sum of their individual ranges. It therefore follows from equation (6–192) that the range of the matrix $\mathbf{W}(t)$ is the space spanned by all the excited modes of the system.

For the multiple input case we have

$$\dot{\mathbf{x}} = \mathbf{A}\mathbf{x} + \mathbf{B}\mathbf{u}$$
$$= \mathbf{A}\mathbf{x} + \sum_{i=1}^{r} \mathbf{b}_i u_i \qquad (6\text{--}193)$$

where $\{\mathbf{b}_i : i = 1, 2, \ldots, r\}$ denotes the columns of $\mathbf{B}$. We may then define, as above, a matrix $\mathbf{W}_i(t)$ corresponding to each of the r inputs. If the matrix $\mathbf{W}(t)$ for such a system is then defined as

$$\mathbf{W}(t) = \sum_{i=1}^{r} \mathbf{W}_i(t) \qquad (6\text{--}194)$$

we will again have, using the fact that the range of a sum of matrices is the direct sum of the individual ranges, that the range of $\mathbf{W}(t)$ is the subspace spanned by all excited modes of the system. There is thus a close relationship between the matrix $\mathbf{W}(t)$ and the controllability of the system; for this reason $\mathbf{W}(t)$ is called the *controllability matrix* of the system.

For computational purposes, the integral form of equation (6–189) is inconvenient; it is often much more useful to have a set of differential equations which can be solved for the elements of $\mathbf{W}$. To obtain such a set of equations we start by differentiating equation (6–190). This gives

$$\frac{dG_{ij}}{dt} = \beta_i \beta_j \exp(\lambda_i + \lambda_j) t$$

Substituting this into equation (6–191) then gives

$$\frac{dG_{ij}}{dt} = \lambda_i G_{ij} + \lambda_j G_{ij} + \beta_i \beta_j \qquad i, j = 1, 2, \ldots, n$$

Expressing this set of equations in matrix form then gives

$$\frac{d\mathbf{G}}{dt} = \mathbf{\Lambda}\mathbf{G} + \mathbf{G}\mathbf{\Lambda} + \boldsymbol{\beta}\boldsymbol{\beta}^t \qquad (6\text{--}195)$$

where $\mathbf{\Lambda}$ is a diagonal matrix whose elements are the eigenvalues of the matrix $\mathbf{A}$ and $\boldsymbol{\beta}$ is the vector previously defined. Now we have, using equation (6–189),

$$\frac{d\mathbf{W}}{dt} = \mathbf{U}\frac{d\mathbf{G}}{dt}\mathbf{U}^t$$

so that

$$\frac{d\mathbf{W}}{dt} = \mathbf{U}[\mathbf{\Lambda G}+\mathbf{G\Lambda}+\boldsymbol{\beta\beta}^t]\mathbf{U}^t$$
$$= \mathbf{U\Lambda G U}^t + \mathbf{UG\Lambda U}^t + \mathbf{U\boldsymbol{\beta\beta}}^t\mathbf{U}^t$$
$$= \mathbf{U\Lambda U}^{-1}\mathbf{UG U}^t + \mathbf{UG U}^t(\mathbf{U}^t)^{-1}\mathbf{\Lambda U}^t + \mathbf{BB}^t \quad (6\text{-}196)$$
$$= \mathbf{AW}+\mathbf{WA}^t+\mathbf{BB}^t$$

The order of a dynamical system may be reduced by projecting an original system trajectory on to some subspace of the system state space, as illustrated by Figure 6–16, and then devising a lower-order dynamical system to generate this subspace trajectory. We therefore seek ways of determining, in some defined sense, the optimal subspace

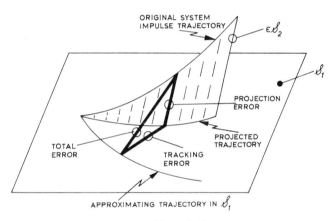

Figure 6–16

of a given order on which to project so that some functional of the error involved in the approximation is minimized. Consider the situation when the system state space trajectory to be approximated is projected *on to* some subspace $\mathscr{S}_1$ along some subspace $\mathscr{S}_2$. The total error incurred may be conveniently expressed as a sum of component errors, which are termed the projection error and the tracking error respectively, as shown in Figure 6–16. If we denote the projection error by $\mathbf{e}_p(t)$ and the tracking error by $\mathbf{e}_T(t)$ then the norm of the instantaneous total error, $\mathbf{e}(t)$ say, will be such that

$$\|\mathbf{e}(t)\|_E^2 = \|\mathbf{e}_p(t)\|_E^2 + \|\mathbf{e}_T(t)\|_E^2 + 2\langle\mathbf{e}_p,\mathbf{e}_T\rangle \quad (6\text{-}197)$$

In the particular case when the projection is an orthogonal projection we will have

$$\|\mathbf{e}(t)\|_E^2 = \|\mathbf{e}_p(t)\|_E^2 + \|\mathbf{e}_T(t)\|_E^2 \qquad (6\text{–}198)$$

The optimal subspace associated with any particular method of projection may be defined as that which makes an error functional

$$E(T) = \int_0^T \|\mathbf{e}(\tau)\|_E^2 \, d\tau \qquad (6\text{–}199)$$

a minimum over some specified time interval $[0, T]$.

In what follows we consider the particular case of the approximation of the response of the single-input system

$$\dot{\mathbf{x}} = \mathbf{A}\mathbf{x} + \mathbf{b}\delta(t) \qquad (6\text{–}200)$$

where $\mathbf{A}$ is an $n \times n$ matrix of constant coefficient, $\mathbf{b}$ an $n \times 1$ constant matrix, and the system input is a unit Dirac impulse. For such an input, let $\mathbf{c}$ be some given constant vector in the state space. Then the component of the response of system (6–200) along $\mathbf{c}$ is

$$y(t) = \frac{\langle \mathbf{c}, \mathbf{r}(t) \rangle}{\|\mathbf{c}\|_E} \qquad (6\text{–}201)$$

where $\mathbf{r}(t)$ is the system zero-initial-state response to a unit Dirac impulse; $\langle \mathbf{c}, \mathbf{r} \rangle$ denotes the scalar product of $\mathbf{c}$ and $\mathbf{r}$, and $\|\mathbf{c}\|_E$ denotes the Euclidean norm of $\mathbf{c}$. Let

$$Y_c = \int_0^t y^2(\tau) \, d\tau$$

then, using equation (6–201) we have

$$\begin{aligned}\|\mathbf{c}\|_E^2 Y_c &= \int_0^t \langle \mathbf{c}, \mathbf{r}(\tau) \rangle^2 \, d\tau \\ &= \mathbf{c}^t \left[\int_0^t \mathbf{r}(\tau) \mathbf{r}^t(\tau) \, d\tau \right] \mathbf{c} \qquad (6\text{–}202) \\ &= \mathbf{c}^t \mathbf{W}(t) \mathbf{c}\end{aligned}$$

from which we conclude that $\mathbf{W}(t)$ must be positive semi-definite, since the term on the left-hand side of equation (6–202) can never be negative.

6.8.1 Optimal Orthogonal Projection on to a Subspace [M11]

Let the eigenvalues and corresponding eigenvectors of the system controllability matrix $\mathbf{W}(T)$ be denoted by

$$\lambda_i(\mathbf{W}) \qquad i = 1, 2, \ldots, n$$

and $\mathbf{u}_i(\mathbf{W})$ $i = 1, 2, \ldots, n$
respectively, and let the eigenvalues be arranged so that

$$\lambda_1(\mathbf{W}) \leqslant \lambda_2(\mathbf{W}) \leqslant \cdots \leqslant \lambda_n(\mathbf{W})$$

with a corresponding set of *orthonormalized* eigenvectors $\{\mathbf{u}_1(w), \ldots, \mathbf{u}_n(w)\}$. Let $\mathbf{p}(\tau)$ be the orthogonal projection of the impulse response $\mathbf{r}(\tau)$ on some subspace $\mathscr{S}_1$ of dimension m. We then have

$$\|\mathbf{r}(\tau)\|_E^2 - \|\mathbf{p}(\tau)\|_E^2 = \|\mathbf{e}(\tau)\|_E^2 \qquad (6\text{-}203)$$

where $\mathbf{e}(\tau)$, the orthogonal complement of $\mathbf{p}(\tau)$, is the error involved in approximating $\mathbf{r}(\tau)$ by $\mathbf{p}(\tau)$. We now investigate under what conditions, for the dimension of $\mathscr{S}_1$ less than n, we can choose $\mathscr{S}_1$ such that the error functional

$$E = \int_0^T \|\mathbf{r}(\tau)\|_E^2 \, d\tau - \int_0^T \|\mathbf{p}(\tau)\|_E^2 \, d\tau \qquad (6\text{-}204)$$

is minimized. Suppose we select an orthonormal basis for $\mathscr{S}_1$ and denote the constituent vectors of this basis set by

$$\{\mathbf{s}_i : i = 1, 2, \ldots, m\}$$

with
$$\langle \mathbf{s}_i, \mathbf{s}_j \rangle = \delta_{ij}$$

where δ_{ij} is the Kronecker delta. We may then represent $\mathbf{p}(\tau)$ as

$$\mathbf{p}(\tau) = \sum_{i=1}^m \langle \mathbf{r}(\tau), \mathbf{s}_i \rangle \mathbf{s}_i$$

so that
$$\int_0^T \|\mathbf{p}(\tau)\|_E^2 \, d\tau = \sum_{i=1}^m \int_0^T \langle \mathbf{r}(\tau), \mathbf{s}_i \rangle^2 \, d\tau \qquad (6\text{-}205)$$

The error functional E given by equation (6-204) will be minimized if we choose the set of vectors $\{\mathbf{s}_i : i = 1, 2, \ldots, m\}$ so as to maximize the right-hand side term of equation (6-205). Denote the ith term of the sum in equation (6-205) by I_i. Then

$$I_1 = \int_0^\tau \langle \mathbf{r}(\tau), \mathbf{s}_1 \rangle^2 \, d\tau = \mathbf{s}_1^t \left[\int_0^T \mathbf{r}(\tau)\mathbf{r}^t(\tau) \, d\tau \right] \mathbf{s}_1$$

$$= \mathbf{s}_1^t \mathbf{W}(T) \mathbf{s}_1 \qquad (6\text{-}206)$$

The problem of maximizing I_1 is thus the problem of finding the maximum of the quadratic form $\mathbf{s}_1^t [\mathbf{W}(T)] \mathbf{s}_1$ under the constraint

$$\|\mathbf{s}_1\|_E = 1$$

This maximum has been shown in Section 2.18 of Chapter 2 to be equal to the largest eigenvalue of $\mathbf{W}(T)$, $\lambda_n(\mathbf{W})$, and to be attained for the eigenvector $\mathbf{u}_n(\mathbf{W})$. To satisfy the condition that the succeeding term

$$I_2 = \int_0^T \langle \mathbf{r}(\tau), \mathbf{s}_2 \rangle^2 \, d\tau = \mathbf{s}_2^t \mathbf{W}(T) \mathbf{s}_2$$

be a maximum under the two subsidiary conditions

$$\|\mathbf{s}_2\|_E = 1 \quad \text{and} \quad \langle \mathbf{s}_1, \mathbf{s}_2 \rangle = 0$$

the extremal theory of quadratic forms of Section 2.16 of Chapter 2 tells us that the corresponding maximum value of I_2 will be $\lambda_{n-1}(\mathbf{W})$ and be attained for $\mathbf{u}_{n-1}(\mathbf{W})$. (If $\lambda_n(\mathbf{W}) = \lambda_{n-1}(\mathbf{W})$ then we may choose, from the infinite set of $\mathbf{u}_{n-1}(\mathbf{W})$ then available, that one which is orthogonal to $\mathbf{s}_1$.)

Proceeding in this way, we find that the appropriate choice for the basis set $\{\mathbf{s}_i : i = 1, 2, \ldots, m\}$ consists of the first m eigenvectors of $\mathbf{W}(T)$ corresponding to the decreasing order set of eigenvalues $\lambda_n(\mathbf{W}), \ldots, \lambda_{(n-m)}(\mathbf{W})$. This gives

$$\underset{\mathbf{s}_i}{\text{maximum}} \sum_{i=1}^m I_i = \sum_{i=n-m}^n \lambda_i(\mathbf{W})$$

Now

$$\int_0^T \|\mathbf{r}(\tau)\|_E^2 \, d\tau = \sum_{i=1}^n \int_0^T r_i^2(\tau) \, d\tau$$

$$= \text{trace}\,[\mathbf{W}(T)]$$

$$= \sum_{i=1}^n \lambda_i(\mathbf{W})$$

and so we have that

$$E = \sum_{i=1}^n \lambda_i(\mathbf{W}) - \sum_{i=n-m}^n \lambda_i(\mathbf{W}) = \sum_{i=1}^m \lambda_i(\mathbf{W}) \qquad (6\text{--}207)$$

The above argument has shown that the particular m-dimensional subspace $\mathscr{S}_1$ on to which we may orthogonally project with least error in the sense of minimizing the error functional E, is the subspace spanned by the set of eigenvectors corresponding to the ordered set of eigenvalues of the controllability matrix

$$\lambda_1(\mathbf{W}) \leqslant \lambda_2(\mathbf{W}) \leqslant \cdots \leqslant \lambda_m(\mathbf{W})$$

It further shows that the integral-square error of the corresponding approximation is given by the sum of the $(n-m)$ smallest eigenvalues of the controllability matrix.

Reduction 483

If $\det \mathbf{W}(T) = 0$ due to the existence of a zero eigenvalue, $\lambda_1(\mathbf{W})$ say, then the zero-initial-state response will lie in the $(n-1)$-dimensional subspace spanned by $\mathbf{u}_2(\mathbf{W}), \ldots, \mathbf{u}_n(\mathbf{W})$, and no approximation is involved in the projection for such a reducible system. Similar considerations apply when the zero eigenvalue has a multiplicity q, the trajectory then lying in the $(n-q)$-dimensional subspace spanned by the eigenvectors corresponding to the non-zero eigenvalues.

6.8.2 Optimal Projection Along an Invariant Subspace [M11]

If we project along a subspace $\mathscr{S}_2$ which is an invariant subspace under $\mathbf{A}$, that is one spanned by a set of eigenvectors of $\mathbf{A}$, then the tracking error will be zero. Suppose that we have selected some m-dimensional invariant subspace along which to project, on the grounds of their comprising a set of fastest-decaying modes, least-weighted modes or some equivalent criterion. Consider the problem of finding the optimal $(n-m)$-dimensional subspace $\mathscr{S}_1$ on which to project so that the projection-error functional

$$\int_0^T \|\mathbf{e}(\tau)\|_E^2 \, d\tau$$

is a minimum.

From the eigenvector set spanning $\mathscr{S}_2$ we may form, using the Gram–Schmidt orthonormalizing process of Section 2.5 of Chapter 2, an orthonormal set of vectors

$$\{\mathbf{t}_i : i = 1, 2, \ldots, m\}$$

which span $\mathscr{S}_2$. We take the required $(n-m)$-dimensional optimal subspace on which to project to be defined by some, as yet unknown, set of vectors

$$\{\mathbf{p}_j : j = 1, 2, \ldots, (n-m)\}$$

which span it. Now the state space is the direct sum of the subspaces $\mathscr{S}_1$ and $\mathscr{S}_2$ and thus the sets

$$\{\mathbf{t}_i : i = 1, 2, \ldots, m\} \quad \text{and} \quad \{\mathbf{p}_j : j = 1, 2, \ldots, (n-m)\}$$

together span the whole state space. Let

$$\{\mathbf{s}_i : i = 1, 2, \ldots, n\}$$

be the dual basis set for the *combined* basis set $\{\mathbf{t}_i : 1 \leqslant i \leqslant m\}$ and $\{\mathbf{p}_j : 1 \leqslant j \leqslant n-m\}$. Since the error is wholly a projection error it will have zero components in $\mathscr{S}_1$ and is thus given by

$$\mathbf{e}(t) = \sum_{i=1}^m \langle \mathbf{r}(t), \mathbf{s}_i \rangle \mathbf{t}_i$$

so that
$$\|\mathbf{e}(t)\|_E^2 = \sum_{i=1}^m \langle \mathbf{r}(t), \mathbf{s}_i \rangle^2$$

and thus
$$\int_0^T \|\mathbf{e}(\tau)\|_E^2 \, d\tau = \sum_{i=1}^m \mathbf{s}_i^t \left[\int_0^T \mathbf{r}(\tau)\mathbf{r}^t(\tau) \, d\tau \right] \mathbf{s}_i$$

$$= \sum_{i=1}^m \mathbf{s}_i^t \mathbf{W}(T)\mathbf{s}_i$$

where $\mathbf{W}(T)$ is the controllability matrix.

Put
$$\int_0^T \|\mathbf{e}(\tau)\|_E^2 \, d\tau = \sum_{i=1}^n E_i$$

where
$$E_i = \mathbf{s}_i^t \mathbf{W}(T)\mathbf{s}_i$$

and consider the minimization of E_i subject to the constraints

$$\mathbf{T}^t \mathbf{s}_i = \mathbf{e}_i \qquad i = 1, 2, \ldots, n$$

where $\mathbf{T}$ is a matrix whose ith column is the vector $\mathbf{t}_i$ and $\mathbf{e}_i$ is the ith standard basis vector. These constraints ensure that the minimizing vectors selected will be suitable members of the defined dual vector set. To handle this set of constraints we may introduce an m-dimensional column vector of Lagrangian multipliers $\boldsymbol{\mu}_i$ and thus consider

$$\operatorname*{minimum}_{\mathbf{s}_i} [\mathbf{s}_i^t \mathbf{W}(T)\mathbf{s}_i - 2\boldsymbol{\mu}_i^t(\mathbf{T}^t \mathbf{s}_i - \mathbf{e}_i)] \triangleq \min_{\mathbf{s}_i} \phi$$

We then have
$$\left(\frac{\partial \phi}{\partial \mathbf{s}_i} \right)^t = 2\mathbf{W}\mathbf{s}_i - 2\mathbf{T}\boldsymbol{\mu}_i$$

and
$$\frac{\partial}{\partial \mathbf{s}_i} \left(\frac{\partial \phi}{\partial \mathbf{s}_i} \right)^t = 2\mathbf{W}$$

If we assume that $\mathbf{W}$ is positive-definite then we will have a minimum given by

$$2\mathbf{W}\mathbf{s}_i - 2\mathbf{T}\boldsymbol{\mu}_i = \mathbf{0} \qquad i = 1, 2, \ldots, m$$

so that
$$\mathbf{s}_i = \mathbf{W}^{-1} \mathbf{T} \boldsymbol{\mu}_i \qquad i = 1, 2, \ldots, m$$

The satisfaction of the constraint at the minimum gives

$$\mathbf{T}^t \mathbf{W}^{-1} \mathbf{T} \boldsymbol{\mu}_i = \mathbf{e}_i \qquad i = 1, 2, \ldots, m$$

so that
$$\boldsymbol{\mu}_i = [\mathbf{T}^t \mathbf{W}^{-1} \mathbf{T}]^{-1} \mathbf{e}_i \qquad i = 1, 2, \ldots, m$$

and thus for a minimum we get that the set of vectors $\mathbf{s}_i$ is given by

$$\mathbf{s}_i = \mathbf{W}^{-1} \mathbf{T} [\mathbf{T}^t \mathbf{W}^{-1} \mathbf{T}]^{-1} \mathbf{e}_i \qquad i = 1, 2, \ldots, m \qquad (6\text{–}208)$$

The set of vectors $\{\mathbf{s}_i : i = 1, 2, \ldots, m\}$ given by equation (6–208) span that portion of the state space which we wish to reject, and we must construct our required set of optimal-subspace-defining vectors

$$\{\mathbf{p}_j : j = 1, 2, \ldots, n - m\}$$

which span $\mathscr{S}_2$ from them. The simplest way to achieve this is to add to the set $\{\mathbf{s}_1, \mathbf{s}_2, \ldots, \mathbf{s}_m\}$ some set of vectors, $\{\mathbf{q}_1, \mathbf{q}_2, \ldots, \mathbf{q}_{n-m}\}$ say, such that the two sets together span the whole state space. An obvious choice would be a suitable set of $(n-m)$ standard basis vectors. The construction of the corresponding dual vector set

$$\{\mathbf{r}_1, \mathbf{r}_2, \ldots, \mathbf{r}_m; \mathbf{p}_1, \mathbf{p}_2, \ldots, \mathbf{p}_{n-m}\}$$

will then automatically generate the required set of vectors which span the optimal subspace $\mathscr{S}_1$. Thus the controllability matrix serves to define the optimal subspace on which to project when studying the reduction of the order of a dynamical system model.

Appendix A

Table of Functional Matrices

$$\mathbf{M}_{m,n} \quad \begin{array}{l} n = \text{System order} \\ m = \text{Order of algebraic form} \end{array}$$

M-matrices for Second-order Dynamical Systems

$$\mathbf{M}_{1,2} = \begin{bmatrix} a_{11} & a_{12} \\ a_{21} & a_{22} \end{bmatrix} \quad \mathbf{M}_{2,2} = \begin{bmatrix} 2a_{11} & a_{12} & 0 \\ 2a_{21} & (a_{11}+a_{22}) & 2a_{12} \\ 0 & a_{21} & 2a_{22} \end{bmatrix}$$

$$\mathbf{M}_{3,2} = \begin{bmatrix} 3a_{11} & a_{12} & 0 & 0 \\ 3a_{21} & (2a_{11}+a_{22}) & 2a_{12} & 0 \\ 0 & 2a_{21} & (a_{11}+2a_{22}) & 3a_{12} \\ 0 & 0 & a_{21} & 3a_{22} \end{bmatrix}$$

$$\mathbf{M}_{4,2} = \begin{bmatrix} 4a_{11} & a_{12} & 0 & 0 & 0 \\ 4a_{21} & (3a_{11}+a_{22}) & 2a_{12} & 0 & 0 \\ 0 & 3a_{21} & (2a_{11}+2a_{22}) & 3a_{12} & 0 \\ 0 & 0 & 2a_{21} & (a_{11}+3a_{22}) & 4a_{12} \\ 0 & 0 & 0 & a_{21} & 4a_{22} \end{bmatrix}$$

$$\mathbf{M}_{5,2} = \begin{bmatrix} 5a_{11} & a_{12} & 0 & 0 & 0 & 0 \\ 5a_{21} & (4a_{11}+a_{22}) & 2a_{12} & 0 & 0 & 0 \\ 0 & 4a_{21} & (3a_{11}+2a_{22}) & 3a_{12} & 0 & 0 \\ 0 & 0 & 3a_{21} & (2a_{11}+3a_{22}) & 4a_{12} & 0 \\ 0 & 0 & 0 & 2a_{21} & (a_{11}+4a_{22}) & 5a_{12} \\ 0 & 0 & 0 & 0 & a_{21} & 5a_{22} \end{bmatrix}$$

M-matrices for Third-order Dynamical Systems

$$\mathbf{M}_{1,3} = \begin{bmatrix} a_{11} & a_{12} & a_{13} \\ a_{21} & a_{22} & a_{23} \\ a_{31} & a_{32} & a_{33} \end{bmatrix}$$

$$\mathbf{M}_{2,3} = \begin{bmatrix} 2a_{11} & a_{12} & 0 & a_{13} & 0 & 0 \\ 2a_{21} & (a_{11}+a_{22}) & 2a_{12} & a_{23} & a_{13} & 0 \\ 0 & a_{21} & 2a_{22} & 0 & a_{23} & 0 \\ 2a_{31} & a_{32} & 0 & (a_{11}+a_{33}) & a_{12} & 2a_{13} \\ 0 & a_{31} & 2a_{32} & a_{21} & (a_{22}+a_{33}) & 2a_{23} \\ 0 & 0 & 0 & a_{31} & a_{32} & 2a_{33} \end{bmatrix}$$

$$\mathbf{M}_{3,3} = \begin{bmatrix} 3a_{11} & a_{12} & 0 & 0 & a_{13} & 0 & 0 & 0 & 0 & 0 \\ 3a_{21} & \left(\begin{array}{c}2a_{11}\\+\\a_{22}\end{array}\right) & 2a_{12} & 0 & a_{23} & a_{13} & 0 & 0 & 0 & 0 \\ 0 & 2a_{21} & \left(\begin{array}{c}a_{11}\\+\\2a_{22}\end{array}\right) & 3a_{12} & 0 & a_{23} & a_{13} & 0 & 0 & 0 \\ 0 & 0 & a_{21} & 3a_{22} & 0 & 0 & a_{23} & 0 & 0 & 0 \\ 3a_{31} & a_{32} & 0 & 0 & \left(\begin{array}{c}2a_{11}\\+\\a_{33}\end{array}\right) & a_{12} & 0 & 2a_{13} & 0 & 0 \\ 0 & 2a_{31} & 2a_{32} & 0 & 2a_{21} & \left(\begin{array}{c}a_{11}\\+\\a_{22}\\+\\a_{33}\end{array}\right) & 2a_{12} & 2a_{23} & 2a_{13} & 0 \\ 0 & 0 & a_{31} & 3a_{32} & 0 & a_{21} & \left(\begin{array}{c}2a_{22}\\+\\a_{33}\end{array}\right) & 0 & 2a_{23} & 0 \\ 0 & 0 & 0 & 0 & 2a_{31} & a_{32} & 0 & \left(\begin{array}{c}a_{11}\\+\\2a_{33}\end{array}\right) & a_{12} & 3a_{13} \\ 0 & 0 & 0 & 0 & 0 & a_{31} & 2a_{32} & a_{21} & \left(\begin{array}{c}a_{22}\\+\\2a_{33}\end{array}\right) & 3a_{23} \\ 0 & 0 & 0 & 0 & 0 & 0 & 0 & a_{31} & a_{32} & 3a_{33} \end{bmatrix}$$

M-matrices for Fourth-order Dynamical System

$$M_{1,4} = \begin{bmatrix} a_{11} & a_{12} & a_{13} & a_{14} \\ a_{21} & a_{22} & a_{23} & a_{24} \\ a_{31} & a_{32} & a_{33} & a_{34} \\ a_{41} & a_{42} & a_{43} & a_{44} \end{bmatrix}$$

$$M_{2,4} = \begin{bmatrix} 2a_{11} & a_{12} & 0 & a_{13} & 0 & 0 & a_{14} & 0 & 0 & 0 \\ 2a_{21} & \left(a_{11}+a_{22}\right) & 2a_{12} & a_{23} & a_{13} & 0 & a_{24} & a_{14} & 0 & 0 \\ 0 & a_{21} & 2a_{22} & 0 & a_{23} & 0 & 0 & a_{24} & 0 & 0 \\ 2a_{31} & a_{32} & 0 & \left(a_{11}+a_{33}\right) & a_{12} & 2a_{13} & a_{34} & 0 & a_{14} & 0 \\ 0 & a_{31} & 2a_{32} & a_{21} & \left(a_{22}+a_{33}\right) & 2a_{23} & 0 & a_{34} & a_{24} & 0 \\ 0 & 0 & 0 & a_{31} & a_{32} & 2a_{33} & 0 & 0 & a_{34} & 0 \\ 2a_{41} & a_{42} & 0 & a_{43} & 0 & 0 & \left(a_{11}+a_{44}\right) & a_{12} & a_{13} & 2a_{14} \\ 0 & a_{41} & 2a_{42} & 0 & a_{43} & 0 & a_{21} & \left(a_{22}+a_{44}\right) & a_{23} & 2a_{24} \\ 0 & 0 & 0 & a_{41} & a_{42} & 2a_{43} & a_{31} & a_{32} & \left(a_{33}+a_{44}\right) & 2a_{34} \\ 0 & 0 & 0 & 0 & 0 & 0 & a_{41} & a_{42} & a_{43} & 2a_{44} \end{bmatrix}$$

$M_{2,5}$ for Fifth-order Dynamical System

$$M_{2,5} = \begin{bmatrix}
2a_{11} & a_{12}+a_{21} & 0 & 0 & 0 & a_{14} & 0 & 0 & a_{15} & 0 & 0 & 0 & 0 & 0 & 0 \\
2a_{21} & a_{11}+a_{22} & 2a_{12} & a_{13} & 0 & a_{24} & a_{14} & 0 & a_{25} & a_{15} & 0 & 0 & 0 & 0 & 0 \\
0 & a_{21} & 2a_{22} & a_{23} & 0 & 0 & a_{24} & 0 & 0 & a_{25} & 0 & 0 & 0 & 0 & 0 \\
2a_{31} & a_{32} & 0 & a_{11}+a_{33} & a_{21} & a_{34} & 0 & a_{14} & a_{35} & 0 & a_{15} & 0 & 0 & 0 & 0 \\
0 & a_{31} & 0 & a_{12}+a_{32} & a_{22} & 0 & a_{34} & a_{24} & 0 & a_{35} & a_{25} & 0 & 0 & 0 & 0 \\
0 & 0 & 0 & 2a_{13} & 2a_{23} & 2a_{33} & 0 & 0 & 0 & 0 & a_{35} & 0 & 0 & 0 & 0 \\
2a_{41} & a_{42} & 0 & 0 & 0 & a_{11}+a_{44} & a_{12}+a_{22} & a_{13} & a_{45} & 0 & 0 & a_{15} & 0 & 0 & 0 \\
0 & a_{41} & 2a_{42} & a_{43} & 0 & a_{21} & a_{11}+a_{22} & a_{23} & 0 & a_{45} & 0 & a_{25} & 0 & 0 & 0 \\
0 & 0 & 0 & a_{41} & 0 & a_{31} & a_{32} & a_{33}+a_{44} & 0 & 0 & a_{45} & a_{35} & 0 & 0 & 0 \\
0 & 0 & 0 & 0 & 0 & 2a_{14} & 2a_{24} & 2a_{34} & 2a_{44} & 0 & 0 & 0 & 0 & 0 & 0 \\
0 & 0 & 0 & 0 & 0 & 0 & 0 & 0 & 0 & 0 & 0 & 0 & a_{15} & 0 & 2a_{15} \\
2a_{51} & a_{52} & 0 & a_{53} & 0 & 0 & 0 & 0 & a_{54} & 0 & 0 & a_{11} & a_{12} & a_{13} & 2a_{25} \\
0 & a_{51} & 2a_{52} & 0 & a_{53} & 0 & a_{54} & 0 & 0 & 0 & 0 & a_{21} & a_{22}+a_{55} & a_{23} & 2a_{35} \\
0 & 0 & 0 & a_{51} & a_{52} & 0 & 0 & a_{54} & 0 & 0 & 0 & a_{31} & a_{32} & a_{33}+a_{55} & 2a_{45} \\
0 & 0 & 0 & 0 & 0 & 0 & 0 & 0 & 2a_{54} & 0 & 0 & a_{41} & a_{42} & a_{43} & a_{44}+a_{55}
\end{bmatrix}$$

References

A1 ANTOSIEWICZ, H. A., 'Linear control systems', *Archive for Rational Mechanics and Analysis*, Vol. 12, No. 4, March 1963
A2 ATHANS, M. and FALB, P. L., *Optimal Control*. McGraw-Hill, 1966
B1 BASHKOW, T. R., 'The A-matrix, network description', *Transactions of Institute of Radio Engineers*, Vol. CT-4, p. 117, 1957
B2 BELLMAN, R., *Introduction to Matrix Analysis*. McGraw-Hill, 1960
B3 BIRKHOFF, G. and MACLANE, S., *A Survey of Modern Algebra*. Macmillan, New York, 1941
B4 BLACKWELL, W. A., *Mathematical Modeling of Physical Networks*. Macmillan, New York, 1968
B5 BLAQUIERE, A., *Nonlinear System Analysis*. Academic Press, 1966
B6 BRAYTON, R. K. and MOSER, J. K., 'A theory of nonlinear networks—1', *Quarterly of Applied Mathematics*, Vol. 22, No. 1, p. 1, 1964
B7 BRAYTON, R. K. and MOSER, J. K., 'A theory of nonlinear networks—2', *Quarterly of Applied Mathematics*, Vol. 22, No. 2, p. 81, 1964
B8 BROWN, B. M., *The Mathematical Theory of Linear Systems*. Chapman and Hall, 1961.
B9 BRYANT, P. R., 'The order of complexity of electrical networks', *Proceedings I.E.E.*, Vol. 106C, p. 174, 1959
B10 BRYANT, P. R., 'The algebra and topology of electrical networks', *Proceedings I.E.E.*, Vol. 108C, p. 215, 1961
B11 BRYANT, P. R., 'The explicit form of Bashkow's A-matrix', *Transactions I.R.E.*, Vol. CT-9, p. 303, 1962
B12 BROMWICH, T. J. I'A., 'Normal Co-ordinates in Dynamical Systems', *Proc. Lond. Math. Soc.* (2), Vol. 15, pp. 401–448, 1916
C1 CAMPBELL, D. P., *Process Dynamics*. Wiley, 1958
C2 CARLIN, H. J. and GIORDANO, *Network Theory*. Prentice-Hall, 1964
C3 CHERRY, C., 'Some general theorems for nonlinear systems possessing reactance', *Philosophical Magazine*, Vol. 42, pp. 1161–1177, 1951

References

C4 CRANDELL, S. H., KARNOPP, D. C., KURTZ, E. F., and PRIDMORE-BROWN, D. C., *Dynamics of Mechanical and Electromechanical Systems*. McGraw-Hill, 1968
D1 DAVIS, H. T., *Introduction to Nonlinear Differential and Integral Equations*. Dover, 1962
D2 DESOER, C. A., 'Modes in Linear Circuits', *Transactions I.R.E.*, Vol. CT-7, p. 211, 1960
D3 DESOER, C. A. and KATZENELSON, J., 'Nonlinear R.L.C. networks', *Bell System Technical Journal*, Vol. 44, pp. 161–198, 1965
D4 DETTMAN, J. W., *Mathematical Methods in Physics and Engineering*. McGraw-Hill, 1962
F1 FREEMAN, H., *Discrete Time Systems*. Wiley, 1965 (p. 51)
F2 FIRESTONE, F. A., *J. Acoust. Soc. Amer.*, Vol. 4, pp. 249–67, 1933
G1 GANTMACHER, F. R., *Theory of Matrices*. Chelsea, 1959
G2 GARDNER, M. F. and BARNES, J. L., *Transients in Linear Systems*. Wiley, 1942
G3 GILBERT, E. G., 'Controllability and Observability in multivariable control systems', *Journal of Society Industrial and Applied Mathematics*, Series A, Vol. 1, pp. 128–151, 1963
G4 GILLE, J. C., PELEGRIN, M. J., and DECAULNE, P., *Feedback Control Systems*. McGraw-Hill, 1959
G5 GUILLEMIN, E. A., *Introductory Circuit Theory*. Wiley, 1953.
H1 HALKIN, H., 'The principle of optimal evolution', *Proceedings of the Second International Symposium on Nonlinear Mechanics and Nonlinear Differential Equations*. Academic Press, 1962
H2 HALMOS, P. R., *Finite Dimensional Vector Spaces*. Van Nostrand, 1950
H3 HAUSER, W., *Introduction to the Principles of Mechanics*. Addison-Wesley, 1965
H4 HILL, R., *Principles of Dynamics*. Pergamon, 1964
H5 HUREWICZ, W., *Lectures on Ordinary Differential Equations*. Wiley, 1958
H6 HUREWICZ, W., *Math. Ann.*, Vol. 46, p. 273, 1895
H7 HERMITE, C., *J. Reine Angew Math.*, 52, p. 39, 1856
J1 JEFFREYS, H. and JEFFREYS, B. S., *Methods of Mathematical Physics*. Cambridge, 1966
K1 KARNOPP, D. C. and ROSENBERG, R. C., *Analysis and Simulation of Multiport Systems*. M.I.T. Press, 1968
K2 KIM, W. H. and CHIEN, R. T-W., *Topological Analysis and Synthesis of Communication Networks*. Columbia University Press, 1963
K3 KOENIG, H. E., TOKAD, Y., and KESAVAN, H. K., *Analysis of Discrete Physical Systems* McGraw-Hill, 1967

K4 KOENIG, H. E. and BLACKWELL, W. A., *Electromechanical System Theory*. McGraw-Hill, 1961
K5 KALMAN, R. E. and BERTRAM, J. E., 'Control Systems analysis and design via the second method of Liapunov', *Transactions A.S.M.E., Journal of Basic Engineering*, Series D, Vol. 82, p. 371, 1960
K6 KALMAN, R. E., H. O., Y. C., and NARENDRA, K. S., 'Controllability of linear dynamical systems', *Contributions to Differential Equations*, Vol. 1, No. 2, pp. 189–213
K7 KALMAN, R. E., 'Mathematical description of linear dynamical systems', *Journal of Society of Industrial and Applied Mathematics on Control*, Series A, Vol. 1, No. 2, p. 152, 1963
K8 KALMAN, R. E., 'Canonical structure of linear dynamical systems', *Proceedings of the National Academy of Sciences*, Vol. 48, No. 4, p. 596, 1962
K9 KRON, G., *Tensor Analysis of Networks*. Wiley, 1939
L1 LANCZOS, C., *The Variational Principles of Mechanics*. University of Toronto Press, 1960
L2 LASALLE, J. P. and LEFSCHETZ, S., *Stability by Lyapunov's Direct Method with Applications*. Academic Press, 1961
L3 LIGHTHILL, M. J., *Fourier Analysis and Generalised Functions*. Cambridge, 1962
L4 LORENS, C. S., *Flowgraphs*. McGraw-Hill, 1964
L5 LIENARD, A. and CHIPART, M. H., *J. Math. Pures. Appl.*, Vol. 10, p. 29, 1914
L6 LIAPUNOV, A. M., *Problème Générale de la Stabilité du Mouvement*. Annals de la Faculté des Sciences, Toulouse, 1907
M1 MACFARLANE, A. G. J., *Engineering Systems Analysis*. Harrap, 1964
M2 MACFARLANE, A. G. J., 'Generalized Block Diagrams', *International Journal of Control*, Vol. 5, No. 3, pp. 245–267, 1967
M3 MACFARLANE, A. G. J., 'Functional-matrix theory for the general linear electrical network', *Proceedings I.E.E.* Vol. 112, No. 4, pp. 754–770, 1965
M4 MACFARLANE, A. G. J., 'Formulation of the state-space equations for nonlinear networks', *International Journal of Control*, Vol. 5, No. 2, pp. 145–161, 1967.
M5 MACFARLANE, A. G. J., 'Systems Matrices', *Proceedings Institution of Electrical Engineers*, Vol. 115, No. 5, pp. 749–754, 1968
M6 MACFARLANE, A. G. J. and MUNRO, N., 'Mappings of the state space into the complex plane and their use in multivariable systems analysis', *International Journal of Control*, Vol. 7, No. 6, pp. 501–555, 1968

M7 MACFARLANE, A. G. J., 'Representation of state-space flows by circles in the complex plane', *Proceedings Institution of Electrical Engineers*, Vol. 115, No. 8, pp. 1195–1199, 1968
M8 MAYNE, D. Q., 'A computational procedure for the minimal realisation of transfer function matrices', *Centre for Computation and Automation Report*. Imperial College, London, 1968
M9 MILLAR, W., 'Some general theorems for nonlinear systems possessing resistance', *Philosophical Magazine*, Vol. 42, pp. 1150–1160, 1951
M10 MIRSKY, L., *An Introduction to linear algebra*. Oxford, 1955
M11 MITRA, D., 'The equivalence and reduction of linear dynamical systems', Ph.D. Thesis, University of London, 1967
M12 MOORE, R. K., *Wave and Diffusion Analogies*. McGraw-Hill, 1964
M13 MURPHY, G. J., *Basic Automatic Control Theory*. Van Nostrand, 1966
M14 MORSE, P. M. and FESHBACH, H., *Methods of Theoretical Physics*. Vol. I, McGraw-Hill, 1953 (p. 847)
M15 MAXWELL, J. C., *Electricity and Magnetism*. Oxford University Press, 1892
M16 MOIGNO, F. L. N. M., *Leçons de Calcul d'après Cauchy*. Paris, 1844
M17 MAXWELL, J. C., 'On Governors', *Proc. Roy. Soc. Lond.*, Vol. 16, p. 270, 1868
M18 MOHR, O., *Civilingenieure*, Vol. 28, p. 112, 1882.
N1 NYQUIST, H., 'Regeneration Theory', *Bell System Technical Journal*, Vol. II, pp. 126–47, 1932
O1 ORE, O., *Graphs and their uses*, New Mathematical Library Number 10. Random House, 1963
P1 PARS, L. A., *Introduction to Calculus of Variations*. Wiley, 1962.
P2 PARS, L. A., *A Treatise on Analytical Dynamics*. Heinemann, 1965
P3 PAYNTER, H. M., *Analysis and Design of Engineering Systems*. M.I.T. Press, 1960
P4 PIPES, L. A., *Applied Mathematics for Engineers and Physicists*. McGraw-Hill, 1946
P5 PONTRYAGIN, L. S., *Ordinary Differential Equations*. Addison-Wesley, 1962
P6 PONTRYAGIN, L. S., BOLTYANSKY, V. G., GAMKRELIDZE, R. V., and MISCHENKO, YE., F., *The Mathematical Theory of Optimal Processes*. Interscience, 1963
P7 PORTER, W. A., *Modern Foundations of Systems Engineering*, Macmillan, New York, 1966

P8 PICARD, E., *Rend. Circ. Math. Palermo*, Vol. 29, p. 79, 1910
R1 RAGAZZINI, J. R. and FRANKLIN, G. F., *Sampled-data Control Systems*. McGraw-Hill, 1958
R2 ROSENBROCK, H. H., 'On the transformation of the linear constant system equations', *Proceedings I.E.E.*, Vol. 114, No. 4, pp. 541–544, 1967
R3 ROSENBROCK, H. H., 'On linear system theory', *Proceedings I.E.E.*, Vol. 114, pp. 1353–1359, 1967
R4 ROSENBROCK, H. H., 'Connection between network theory and the theory of linear dynamical systems'. *Electronics Letters*, Vol. 3, p. 296, 1967
R5 ROUTH, J. C., *Rigid Dynamics*. Vol. 2, Macmillan, 1897 (p. 192)
S1 SCHWARTZ, R. J. and FRIEDLAND, B., *Linear Systems*. McGraw-Hill, 1965
S2 SESHU, S. and REED, M. B., *Linear Graphs and Electrical Networks*. Addison-Wesley, 1961
S3 SHEARER, J. L., MURPHY, A. T., and RICHARDSON, H. H., *Introduction to System Dynamics*. Addison-Wesley, 1967
S4 SOMMERFIELD, A., *Mechanics, Lectures on Theoretical Physics, Vol. 1*. Academic Press, 1964
S5 SOMMERFIELD, A., *Electrodynamics, Lectures on Theoretical Physics, Vol. 3*. Academic Press, 1964
S6 STERN, T., *Theory of nonlinear networks and systems: an introduction*. Addison-Wesley, 1966
S7 STOUT, T. M., 'A Block Diagram Approach to Network Analysis', *A.I.E.E. Transactions*, Vol. 71, Part II, pp. 255–260, 1952
S8 SYNGE, J. L., 'The fundamental theorem of electrical networks', *Quarterly of Applied Mathematics*, Vol. 9, pp. 113–127, 1951
S9 SOMMERFIELD, A., *Thermodynamics and Statistical Mechanics*. Academic Press, 1964 (p. 30)
S10 SCHWARZ, L., *Théorie des Distributions*. Vols 1 and 2, Hermann, Paris, 1950–51
S11 SMITH, O. J. M., *Feedback Control Systems*. McGraw-Hill, 1958
S12 SALZER, C., *Quart. App. Maths.*, Vol. II, p. 119, 1953
T1 TRENT, H. M., 'Isomorphisms between oriented linear graphs and lumped physical systems', *Journal of Acoustical Society of America*, Vol. 27, No. 3, pp. 500–527, 1955
T2 TRUXAL, J. G., *Automatic Feedback Control System Synthesis*. McGraw-Hill, 1955
V1 VOLTERRA, V., *Theory of Functionals*. Blackie, 1929
W1 WHITE, D. C. and WOODSON, H. H., *Electromechanical Energy Conversion*. McGraw-Hill, 1959
W2 WHITTAKER, E. T., *Analytical Dynamics of Particles and Rigid Bodies*. Cambridge, 1927

W3 WEST, J. C., *Analytical techniques for nonlinear control systems.* English Universities Press, 1960

Z1 ZADEH, L. A. and DESOER, C. A., *Linear System Theory: The State Space Approach.* McGraw-Hill, 1963

Index

Abscissa of absolute convergence, 135
across
   stores, 280
   variables, 279
adjoint systems, 462–3
amplitude scaling, 419–20
analogues, 301
analytical
   aspects of state equations, 385–420
   solutions of linear equations, 396–420
angle
   between vectors, 97
   condition, 195
angular
   displacement, 43
   momentum, 41
   velocity, 43
arc, 285
asymptotic
   relations for transfer function, 149
   stability of a singular point, 426
axial vector convention, 48

Backward
   difference, 207
   shift, 207
bang-bang system, 270
basic closed-loop stability theorem, 188
basis
   change of, 101
   definition of, 95
   dual, 100
   orthonormal, 97
   reciprocal, 100
   sequences, 206
   standard, 95
Bendixon's
   first theorem, 394
   second theorem, 395
block diagram, 159–69
   conventions, 159
   definition, 159
   manipulations, 163
Bolza problem, 251

bounded variation, 135
Brayton–Moser equations, 373, 377
Bromwich, T. J. I'A., 138
Bromwich contour, 138
bulk modulus of elasticity, 20

Calculus of variations, 217–26
canonical
   equations for linear networks, 338–62
   equations for nonlinear networks, 363–82
capacitance
   electrical, 79–80
   fluid, 53–9
capacitor
   co-energy, 80
   charge-controlled, 78
   energy, 79
   fluid, 53
   one-to-one, 79
   pneumatic, 55
Cartesian product, 91
cascade
   block diagram, 164
   flowgraph, 172
   reduction rule, 164, 174
Cauchy, A.-L., 383
   integral formula, 140
Cayley–Hamilton theorem, 417
characteristic equation, 103, 420
charge, 74–7
chords, 286
circuit
   consistently-oriented, 286
   matrix, 293
   method, 308
   of m variables, 177
   postulate for across-variables, 289
   product, 177
   rank, 287
   rule, 183
   transmittance, 177
co-content
   electrical, 85

fluid, 62
  mechanical, 32, 46
co-energy
  electrical, 80–1
  fluid, 58, 61
  rotational mechanical, 44–5
  translational mechanical, 26, 29
co-factor, 178
co-Lagrangian
  definition, 239
  equations of motion, 239–41
  network equations, 325–8
commutative properties of variation, 218
complement of set, 91
complex
  conjugate eigenvalues, 402–11
  conjugate eigenvectors, 407–11
  plane mappings, 184–6
components, 1
conditionally stable system, 193
connected graph, 285
connectivity, 285
conservation
  of energy, 241
  of momentum, 241
constant coefficient linear system, 400
constitutive relationship, 7, 10
contact curve, 393
content
  electrical, 85
  fluid, 62
  rotational mechanical, 46
  translational mechanical, 31
continuity of space law, 11
contour integral, 138
controllability, 455–8
  matrix, 478
converter, 83, 280
convolution
  integral, 145
  sequence, 214
coordinate
  definition, 95
  transformation, 102
co-state vector, 255
co-tree, 286
critical
  locus, 205
  point, 190
current, 74, 77
  source, 85

cut-set, 286
cyclomatic number, 287

D'Alembert's principle, 11
decomposition of state space systems, 461
describing function, 201–05
determinant
  differentiation of, 114
  Jacobian, 118
  of flowgraph, 177
diagonal matrix, 125
difference
  backward, 207
  forward, 207
  of sets, 91
  operation, 209
dimension of vector space, 95
Dirac impulse function, 130
direct
  method of Liapunov, 424, 433
  sum of matrices, 115
discrete
  model approximation, 437
  operators, 206
disjoint circuits, 177
displacement, 7
  controlled spring, 25
dissipator, 31
  co-content, 32
  content, 31
distribution theory, 130
division vertex, 162
dominant poles, 150
double pendulum, 236
dual
  basis, 100
  form of Hamilton's postulate, 239
  interaction experiment, 8
  systems, 301
duality and adjoint systems, 462
dualogues, 301
dynamical
  relationships, 7, 10
  transformation matrix, 298

Eigenvalue, 103
  complex-conjugate, 402–11
  extremum characterization, 120
  invariance, 106
  location bounds, 106
  shift theorem, 106

Index

eigenvector
  complex-conjugate, 407–11
  definition, 103
  left-hand, 105
  right-hand, 105
elasticity, 11
electrical
  converter, 83
  system components, 74–89
  system wave propagation, 86–9
elimination
  of intermediate vertex, 176
  of self-circuit on vertex, 174
energy, 16
Euclidean space, 97
Euler–Lagrange equation, 222
evaluation of inverse transforms, 140, 212
event space, vector, 251
excess elements, 345
existence of equation solutions, 385–9
expanding wavefronts in state space, 253
extremum values of constrained functions, 116–20

Feedback-controlled systems, 187
field of optimal trajectories, 272
final value theorem, 132, 209
first variation, 220
fluid systems, 52–66
flux-linkage, 81
flyball governor, 234
force, 3, 7
  orientation conventions, 3
  controlled spring, 25
forest, 286
forward
  difference, 207
  shift, 206
functional, 2, 217, 249
  matrix, 439–46
  relationship vertex, 160
fundamental
  circuit, 295
  circuit matrix, 295
  lemma of calculus of variations, 219
  matrix, 399
  processes of calculus of variations, 217

Gain margin, 192

general
  form of Nyquist criterion, 191
  rules for root loci, 197
generalized
  admittance matrix, 307
  coordinates, 226
  forces, 227
  functional matrix, 442
  impedance matrix, 306
  Mohr circle, 446
  momenta, 238
  velocity, 226
Gershgorin, S. A., 107
governor, 234
Gram–Schmidt procedure, 98
graph
  completely connected, 286
  linear, 33, 285
gravitational postulate, 11
group postulate, 6

Hamilton's equations, 243–4
Hamilton–Jacobi equation, 244
Hamiltonian
  definition, 243
  dual postulate, 239
  primal postulate, 228
  principles for networks, 246–8
heat, 69
  flow, 69
  transfer coefficient, 71
Heaviside
  expansion formula, 142
  step function, 129
Hermite's stability criterion, 423
Hill, R., 11
Hurwitz criterion, 422
Huygens' principle, 254
hydraulic capacitance, 55
hyperplane, 256

Ideal
  delay vertex, 161
  linear dissipator, 32, 47
  linear spring, 43
  rotational spring, 42
  spring postulate, 7
idempotency, 125
identity
  convention, 171
  vertex, 162
incidence matrix, 292
inclusion, 90

independent
  capacitors, 344
  inductors, 344
  vectors, 95
index of closed curve, 395
indirect method of Liapunov, 424, 432
inductive
  co-energy, 82
  energy, 82
inductor, 80, 82
inertia, 11
inertial
  mass, 6, 18
  reference system, 34
initial value theorem, 132, 209
inner product, 96
input variables, 2, 455
instability, 426
integral invariants, 369–71
integrated
  across-variables, 279
  through-variables, 279
integrator vertex, 161
interaction, 3
  coefficient, 6
interconnective
  constraints on power variables, 287
  relations, 11
intersection of sets, 91
invariance of eigenvalues, 106
inversion of transforms, 135–43, 212–14
isoclines, 394

Jacobian determinant, 118

Kalman, R. E., 462, 491–2
kinematical variables, 3
kinetic energy, 29
Kirchhoff's
  current law, 288
  voltage law, 289
Kron, G., 310, 314
Kronecker delta, 100

Lagrange multiplier, 116, 118
Lagrangian
  definition, 227
  equations, 227–38
  network equations, 323–5
  problems, 250
Laplace transform
  definition, 126

of derivatives, 128
of integrals, 128
solution of equations, 143, 413
table, 133
laws
  of motion, 3
  of variation, 218
left-hand
  eigenvectors, 105
  rule, 191
Legendre dual transformation, 27, 244
level, 52
Liapunov
  direct method, 433
  first-approximation matrix, 389–90
  function, 427
  indirect method, 429
  stability theory, 424–34
Lienard–Chipart criteria, 423
limit cycles, 394, 433
linear
  functional matrix, 440
  graph, 33, 285
  graphs for electrical systems, 85
  graphs for fluid systems, 63
  graphs for mechanical systems, 33, 48
  graphs for thermal systems, 72
  graph summary of conventions, 88
  independence, 94
  vector spaces, 92
Liouville's theorem, 245
Lipschitz condition, 386
location bounds on eigenvalues, 106

Mason's circuit rule, 177–84
mass symbol, 33
matrix, 92
  differentiation, 114
  direct sum, 115
  functions, 124
  representation of basis change, 101
  representation of operator, 98
matrizant, 400
matter, 3
Maxwell
  Vyschnegradskii criterion, 421
  stationary heat theorem, 330
maximal effort system, 270
Mayer problem, 250
mechanical
  circuit, 37
  circuit postulate, 38

duality, 10
systems, 3–52
terminal, 23
transformer, 51
vertex, 23
vertex postulate, 38
metric space, 92
minimization of integral functional, 265
minimum-phase transfer function, 158
mixed transform network analysis method, 318
modality, 434–7
modulus condition, 195
Mohr, O., 446
Moigno, F. N. M., 383
momentum, 7

Negative
  definite function, 427
  semi-definite function, 427
network
  analysis, 278–382
  diagram, 36
  elements, 33
  models, 278–382
  nonlinear, 363–82
Newton's laws, 7
nilpotency, 125
nonlinear
  capacitor, 58, 77–80
  inductor, 80–2
  spring, 25
  networks, 363–82
norm, 97
non-state elements, 345
n-tuple number, 92
null
  graph, 286
  matrix, 125
  space, 110
nullity, 110
Nyquist
  contour, 185
  diagram, 190
  stability criterion, 184–94

Observability, 455–76
one-to-one
  capacitor, 79
  spring, 25
open-loop transfer function, 188
operations on sets, 91

optimal
  control problem, 251–77
  linearization, 201
  orthogonal projection onto subspace, 481
  projection along invariant subspace, 483
oriented line segment, 34, 49, 286
oriented linear graph, 286
orthonormal basis, 97
output variables, 2, 455

Path, 285
performance index, 250
phase margin, 192
physical variable, 3
pneumatic tank capacitance, 55
Poincaré, H., 383
  index theorems, 396
Poisson ratio, 42
polar vector convention, 48
poles, 151
  dominant, 150
  residue at, 140
Pontryagin
  equations, 248–77
  maximum principle, 255
positive
  definite function, 427
  semi-definite function, 427
potential
  analogy, 154
  co-energy, 26
  difference, 76
  energy, 26
power, 16
pressure-momentum variable, 60
pressurized tank, 55
primal
  form of Hamilton's postulate, 227
  interaction experiment, 4
projection
  along invariant subspace, 483
  on to subspace, 481
  theorem, 100–01
propagation of longitudinal waves, 23

Range space, 110
rank, 110
reciprocal
  basis, 100
  inductance, 82
  matrix spectrum, 106

rectangular pulse function, 130
reduced incidence matrix, 292
reduction, 476–85
region
　of absolute convergence, 134
　of asymptotic stability, 426
relationships, 2
relative stability criteria, 192
removal
　of intermediate vertex, 174
　of self-circuit, 175
　of simple circuit, 175
repeated eigenvalues, 411
representation
　of electrical systems by graphs, 85
　of fluid systems by graphs, 63
　of mechanical systems by graphs, 33, 48
　of operators in new basis, 102
　of thermal systems by graphs, 72
　of systems by networks, 36
residue, 140
resistor, 83–5
　co-content, 85
　content, 85
right-hand eigenvector, 105
root-locus method, 194–200
rotational
　compliance, 43
　dissipator co-content, 46
　dissipator content, 46
　inertia, 41, 44
　kinetic co-energy, 45
　kinetic energy, 45
　mechanical components, 19, 41
　mechanical potential co-energy, 44
　mechanical potential energy, 43
　spring, 42
　stiffness, 42–3
　viscous damping coefficient, 47
Routh's stability criterion, 421

Saddle point, 391
Salzer, C., 314
scalar-valued vector functions, 109
scaling, 419
　vertex, 162
schematic diagrams, 33
Schwarz, L., 130
second variation, 220
self-circuit, 161, 177
set, 90
　of possible events, 252
　operations, 91
shift
　forward, 206
　backward, 207
shifting scaling vertex, 165
signal-flow graphs, 170–84
　conventions, 170
　manipulations, 172
simple
　cascade rule, 163, 172
　form of Nyquist criterion, 189
　parallel combination rule, 164, 172
singular points, 389, 425
　classification, 390
　stability, 428
sinusoidal input response, 150
sources, 31
space
　physical, 3
　vector, 90
spatial relationships, 4
special
　theory of relativity, 28
　types of excitation function, 129
spectral
　representation of matrix, 436
　form for constant coefficient system, 401
spring, 24–7
stability, 420–34
　of singular point, 425
stability criteria, 420–3
stable
　focus, 392
　node, 391
standard
　basis, 95
　dynamic object, 6
state
　elements, 345
　models, 383–485
　space equations, 351, 363–82, 385
　system decomposition, 461
　variables, 2
stationary
　principles for networks, 329–33
　values for integrals, 220
Stodola criterion, 420
stored energy, 26
stores, 280
subgraph, 285
subspace, 94

summation
  conventions, 170
  vertex, 159
susceptance, 79
switching curve, 274
Sylvester expansion, 417
symmetrical matrix, 105

Tellegen's theorem, 296
temperature, 68
thermal systems, 66–73
through store, variable, 279–80
time, 3
  scaling, 419
  shift, 129
topological network relationships, 290–6
torque, 40
torsional waves, 41
trajectory properties in state plane, 393–6
transfer function
  determination of response from, 151
  relationships, 143–58
transform pairs, 133, 213
transformation
  active interpretation, 102
  of gradient vectors, 114
  Laplace, 126
  passive interpretation, 102
  z, 207
transition matrix, 400
translational
  mechanical components, 19
  converters, 29
  mass, 27–8
  mechanical kinetic energy, 29
  mechanical spring, 24
  mechanical system, 3
transmission of power, 19
transversality conditions, 222
travelling waves, 22

tree, 286
Trent, H. M., 304

Unit matrix, 125
uniqueness of equation solutions, 385–9
unstable focus, node, 391–2

Variation, 218
  between fixed end points, 222
  first, second, 220
variational principles for networks, 329–38
vector differentiation, 112
velocity, 34, 49
vertex
  integrator, 161
  mechanical, 23
  method, 313
  multiplication, 160
  summation, 159
  postulate for through-variables, 288
virtual work, 228
viscous damping coefficient, 32, 47
voltage, 76
  controlled capacitor, 79
  source, 85
Volterra, V., 3
volume flow rate, 53, 60
vortex, 392

Wave propagation, 20, 64, 73, 86
weighting function, 144

Zero matrix, 125
zeros of transfer function, 151
z-transfer function, 215
z-transform
  convergence, 210
  definition, 207
  table, 213
  use in discrete systems, 206